PRACTICAL GUIDE to CHEMOMETRICS

SECOND EDITION

Edited by
PAUL GEMPERLINE

Taylor & Francis
Taylor & Francis Group
Boca Raton   London   New York

CRC is an imprint of the Taylor & Francis Group,
an informa business

Published in 2006 by
CRC Press
Taylor & Francis Group
6000 Broken Sound Parkway NW, Suite 300
Boca Raton, FL 33487-2742

© 2006 by Taylor & Francis Group, LLC
CRC Press is an imprint of Taylor & Francis Group

No claim to original U.S. Government works
Printed in the United States of America on acid-free paper
10 9 8 7 6 5 4 3 2 1

International Standard Book Number-10: 1-57444-783-1 (Hardcover)
International Standard Book Number-13: 978-1-57444-783-5 (Hardcover)
Library of Congress Card Number 2005054904

This book contains information obtained from authentic and highly regarded sources. Reprinted material is quoted with permission, and sources are indicated. A wide variety of references are listed. Reasonable efforts have been made to publish reliable data and information, but the author and the publisher cannot assume responsibility for the validity of all materials or for the consequences of their use.

No part of this book may be reprinted, reproduced, transmitted, or utilized in any form by any electronic, mechanical, or other means, now known or hereafter invented, including photocopying, microfilming, and recording, or in any information storage or retrieval system, without written permission from the publishers.

For permission to photocopy or use material electronically from this work, please access www.copyright.com (http://www.copyright.com/) or contact the Copyright Clearance Center, Inc. (CCC) 222 Rosewood Drive, Danvers, MA 01923, 978-750-8400. CCC is a not-for-profit organization that provides licenses and registration for a variety of users. For organizations that have been granted a photocopy license by the CCC, a separate system of payment has been arranged.

Trademark Notice: Product or corporate names may be trademarks or registered trademarks, and are used only for identification and explanation without intent to infringe.

Library of Congress Cataloging-in-Publication Data

Practical guide to chemometrics / edited by Paul Gemperline.--2nd ed.
    p. cm.
Includes bibliographical references and index.
ISBN 1-57444-783-1 (alk. paper)
1. Chemometrics. I. Gemperline, Paul.

QD75.4.C45P73 2006
543.072--dc22
                                                        2005054904

Taylor & Francis Group
is the Academic Division of Informa plc.

Visit the Taylor & Francis Web site at
http://www.taylorandfrancis.com

and the CRC Press Web site at
http://www.crcpress.com

Preface

Chemometrics is an interdisciplinary field that combines statistics and chemistry. From its earliest days, chemometrics has always been a practically oriented subdiscipline of analytical chemistry aimed at solving problems often overlooked by mainstream statisticians. An important example is solving multivariate calibration problems at reduced rank. The method of partial least-squares (PLS) was quickly recognized and embraced by the chemistry community long before many practitioners in statistics considered it worthy of a "second look."

For many chemists, training in data analysis and statistics has been limited to the basic univariate topics covered in undergraduate analytical chemistry courses such as univariate hypothesis testing, for example, comparison of means. A few more details may have been covered in some senior-level courses on instrumental methods of analysis where topics such as univariate linear regression and prediction confidence intervals might be examined. In graduate school, perhaps a review of error propagation and analysis of variance (ANOVA) may have been encountered in a core course in analytical chemistry. These tools were typically introduced on a very practical level without a lot of the underlying theory. The chemistry curriculum simply did not allow sufficient time for more in-depth coverage. However, during the past two decades, chemometrics has emerged as an important subdiscipline, and the analytical chemistry curriculum has evolved at many universities to the point where a small amount of time is devoted to practical application-oriented introductions to some multivariate methods of data analysis.

This book continues in the practical tradition of chemometrics. Multivariate methods and procedures that have been found to be extraordinarily useful in analytical chemistry applications are introduced with a minimum of theoretical background. The aim of the book is to illustrate these methods through practical examples in a style that makes the material accessible to a broad audience of nonexperts.

Editor

Paul J. Gemperline, Ph.D., ECU distinguished professor of research and Harriot College distinguished professor of chemistry, has more than 20 years of experience in chemometrics, a subdiscipline of analytical chemistry that utilizes multivariate statistical and numerical analysis of chemical measurements to provide information for understanding, modeling, and controlling industrial processes. Dr. Gemperline's achievements include more than 50 publications in the field of chemometrics and more than $1.5 million in external grant funds. Most recently, he was named recipient of the 2003 Eastern Analytical Symposium's Award in Chemometrics, the highest international award in the field of chemometrics.

Dr. Gemperline's training in scientific computing began in the late 1970s in graduate school and developed into his main line of research in the early 1980s. He collaborated with pharmaceutical company Burroughs Wellcome in the early 1980s to develop software for multivariate pattern-recognition analysis of near-infrared reflectance spectra for rapid, nondestructive testing of pharmaceutical ingredients and products. His research and publications in this area gained international recognition. He is a sought-after lecturer and has given numerous invited lectures at universities and international conferences outside the United States. Most recently, Dr. Gemperline participated with a team of researchers to develop and conduct training on chemometrics for U.S. Food and Drug Administration (FDA) scientists, inspectors, and regulators of the pharmaceutical industry in support of their new Process Analytical Technology initiative.

The main theme of Dr. Gemperline's research in chemometrics is focused on development of new algorithms and software tools for analysis of multivariate spectroscopic measurements using pattern-recognition methods, artificial neural networks, multivariate statistical methods, multivariate calibration, and nonlinear model estimation. His work has focused on applications of process analysis in the pharmaceutical industry, with collaborations and funding from scientists at Pfizer, Inc. and GlaxoSmithKline. Several of his students are now employed as chemometricians and programmers at pharmaceutical and scientific instrument companies. Dr. Gemperline has also received significant funding from the National Science Foundation and the Measurement and Control Engineering Center (MCEC), an NSF-sponsored University/Industry Cooperative Research Center at the University of Tennessee, Knoxville.

Contributors

Karl S. Booksh
Department of Chemistry and Biochemistry
Arizona State University
Tempe, Arizona

Steven D. Brown
Department of Chemistry and Biochemistry
University of Delaware
Newark, Delaware

Charles E. Davidson
Department of Chemistry
Clarkson University
Potsdam, New York

Anna de Juan
Department of Analytical Chemistry
University of Barcelona
Barcelona, Spain

Paul J. Gemperline
Department of Chemistry
East Carolina University
Greenville, North Carolina

Mia Hubert
Department of Mathematics
Katholieke Universiteit Leuven
Leuven, Belgium

John H. Kalivas
Department of Chemistry
Idaho State University
Pocatello, Idaho

Barry K. Lavine
Department of Chemistry
Oklahoma State University
Stillwater, Oklahoma

Marcel Maeder
Department of Chemistry
University of Newcastle
Newcastle, Australia

Yorck-Michael Neuhold
Department of Chemistry
University of Newcastle
Newcastle, Australia

Kalin Stoyanov
Sofia, Bulgaria

Romà Tauler
Institute of Chemical and Environmental Research
Barcelona, Spain

Anthony D. Walmsley
Department of Chemistry
University of Hull
Hull, England

Contents

Chapter 1
Introduction to Chemometrics 1
Paul J. Gemperline

Chapter 2
Statistical Evaluation of Data 7
Anthony D. Walmsley

Chapter 3
Sampling Theory, Distribution Functions, and the Multivariate
Normal Distribution 41
Paul J. Gemperline and John H. Kalivas

Chapter 4
Principal Component Analysis 69
Paul J. Gemperline

Chapter 5
Calibration 105
John H. Kalivas and Paul J. Gemperline

Chapter 6
Robust Calibration 167
Mia Hubert

Chapter 7
Kinetic Modeling of Multivariate Measurements with
Nonlinear Regression 217
Marcel Maeder and Yorck-Michael Neuhold

Chapter 8
Response-Surface Modeling and Experimental Design 263
Kalin Stoyanov and Anthony D. Walmsley

Chapter 9
Classification and Pattern Recognition 339
Barry K. Lavine and Charles E. Davidson

Chapter 10
Signal Processing and Digital Filtering 379
Steven D. Brown

Chapter 11
Multivariate Curve Resolution 417
Romà Tauler and Anna de Juan

Chapter 12
Three-Way Calibration with Hyphenated Data..475
Karl S. Booksh

Chapter 13
Future Trends in Chemometrics 509
Paul J. Gemperline

Index..521

1 Introduction to Chemometrics

Paul J. Gemperline

CONTENTS

1.1 Chemical Measurements — A Basis for Decision Making 1
1.2 Chemical Measurements — The Three-Legged Platform 2
1.3 Chemometrics 2
1.4 How to Use This Book 3
    1.4.1 Software Applications 4
1.5 General Reading on Chemometrics 5
References 6

1.1 CHEMICAL MEASUREMENTS — A BASIS FOR DECISION MAKING

Chemical measurements often form the basis for important decision-making activities in today's society. For example, prior to medical treatment of an individual, extensive sets of tests are performed that often form the basis of medical treatment, including an analysis of the individual's blood chemistry. An incorrect result can have life-or-death consequences for the person receiving medical treatment. In industrial settings, safe and efficient control and operation of high energy chemical processes, for example, ethylene production, are based on on-line chemical analysis. An incorrect result for the amount of oxygen in an ethylene process stream could result in the introduction of too much oxygen, causing a catastrophic explosion that could endanger the lives of workers and local residents alike. Protection of our environment is based on chemical methods of analysis, and governmental policymakers depend upon reliable measurements to make cost-effective decisions to protect the health and safety of millions of people living now and in the future. Clearly, the information provided by chemical measurements must be reliable if it is to form the basis of important decision-making processes like the ones described above.

1.2 CHEMICAL MEASUREMENTS — THE THREE-LEGGED PLATFORM

Sound chemical information that forms the basis of many of humanity's important decision-making processes depends on three critical properties of the measurement process, including its (1) chemical properties, (2) physical properties, and (3) statistical properties. The conditions that support sound chemical measurements are like a platform supported by three legs. Credible information can be provided only in an environment that permits a *thorough understanding and control* of these three critical properties of a chemical measurement:

1. Chemical properties, including stoichiometry, mass balance, chemical equilibria, kinetics, etc.
2. Physical properties, including temperature, energy transfer, phase transitions, etc.
3. Statistical properties, including sources of errors in the measurement process, control of interfering factors, calibration of response signals, modeling of complex multivariate signals, etc.

If any one of these three legs is missing or absent, the platform will be unstable and the measurement system will fail to provide reliable results, sometimes with catastrophic consequences. It is the role of statistics and chemometrics to address the third critical property. It is this fundamental role that provides the primary motivation for developments in the field of chemometrics. Sound chemometric methods and a well-trained work force are necessary for providing reliable chemical information for humanity's decision-making activities. In the subsequent sections, we begin our presentation of the topic of chemometrics by defining the term.

1.3 CHEMOMETRICS

The term chemometrics was first coined in 1971 to describe the growing use of mathematical models, statistical principles, and other logic-based methods in the field of chemistry and, in particular, the field of analytical chemistry. Chemometrics is an interdisciplinary field that involves multivariate statistics, mathematical modeling, computer science, and analytical chemistry. Some major application areas of chemometrics include (1) calibration, validation, and significance testing; (2) optimization of chemical measurements and experimental procedures; and (3) the extraction of the maximum of chemical information from analytical data.

In many respects, the field of chemometrics is the child of statistics, computers, and the "information age." Rapid technological advances, especially in the area of computerized instruments for analytical chemistry, have enabled and necessitated phenomenal growth in the field of chemometrics over the past 30 years. For most of this period, developments have focused on multivariate methods. Since the world around us is inherently multivariate, it makes sense to treat multiple measurements simultaneously in any data analysis procedure. For example, when we measure the ultraviolet (UV) absorbance of a solution, it is easy to measure its entire spectrum quickly and rapidly with low noise, rather than measuring its absorbance at a single wavelength. By properly considering the distribution of multiple variables simultaneously, we obtain *more information* than could be obtained by considering each variable individually. This is one of the so-called *multivariate advantages*. The additional information comes to us in the form of correlation. When we look at one variable at a time, we neglect correlation between variables, and hence miss part of the picture.

A recent paper by Bro described four additional advantages of multivariate methods compared with univariate methods [1]. Noise reduction is possible when multiple redundant variables are analyzed simultaneously by proper multivariate methods. For example, low-noise factors can be obtained when principal component analysis is used to extract a few meaningful factors from UV spectra measured at hundreds of wavelengths. Another important multivariate advantage is that partially selective measurements can be used, and by use of proper multivariate methods, results can be obtained free of the effects of interfering signals. A third advantage is that false samples can be easily discovered, for example in spectroscopic analysis. For any well characterized chemometric method, aliquots of material measured in the future should be properly explained by linear combinations of the training set or calibration spectra. If new, foreign materials are present that give spectroscopic signals slightly different from the expected ingredients, these can be detected in the spectral residuals and the corresponding aliquot flagged as an outlier or "false sample." The advantages of chemometrics are often the consequence of using multivariate methods. The reader will find these and other advantages highlighted throughout the book.

1.4 HOW TO USE THIS BOOK

This book is suitable for use as an introductory textbook in chemometrics or for use as a self-study guide. Each of the chapters is self-contained, and together they cover many of the main areas of chemometrics. The early chapters cover tutorial topics and fundamental concepts, starting with a review of basic statistics in Chapter 2, including hypothesis testing. The aim of Chapter 2 is to review suitable protocols for the planning of experiments and the analysis of the data, primarily from a univariate point of view. Topics covered include defining a research hypothesis, and then implementing statistical tools that can be used to determine whether the stated hypothesis is found to be true. Chapter 3 builds on the concept of the univariate normal distribution and extends it to the multivariate normal distribution. An example is given showing the analysis of near infrared spectral data for raw material testing, where two degradation products were detected at 0.5% to 1% by weight. Chapter 4 covers principal component analysis (PCA), one of the workhorse methods of chemometrics. This is a topic that all basic or introductory courses in chemometrics should cover. Chapter 5 covers the topic of multivariate calibration, including partial least-squares, one of the single most common application areas for chemometrics. Multivariate calibration refers generally to mathematical methods that transform and instrument's response to give an estimate of a more informative chemical or physical variable, e.g., the target analyte. Together, Chapters 3, 4, and 5 form the introductory core material of this book.

The remaining chapters of the book introduce some of the advanced topics of chemometrics. The coverage is fairly comprehensive, in that these chapters cover some of the most important advanced topics. Chapter 6 presents the concept of robust multivariate methods. Robust methods are insensitive to the presence of outliers. Most of the methods described in Chapter 6 can tolerate data sets contaminated with up to 50% outliers without detrimental effects. Descriptions of algorithms and examples are provided for robust estimators of the multivariate normal distribution, robust PCA, and robust multivariate calibration, including robust PLS. As such, Chapter 6 provides an excellent follow-up to Chapters 3, 4, and 5.

Chapter 7 covers the advanced topic of nonlinear multivariate model estimation, with its primary examples taken from chemical kinetics. Chapter 8 covers the important topic of experimental design. While its position in the arrangement of this book comes somewhat late, we feel it will be much easier for the reader or student to recognize important applications of experimental design by following chapters on calibration and nonlinear model estimation. Chapter 9 covers the topic of multivariate classification and pattern recognition. These types of methods are designed to seek relationships that describe the similarity or dissimilarity between diverse groups of data, thereby revealing common properties among the objects in a data set. With proper multivariate approaches, a large number of features can be studied simultaneously. Examples of applications in this area of chemometrics include the identification of the source of pollutants, detection of unacceptable raw materials, intact classification of unlabeled pharmaceutical products for clinical trials through blister packs, detection of the presence or absence of disease in a patient, and food quality testing, to name a few.

Chapter 10, Signal Processing and Digital Filtering, is concerned with mathematical methods that are intended to enhance signals by decreasing the contribution of noise. In this way, the "true" signal can be recovered from a signal distorted by other effects. Chapter 11, Multivariate Curve Resolution, describes methods for the mathematical resolution of multivariate data sets from evolving systems into descriptive models showing the contributions of pure constituents. The ability to correctly recover pure concentration profiles and spectra for each of the components in the system depends on the degree of overlap among the pure profiles of the different components and the specific way in which the regions of these profiles are overlapped. Chapter 12 describes three-way calibration methods, an active area of research in chemometrics. Chapter 12 includes descriptions of methods such as the generalized rank annihilation method (GRAM) and parallel factor analysis (PARAFAC). The main advantage of three-way calibration methods is their ability to estimate analyte concentrations in the presence of unknown, uncalibrated spectral interferents. Chapter 13 reviews some of the most active areas of research in chemometrics.

1.4.1 SOFTWARE APPLICATIONS

Our experience in learning chemometrics and teaching it to others has demonstrated repeatedly that people learn new techniques by using them to solve interesting problems. For this reason, many of the contributing authors to this book have chosen to illustrate their chemometric methods with examples using

Microsoft® Excel, MATLAB, or other powerful computer applications. For many research groups in chemometrics, MATLAB has become a workhorse research tool, and numerous public-domain MATLAB software packages for doing chemometrics can be found on the World Wide Web. MATLAB is an interactive computing environment that takes the drudgery out of using linear algebra to solve complicated problems. It integrates computer graphics, numerical analysis, and matrix computations into one simple-to-use package. The package is available on a wide range of personal computers and workstations, including IBM-compatible and Macintosh computers. It is especially well-suited to solving complicated matrix equations using a simple "algebra-like" notation. Because some of the authors have chosen to use MATLAB, we are able to provide you with some example programs. The equivalent programs in BASIC, Pascal, FORTRAN, or C would be too long and complex for illustrating the examples in this book. It will also be much easier for you to experiment with the methods presented in this book by trying them out on your data sets and modifying them to suit your special needs. Those who want to learn more about MATLAB should consult the manuals shipped with the program and numerous web sites that present tutorials describing its use.

1.5 GENERAL READING ON CHEMOMETRICS

A growing number of books, some of a specialized nature, are available on chemometrics. A brief summary of the more general texts is given here as guidance for the reader. Each chapter, however, has its own list of selected references.

JOURNALS

1. *Journal of Chemometrics* (Wiley) — Good for fundamental papers and applications of advanced algorithms.
2. *Journal of Chemometrics and Intelligent Laboratory Systems* (Elsevier) — Good for conference information; has a tutorial approach and is not too mathematically heavy.
3. Papers on chemometrics can also be found in many of the more general analytical journals, including: *Analytica Chimica Acta, Analytical Chemistry, Applied Spectroscopy, Journal of Near Infrared Spectroscopy, Journal of Process Control,* and *Technometrics.*

BOOKS

1. Adams, M.J., *Chemometrics in Analytical Spectroscopy*, 2nd ed., The Royal Society of Chemistry: Cambridge. 2004.
2. Beebe, K.R., Pell, R.J., and Seasholtz, M.B. *Chemometrics: A Practical Guide.*, John Wiley & Sons: New York. 1998.
3. Box, G.E.P., Hunter, W.G., and Hunter, J.S. *Statistics for Experimenters.* John Wiley & Sons: New York. 1978.
4. Brereton, R.G. *Chemometrics: Data Analysis for the Laboratory and Chemical Plant.* John Wiley & Sons: Chichester, U.K. 2002.
5. Draper, N.R. and Smith, H.S. *Applied Regression Analysis*, 2nd ed., John Wiley & Sons: New York. 1981.

6. Jackson, J.E. *A User's Guide to Principal Components.* John Wiley & Sons: New York. 1991.
7. Jollife, I.T. *Principal Component Analysis.* Springer-Verlag: New York. 1986.
8. Kowalski, B.R., Ed. *NATO ASI Series. Series C, Mathematical and Physical Sciences, Vol. 138: Chemometrics, Mathematics, and Statistics in Chemistry.* Dordrecht; Lancaster: Published in cooperation with NATO Scientific Affairs Division [by] Reidel, 1984.
9. Kowalski, B.R., Ed. *Chemometrics: Theory and Application.* ACS Symposium Series 52. American Chemical Society: Washington, DC. 1977.
10. Malinowski, E.R. *Factor Analysis of Chemistry.* 2nd ed., John Wiley & Sons: New York. 1991.
11. Martens, H. and Næs, T. *Multivariate Calibration.* John Wiley & Sons: Chichester, U.K. 1989.
12. Massart, D.L., Vandeginste, B.G.M., Buyden, L.M.C., De Jong, S., Lewi, P.J., and Smeyers-Verbeke, J. *Handbook of Chemometrics and Qualimetrics*, Part A and B. Elsevier: Amsterdam. 1997.
13. Miller, J.C. and Miller, J.N. *Statistics and Chemometrics for Analytical Chemistry,* 4th ed., Prentice Hall: Upper Saddle River N.J. 2000.
14. Otto, M. *Chemometrics: Statistics and Computer Application in Analytical Chemistry.* John Wiley & Sons-VCH: New York. 1999.
15. Press, W.H.; Teukolsky, S.A., Flannery, B.P., and Vetterling, W.T. *Numerical Recipes in C. The Art of Scientific Computing*, 2nd ed., Cambridge University Press: New York. 1992.
16. Sharaf, M.A., Illman, D.L., and Kowalski, B.R. *Chemical Analysis, Vol. 82: Chemometrics.* John Wiley & Sons: New York. 1986.

REFERENCES

1. Bro, R., Multivariate calibration. What is in chemometrics for the analytical chemist? *Analytica Chimica Acta*, 2003. 500(1-2): 185–194.

2 Statistical Evaluation of Data

Anthony D. Walmsley

CONTENTS

Introduction 8
2.1 Sources of Error 9
    2.1.1 Some Common Terms 10
2.2 Precision and Accuracy 12
2.3 Properties of the Normal Distribution 14
2.4 Significance Testing 18
    2.4.1 The F-test for Comparison of Variance (Precision) 19
    2.4.2 The Student t-Test 22
    2.4.3 One-Tailed or Two-Tailed Tests 24
    2.4.4 Comparison of a Sample Mean with a Certified Value 24
    2.4.5 Comparison of the Means from Two Samples 25
    2.4.6 Comparison of Two Methods with Different Test Objects or Specimens 26
2.5 Analysis of Variance 27
    2.5.1 ANOVA to Test for Differences Between Means 28
    2.5.2 The Within-Sample Variation (Within-Treatment Variation) 29
    2.5.3 Between-Sample Variation (Between-Treatment Variation) 29
    2.5.4 Analysis of Residuals 30
2.6 Outliers 33
2.7 Robust Estimates of Central Tendency and Spread 36
2.8 Software 38
    2.8.1 ANOVA Using Excel 39
Recommended Reading 40
References 40

INTRODUCTION

Typically, one of the main errors made in analytical chemistry and chemometrics is that the chemical experiments are performed with no prior plan or design. It is often the case that a researcher arrives with a pile of data and asks "what does it mean?" to which the answer is usually "well what do you think it means?" The weakness in collecting data without a plan is that one can quite easily acquire data that are simply not relevant. For example, one may wish to compare a new method with a traditional method, which is common practice, and so aliquots or test materials are tested with both methods and then the data are used to test which method is the best (Note: for "best" we mean the most suitable for a particular task, in most cases "best" can cover many aspects of a method from highest purity, lowest error, smallest limit of detection, speed of analysis, etc. The "best" method can be defined for each case). However, this is not a direct comparison, as the new method will typically be one in which the researchers have a high degree of domain experience (as they have been developing it), meaning that it is an optimized method, but the traditional method may be one they have little experience with, and so is more likely to be nonoptimized. Therefore, the question you have to ask is, "Will simply testing objects with both methods result in data that can be used to compare which is the better method, or will the data simply infer that the researchers are able to get better results with their method than the traditional one?" Without some design and planning, a great deal of effort can be wasted and mistakes can be easily made. It is unfortunately very easy to compare an optimized method with a nonoptimized method and hail the new technique as superior, when in fact, all that has been deduced is an inability to perform both techniques to the same standard.

Practical science should not start with collecting data; it should start with a hypothesis (or several hypotheses) about a problem or technique, etc. With a set of questions, one can plan experiments to ensure that the data collected is useful in answering those questions. Prior to any experimentation, there needs to be a consideration of the analysis of the results, to ensure that the data being collected are relevant to the questions being asked. One of the desirable outcomes of a structured approach is that one may find that some variables in a technique have little influence on the results obtained, and as such, can be left out of any subsequent experimental plan, which results in the necessity for less rather than more work.

Traditionally, data was a single numerical result from a procedure or assay; for example, the concentration of the active component in a tablet. However, with modern analytical equipment, these results are more often a spectrum, such as a mid-infrared spectrum for example, and so the use of multivariate calibration models has flourished. This has led to more complex statistical treatments because the result from a calibration needs to be validated rather than just a single value recorded. The quality of calibration models needs to be tested, as does the robustness, all adding to the complexity of the data analysis. In the same way that the spectroscopist relies on the spectra obtained from an instrument, the analyst must rely on the results obtained from the calibration model (which may be based on spectral data); therefore, the rigor of testing must be at the same high standard as that of the instrument manufacturer. The quality of any model is very dependent on the test specimens used to build it, and so sampling plays a very important part in analytical methodology. Obtaining a good representative sample or set of test specimens is not easy without some prior planning, and in cases where natural products or natural materials are used or where no design is applicable, it is critical to obtain a representative sample of the system.

The aim of this chapter is to demonstrate suitable protocols for the planning of experiments and the analysis of the data. The important question to keep in mind is, "What is the purpose of the experiment and what do I propose as the outcome?" Usually, defining the question takes greater effort than performing any analysis. Defining the question is more technically termed defining the research hypothesis, following which the statistical tools can be used to determine whether the stated hypothesis is found to be true.

One can consider the application of statistical tests and chemometric tools to be somewhat akin to torture—if you perform it long enough your data will tell you anything you wish to know—but most results obtained from torturing your data are likely to be very unstable. A light touch with the correct tools will produce a much more robust and useable result then heavy-handed tactics ever will. Statistics, like torture, benefit from the correct use of the appropriate tool.

2.1 SOURCES OF ERROR

Experimental science is in many cases a quantitative subject that depends on numerical measurements. A numerical measurement is almost totally useless unless it is accompanied by some estimate of the error or uncertainty in the measurement. Therefore, one must get into the habit of estimating the error or degree of uncertainty each time a measurement is made. Statistics are a good way to describe some types of error and uncertainty in our data. Generally, one can consider that simple statistics are a numerical measure of "common sense" when it comes to describing errors in data. If a measurement seems rather high compared with the rest of the measurements in the set, statistics can be employed to give a numerical estimate as to how high. This means that one must not use statistics blindly, but must always relate the results from the given statistical test to the data to which the data has been applied, and relate the results to given knowledge of the measurement. For example, if you calculate the mean height of a group of students, and the mean is returned as 296 cm, or more than 8 ft, then you must consider that unless your class is a basketball team, the mean should not be so high. The outcome should thus lead you to consider the original data, or that an error has occurred in the calculation of the mean.

One needs to be extremely careful about errors in data, as the largest error will always dominate. If there is a large error in a reference method, for example, small measurement errors will be superseded by the reference errors. For example, if one used a bench-top balance accurate to one hundredth of a gram to weigh out one gram of substance to standardize a reagent, the resultant standard will have an accuracy of only one part per hundredth, which is usually considered to be poor for analytical data.

Statistics must not be viewed as a method of making sense out of bad data, as the results of any statistical test are only as good as the data to which they are applied. If the data are poor, then any statistical conclusion that can be made will also be poor.

Experimental scientists generally consider there to be three types of error:

1. *Gross error* is caused, for example, by an instrumental breakdown such as a power failure, a lamp failing, severe contamination of the specimen or a simple mislabeling of a specimen (in which the bottle's contents are not as recorded on the label). The presence of gross errors renders an experiment useless. The most easily applied remedy is to repeat the experiment. However, it can be quite difficult to detect these errors, especially if no replicate measurements have been made.
2. *Systematic error* arises from imperfections in an experimental procedure, leading to a bias in the data, i.e., the errors all lie in the same direction for all measurements (the values are all too high or all too low). These errors can arise due to a poorly calibrated instrument or by the incorrect use of volumetric glassware. The errors that are generated in this way can be either constant or proportional. When the data are plotted and viewed, this type of error can usually be discovered, i.e., the intercept on the y-axis for a calibration is much greater than zero.
3. *Random error* (commonly referred to as noise) produces results that are spread about the average value. The greater the degree of randomness, the larger the spread. Statistics are often used to describe random errors. Random errors are typically ones that we have no control over, such as electrical noise in a transducer. These errors affect the precision or reproducibility of the experimental results. The goal is to have small random errors that lead to good precision in our measurements. The precision of a method is determined from replicate measurements taken at a similar time.

2.1.1 SOME COMMON TERMS

Accuracy: An experiment that has small systematic error is said to be accurate, i.e., the measurements obtained are close to the true values.

Precision: An experiment that has small random errors is said to be precise, i.e., the measurements have a small spread of values.

Within-run: This refers to a set of measurements made in succession in the same laboratory using the same equipment.

Between-run: This refers to a set of measurements made at different times, possibly in different laboratories and under different circumstances.

Repeatability: This is a measure of within-run precision.

Reproducibility: This is a measure of between-run precision.

Mean, Variance, and *Standard Deviation:* Three common statistics can be calculated very easily to give a quick understanding of the quality of a dataset and can also be used for a quick comparison of new data with some prior datasets. For example, one can compare the mean of the dataset with the mean from a standard set. These are very useful exploratory statistics, they are easy to calculate, and can also be used in subsequent data analysis tools. The *arithmetic mean* is a measure of the average or central tendency of a set of data and is usually denoted by the symbol $\bar{x}$. The value for the mean is calculated by summing the data and then dividing this sum by the number of values (n).

$$\bar{x} = \frac{\sum x_i}{n} \qquad (2.1)$$

The *variance* in the data, a measure of the spread of a set of data, is related to the precision of the data. For example, the larger the variance, the larger the spread of data and the lower the precision of the data. Variance is usually given the symbol s^2 and is defined by the formula:

$$s^2 = \frac{\sum (x_i - \bar{x})^2}{n} \qquad (2.2)$$

The *standard deviation* of a set of data, usually given the symbol s, is the square root of the variance. The difference between standard deviation and variance is that the standard deviation has the same units as the data, whereas the variance is in units squared. For example, if the measured unit for a collection of data is in meters (m) then the units for the standard deviation is m and the unit for the variance is m^2. For large values of n, the *population standard deviation* is calculated using the formula:

$$s = \sqrt{\frac{\sum (x_i - \bar{x})^2}{n}} \qquad (2.3)$$

If the standard deviation is to be estimated from a small set of data, it is more appropriate to calculate the *sample standard deviation*, denoted by the symbol $\hat{s}$, which is calculated using the following equation:

$$\hat{s} = \sqrt{\frac{\sum (x_i - \bar{x})^2}{n-1}} \qquad (2.4)$$

The *relative standard deviation* (or *coefficient of variation*), a dimensionless quantity (often expressed as a percentage), is a measure of the relative error, or noise in some data. It is calculated by the formula:

$$\text{RSD} = \frac{s}{\bar{x}} \qquad (2.5)$$

When making some analytical measurements of a quantity (x), for example the concentration of lead in drinking water, all the results obtained will contain some random errors; therefore, we need to repeat the measurement a number of times (n). The standard error of the mean, which is a measure of the error in the final answer, is calculated by the formula:

$$s_M = \frac{s}{\sqrt{n}} \tag{2.6}$$

It is good practice when presenting your results to use the following representation:

$$\bar{x} \pm \frac{s}{\sqrt{n}} \tag{2.7}$$

Suppose the boiling points of six impure ethanol specimens were measured using a digital thermometer and found to be: 78.9, 79.2, 79.4, 80.1, 80.3, and 80.9°C. The mean of the data, $\bar{x}$, is 79.8°C, the standard deviation, s, is 0.692°C. With the value of $n = 6$, the standard error, s_m, is found to be 0.282°C, thus the true temperature of the impure ethanol is in the range 79.8 ± 0.282°C ($n = 6$).

2.2 PRECISION AND ACCURACY

The ability to perform the same analytical measurements to provide precise and accurate results is critical in analytical chemistry. The quality of the data can be determined by calculating the precision and accuracy of the data. Various bodies have attempted to define *precision*. One commonly cited definition is from the International Union of Pure and Applied Chemistry (IUPAC), which defines precision as "relating to the variations between variates, i.e., the scatter between variates."[1] *Accuracy* can be defined as the ability of the measured results to match the true value for the data. From this point of view, the standard deviation is a measure of precision and the mean is a measure of the accuracy of the collected data. In an ideal situation, the data would have both high accuracy and precision (i.e., very close to the true value and with a very small spread). The four common scenarios that relate to accuracy and precision are illustrated in Figure 2.1. In many cases, it is not possible to obtain high precision and accuracy simultaneously, so common practice is to be more concerned with the precision of the data rather than the accuracy. Accuracy, or the lack of it, can be compensated in other ways, for example by using aliquots of a reference material, but low precision cannot be corrected once the data has been collected.

To determine precision, we need to know something about the manner in which data is customarily distributed. For example, high precision (i.e., the data are very close together) produces a very narrow distribution, while low precision (i.e., the data are spread far apart) produces a wide distribution. Assuming that the data are normally distributed (which holds true for many cases and can be used as an approximation in many other cases) allows us to use the well understood mathematical distribution known as the normal or Gaussian error distribution. The advantage to using such a model is that we can compare the collected data with a well understood statistical model to determine the precision of the data.

Statistical Evaluation of Data

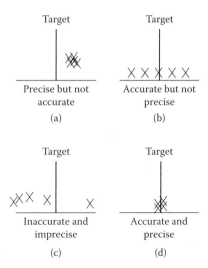

FIGURE 2.1 The four common scenarios that illustrate accuracy and precision in data: (a) precise but not accurate, (b) accurate but not precise, (c) inaccurate and imprecise, and (d) accurate and precise.

Although the standard deviation gives a measure of the spread of a set of results about the mean value, it does not indicate the way in which the results are distributed. To understand this, a large number of results are needed to characterize the distribution. Rather than think in terms of a few data points (for example, six data points) we need to consider, say 500 data points, so the mean, $\bar{x}$, is an excellent estimate of the true mean or population mean, μ. The spread of a large number of collected data points will be affected by the random errors in the measurement (i.e., the sampling error and the measurement error) and this will cause the data to follow the normal distribution. This distribution is shown in Equation 2.8:

$$y = \frac{\exp[-(x-\mu)^2/2\sigma^2]}{\sigma\sqrt{2\pi}} \qquad (2.8)$$

where μ is the true mean (or *population mean*), x is the measured data, and σ is the true standard deviation (or the *population standard deviation*). The shape of the distribution can be seen in Figure 2.2, where it can be clearly seen that the smaller the spread of the data, the narrower the distribution curve.

It is common to measure only a small number of objects or aliquots, and so one has to rely upon the *central limit theorem* to see that a small set of data will behave in the same manner as a large set of data. The central limit theorem states that "as the size of a sample increases (number of objects or aliquots measured), the data will tend towards a normal distribution." If we consider the following case:

$$y = x_1 + x_2 + \ldots + x_n \qquad (2.9)$$

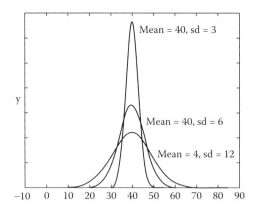

FIGURE 2.2 The normal distribution showing the effect of the spread of the data with a mean of 40 and standard deviations of 3, 6, and 12.

where n is the number of independent variables, x_i, that have mean, μ, and variance, σ^2, then for a large number of variables, the distribution of y is approximately normal, with mean $\Sigma\mu$ and variance $\Sigma\mu^2$, despite whatever the distribution of the independent variable x might be.

2.3 PROPERTIES OF THE NORMAL DISTRIBUTION

The actual shape of the curve for the normal distribution and its symmetry around the mean is a function of the standard deviation. From statistics, it has been shown that 68% of the observations will lie within ±1 standard deviation, 95% lie within ±2 standard deviations, and 99.7% lie within ±3 standard deviations of the mean (see Figure 2.3). We can easily demonstrate how the normal distribution can be

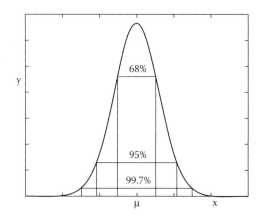

FIGURE 2.3 A plot of the normal distribution showing that approximately 68% of the data lie within ±1 standard deviation, 95% lie within ±2 standard deviation, and 99.7% lie within ±3 standard deviations.

Statistical Evaluation of Data

populated using two six-sided dice. If both dice are thrown together, there is only a small range of possible results: 2, 3, 4, 5, 6, 7, 8, 9, 10, 11 and 12. However, some results have a higher frequency of occurrence due to the number of possible combinations of values from each single die. For example, one possible event that will result in a 2 or a 12 being the total from the dice is a 1 or a 6 on both dice. To obtain a sum of 7 on one roll of the dice there are a number of possible combinations (1 and 6, 2 and 5, 3 and 4, 4 and 3, 5 and 2, 6 and 1). If you throw the two dice a small number of times, it is unlikely that every possible result will be obtained, but as the number of throws increases, the population will slowly fill out and become normal. Try this yourself.

(*Note*: The branch of statistics concerned with measurements that follow the normal distribution are known as parametric statistics. Because many types of measurements follow the normal distribution, these are the most common statistics used. Another branch of statistics designed for measurements that do not follow the normal distribution is known as nonparametric statistics.)

The *confidence interval* is the range within which we can reasonably assume a true value lies. The extreme values of this range are called the confidence limits. The term "confidence" implies that we can assert a result with a given degree of confidence, i.e., a certain probability. Assuming that the distribution is normal, then 95% of the sample means will lie in the range given by:

$$\mu - 1.96 \frac{\sigma}{\sqrt{n}} < x < \mu + 1.96 \frac{\sigma}{\sqrt{n}} \quad (2.10)$$

However, in practice we usually have a measurement of one specimen or aliquot of known mean, and we require a range for μ. Thus, by rearrangement:

$$x - 1.96 \frac{\sigma}{\sqrt{n}} < \mu < x + 1.96 \frac{\sigma}{\sqrt{n}} \quad (2.11)$$

Thus,

$$\mu = x \pm t \frac{\sigma}{\sqrt{n}} \quad (2.12)$$

The appropriate value of t (which is found in the statistical tables) depends both on ($n - 1$), which is the number of degrees of freedom and the degree of confidence required (the term "degrees of freedom" refers to the number of independent deviations used in calculating σ). The value of 1.96 is the t value for an infinite number of degrees of freedom and the 95% confidence limit.

For example, consider a set of data where:

$\bar{x} = 100.5$
$s = 3.27$
$n = 6$

TABLE 2.1
The *t*-Distribution

Value of *t* for a Confidence Interval of	90%	95%	98%	99%
Critical value of $|t|$ for *P* values of	0.10	0.05	0.02	0.01
Number of degrees of freedom				
1	6.31	12.71	31.82	63.66
2	2.92	4.30	6.96	9.92
3	2.35	3.18	4.54	5.84
4	2.13	2.78	3.75	4.60
5	2.02	2.57	3.36	4.03
6	1.94	2.45	3.14	3.71
7	1.89	2.36	3.00	3.50
8	1.86	2.31	2.90	3.36
9	1.83	2.26	2.82	3.25
10	1.81	2.23	2.76	3.17
12	1.78	2.18	2.68	3.05
14	1.76	2.14	2.62	2.98
16	1.75	2.12	2.58	2.92
18	1.73	2.10	2.55	2.88
20	1.72	2.09	2.53	2.85
30	1.70	2.04	2.46	2.75
50	1.68	2.01	2.40	2.68
∞	1.64	1.96	2.33	2.58

Note: The critical values of $|t|$ are appropriate for a two-tailed test. For a one-tailed test, use the $|t|$ value from the column with twice the *P* value.

The 95% confidence interval is computed using $t = 2.57$ (from Table 2.1)

$$\mu = 100.5 \pm 2.57 \frac{3.27}{\sqrt{6}}$$

$$\mu = 100.5 \pm 3.4$$

Summary statistics are very useful when comparing two sets of data, as we can compare the quality of the analytical measurement technique used. For example, a pH meter is used to determine the pH of two solutions, one acidic and one alkaline. The data are shown below.

pH Meter Results for the pH of Two Solutions, One Acidic and One Alkaline

| Acidic solution | 5.2 | 6.0 | 5.2 | 5.9 | 6.1 | 5.5 | 5.8 | 5.7 | 5.7 | 6.0 |
| Alkaline solution | 11.2 | 10.7 | 10.9 | 11.3 | 11.5 | 10.5 | 10.8 | 11.1 | 11.2 | 11.0 |

For the acidic solution, the mean is found to be 5.6, with a standard deviation of 0.341 and a relative standard deviation of 6.0%. The alkaline solution results give a mean of 11.0, a standard deviation of 0.301 and a relative standard deviation of 2.7%. Clearly, the precision for the alkaline solution is higher (RSD 2.7% compared with 6.0%), indicating that the method used to calibrate the pH meter worked better with higher pH. Because we expect the same pH meter to give the same random error at all levels of pH, the low precision indicates that there is a source of systematic error in the data. Clearly, the data can be very useful to indicate the presence of any bias in an analytical measurement. However, what is good or bad precision? The RSD for a single set of data does not give the scientist much of an idea of whether it is the experiment that has a large error, or whether the error lies with the specimens used. Some crude rules of thumb can be employed: a RSD of less than 2% is considered acceptable, whereas an RSD of more than 5% might indicate error with the analytical method used and would warrant further investigation of the method.

Where possible, we can employ methods such as experimental design to allow for an examination of the precision of the data. One key requirement is that the analyst must make more than a few measurements when collecting data and these should be true replicates, meaning that a set of specimens or aliquots are prepared using exactly the same methodology, i.e., it is not sufficient to make up one solution and then measure it ten times. Rather, we should make up ten solutions to ensure that the errors introduced in preparing the solutions are taken into account as well as the measurement error. Modern instruments have very small measurement errors, and the variance between replicated measurements is usually very low. The largest source of error will most likely lie with the sampling and the preparation of solutions and specimens for measuring.

The accuracy of a measurement is a parameter used to determine just how close the determined value is to the true value for the test specimens. One problem with experimental science is that the true value is often not known. For example, the concentration of lead in the Humber Estuary is not a constant value and will vary depending upon the time of year and the sites from which the test specimens s are taken. Therefore, the true value can only be estimated, and of course will also contain measurement and sampling errors. The formal definition of accuracy is the difference between the experimentally determined mean of a set of test specimens, $\bar{x}$, and the value that is accepted as the true or correct value for that measured analyte, μ_o. The difference is known statistically as the error (e) of x, so we can write a simple equation for the error:

$$e = x - \mu_o \tag{2.13}$$

The larger the number of aliquots or specimens that are determined, the greater the tendency of $\bar{x}$ toward the true value μ_0 (which is obtained from an infinite number of measurements). The absolute difference between μ and the true value is called the *systematic error* or *bias*. The error can now be written as:

$$e = x - \mu + \mu - \mu_o \tag{2.14}$$

The results obtained by experimentation ($\bar{x}$ and σ) will be uncertain due to random errors, which will affect the systematic error or bias. These random errors should be minimized, as they affect the precision of the method.

Several types of bias are common in analytical methodology, including laboratory bias and method bias. *Laboratory bias* can occur in specific laboratories, due to an uncalibrated balance or contaminated water supply, for example. This source of bias is discovered when results of interlaboratory studies are compared and statistically evaluated. *Method bias* is not readily distinguishable between laboratories following a standard procedure, but can be identified when reference materials are used to compare the accuracy of different methods. The use of interlaboratory studies and reference materials allows experimentalists to evaluate the accuracy of their analysis.

2.4 SIGNIFICANCE TESTING

To decide whether the difference between the measured values and standard or references values can be attributable to random errors, a statistical test known as a *significance test* can be employed. This approach is used to investigate whether the difference between the two results is significant or can be explained solely by the effect of random variations. Significance tests are widely used in the evaluation of experimental results. The term "significance" has a real statistical meaning and can be determined only by using the appropriate statistical tools. One can visually estimate that the results from two methods produce similar results, but without the use of a statistical test, a judgment on this approach is purely empirical. We could use the empirical statement "there is no difference between the two methods," but this conveys no quantification of the results. If we employ a significance test, we can report that "there is no significant difference between the two methods." In these cases, the use of a statistical tool simply enables the scientist to quantify the difference or similarity between methods. Summary statistics can be used to provide empirical conclusions, but no quantitative result. Quantification of the results allows for a better understanding of the variables impacting on our data, better design of experiments, and also for knowledge transfer. For example, an analyst with little experimental experience can use significance testing to evaluate the data and then incorporate these quantified results with empirical judgment. It is always a good idea to use one's common sense when applying statistics. If the statistical result flies in the face of the expected result, one should check that the correct method has been used with the correct significance level and that the calculation has been performed correctly. If the statistical result does not confirm the expected result, one must be sure that no errors have occurred, as the use of a significance test will usually confirm the expected result.

The obligation lies with the analyst to evaluate the significance of the results and report them in a correct and unambiguous manner. Thus, significance testing is used to evaluate the quality of results by estimating the accuracy and precision errors in the experimental data.

The simplest way to estimate the accuracy of a method is to analyze reference materials for which there are known values of µ for the analyte. Thus, the difference

Statistical Evaluation of Data

between the experimentally determined mean and the true value will be due to both the method bias and random errors.

Before we jump in and use significance tests, we need to first understand the *null hypothesis*. In making a significance test we test the truth of a hypothesis, which is known as a null hypothesis. The term "null" is used to imply that there is no difference between the observed and known values other than that which can be attributed to random variation. Usually, the null hypothesis is rejected if the probability of the observed difference's occurring by chance is less than 1 in 20 (i.e., 0.05 or 5%). In such a case, the difference is said to be significant at the 0.05 (or 5%) level. This also means there will be a 1 in 20 chance of making an incorrect conclusion from the test results. Test objects at the extremes of the distribution can be incorrectly classified as being from outside the true population, and objects that are, in fact, outside the true population can be incorrectly classified as being from the sample population. We must be aware that statistical tests will sometimes highlight occasional anomalies that must be investigated further rather than rejected outright. These effects are most commonly seen when using statistical control charts, where there are a number of specimens measured over a long period of time. For example, in a control chart used to monitor a process over a period of 100 sample intervals, we expect to find, on the average, five test objects outside the statistical bounds. Significance testing falls into two main sections: testing for accuracy (using the student t-test) and testing for precision (using the F-test).

2.4.1 THE F-TEST FOR COMPARISON OF VARIANCE (PRECISION)

The F-test is a very simple ratio of two sample variances (the squared standard deviations), as shown in Equation 2.15

$$F = \frac{s_1^2}{s_2^2} \qquad (2.15)$$

where s_1^2 is the variance for the first set of data and s_2^2 is the variance for the second data set. Remember that the ratio must return an F value such that $F \geq 1$, so the numerator and denominator must be arranged appropriately. Care must be taken to use the correct degrees of freedom when reading the F table value to ensure that they are matched to the denominator and numerator.

If the null hypothesis is retained, i.e., there is no statistical significant difference between the two variances, then the calculated F value will approach 1. Some critical values of F can be found in Table 2.2a and Table 2.2b. The test can be used in two ways; to test for a significant difference in the variances of the two samples or to test whether the variance is significantly higher or lower for either of the two data sets, hence two tables are shown, one for the one-tailed test and one for the two-tailed test.

TABLE 2.2a
Critical Values for F for a One-Tailed Test ($p = 0.05$)

v_2	\\ v_1	1	2	3	4	5	6	7	8	9	10	20	30	∞
1		161.4	199.5	215.7	224.6	230.2	234	236.8	238.9	240.5	241.9	248	250.1	254.3
2		18.51	19	19.16	19.25	19.3	19.33	19.35	19.37	19.38	19.4	19.45	19.46	19.5
3		10.13	9.55	9.28	9.12	9.01	8.94	8.89	8.85	8.81	8.79	8.66	8.62	8.53
4		7.71	6.94	6.59	6.39	6.26	6.16	6.09	6.04	6	5.96	5.8	5.75	5.63
5		6.61	5.79	5.41	5.19	5.05	4.95	4.88	4.82	4.77	4.74	4.56	4.5	4.36
6		5.99	5.14	4.76	4.53	4.39	4.28	4.21	4.15	4.1	4.06	3.87	3.81	3.67
7		5.59	4.74	4.35	4.12	3.97	3.87	3.79	3.73	3.68	3.64	3.44	3.38	3.23
8		5.32	4.46	4.07	3.84	3.69	3.58	3.5	3.44	3.39	3.35	3.15	3.08	2.93
9		5.12	4.26	3.86	3.63	3.48	3.37	3.29	3.23	3.18	3.14	2.94	2.86	2.71
10		4.96	4.1	3.71	3.48	3.33	3.22	3.14	3.07	3.02	2.98	2.77	2.7	2.54
20		4.35	3.49	3.1	2.87	2.71	2.6	2.51	2.45	2.39	2.35	2.12	2.04	1.84
30		4.17	3.32	2.92	2.69	2.53	2.42	2.33	2.27	2.21	2.16	1.93	1.84	1.62
∞		3.84	3	2.6	2.37	2.21	2.1	2.01	1.94	1.88	1.83	1.57	1.46	1

Note: v_1 = number of degrees of freedom of the numerator; v_2 = number of degrees of freedom of the denominator.

TABLE 2.2b
Critical Values for F for a Two-Tailed Test ($p = 0.05$)

v_2	1	2	3	4	5	6	7	8	9	10
1	647.7890	799.5000	864.1630	899.5833	921.8479	937.1111	948.2169	956.6562	963.2846	968.6274
2	38.5063	39.0000	39.1655	39.2484	39.2982	39.3315	39.3552	39.3730	39.3869	39.3980
3	17.4434	16.0441	15.4392	15.1010	14.8848	14.7347	14.6244	14.5399	14.4731	14.4189
4	12.2179	10.6491	9.9792	9.6045	9.3645	9.1973	9.0741	8.9796	8.9047	8.8439
5	10.0070	8.4336	7.7636	7.3879	7.1464	6.9777	6.8531	6.7572	6.6811	6.6192
6	8.8131	7.2599	6.5988	6.2272	5.9876	5.8198	5.6955	5.5996	5.5234	5.4613
7	8.0727	6.5415	5.8898	5.5226	5.2852	5.1186	4.9949	4.8993	4.8232	4.7611
8	7.5709	6.0595	5.4160	5.0526	4.8173	4.6517	4.5286	4.4333	4.3572	4.2951
9	7.2093	5.7147	5.0781	4.7181	4.4844	4.3197	4.1970	4.1020	4.0260	3.9639
10	6.9367	5.4564	4.8256	4.4683	4.2361	4.0721	3.9498	3.8549	3.7790	3.7168
20	5.8715	4.4613	3.8587	3.5147	3.2891	3.1283	3.0074	2.9128	2.8365	2.7737
30	5.5675	4.1821	3.5894	3.2499	3.0265	2.8667	2.7460	2.6513	2.5746	2.5112
∞	5.0239	3.6889	3.1161	2.7858	2.5665	2.4082	2.2875	2.1918	2.1136	2.0483

Note: v_1 = number of degrees of freedom of the numerator; v_2 = number of degrees of freedom of the denominator.

For example, suppose we wish to determine whether two synthetic routes for producing the same product have the same precision. The data for the two routes are shown below.

Synthetic Route 1 (% yield)	Synthetic Route 2 (% yield)
79.4	78.0
77.1	81.2
76.2	80.5
77.5	78.2
78.6	79.8
77.7	79.5
$\bar{x} = 77.7$	$\bar{x} = 79.5$
$s_1 = 1.12$	$s_2 = 1.26$
$s_1^2 = 1.25$	$s_2^2 = 1.58$
$n = 6$	$n = 6$

To test that the precision of the two routes is the same, we use the F test, so

$$F = \frac{1.58}{1.25} = 1.26$$

As we are testing for a significant difference in the precision of the two routes, the two-tailed test value for F is required. In this case, at the 95% significance level, for 5 degrees of freedom for both the numerator and denominator, the critical value of F is 7.146. As the calculated value from the data is smaller than the critical value of F, we can see that the null hypothesis is accepted and that there is no significant difference in the precision of the two synthetic routes.

The F test is a very simple but powerful statistical test, as many other tests require the variances of the data or populations to be similar (i.e., not significantly different). This is quite logical; it would be rather inappropriate to test the means of two data sets if the precisions of the data were significantly different. As mentioned previously, the precision is related to the reproducibility of the data collected. If we have poor reproducibility, then the power and the significance of further testing are somewhat limited.

2.4.2 THE STUDENT T-TEST

This test is employed to estimate whether an experimental mean, $\bar{x}$, differs significantly from the true value of the mean, μ. This test, commonly known as the t-test, has several possible variations: the standard t-test, the paired t-test, and the t-test with nonequal variance. The computation of each test is quite simple, but the analyst must ensure that the correct test procedure is used.

In the case where the deviation between the known and the experimental values is considered to be due to random errors, the method can be used to assess accuracy. If this assumption is not made, the deviation becomes a measure of the systematic error or bias. The approach to accuracy is limited to where test objects can be compared with reference materials, which is not always the case, for example, where

Statistical Evaluation of Data

an unusual sample matrix is present. In most cases, when a reference material is not available, a standard reference method is used. Of course the reference method gives only an estimate of the true value and may itself be incorrect (i.e., not the true value) but the methodology does provide a procedural standard that can be used for comparison. It is very important to be able to perform the standard reference method as well as any new method, as poor accuracy and precision in the reference results will invalidate any statistical test results.

The numerical value of the t-test to be compared with critical values of t from tables is calculated from experimental results using the following formula:

$$t = \frac{\bar{x} - \mu}{\sigma/\sqrt{n}} \tag{2.16}$$

If the calculated value of t (without regard to sign) exceeds a certain critical value (defined by the required confidence limit and the number of degrees of freedom) then the null hypothesis is rejected. For example, a method for determining lead by atomic absorption returned the following values for a standard reference material containing 38% Pb: 38.9, 37.4, and 37.1%

Let us test the result for any evidence of systematic error. We calculate the appropriate summary statistics and the critical value of t:

$$\bar{x} = 37.8\%$$

$$\sigma = 0.964\%$$

$$\mu = 38.9\%$$

$$t = \frac{37.8 - 38.9}{0.964/\sqrt{3}} = 1.98$$

$$t_{tables, 95\%} = 4.30$$

Comparing the calculated value of t with the critical value at the 95% confidence level (obtained from Table 2.1) we observe that the calculated value is less than the critical level at the desired confidence level, so the null hypothesis is retained and there is no evidence of systematic error in these data.

It is worth noting that the critical t value for an infinite number of test objects at the 95% confidence limit is 1.96 and here, with a sample size of $n = 3$, the value is 4.3, so clearly the larger the number of test objects, the smaller the t critical value becomes. For example, for a sample size of $n = 6$ (and therefore 5 degrees of freedom), the t critical value is 2.57. This is useful, as $n = 6$ is a very common number of test objects to run in an analytical test, and so remembering the critical value saves one from hunting statistical tables. If the calculated value for a data set is less than 2.57, the null hypothesis is retained, and if it is greater than 2.57 the null hypothesis is rejected.

2.4.3 ONE-TAILED OR TWO-TAILED TESTS

This may seem like the introduction of jargon into the hypothesis testing but actually it is just more common sense. As has been mentioned, the sign of the value used in the t-test is meaningless—it can be positive or negative. This is because the mean of a sample set may be lower (negative sign) or higher (positive sign) than the accepted true value. The normal distribution is symmetrical about the mean, so if all one wishes to determine is whether the two means are from the same population (i.e., there is no significant difference in the means) then you can use a two-tailed test, as the value can be either higher or lower than the true mean. However, if one wishes to determine whether a sample mean is either higher or lower, a one-tailed test must be used. This is very useful, especially when one wants to compare limits of detection. For example, this approach can be used to determine whether a new method has a significantly lower limit of detection, rather than just a lower limit of detection.

As mentioned previously, experimentalists need to consider the questions they want to have answered before using statistical tests, as the quality of the results is dependent upon the right question's being asked. As with the F test, care must be taken when using these tests to ensure that the correct values are used. Many spreadsheets will also perform statistical testing, but the answers are often not as clear (they often return a probability of significance that some find is somewhat less clear than a simple comparison). The other piece of jargon that one will come across when using significance testing is the number of degrees of freedom (d.o.f.), usually given the notation $(n - 1)$, where n is the number of objects (or the number of experiments performed). The best way to understand d.o.f. is to think of the number of things that have varied during the collecting of the data. If you run one experiment or take one sample there is no possible variation, therefore d.o.f. will be equal to zero. However, if you measure six objects (or perform six experiments), there are five possible sources of variation. The correction for d.o.f is very important, especially when comparing data sets with different numbers of experiments, but the rule for calculating it remains the same; d.o.f. is the number of possible variations within the data collected.

There are three major uses for the t-test:

1. Comparison of a sample mean with a certified value
2. Comparison of the means from two samples
3. Comparison of the means of two methods with different samples

All three of these situations can be tested for statistical significance, the only difference is the type of test used in each case. In most cases in the real analytical world, the first and last cases are the most commonly encountered.

2.4.4 COMPARISON OF A SAMPLE MEAN WITH A CERTIFIED VALUE

A common situation is one in which we wish to test the accuracy of an analytical method by comparing the results obtained from it with the accepted or true value of an available reference sample. The utilization of the test is illustrated in the following example:

$\bar{x} = 85$ (obtained from 10 replicate test objects)
$s = 0.6$
$\mu = 83$ (the accepted value or true value of the reference material)

Using the general form of Student's *t*-test, we calculate a value of *t* from the experimental data.

$$t = \frac{\bar{x} - \mu}{s/\sqrt{n}} = \frac{85 - 83}{0.6/\sqrt{10}} = 10.5$$

From tables, we obtain the critical value of $t = 1.83$ for a one-tailed test (i.e., the result from our method is significantly higher than the reference sample value at the 95% confidence limit). Comparing the calculated value of *t* with the critical value of *t*, we observe that the null hypothesis is rejected and there is a significant difference between the experimentally determined mean compared with the reference result. Clearly, the high precision of the method (0.6) compared with the deviation between mean result and the accepted or true value (85–83), contributes to the rejection of the null hypothesis.

2.4.5 COMPARISON OF THE MEANS FROM TWO SAMPLES

This version of the *t*-test is used when comparing two methods. Usually a new method that is under development is compared with an existing approved method. Cases like this exist when there is no suitable reference sample available for testing the new method. This situation is quite common, as there are many possible sample matrices and only limited availability of reference materials. This test is slightly different from the one previously described because in this case there will be two standard deviations (one for each method) as well as the two means. Prior to conducting the test, we first need to ensure that the variances for both methods are statistically similar prior to performing any analysis on the sample means. Hence, we perform the *F* test first. The following example is used to illustrate the comparison.

Reference Method	New Method
$\bar{x}_1 = 6.40$	$\bar{x}_2 = 6.56$
$s_1 = 0.126$	$s_2 = 0.179$
$s_1^2 = 0.015$	$s_2^2 = 0.032$
$n = 10$	$n = 10$

First, we will perform the *F* test to ensure that the variances from each method are statistically similar.

$$F = \frac{s_1^2}{s_2^2} = \frac{0.032}{0.015} = 2.13$$

For a two-tailed test (as we are testing to determine that the two variances are statistically similar), the critical value from tables gives $F_{9,9} = 4.026$ at the 95% confidence limit. As the calculated value for F is lower than the critical value, we accept the null hypothesis that there is no significant difference in the variances from the two methods.

We can now apply the t-test. However, as there are two standard deviations, we must first calculate the pooled estimate of the standard deviation, which is based on their individual standard deviations. To do this we use Equation 2.17:

$$s^2 = \frac{(n_1 - 1)s_1^2 + (n_2 - 1)s_2^2}{n_1 + n_2 - 2} \qquad (2.17)$$

giving the following result

$$s^2 = \frac{9 \times 0.015 + 9 \times 0.032}{18} = 0.0235$$

$$s = 0.153$$

The calculated value of t is computed using Equation 2.18:

$$t = \frac{\bar{x}_1 - \bar{x}_2}{S\left(\dfrac{1}{n_1} + \dfrac{1}{n_2}\right)^{1/2}} \qquad (2.18)$$

giving the result, $t = 2.35$.

$$t = \frac{6.40 - 6.56}{0.153\left(\dfrac{1}{10} + \dfrac{1}{10}\right)^{1/2}} = -2.35$$

As there are 18 degrees of freedom, the critical value for t at the 95% confidence limit for a two-tailed test is 2.10. Given that the calculated value for t is greater than the critical value, the null hypothesis is rejected, and we conclude there is a statistical significant difference between the new method and the reference method. We conclude that the two methods have similar precision but significantly different accuracy.

2.4.6 COMPARISON OF TWO METHODS WITH DIFFERENT TEST OBJECTS OR SPECIMENS

Sometimes when comparing two methods in analytical chemistry we are unable to obtain true replicates of each specimen or aliquot, due to limited availability of the test material or the requirements of the analytical method. In these cases, each test object or specimen has to be treated independently for the two methods, i.e., it is not possible to calculate a mean and standard deviation for the samples as each specimen analyzed is different. It is worth remembering that the *t*-test is not for testing the specimens but the methods that have been used to analyze them. The type of test that is used for this analysis of this kind of data is known as the paired *t*-test. As the name implies, test objects or specimens are treated in pairs for the two methods under observation. Each specimen or test object is analyzed twice, once by each method. Instead of calculating the mean of method one and method two, we need to calculate the differences between method one and method two for each sample, and use the resulting data to calculate the mean of the differences $\bar{x}_d$ and the standard deviation of these differences, s_d. The use of the paired *t*-test is illustrated using the data shown below as an example.

Aliquot	Method One	Method Two	Difference
1	90	87	3
2	30	34	−4
3	62	60	2
4	47	50	−3
5	61	63	−2
6	53	48	5
7	40	38	2
8	88	80	8
9	76	78	−2
10	10	15	−5

We calculate the *t* statistic using Equation 2.19:

$$t = \frac{\bar{x}_d \sqrt{n}}{s_d} \tag{2.19}$$

Using the results from Equation 2.19, we obtain the following value for the calculated value of the *t*-statistic:

$$t = \frac{0.4\sqrt{10}}{4.029} = 0.31$$

The critical value of *t* at the 95% confidence limit is 2.26, so we accept the null hypothesis that there is no significant difference in the accuracy of the two methods. The paired *t*-test is a common type of test to use, as it is often the availability of test objects or specimens that is the critical factor in analysis.

2.5 ANALYSIS OF VARIANCE

In the previous section, Student's *t*-test was used to compare the statistical significance of mean results obtained by two different methods. When we wish to compare more than two methods or sample treatments, we have to consider two possible sources of variation, those associated with systematic errors and those arising from random errors. To conform to standard nomenclature used with design of experiments, it is useful to state here that the sample means are the same as the treatment means, which is the term normally associated with design of experiments (DoE). Subsequently, we will use both sample and treatment means synonymously.

Analysis of variance (ANOVA) is a useful technique for comparing more than two methods or treatments. The variation in the sample responses (treatments) is used to decide whether the sample treatment effect is significant. In this way, the data can be treated as random samples from h normal populations having the same variance, σ^2, and differing only by their means. The null hypothesis in this case is that the sample means (treatment means) are not different and that they are from the same population of sample means (treatments). Thus, the variance in the data can be assessed in two ways, namely the *between-sample means* (treatment means) and the *within-sample means* (treatment means).

A common example where ANOVA can be applied is in interlaboratory trials or method comparison. For example, one may wish to compare the results from four laboratories, or perhaps to evaluate three different methods performed in the same laboratory. With inter-laboratory data, there is clearly variation between the laboratories (between sample/treatment means) and within the laboratory samples (treatment means). ANOVA is used in practice to separate the *between-laboratories variation* (the treatment variation) from the random *within-sample variation*. Using ANOVA in this way is known as one-way (or one factor) ANOVA.

2.5.1 ANOVA TO TEST FOR DIFFERENCES BETWEEN MEANS

Let us use an example to illustrate how the ANOVA calculations are performed on some test data. A chemist wishes to evaluate four different extraction procedures that can be used to determine an organic compound in river water (the quantitative determination is obtained using ultraviolet [UV] absorbance spectroscopy). To achieve this goal, the analyst will prepare a test solution of the organic compound in river water and will perform each of the four different extraction procedures in replicate. In this case, there are three replicates for each extraction procedure. The quantitative data is shown below.

Extraction Method	Replicate Measurements (arbitrary units)	Mean Value (arbitrary units)
A	300, 294, 304	299
B	299, 291, 300	296
C	280, 281, 289	283
D	305, 310, 300	305
Overall Mean		296

From the data we can see that the mean values obtained for each extraction procedure are different; however, we have not yet included an estimate of the effect of random error that may cause variation between the sample means. ANOVA is used to test whether the differences between the extraction procedures are simply due to random errors. To do this we will use the null hypothesis, which assumes that the data are drawn from a population μ and have a variance of σ^2.

2.5.2 THE WITHIN-SAMPLE VARIATION (WITHIN-TREATMENT VARIATION)

The variance, s^2, can be determined for each extraction method using the following equation:

$$s^2 = \frac{\sum (x_1 - \bar{x})^2}{n-1} \qquad (2.20)$$

Using the data from above, we obtain the following result:

$$\text{Variance for method } A = \frac{(300-299)^2 + (294-299)^2 + (304-299)^2}{3-1} = 25.5$$

$$\text{Variance for method } B = \frac{(299-296)^2 + (291-296)^2 + (300-296)^2}{3-1} = 25.0$$

$$\text{Variance for method } C = \frac{(280-283)^2 + (281-283)^2 + (289-283)^2}{3-1} = 24.5$$

$$\text{Variance for method } D = \frac{(305-305)^2 + (310-305)^2 + (300-305)^2}{3-1} = 25.0$$

If we now take the average of these method variances, we obtain the within-sample (within-treatment) estimate of the variance.

$$\frac{(25.5 + 25.0 + 24.5 + 25.0)}{4} = 25$$

This is known as the *mean square* because it is a sum of the squared terms (SS) divided by the number of degrees of freedom. This estimate has 8 degrees of freedom; each sample estimate (treatment) has 2 degrees of freedom and there are four samples (treatments). One is then able to calculate the sum of squared terms by multiplying the mean square (MS) by the number of degrees of freedom.

2.5.3 BETWEEN-SAMPLE VARIATION (BETWEEN-TREATMENT VARIATION)

The between-treatment variation is calculated in the same manner as the within-treatment variation.

Method mean variance

$$= \frac{(299-296)^2 + (296-296)^2 + (283-296)^2 + (305-296)^2}{4-1} = 86$$

Summarizing these results, we have:

Within-sample mean square = 25 with 8 d.o.f.
Between-sample mean square = 86 with 3 d.o.f.

The one-tailed F test is used to test whether the between-sample variance is significantly greater than the within-sample variance. Applying the F test we obtain:

$$F_{calc} = \frac{86}{25} = 3.44$$

From Table 2.2b, we obtain a value of $F_{crit} = 4.006$ ($p = 0.05$). As the calculated value of F is less than the critical value of F, the null hypothesis is accepted and we conclude there is no significant difference in the method means.

A result indicating the differences are significantly different in a one-way ANOVA would be indicative of various problems, which could range from one mean being very different, to all means being different, and as such it is important to use a simple method to estimate the source of this variation. The simplest method of estimating the difference between different mean values is to calculate the least significant difference (l.s.d.). A simple method for deciding the cause of a significant result is to arrange the means in increasing order and compare the difference between adjacent means with the least significant difference. The l.s.d. is calculated using the following formula:

$$\text{l.s.d.} = s \sqrt{\left(\frac{2}{n}\right)} \times t_{h(n-1)} \qquad (2.21)$$

where s is the within-sample estimate of variance and $h(n-1)$ is the number of degrees of freedom. For the data used previously, the least significant difference is

$$\sqrt{25} \times \sqrt{\left(\frac{2}{3}\right)} \times 2.36 = 9.63$$

which, when compared with the data for the adjacent means, gives:

$$\bar{x}_A = 25.5, \quad \bar{x}_B = 25.0, \quad \bar{x}_C = 24.5, \quad \bar{x}_D = 25.0$$

The difference between adjacent values clearly shows that there are no significance differences in the means, as the least significant difference, 9.63, is much larger than any of the differences between the pairs of results (the largest difference is between A and C is only 1.0 in magnitude).

2.5.4 ANALYSIS OF RESIDUALS

Results for which the mean values of the samples (treatments) are different, but which have the same variance, is said to be homoscedastic, as opposed to having different variance, which is said to be heteroscedastic. Thus, in the case of homoscedastic variation, the variance is constant with increasing mean response, whereas with heteroscedastic variation the variance increases with the mean response. ANOVA is quite sensitive to

Statistical Evaluation of Data

heteroscedasticity because it attempts to use a comparison of the estimates of variance from different sources to infer whether the treatments have a significant effect. If the data tends to be heteroscedastic, it might be necessary to transform the data to stabilize the variance and repeat the ANOVA. Typical transformations for experimental data would be taking the square root, logarithm, reciprocal, or reciprocal square root.

It is common to use a shorter calculation than the one described previously to achieve the results from ANOVA (or to simply use software). The shortened form of the calculation involves summing the squares of the deviations from the overall mean and dividing by the number of degrees of freedom. Assessing total variance this way takes into account both the within- and between-treatment variations. There is a direct relationship between the sum of between- and within-treatment variations, so by calculating the between-treatment variation and the total variation, one can obtain the within-treatment variation using subtraction.

The table below shows the summary approach for the ANOVA calculations.

Source of Variation	Sum of Squares	Degrees of Freedom
Between-samples (treatments)	$\left(\sum_{t} T_i^2/n\right) - \left(T^2/N\right)$	$h - 1$
Within-samples (treatments)	By subtraction	By subtraction
Total	$\left(\sum_i \sum_j x_{ij}^2\right) - \left(T^2/N\right)$	$N - 1$

In the above table, N is the total number of measurements, n is the number of replicate measurements for each sample or treatment, h is the number of treatments, T_i is the sum of the measurements for the ith sample or treatment, T is the grand total of all measurements, and Σx^2 is the sum of squares of all the data points.

We can illustrate this approach with some new data that were collected to determine whether there is a random sampling effect, rather than a fixed effect, which is the source of variation in the data. The data collected were for the determination of arsenic in coal. The data consisted of five samples of coal and each sample was analyzed four times. The data for arsenic content (ng/g) is shown below:

Coal Sample	Arsenic Content (ng/g)	Mean
A	72, 73, 72, 71	72
B	73, 74, 75, 73	74
C	74, 75, 74, 76	75
D	71, 72, 71, 73	72
E	76, 75, 71, 76	75

The first step is to calculate the mean squares. It is worth remembering that, as the calculation is based upon variance in the data, one can always subtract a common value from the data to make the longhand calculation easier. This will have no effect on the result (and if the calculations are performed by computer, it is not relevant). For this data, the common scalar 70 has been removed from all the data.

Sample	Data (Original Values — 70)	T	T^2
A	2, 3, 2, 1	9	81
B	3, 4, 5, 3	15	225
C	4, 5, 4, 6	19	361
D	1, 2, 1, 3	7	49
E	6, 5, 7, 6	24	576
		$\sum T = 74$	$\sum T^2 = 1292$

For the arsenic data, $n = 4$, $h = 5$, $N = 20$, and $\Sigma x^2 = 331$. We can now use this data to set up an ANOVA table:

	Sum of Squares	Degrees of Freedom	Mean Squares
Between-sample (treatment)	$1291/4 - 74^2/20 = 49$	4	$49/4 = 12$
Within-sample (treatment)	8	15	$8/15 = 0.53$
Total	$331 - 74^2/20 = 57$	19	

From the ANOVA table we can see that the between-treatment mean square is much larger than the within-treatment mean square, but to test the significance of this we use the F-test:

$$F_{calc} = 12/0.53 = 5.446$$

The $F_{crit(4,15)}$ value from tables at the 95% confidence limit for a two-tailed test is 3.804. Therefore, the F_{calc} value is larger then the critical value, which means there is a significant difference in the sampling (treatment) error compared with the analytical error. This finding is very important, especially for environmental data where sampling is very much a part of the analytical methodology, as there is a drive among analysts to gain better and better analytical precision by the employment of higher-cost, high-resolution instruments, but without proper attention to the precision in the sampling procedure. This fact is often borne out in the field of process analysis, where instead of sampling a process stream at regular intervals and then analyzing the samples in a dedicated laboratory (with high precision), analyzers are employed on-line to analyze the process stream continuously, hence, reducing the sampling error. Often these instruments have poorer precision than the laboratory instruments, but the lower sampling error means that the confidence in the result is high.

One can conclude that ANOVA can be a very useful test for evaluating both systematic and random errors in data, and is a useful addition to the basic statistical tests mentioned previously in this chapter. It is important to note, however, there are other factors that can greatly influence the outcome of any statistical test, as any result obtained is directly affected by the quality of the data used. It is therefore important to assess the quality of the input data, to ensure that it is free from errors. One of the most commonly encountered errors is that of outliers.

2.6 OUTLIERS

The inclusion of bad data in any statistical calculation can lead the unwary to false conclusions. The effect of one or a few erroneous data points can totally obscure underlying trends in the data, and it is especially true in experimental science when the number of samples used is few and the cost of the experimentation is high. Clearly, the best manner in which to avoid including outliers in your data is to have sufficient replicates for all samples, but often this is not possible in practice. There are two common sources of outliers in data, the first being outliers in the analytical measurement or samples. These are called *experimental outliers*. The second case is where the error lies not with the measurement but with the reference value either being incorrectly entered into a data book or the standard being made up incorrectly. These kinds of errors are not that uncommon, but they are too easily ignored. One cannot simply remove data that does not seem to fit the original hypothesis; the data must be systematically scrutinized to ensure that any suspected outliers can be proven to lie outside the expected range for that data.

For a quick investigation of a small number of data (less than 20 values), one can use the Dixon Q test, which is ready-made for testing small sets of experimental data. The test is performed by comparing the difference between a suspected outlier and its nearest data point with the range of the data, producing a ratio of the two (i.e., a Q_{calc} value, see Equation 2.22), which is then compared with critical values of Q from tables (see Table 2.3).

$$Q = \frac{|\text{suspect value} - \text{nearest value}|}{|\text{largest value} - \text{smallest value}|} \quad (2.22)$$

TABLE 2.3
Critical Values of Dixon's Q Test for a Two-Tailed Test at the 95% Confidence Level[2]

3	0.970
4	0.829
5	0.710
6	0.625
7	0.568
8	0.526
9	0.493
10	0.466
11	0.444
12	0.426
13	0.410
14	0.396
15	0.384
16	0.374
17	0.365
18	0.356
19	0.349
20	0.342

As is common with all other hypothesis tests covered in this chapter, the calculated value of Q is compared with the appropriate critical value (shown in Table 2.3), and if the calculated value is greater than the critical value, the null hypothesis is rejected and the suspect data is treated as an outlier. Note that the result from the calculation is the modulus result (all negatives are ignored).

If we examine the data used in the arsenic in coal example (for sample or treatment E), we have the following results: arsenic content (ng/g) 76, 75, 71, 76.

The hypothesis we propose to test is that 71 ng/g is not an outlier in this data. Using the Dixon Q test, we obtain the following result:

$$Q = \frac{71-75}{76-71} = \frac{-4}{5} = |0.8|$$

Comparing the calculated value with the critical value for the 95% level, $Q_{crit} = 0.829$, we observe that the calculated value is lower, so the suspect data point is retained, e.g., 71 ng/g arsenic is not an outlier.

It is useful to see what effect retaining or not retaining a data point has on the mean and standard deviation for a set of data. The table below shows descriptive statistics for this data.

	$\bar{x}$	s
Retaining 71 ng/g	74.5	2.4
Rejecting 71 ng/g	75.6	0.58

From the above table it is clear that the main effect is on of the standard deviation, which is an order of magnitude smaller when the suspected point is rejected. This example illustrates how it is important to apply a suitable statistical test, as simply examining the effect of deleting a suspected outlier on the standard deviation may have led us to incorrectly reject the data point. The result from the Q test is clearly quite close (the calculated value is very similar to the critical value), but the data point is not rejected. It is important to get a "feel" of how far a data point must be away from the main group of data before it will be rejected. Clearly, if the main group of data has a small spread (or range from highest to lowest value) then the suspect value will not have to lie very far away from the main group of data before it is rejected. If the main group of data has a wide spread or range, then the outlier will have to lie far outside the range of the main group of data before it will be seen as an outlier.

For the data from the arsenic in the coal example, if we replaced the 71 ng/g value with 70 ng/g, we would obtain $Q_{calc} = 0.833$, which is greater than the $Q_{crit} = 0.829$, and so it would now be rejected as an outlier.

Typically, two types of "extreme values" can exist in our experimentally measured results, namely *stragglers* and *outliers*. The difference between the two is the confidence level required to distinguish between them. Statistically, *stragglers* are detected between the 95% and 99% confidence levels; whereas *outliers* are detected at >99% confidence limit. It is always important to note that no matter how extreme a data point may be in our results, the data point could in fact be correct, and we need to remember that, when using the 95% confidence limit, one in every 20 samples we examine will be classified incorrectly.

Statistical Evaluation of Data

A second method that can be employed for testing for outliers (or extreme values) in experimental data are the Grubbs' tests (Grubbs' 1, 2, and 3). The formulae can be found in Equation 2.23, Equation 2.24, and Equation 2.25, respectively,

$$G_1 = \frac{|\bar{x} - x_i|}{s} \qquad (2.23)$$

$$G_2 = \frac{x_n - x_1}{s} \qquad (2.24)$$

$$G_3 = 1 - \left(\frac{(n-3) \times s_{n-2}^2}{(n-1) \times s^2} \right) \qquad (2.25)$$

where s is the standard deviation, x_i is the suspected extreme value, x_n and x_1 are the most extreme values, and s_{n-2} is the standard deviation for the data excluding the two most extreme values.

What is unique about the use of the Grubbs' tests is that, before the tests are applied, data are sorted into ascending order. The test values for G_1, G_2, and G_3 are compared with values obtained from tables (see Table 2.4), as has been common with all the tests discussed previously. If the test values are greater than the tabulated values, we reject the null hypothesis that they are from the same population and reject the suspected values as outliers. Again, the level of confidence that is used in outlier rejection is usually at the 95 and 99% limits.

TABLE 2.4
Critical Values of G for the Grubbs' Test

	95% Confidence Limit			99% Confidence Limit		
	G_1	G_2	G_3	G_1	G_2	G_3
3	1.153	2.00	—	1.155	2.00	—
4	1.463	2.43	0.9992	1.492	2.44	1.0000
5	1.672	2.75	0.9817	1.749	2.80	0.9965
6	1.822	3.01	0.9436	1.944	3.10	0.9814
7	1.938	3.22	0.8980	2.097	3.34	0.9560
8	2.032	3.40	0.8522	2.221	3.54	0.9250
9	2.110	3.55	0.8091	2.323	3.72	0.8918
10	2.176	3.68	0.7695	2.410	3.88	0.8586
12	2.285	3.91	0.7004	2.550	4.13	0.7957
13	2.331	4.00	0.6705	2.607	4.24	0.7667
15	2.409	4.17	0.6182	2.705	4.43	0.7141
20	2.557	4.49	0.5196	2.884	4.79	0.6091
25	2.663	4.73	0.4505	3.009	5.03	0.5320
30	2.745	4.89	0.3992	3.103	5.19	0.4732
35	2.811	5.026	0.3595	3.178	5.326	0.4270
40	2.866	5.150	0.3276	3.240	5.450	0.3896
50	2.956	5.350	0.2797	3.336	5.650	0.3328

The following example shows how the Grubbs' test is applied to chemical data. The results obtained for the determination of cadmium in human hair by total reflection x-ray fluorescence (TXRF) are shown below:

Cadmium (ng/g)

1.574 1.275 1.999 1.851 1.924 2.421 2.969 1.249

1.810 1.425 2.914 2.217 1.059 2.187 1.876 2.002

First, we arrange the data in ascending order of magnitude:

x_1 x_n
1.059 1.249 1.275 1.425 1.574 1.81 1.851 1.876 1.924 1.999 2.002 2.187 2.217 2.421 2.914 2.969

$n = 16$, mean = 1.922, $s = 0.548$, $s_{n-2}^2 = 0.2025$

$$G_1 = \frac{2.969 - 1.922}{0.548} = 1.91$$

$$G_2 = \frac{2.969 - 1.059}{0.548} = 3.485$$

$$G_3 = 1 - \left(\frac{13 \times 0.2025}{15 \times 0.548^2}\right) = 0.584$$

The 95% confidence limit for Grubbs' critical values for the data are:

$G_1 = 2.409$, $G_2 = 4.17$, and $G_3 = 0.6182$.

Comparing the calculated values of G with the critical values, we observe that there are no outliers in this data. A comparison at the 99% confidence limit also returns the same result. Again, this result is worth commenting upon, as the range of values for these data seem quite large (from 1.059 to 2.969), which might indicate that there are extreme values in the data. There is no statistical evidence of this from any of the three Grubbs' tests applied. This is because the data have quite a large spread, which is indicative of a lack of analytical precision in the results, rather than the presence of any extreme values. Simply looking at data and seeing high or low values is not a robust method for determining extreme values.

It is worth noting that a useful rule of thumb can be applied to the rejection of outliers in data. If more than 20% of the data are rejected as outliers, then one should examine the quality of the collected data and the distribution of the results.

2.7 ROBUST ESTIMATES OF CENTRAL TENDENCY AND SPREAD

Most of the methods discussed previously were based on the assumption that the data were normally distributed, however there are numerous other possible distributions. If the number of objects or specimens measured is small, it is often not possible to determine whether a set of data conform to any known distribution.

Statistical Evaluation of Data

Robust statistics are a set of methods that are largely unaffected by the presence of extreme values. Commonly used statistics of this type are the median and the median absolute deviation (MAD). The median is a measure of the central tendency of the data and can be used to replace the mean value. If the data are normally distributed (i.e., symmetrical about the mean) then the mean and median will have the same value. The calculation of the median is very simple. It is calculated by arranging the data in ascending order. From the series of sorted data, the median is simply the central number in this series (or the mean of the two center numbers if the number of points are even).

Below is another example from the analysis of human hairs. The analytical results are for the concentration of copper (in ng/g):

Copper (ng/g)

48.81   30.61   39.01   65.42   44.19   51.44   46.29   50.91   48.47   41.83   29.27   79.34

Arranging this data in order, we have:

29.27   30.61   39.01   41.83   44.19   46.29   48.47   48.81   50.91   51.44   65.42   89.34

The median is then 47.38, and the mean is 48.79.

The median absolute deviation, MAD, is a robust estimate for gauging the spread of the data and similar to the standard deviation. To calculate the MAD value, we use Equation 2.26,

$$\text{MAD} = \text{median}(|x_i - \tilde{x}|) \quad (2.26)$$

where $\tilde{x}$ is the median of the data. Using the copper in human hair data as an example, we obtain the following,

$$\text{MAD} = \text{median}(|29.27 - 47.38|, |30.67 - 47.38|, \ldots)$$

MAD = median (1.09 1.09 1.43 3.19 3.53 4.06 5.55 8.37 16.77 18.04 18.11 41.96)

$$\text{MAD} = \frac{4.06 + 5.55}{2} = 4.805$$

whereas the standard deviation = 15.99.

If the MAD value is scaled by a factor of 1.483, it becomes comparable to the standard deviation (MAD_E). In this example, $\text{MAD}_E = 7.13$, which is less than half the standard deviation. We can clearly see that although the mean and the median values are quite similar, the spread of the data is quite different (i.e., there is a large difference between the MAD and standard deviation).

Of course, we could use the tests described previously to see whether there are extreme values in the data, but then we cannot be certain that those extreme values are outliers or that the data simply does not follow a normal distribution. To answer this question, we have to look at the origin of the data to try to understand which tests to apply and for what reason. For example, we would expect replicates of an analysis to follow a normal distribution, as the errors that are expected would be random errors. However, in the case of copper in human hair samples, the hair comes from different people, thus different environments, colors, hair products, etc., so the distribution of the data is not so easy to estimate.

2.8 SOFTWARE

Most of the methods detailed in this chapter can be performed using computer software, even nonspecialized software such as a spreadsheet. This saves time and reduces errors in the data processing stage, but the results are only as good as the data from which they are derived. There still remains the problem in choosing the correct test to use for your data and ensuring that the data are in the correct format for the software, which sometime is not quite as straightforward as one would hope. It is also worth noting that the output from the software is not as clear as the comparison of a test result with a tabulated result, as most software packages commonly estimate the probability level for the calculated test statistic in question ($0 < p \leq 1$), rather than comparing the value of the calculated test statistic to a tabulated value. Some users can find this confusing. One of the best methods to ensure that you understand the output from the software is to use data for which you know the correct answer worked out longhand (one of the previous examples would suffice) and then use that data with the software package of your choice to compare the output. Also, be aware that some of the statistical tools available with many software packages are often not installed by default and have to be installed when first used.

For example, using Microsoft Excel™ to perform the t-test, one would use the following syntax:

$$= \text{TTEST(ARRAY1, ARRAY2, tails, type)}$$

where *Array1* and *Array2* are the data you wish to use, *tails* is 1 for one-tailed tests and 2 for two-tailed tests, and *type* is 1 for a paired test, 2 for two samples with equal variance, and 3 for two samples with unequal variance. The output or result is not a t_{calc} value but a probability. If we use the data from the previous paired t-test, the probability returned is 0.77, which is less than the 0.95 probability level, and, as such, we accept the null hypothesis, which is the same result we obtained using the longhand method and the t-test tables.

One can also use the Data Analysis Toolbox feature of Microsoft Excel. If this feature does not appear in the *Tool* menu, you will need to install it. To perform the same test, select Tools\Data Analysis Toolbox\t test: Paired two Sample for Means.

Statistical Evaluation of Data

Select the input range for each variable and then output range as a new workbook. The output is shown below:

	t-Test: Paired Two Sample for Means	
	Variable 1	Variable 2
Mean	55.7	55.3
Variance	642.0111111	518.9
Observations	10	10
Pearson correlation	0.990039365	
Hypothesized mean difference	0	
d.o.f.	9	
t Stat	0.297775	
P(T ≤ t) one-tail	0.386317127	
t Critical one-tail	1.833113856	
P(T ≤ t) two-tail	0.772634254	
t Critical two-tail	2.262158887	

2.8.1 ANOVA USING EXCEL

The ANOVA calculation can be performed in Excel only if the Data Analysis Toolbox has been installed. To perform the calculation, you can select the ANOVA tool under Tools\Data Analysis\ANOVA Single Factor from the Excel toolbar. Using the following example of arsenic content of coal taken from different parts of a ship's hold, where there are five sampling points and four aliquots or specimens taken at each point, we have the data as shown below:

Sample	Arsenic Content (ng/g)			
A	72	73	72	71
B	73	74	75	73
C	74	75	74	76
D	71	72	71	73
E	76	75	71	76

We wish to determine whether there is a statistically significant difference in the sampling error vs. the error of the analytical method. It has been previously mentioned in this chapter that the sampling errors are often much greater than the analytical errors, and so now we can use ANOVA to illustrate this example.

To perform the analysis, enter the data into an Excel spreadsheet (start at the top left-hand corner cell A1), then select the ANOVA : Single Factor option from the Tool Menu. Select all the data by entering $B$2:$E$6 in the input range box (or select the data using the mouse). Now ensure that you select the Grouped By Rows Radio Button, as the default is to assume the data are grouped in columns (remember we want to determine whether there is a difference in the sampling error over the analytical error). Check that the probability level (alpha) is set to 0.05 (this is the default, but it might have been changed by another user) to ensure that one is testing at the 95% confidence level. The output is best saved into another workbook and is shown below:

ANOVA: Single Factor

SUMMARY

Groups	Count	Sum	Average	Variance
A	4	288	72	0.666667
B	4	295	73.75	0.916667
C	4	299	74.75	0.916667
D	4	287	71.75	0.916667
E	4	298	74.5	5.666667

ANOVA

Source of Variation	SS	df	MS	F	P-value	F crit
Between groups	31.3	4	7.825	4.307339	0.016165	3.055568
Within groups	27.25	15	1.816667			
Total	58.55	19				

From these results, it is clear that the random sampling error (the between-group variance) is statistically significantly different compared with the random analytical error (the within-groups variance).

Excel, a very powerful tool for many statistical calculations, is widely available. The routine use of a spreadsheet will dramatically reduce any errors due to incorrect calculations performed by hand as the data you are using are always visible on screen and so any errors are easily spotted. Also, saving the workbooks allows one to review any calculations over time to ensure no errors have occurred.

RECOMMENDED READING

Miller, J.C. and Miller, J.N., *Statistics and Chemometrics for Analytical Chemistry*, 4th ed., Prentice Hall, New York, 2000.

REFERENCES

1. IUPAC, Compendium of Analytical Nomenclature, 1997; http://www.iupac.org/publications/analytical_compendium/.
2. Rorabacher, D.B., Statistical method for rejection of deviant values: Critical values of Dixon's "Q" parameter and related subrange ratio of the 95% confidence level. *Anal. Chem.*, 63, 139–146, 1991.

3 Sampling Theory, Distribution Functions, and the Multivariate Normal Distribution

Paul J. Gemperline and John H. Kalivas

CONTENTS

3.1 Sampling and Sampling Distributions 42
    3.1.1 The Normal Distribution 43
    3.1.2 Standard Normal Distribution 45
3.2 Central Limit Theorem 45
    3.2.1 Implications of the Central Limit Theorem 45
3.3 Small Sample Distributions 46
    3.3.1 The t-Distribution 46
    3.3.2 Chi-Square Distribution 47
3.4 Univariate Hypothesis Testing 48
    3.4.1 Inferences about Means 49
    3.4.2 Inferences about Variance and the F-Distribution 51
3.5 The Multivariate Normal Distribution 51
    3.5.1 Generalized or Mahalanobis Distances 52
    3.5.2 The Variance–Covariance Matrix 53
    3.5.3 Estimation of Population Parameters from Small Samples 54
    3.5.4 Comments on Assumptions 55
    3.5.5 Generalized Sample Variance 55
    3.5.6 Graphical Illustration of Selected Bivariate Normal Distributions 56
    3.5.7 Chi-Square Distribution 58
3.6 Hypothesis Test for Comparison of Multivariate Means 59
3.7 Example: Multivariate Distances 59
    3.7.1 Step 1: Graphical Review of smx.mat Data File 60
    3.7.2 Step 2: Selection of Variables (Wavelengths) 61
    3.7.3 Step 3: View Histograms of Selected Variables 61
    3.7.4 Step 4: Compute the Training Set Mean and Variance–Covariance Matrix 62

3.7.5   Step 5: Calculate Mahalanobis Distances and
         Probability Densities 64
3.7.6   Step 6: Find "Acceptable" and
         "Unacceptable" Objects 65
Recommended Reading 66
References 67

In this chapter we introduce the multivariate normal distribution and its use for hypothesis testing in chemometrics. The world around us is inherently multivariate, and it makes sense to consider multiple measurements on a single object simultaneously. For example, when we measure the ultraviolet (UV) absorbance of solution, it is easy to measure its entire spectrum quickly and rapidly rather than measuring its absorbance at a single wavelength. We will learn that by properly considering the distribution of multiple variables simultaneously, we get *more information* than is obtained by considering each variable individually. This is the so-called *multivariate advantage*. The additional information comes to us in the form of correlation. When we look at one variable at a time, we neglect the correlation between variables, and hence we miss part of the picture.

Before launching into the topic of multivariate distributions, a review of univariate sampling distributions is presented as well as a review of univariate hypothesis testing. We begin with a discussion of the familiar Gaussian or normal probability distribution function, the chi-square distribution, and the F-distribution. A review of these distribution functions and their application to univariate descriptive statistics provides the necessary background and sets the stage for an introduction to multivariate descriptive statistics and their sampling distributions.

3.1   SAMPLING AND SAMPLING DISTRIBUTIONS

Any statistical study must begin with a representative sample. By making measurements on a small representative set of objects, we can learn something about the characteristics of the whole group. The statistical descriptions we develop can then be used for making inferences and decisions. There are many methods for selecting a sample, but the most common is *simple random sampling*. When the population being sampled is an infinite population (the usual case in chemistry), each object selected for measurement must be (1) selected at random from the same population and (2) selected independently from the other objects.

To a statistician, a population is a *complete collection of measurements on objects that share one or more common features*. For example, one might be interested in determining the average age of male freshman students at a university. Such a group represents a *finite population* of size n, and one could characterize this population by calculating the *population mean*, μ, and the *population standard deviation*, σ, for every individual.

$$\mu = \frac{1}{n}\sum_{i=1}^{n} x_i, \quad \sigma = \left[\frac{\sum_{i=1}^{n}(x_i - \mu)^2}{n}\right] \quad (3.1)$$

In chemistry, a more relevant example might be the determination of the composition of ingredients like pseudoephedrine hydrochloride, microcrystalline cellulose, and magnesium stearate in granules of a pharmaceutical preparation. This example represents an *infinite population*, because the concentration of an ingredient in an aliquot of material can take on any conceivable value.

The goal of the pharmaceutical manufacturing process is to produce a mixture of granules having a homogeneous distribution of ingredients that are then fed to a tablet press. Provided the composition of granules is homogeneous, the tablets so produced would have uniform potency. In this example, it would not be practical to collect a comprehensive set of information for the entire manufactured lot. Alternatively, one might take several small aliquots of granules and assume that all chemical species have an equal chance of appearing at identical concentrations in the individual aliquots. Unfortunately, the composition of ingredients is most likely not distributed homogeneously throughout the granules, which therefore presents the analyst with a problem when trying to obtain a representative sample. An inherent sampling error always transpires owing to the heterogeneity of the population. In addition, an analysis error exists that is caused by random error present in the measurements used for the analysis. The resulting final variance, s^2, can be represented as

$$s^2 = s_s^2 + s_a^2 \qquad (3.2)$$

where s_s^2 denotes the sampling variance and s_a^2 represents the variance due to analysis. Here, the definition of the word "sample" means a *subset of measurements selected from the population of interest*. Notice that chemists often refer to a sample as a representative aliquot of substance that is to be measured, which is different than the definition used by statisticians. The discussion that follows pertains primarily to the sampling of homogeneous populations; a discussion of sampling heterogeneous populations can be found in more specialized texts.

3.1.1 THE NORMAL DISTRIBUTION

Consider the situation in which a chemist randomly samples a bin of pharmaceutical granules by taking n aliquots of equal convenient sizes. Chemical analysis is then performed on each aliquot to determine the concentration (percent by weight) of pseudoephedrine hydrochloride. In this example, measurement of concentration is referred to as a *continuous random variable* as opposed to a *discrete random variable*. Discrete random variables include counted or enumerated items like the roll of a pair of dice. In chemistry we are interested primarily in the measurement of continuous properties and limit our discussion to continuous random variables.

A *probability distribution function* for a continuous random variable, denoted by $f(x)$, describes how the frequency of repeated measurements is distributed over the range of observed values for the measurement. When considering the probability distribution of a continuous random variable, we can imagine that a set of such measurements will lie within a specific interval. The area under the curve of a graph of a probability distribution for a selected interval gives the probability that a measurement will take on a value in that interval.

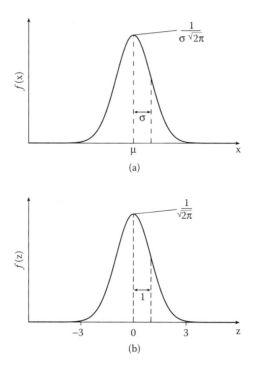

FIGURE 3.1 Distribution curves: (a) normal and (b) standard normal.

If normal distributions are followed, the probability function curves for the concentration of pseudoephedrine hydrochloride from the previous example should follow the familiar bell-shaped curve as shown in Figure 3.1, where μ specifies the *population mean* concentration for a species and x represents an individual concentration value for that species. The probability function for the normal distribution is given by the function,

$$f(x) = \frac{1}{\sigma\sqrt{2\pi}} \exp\left[-\frac{(x-\mu)^2}{2\sigma^2}\right] \tag{3.3}$$

where σ is the population standard deviation.

The highest point in the curve is represented by the mean because the measurements tend to cluster around some central or average value. Small deviations from the mean are more likely than large deviations, thus the curve is highest at the mean, and the tails of the curve asymptotically approach zero as the axes extend to infinity in both directions. The shape of the curve is symmetrical because negative deviations from the mean value are just as likely as positive deviations.

In this example, the normal distribution for pseudoephedrine hydrochloride can be described as $x = N(\mu, \sigma^2)$, where σ^2 is termed the *variance*. When sampling an infinite population, as is the case in this example, it is impossible to determine the true population mean, μ, and standard deviation, σ. A reasonable, more feasible approach is to use an assembly of n aliquots. In this case, $\bar{x}$ (the mean of the n aliquots taken) is an estimate of μ, and σ is estimated by s (the standard deviation for the n aliquots), which is calculated according to Equation 3.4:

$$s = \left[\frac{\sum_{i=1}^{n}(x_i - \bar{x})^2}{n-1} \right]^{1/2} \quad (3.4)$$

The resulting concentration distribution is now characterized by using the notation $x = N(\bar{x}, s^2)$.

3.1.2 Standard Normal Distribution

For convenience, the normal distribution can be transformed to a standard normal distribution where the mean is zero and the standard deviation equals 1. The transformation is achieved using Equation 3.5:

$$z_i = (x_i - \mu)/\sigma \quad (3.5)$$

The probability distribution can now be represented by Equation 3.6

$$f(z) = \frac{1}{\sqrt{2\pi}} \exp\left[-\frac{z^2}{2}\right] \quad (3.6)$$

and represented by the notation $z = N(0,1)$. Figure 3.1 shows a plot of the standard normal distribution. In terms of our pharmaceutical example, the normal concentration distribution for each chemical species with their different means and standard deviations can be transformed to $z = N(0,1)$. A single table of probabilities, which can be found in most statistical books, can now be used.

3.2 CENTRAL LIMIT THEOREM

According to the important theorem known as the *central limit theorem*, if N samples of size n are obtained from a population with mean, μ, and standard deviation, σ, the probability distribution for the means will approach the normal probability distribution as N becomes large *even if the underlying distribution is nonnormal.* For example, as more samples are selected from a bin of pharmaceutical granules, the distribution of N means, $\bar{x}$, will tend toward a normal distribution with mean μ and standard deviation $\sigma_{\bar{x}} = \sigma/\sqrt{n}$, regardless of the underlying distribution.

3.2.1 Implications of the Central Limit Theorem

With the central limit theorem, we have expanded from dealing with individual concentration determinations to concentration means. Each chemical species distribution can be transformed to a standard distribution by

$$z = (\bar{x} - \mu)/\sigma_{\bar{x}} \quad (3.7)$$

with $\bar{x}$ functioning as the random variable. Provided that σ is known, the population mean can now be estimated to lie in the range

$$\mu = \bar{x} \pm z\sigma_{\bar{x}} \qquad (3.8)$$

where z is obtained for the desired level of confidence, $\alpha/2$, from a table of probabilities. Equation 3.8 describes what is commonly called the *confidence interval of the mean* at $100\%(1 - \alpha)$. Using statistical tables to look up values of z, we can estimate the interval in which the true mean lies at any desired confidence level. For example, if we determine the average concentration of pseudoephedrine hydrochloride and its 95% confidence interval in a tablet to be 30.3 ± 0.2 mg, we would say: "There is a 95% probability that the true mean lies in the interval 30.1 to 30.5 mg."

3.3 SMALL SAMPLE DISTRIBUTIONS

The population mean and standard deviation cannot be determined for an infinite population; hence, they must be estimated from a sample of size n. When μ and σ are estimated from small samples, $\mu \approx \bar{x}$ and $\sigma \approx s$, the uncertainty in their estimates may be large, depending on the size of n, thus the confidence interval described in Equation 3.8 must be inflated accordingly by use of the *t*-distribution. When n is small, say 3 to 5, the uncertainty is large, whereas when n is large, say 30 to 50, the uncertainty is much smaller.

3.3.1 THE T-DISTRIBUTION

In order to compensate for the uncertainty incurred by taking small samples of size n, the t probability distribution shown in Figure 3.2 is used in the calculation of confidence intervals, replacing the normal probability distribution based on z values shown in Figure 3.1. When $n \geq 30$, the *t*-distribution approaches the standard normal probability distribution. For small samples of size n, the confidence interval of the mean is inflated and can be estimated using Equation 3.9

$$\mu = \bar{x} \pm t_{\alpha/2} s_{\bar{x}} \qquad (3.9)$$

where t expresses a value for $n - 1$ degrees of freedom at a desired confidence level. The term *degrees of freedom* refers to the number of independent deviations $(x_i - \bar{x})$ that are used in calculating s. For example, if one wished to estimate the $100\%(1 - \alpha) = 95\%$ confidence interval of a mean at $n - 1 = 5$ degrees of freedom, the critical value of $t_{\alpha/2} = 2.571$ at $\alpha/2 = 0.025$ obtained from standard *t*-tables would be used in Equation 3.9. Here, $\alpha/2 = 0.025$ represents the fraction of values in the right-hand tail and the left-hand tail of the *t*-distribution. The selected values in the corresponding *t*-distribution are illustrated graphically in Figure 3.2b. Equation 3.9 does not imply that the sample means are not normally distributed; rather, it suggests that s is a poor estimate of σ except when n is large.

Sampling Theory, Distribution Functions

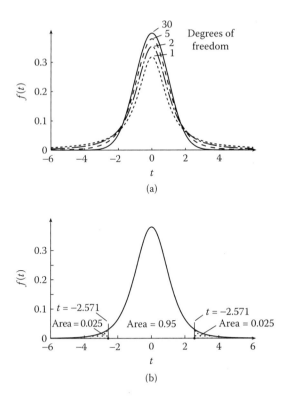

FIGURE 3.2 Illustration of the *t*-distribution (a) for various degrees of freedom and (b) the area under the curve equal to 0.95 at 5 degrees of freedom.

3.3.2 CHI-SQUARE DISTRIBUTION

In the previous section we discussed the ramifications of the uncertainty in estimating means from small samples and described how the sample mean, $\bar{x}$, follows a *t*-distribution. In this section, we discuss the ramifications of the uncertainty in estimating s^2 from small samples. The variable s^2 is called the *sample variance*, which is an estimate of the *population variance*, σ^2. For simple random samples of size n selected from a normal population, the quantity in Equation 3.10

$$\frac{(n-1)s^2}{\sigma^2} \tag{3.10}$$

follows a *chi-square distribution* with $n-1$ degrees of freedom. A graph of the chi-square distribution is shown in Figure 3.3 for selected degrees of freedom. Tables of the areas under the curve are available in standard statistical textbooks

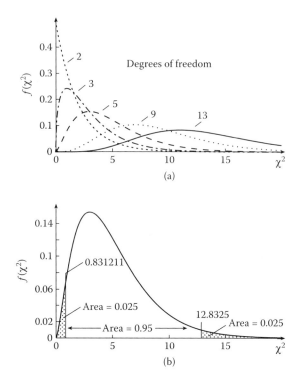

FIGURE 3.3 Illustration of the chi-square distribution (a) for various degrees of freedom and (b) the area under the curve equal to 0.95 at 5 degrees of freedom.

and can be used to estimate the confidence interval for sample variances shown in Equation 3.11

$$\frac{(n-1)s^2}{\chi^2_{\alpha/2}} \leq \sigma^2 \leq \frac{(n-1)s^2}{\chi^2_{(1-\alpha/2)}} \tag{3.11}$$

where α represents the fraction of values between the right-hand tail, $\chi^2_{\alpha/2}$, and left-hand tail, $\chi^2_{(1-\alpha/2)}$, of the chi-square distribution as shown in Figure 3.3a.

3.4 UNIVARIATE HYPOTHESIS TESTING

In the previous sections we discussed probability distributions for the mean and the variance as well as methods for estimating their confidence intervals. In this section we review the principles of *hypothesis testing* and how these principles can be used for *statistical inference*. Hypothesis testing requires the supposition of two hypotheses: (1) the *null hypothesis*, denoted with the symbol H_0, which designates the hypothesis being tested and (2) the *alternative hypothesis* denoted by H_a. If the tested null hypothesis is rejected, the alternative hypothesis must be accepted. For example, if

Sampling Theory, Distribution Functions

we were making a comparison of two means, $\bar{x}_1$ and $\bar{x}_2$, the appropriate null hypothesis and alternative hypothesis would be:

$$H_o : \bar{x}_1 \leq \bar{x}_2$$
$$H_a : \bar{x}_1 > \bar{x}_2$$
(3.12)

In order to test these two competing hypotheses, we calculate a test statistic and attempt to prove the null hypothesis false, thus proving the alternative hypothesis true. It is important to note that we cannot prove the null hypothesis to be true; we can only prove it to be false.

3.4.1 Inferences about Means

Depending on the circumstances at hand, several different types of mean comparisons can be made. In this section we review the method for comparison of two means with independent samples. Other applications, such as a comparison of means with matched samples, can be found in statistical texts. Suppose, for example, we have two methods for the determination of lead (Pb) in orchard leaves. The first method is based on the electrochemical method of potentiometric stripping analysis [1], and the second is based on the method of atomic absorption spectroscopy [2]. We perform replicate analyses of homogeneous aliquots prepared by dissolving the orchard leaves into one homogeneous solution and obtain the data listed in Table 3.1.

We wish to perform a test to determine whether the difference between the two methods is statistically significant. In other words, can the difference between the two means be attributed to random chance alone, or are other significant experimental factors at work? The hypothesis test is performed by formulating an appropriate null hypothesis and an alternative hypothesis:

$$H_0 : \bar{x}_1 = \bar{x}_2 \qquad H_0 : \bar{x}_1 - \bar{x}_2 = 0$$
$$\text{or}$$
$$H_1 : \bar{x}_1 > \bar{x}_2 \qquad H_1 : \bar{x}_1 - \bar{x}_2 > 0$$
(3.13)

In developing the hypothesis, note that a difference of zero between the two means is equivalent to a hypothesis stating that the two means are equal.

To make the test, we compute a test statistic based on small sample measurements such as those described in Table 3.1 and compare it with tabulated values. The result

TABLE 3.1
Summary Data for the Analysis of Pb in Orchard Leaves

Method	Potentiometric Stripping	Atomic Absorption
Sample size, N	5	5
Mean, $\bar{x}$	5.03 ppb	4.93 ppb
Std. dev., s	0.08 ppb	0.12 ppb

of the test can have four possible outcomes. If we (1) accept H_0 when it is true, then we have made the correct decision; however, if we (2) accept H_0 when H_a is true, we have made what is called a *Type II error*. The probability of making such an error is called β. If we (3) reject H_0 when it is false, we have made the correct decision; however, if we (4) reject H_0 when it is true, then we have made what is called a *Type I error*. The probability of making such an error is called α.

Most applications of statistical hypothesis testing require that we specify the maximum allowable probability of making a Type I error, and this is called the *significance level* of the test. Typically, significance levels of 0.01 or 0.05 are used. This implies we have a high degree of confidence in making a decision to reject H_0. For example, when we reject H_0 at the 95% confidence level, 5% of the time we expect to make the wrong decision. In other words, we determine that there is a 5% probability the difference is due to random chance." Consequently, we are very confident we have made the correct decision.

To make the test for the comparison of means described in Equation 3.13, we compute the test statistic, t_{calc},

$$t_{calc} = \frac{\bar{x}_1 - \bar{x}_2}{s_p} \sqrt{\frac{n_1 n_2}{n_1 + n_2}} \quad (3.14)$$

and compare it with tabulated values of t at $n_1 + n_2 - 2$ degrees of freedom at a significance level α, where s_p is the pooled standard deviation.

$$s_p^2 = \frac{(n_1 - 1)s_1^2 + (n_2 - 1)s_2^2}{n_1 + n_2 - 2} \quad (3.15)$$

If $t_{calc} > t_\alpha$, then we reject H_0 at the 100%(1 − α) confidence level. For the data shown in Table 3.1, we have $t_{calc} = 2.451$ and $t_{\alpha=0.05, v=8} = 1.860$, thus we reject H_0 and accept H_1 at the 95% confidence level. We say that there is less than a 5% probability the difference is due to random chance. The language used to describe the significance level of a hypothesis test and the confidence level of the decision making implies a relationship between the two. The formula for calculating the confidence interval of the difference between two means is given in Equation 3.16

$$(\bar{x}_1 - \bar{x}_2) \pm t_{\alpha/2} s_p \sqrt{\frac{n_1 + n_2}{n_1 n_2}} \quad (3.16)$$

where $t_{\alpha/2}$ is obtained from the *t*-distribution at $n_1 + n_2 - 2$ degrees of freedom and s_p is the pooled standard deviation. Note that a simple rearrangement of this equation gives a form similar to Equation 3.14. The *t*-test for the comparison of means is equivalent to estimating the confidence interval for the test statistic and then checking to see if the confidence interval contains the hypothesized value for the test statistic.

Sampling Theory, Distribution Functions

3.4.2 Inferences About Variance and the F-Distribution

It is sometimes desirable to compare the variances of two populations. For example, the data shown in Table 3.1 represent two different populations. Prior to calculating a pooled standard deviation, it might be appropriate to test to see if the variances are equivalent, e.g., we might ask, "Is the difference between the two variances, s_1^2 and s_2^2, statistically significant, or can the difference be explained by random chance alone?" The F-distribution is used for conducting such tests and describes the distribution of the ratio, $F = s_1^2/s_2^2$, for independent random samples of size n_1 and n_2. The ratio is always arranged so that F is greater than one, thus the larger of the two variances, s_1^2 and s_2^2, is placed in the numerator of the ratio. The F-distribution has n_1 degrees of freedom in the numerator and n_2 degrees of freedom in the denominator. To conduct a one-tailed test to compare the variances of two populations, the following set of null and alternative hypotheses is formed:

$$H_0 : s_1^2 \leq s_1^2$$
$$H_1 : s_1^2 > s_1^2 \quad (3.17)$$

The F-test statistic is computed and compared with tabulated values at significance level α.

$$\text{Test statistic}: F = \frac{s_1^2}{s_1^2} \quad \text{Reject } H_0 \text{ if } F > F_\alpha \quad (3.18)$$

Following the example shown in Table 3.1, we have $F = 2.25$ and $F_{\alpha=0.05, v_1=4, v_2=4} = 6.39$ at $\alpha = 0.05$ and degrees of freedom $v_1, v_2 = 4$ in the numerator and denominator. We thus accept H_0 at the 95% confidence level and say that the difference between the two variances is not statistically significant.

3.5 THE MULTIVARIATE NORMAL DISTRIBUTION

As we saw in Section 3.1.1, the familiar bell-shaped curve describes the sampling distributions of many experiments. Many distributions encountered in chemistry are approximately normal [3]. Regardless of the form of the parent population, the central limit theorem tells us that sums and means of samples of random measurements drawn from a population tend to possess approximately bell-shaped distributions in repeated sampling. The functional form of the curve is described by Equation 3.19.

$$f(x) = \frac{1}{\sigma\sqrt{2\pi}} \exp\left[-\frac{(x-\mu)^2}{2\sigma^2}\right] \quad (3.19)$$

The term $1/\sigma\sqrt{2\pi}$ is a normalization constant that sets the total area under the curve to exactly 1.0. The approximate area under the curve within ±1 standard deviation is 0.68, and the approximate area under the curve within ±2 standard deviations is 0.95.

The multivariate normal distribution is a generalization of the univariate normal distribution with $p \geq 2$ dimensions. Consider a $1 \times p$ vector $\mathbf{x}_i^T$ obtained by measuring several variables for the ith observation and the corresponding vector of means for each variable:

$$\mathbf{x}^T = [x_1, x_2, \cdots x_p] \qquad (3.20)$$

$$\boldsymbol{\mu}^T = [\mu_1, \mu_2, \cdots \mu_p] \qquad (3.21)$$

In the example at the beginning of this chapter, we considered the *univariate* distribution of pseudoephedrine hydrochloride in a preparation of pharmaceutical granules. We neglected the other ingredients, including microcrystalline cellulose and magnesium stearate, which were also in the granules. If we wish to properly consider the distribution of all three ingredients *simultaneously*, then we must consider a *multivariate distribution* with $p = 3$ variables. Each object or aliquot of pharmaceutical granules can be assayed for the concentration of each of the three ingredients; thus each object in a sample of size n is represented by a vector of length 3. The resulting sample or set of observations is an $(n \times p)$ matrix, one row per object, with variables arranged in columns.

By properly considering the distribution of all three variables simultaneously, we get more information than is obtained by considering each variable individually. This is the so-called *multivariate advantage*. This extra information is in the form of correlation between the variables.

3.5.1 Generalized or Mahalanobis Distances

For convenience, we normalized the univariate normal distribution so that it had a mean of zero and a standard deviation of one (see Section 3.1.2, Equation 3.5 and Equation 3.6). In a similar fashion, we now define the *generalized multivariate squared distance* of an object's data vector, $\mathbf{x}_i$, from the mean, $\boldsymbol{\mu}$, where $\boldsymbol{\Sigma}$ is the variance–covariance matrix (described later):

$$d_i^2 = (\mathbf{x}_i - \boldsymbol{\mu}) \boldsymbol{\Sigma}^{-1} (\mathbf{x}_i - \boldsymbol{\mu})^T \qquad (3.22)$$

This distance is also called the *Mahalanobis distance* by many practitioners after the famous Indian mathematician, Mahalanobis [4]. The distance in multivariate space is analogous to the normalized univariate squared distance of a single point (in units of standard deviations) from the mean:

$$z_i^2 = \left[\frac{x_i - \mu}{\sigma}\right]^2 = (x_i - \mu)\, \sigma^{-1} (x_i - \mu) \qquad (3.23)$$

3.5.2 THE VARIANCE–COVARIANCE MATRIX

The Σ matrix is the $p \times p$ variance–covariance matrix, which is a measure of the degree of scatter in the multivariate distribution.

$$\Sigma = \begin{bmatrix} \sigma_{1,1}^2 & \sigma_{1,2}^2 & \cdots & \sigma_{1,p}^2 \\ \sigma_{2,1}^2 & \sigma_{2,2}^2 & \cdots & \sigma_{2,p}^2 \\ \vdots & \vdots & \ddots & \vdots \\ \sigma_{p,1}^2 & \sigma_{p,2}^2 & \cdots & \sigma_{p,p}^2 \end{bmatrix} \quad (3.24)$$

The variance and covariance terms $\sigma_{i,i}^2$ and $\sigma_{i,j}^2$ in the variance–covariance matrix are given by Equation 3.25 and Equation 3.26, respectively.

$$\sigma_{i,i}^2 = \frac{1}{n-1} \sum_{k=1}^{n} (x_{k,i} - \bar{x}_i)^2 \quad (3.25)$$

$$\sigma_{i,j}^2 = \frac{1}{n-1} \sum_{k=1}^{n} (x_{k,i} - \bar{x}_i)(x_{k,j} - \bar{x}_j) \quad (3.26)$$

Note that the matrix is symmetrical about the diagonal; variances appear on the diagonal and covariances appear on the off-diagonal. If we were to neglect the covariance terms from the variance–covariance matrix, any resulting statistical analysis that employed it would be equivalent to a univariate analysis in which we consider each variable one at a time. At the beginning of the chapter we noted that considering all variables simultaneously yields more information, and here we see that it is precisely the covariance terms of the variance–covariance matrix that encodes this extra information.

Having described squared distances and the variance–covariance matrix, we are now in a position to introduce the multivariate normal distribution, which is represented in Equation 3.27,

$$f(x) = \frac{1}{(2\pi)^{p/2} |\Sigma|^{1/2}} e^{-1/2[(\mathbf{x}_i - \boldsymbol{\mu})\Sigma^{-1}(\mathbf{x}_i - \boldsymbol{\mu})^T]} \quad (3.27)$$

where the constant

$$(2\pi)^{p/2} |\Sigma|^{1/2} \quad (3.28)$$

normalizes the *volume* of the distribution to 1.00. Comparing Equation 3.27 to Equation 3.19 reveals significant similarities. Each contains a normalization constant, and each contains an exponential term that characterizes the squared normalized distance. In fact, Equation 3.27 is a generalization of Equation 3.19

to more than one variable. If only one variable ($p = 1$) is employed in Equation 3.27, the simpler univariate normal distribution described by Equation 3.19 is obtained.

The variance–covariance matrix can be normalized to give the matrix of correlation coefficients between variables. Recall that the correlation coefficient is the cosine of the angle, ϕ, between two vectors. Because the correlation of any variable with itself is always perfect ($\rho_{i,j} = 1$), the diagonal elements of the correlation matrix, **R**, are always 1.00.

$$\rho_{ij} = \frac{\sigma_{ij}}{\sqrt{\sigma_{ii}}\sqrt{\sigma_{jj}}} \quad (3.29)$$

$$\mathbf{R} = \begin{bmatrix} 1.0 & \rho_{1,2} & \cdots & \rho_{1,p} \\ \rho_{2,1} & 1.0 & \cdots & \rho_{2,p} \\ \vdots & \vdots & 1.0 & \vdots \\ \rho_{p,1} & \rho_{p,2} & \cdots & 1.0 \end{bmatrix} \quad (3.30)$$

3.5.3 ESTIMATION OF POPULATION PARAMETERS FROM SMALL SAMPLES

The population parameters $\mathbf{\Sigma}$ and $\boldsymbol{\mu}$ completely specify the properties of a multivariate distribution. Usually it is impossible determine the population parameters; therefore, one usually tries to estimate them from a small finite sample of size n, where n is the number of observations, The *population mean vector*, $\boldsymbol{\mu}$, is approximated by the *sample mean vector*, $\bar{\mathbf{x}}$, which is simply the mean of each column in the data matrix $\mathbf{X}$ shown in Figure 3.4. As n becomes large, the approximation in Equation 3.31 becomes better.

$$\boldsymbol{\mu}^T \approx \bar{\mathbf{x}}^T = [\bar{x}_1, \bar{x}_2 \ldots \bar{x}_p] = \left[\frac{1}{n}\sum_{i=1}^{n} x_{i,1} \quad \frac{1}{n}\sum_{i=1}^{n} x_{i,2} \quad \cdots \quad \frac{1}{n}\sum_{i=1}^{n} x_{i,p} \right] \quad (3.31)$$

The *population variance-covariance matrix*, $\mathbf{\Sigma}$, is approximated by the *sample variance-covariance matrix*, $\mathbf{S}$, when small samples are used. In order to calculate

$$\mathbf{X} = \begin{bmatrix} x_{1,1} & x_{1,2} & \cdots & x_{1,p} \\ x_{2,1} & x_{2,2} & \cdots & x_{2,p} \\ \vdots & \vdots & \ddots & \vdots \\ x_{n,1} & x_{n,2} & \cdots & x_{n,p} \end{bmatrix}_{(n \times p)}$$

p variables in columns; n objects in rows

FIGURE 3.4 Arrangement of a multivariate data set in matrix form.

Sampling Theory, Distribution Functions

the sample variance–covariance matrix, the sample mean vector must be subtracted row-wise from each row of **X**. As n becomes large, the approximation in Equation 3.32 becomes better.

$$\Sigma \approx \mathbf{S} = \frac{1}{n-1}(\mathbf{X} - \bar{\mathbf{x}})^T(\mathbf{X} - \bar{\mathbf{x}}^T) \tag{3.32}$$

3.5.4 Comments on Assumptions

In the multivariate distribution, it is assumed that measurements of objects or aliquots of material (e.g., a single trial) produces vectors, $\mathbf{x}_i$, having a multivariate normal distribution. The measurements of the p variables in a *single* object, such as $\mathbf{x}_i^T = [x_{i,1}, x_{i,2}, ..., x_{i,p}]$ will usually be correlated. In fact, this is expected to be the case. The measurements from *different* objects, however, are assumed to be independent. The independence of measurements from object to object or from trial to trial may not hold when an instrument drifts over time, as with sets of p wavelengths in a spectrum. Violation of the tentative assumption of independence can have a serious impact on the quality of statistical inferences. As a consequence of these assumptions, we can make the following statements about data sets that meet the above criteria:

- Linear combinations of the columns of **X** are normally distributed.
- All subsets of the components of **X** have a multivariate normal distribution.
- Zero covariance implies that the corresponding variables are independently distributed.
- **S** and $\bar{\mathbf{x}}$ are sufficient statistics when all of the sample information in the data matrix **X** is contained in $\bar{\mathbf{x}}$ and **S**, regardless of the sample size n. Generally, large n leads to better estimates of $\bar{\mathbf{x}}$ and **S**.
- Highly correlated variables should not be included in columns of **X**. In this case, computation of multivariate distances becomes problematic because computation of the inverse of the variance–covariance matrix becomes unstable (see Equation 3.22).

3.5.5 Generalized Sample Variance

The determinant of **S** is often called the *generalized sample variance*. It is proportional to the square of the volume generated by the p deviation vectors, $\mathbf{x} - \bar{\mathbf{x}}$.

$$|\mathbf{S}| = (n-1)^p (\text{volume})^2 \quad \text{or} \quad V = \sqrt{\frac{|\mathbf{S}|}{(n-1)^p}}$$

The generalized sample variance describes the scatter in the multivariate distribution. A large volume indicates a large generalized variance and a large amount of scatter in the multivariate distribution. A small volume indicates a small generalized variance and a small amount of scatter in the multivariate distribution. Note that if there are linear dependencies between variables, then the generalized volume will be zero. In this case, the offending row(s) of variables should be identified and removed from the data set before an analysis is performed.

The total sample variance is the sum of the diagonal elements of the sample variance–covariance matrix, **S**. Total variance = $s^2_{1,1} + s^2_{2,2} + ... + s^2_{p,p}$. Geometrically, the total sample variance is the sum of the squared lengths of the p deviation vectors, $\mathbf{x} - \bar{\mathbf{x}}$.

3.5.6 GRAPHICAL ILLUSTRATION OF SELECTED BIVARIATE NORMAL DISTRIBUTIONS

Some plots of several bivariate distributions are provided in Figure 3.5 to Figure 3.7. In each case, the variance–covariance matrix, **S**, is given, followed by a scatter plot

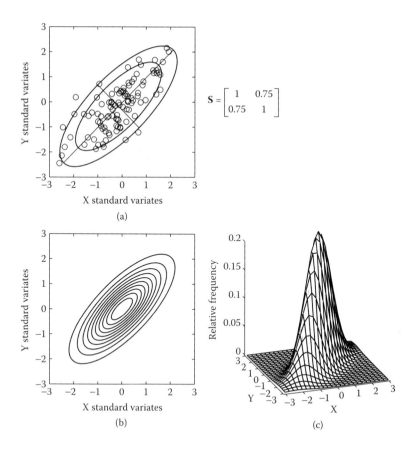

FIGURE 3.5 Scatter plots (a) of a bivariate normal distribution (100 points) with a correlation of 0.75, $|\mathbf{S}| = 0.44$. Ellipses are drawn at 80 and 95% confidence intervals. Contour plots (b) and mesh plots (c) of the corresponding bivariate normal distribution functions are also shown.

Sampling Theory, Distribution Functions

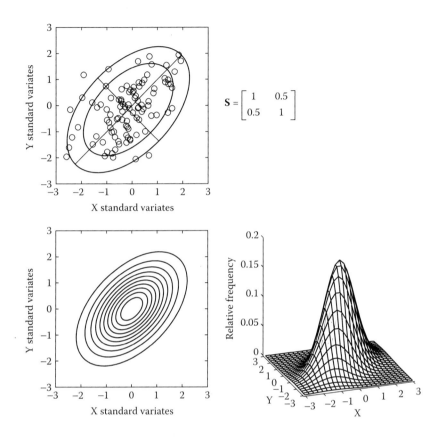

FIGURE 3.6 Scatter plots (a) of a bivariate normal distribution (100 points) with a correlation of 0.50, $|S| = 0.75$. Ellipses are drawn at 80 and 95% confidence intervals. Contour plots (b) and mesh plots (c) of the corresponding bivariate normal distribution functions are also shown.

of the measurements on variable $p = 1$ vs. $p = 2$, as well as contour plots and mesh plots of the corresponding distribution. Bivariate distributions with a high level of correlation have an elongated ellipsoid shape and $|S|$ approaches zero, whereas bivariate distributions with a low level of correlation are shorter and wider and $|S| > 0$. In the limit, as the correlation between the two variables goes to zero, the distribution becomes spherical and $|S|$ approaches its data-dependent upper limit. In the examples that follow, $x_{i,j}$ has been standardized to $z_{i,j} = (x_{i,j} - \bar{x}_j)/s_j$ and, as the correlation varies from 0 to 1, $|S|$ also ranges from 0 to 1.

From the plots of two-dimensional normal distributions in Figure 3.5 through Figure 3.7, it is clear that contours of constant probability density are ellipses centered at the mean. For p-dimensional normal distributions, contours of constant probability density are ellipsoids in $p = 3$ dimensions, or hyperellipsoids in $p > 3$ dimensions, centered about the centroid. The axes of each ellipsoid of constant

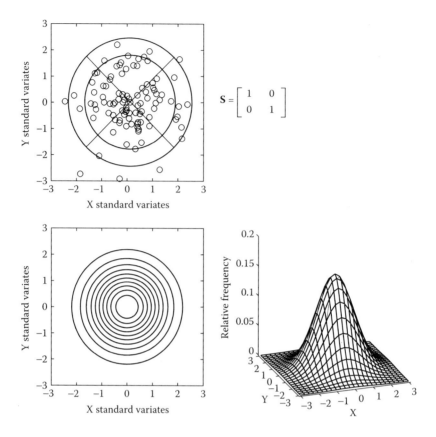

FIGURE 3.7 Scatter plots (a) of a bivariate normal distribution (100 points) with a correlation of 0, $|S| = 1$. Ellipses are drawn at 80 and 95% confidence intervals. Contour plots (b) and mesh plots (c) of the corresponding bivariate normal distribution functions are also shown.

density are in the direction of the eigenvectors, $e_1, e_2, ..., e_n$, of Σ^{-1}, and their lengths are proportional to the reciprocals of the square roots of the eigenvalues, $\lambda_1, \lambda_2, ..., \lambda_n$. For a bivariate distribution, there are two axes:

$$\text{Major axis} = \pm c\sqrt{\lambda_1} \cdot e_1$$
$$\text{Minor axis} = \pm c\sqrt{\lambda_2} \cdot e_2$$
(3.33)

3.5.7 CHI-SQUARE DISTRIBUTION

It can be shown that values of $\chi_p^2(\alpha)$ from the chi-square distribution with p degrees of freedom give contours that contain $(1 - \alpha) \times 100\%$ of the volume in the multivariate normal distribution curve. For example, picking a value of $\chi_p^2(\alpha) = 5.99$ for c with $p = 2$ and $\alpha = 0.05$ gives an ellipse that circumscribes 95% of the sample population in Figure 3.5 through Figure 3.7. Samples having squared distances less than this value will lie *inside* the ellipse. Samples having squared distances greater than this will lie *outside* the ellipse.

3.6 HYPOTHESIS TEST FOR COMPARISON OF MULTIVARIATE MEANS

Often it is useful to compare a multivariate measurement for an object with the mean of a multivariate population [5]. To perform the test, we will determine if a $1 \times p$ vector $\bar{\mathbf{x}}^T$ is an acceptable value for the mean of a multivariate normal distribution, $\boldsymbol{\mu}^T$, according to the null hypothesis and alternative hypothesis shown in Equation 3.34.

$$H_0 : \bar{\mathbf{x}} = \boldsymbol{\mu} \quad \text{vs.} \quad H_1 : \bar{\mathbf{x}} \neq \boldsymbol{\mu} \qquad (3.34)$$

The true values of $\boldsymbol{\Sigma}$ and $\boldsymbol{\mu}$ are usually estimated from a small sample of size n. When n is very large, the estimates $\bar{\mathbf{x}}$ and $\mathbf{S}$ are very good; however, n is usually small, and thus the estimates $\bar{\mathbf{x}}$ and $\mathbf{S}$ have a lot of uncertainty. In this case it is necessary to make an adjustment for the confidence interval, $100\%(1-\alpha)$, of the sample mean and scatter matrix by use of Hotelling's T^2 statistic.

$$T^2 = (\bar{\mathbf{x}} - \boldsymbol{\mu})\mathbf{S}^{-1}(\bar{\mathbf{x}} - \boldsymbol{\mu})^T > \frac{(n-1)}{(n-p)} F_{p, n-p, \alpha} \qquad (3.35)$$

The T^2 statistic is computed and compared with $(n-1)/(n-p)F$ values at significance level α, and we reject the null hypothesis, H_0, when $T^2 > (n-1)/(n-p)F$.

$$\text{Test statistic}: F = \frac{s_1^2}{s_1^2} \quad \text{Reject} \quad H_0 \text{ if } F > F_\alpha \qquad (3.36)$$

3.7 EXAMPLE: MULTIVARIATE DISTANCES

A set of data is provided in the file called "smx.mat." The measurements consist of 83 NIR (near-infrared) reflectance spectra of many different lots of sulfamethoxazole, an active ingredient used in pharmaceutical preparations. The data set has been partitioned into three parts: a training set of 42 spectra, a test set of 13 spectra, and a set of 28 "reject" spectra. The reject samples were intentionally spiked with two degradation products of sulfamethoxazole, sulfanilic acid or sulfanilamide, at 0.5 to 5% by weight. In these exercises, you will inspect the data set, select several wavelengths, and calculate the Mahalanobis distances of samples using the reflectance measurements at the selected wavelengths to determine whether the NIR measurement can be used to detect samples with the above impurities. Sample MATLAB code for performing the analysis is provided at the end of each section.

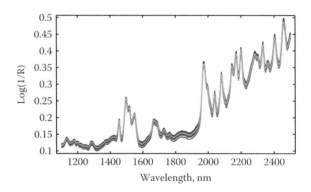

FIGURE 3.8 NIR reflectance spectra of 83 aliquots of sulfamethoxazole powder.

3.7.1 Step 1: Graphical Review of smx.mat Data File

The first step, as in any chemometrics study, begins by plotting the data. Using the MATLAB `plot` command, plot the NIR spectra. As seen in Figure 3.8, there are different baseline offsets between the different spectra, which are typically due to differences in the particle-size distribution of the measured aliquots. The effect of baseline offsets can be removed by taking the first derivative of the spectra. A simple numerical approximation of the first difference can be obtained by taking the difference between adjacent points using the MATLAB `diff` command (Figure 3.9). Use the MATLAB `zoom` command to investigate small regions of the derivative spectra, where a significant amount of spectral variability can be observed. These regions may be useful for finding differences between samples. By zooming in on selected regions, pick at least four wavelengths for subsequent analysis. Good candidates will be uncorrelated (neighbors tend to be highly

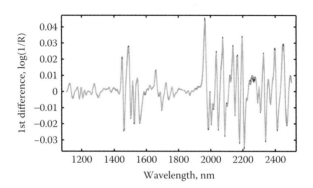

FIGURE 3.9 First-difference NIR reflectance spectra of 83 aliquots of sulfamethoxazole powder.

correlated) and exhibit a lot of variability. In fact, highly correlated variables should be avoided, since then calculation of the inverse of the variance–covariance matrix becomes unstable.

MATLAB Example 3.1

```
load smx
whos
plot(w,a);
da=diff(a');
figure(2);
plot(w(2:end),da);
zoom;
```

3.7.2 STEP 2: SELECTION OF VARIABLES (WAVELENGTHS)

Having selected wavelength regions of interest, it will be necessary to find their respective column indices in the data matrix. Use the MATLAB find command to determine the indices of the wavelengths you have selected. In the example below, variables at 1484, 1692, 1912, and 2264 nm were selected. These are not necessarily the most informative variables for this problem and you are encouraged to try and obtain better results by picking different sets of variables. Once these indices are known, select a submatrix containing 83 rows (spectra) and 4 columns (wavelength variables). The variables trn, tst, and rej in the smx.mat file contain the row indices of objects to be partitioned into a training set, test set, and reject set, respectively. Use the indices trn, tst, and rej to partition the previous submatrix into three new matrices, one containing the training spectra, one containing the test spectra, and one containing the reject spectra.

MATLAB Example 3.2

```
find(w==1484)
find(w==1692)
find(w==1912)
find(w==2264)
% select submatrix of a with 4 wavelengths
wvln_idx=[97 149 204 292]
a4=a(:,wvln_idx);
% Split the data set in 3 parts: training set, test set, and reject set
atrn=a4(trn,:);
atst=a4(tst,:);
arej=a4(rej,:);
```

3.7.3 STEP 3: VIEW HISTOGRAMS OF SELECTED VARIABLES

The training set will be used to calculate the multivariate mean and variance–covariance matrix; however, before calculating these parameters, we will graphically examine the training set to see if it contains measurements that are approximately normally distributed. This can be accomplished by several methods, the simplest being to plot histograms of the individual variables. Use the MATLAB hist command to

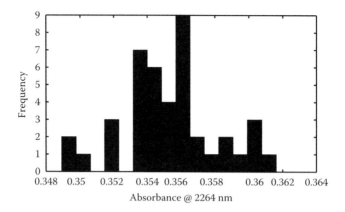

FIGURE 3.10 Histogram showing the absorbance of the sulfamethoxazole training set at 2264 nm.

make a histogram of each selected variable (type help hist) and note which variables tend to show normal behavior or a lack of normal behavior. For example, Figure 3.10 shows a plot of a histogram of the absorbance values at 2264 nm for the training set. The training set contains 42 objects or spectra, thus it is expected that the corresponding histogram will have some gaps and spikes.

MATLAB Example 3.3

```
hist(atrn(:,1),15); % plot a histogram with 15 bins
hist(atrn(:,2),15);
hist(atrn(:,3),15);
hist(atrn(:,4),15);
```

3.7.4 STEP 4: COMPUTE THE TRAINING SET MEAN AND VARIANCE–COVARIANCE MATRIX

In Step 4 we use the MATLAB mean, cov, and corrcoef functions to compute $\bar{x}$, **S**, **R** for the training set. Table 3.2 shows the results of one such calculation, the correlation matrix. It can be used to identify the pair of variables that exhibit the largest correlation and the pair of variables that exhibit the least correlation. After identifying these matrices, use MATLAB to construct scatter plots using these pairs of variables. For example, in Figure 3.11, the absorbance of training samples at 1912 nm is plotted against the absorbance at 2264 nm. There is a total of 42 points in the plot, one for each spectrum or object in the training set. The distribution appears approximately bivariate normal, with the highest density of points near the centroid and a lower density of points at the edges of the cluster. Ellipses are drawn at the 80 and 95% confidence intervals.

TABLE 3.2
Correlation Coefficients between Absorbance at Selected Pairs of Wavelengths for Sulfamethoxazole Training Set

	Wavelengths (nm)			
	1484	1692	1912	2264
1484	1.000	0.747	0.851	0.757
1692	0.747	1.000	0.940	0.714
1912	0.851	0.940	1.000	0.690
2264	0.757	0.714	0.690	1.000

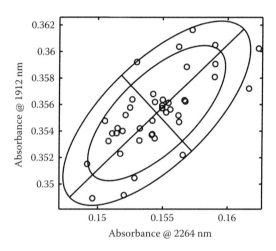

FIGURE 3.11 Scatter plot of absorbance at 1912 nm vs. 2264 nm for the sulfamethoxazole training set. Ellipses are drawn at the 80 and 95% confidence intervals.

MATLAB Example 3.4

```
m=mean(atrn)
format short e
s=cov(atrn)
format short
r=corrcoef(atrn)
% Make scatter plots of pairs with high correlation, low correlation
figure(1); plot(atrn(:,3),atrn(:,4),'o');
figure(2); plot(atrn(:,3),atrn(:,2),'o');
```

3.7.5 Step 5: Calculate Mahalanobis Distances and Probability Densities

A short MATLAB function called `m_dist.m` is provided for computing the Mahalanobis distances or generalized distances. Two data sets must be specified, a training set and a test set. The sample mean vector and sample-covariance matrix are calculated from the training set. Distances from the training set centroid are calculated for each object or row in the test set. If you wish to calculate the distances of the training objects from the centroid of the training set, then the function can be called with the same data set for the training and test sets. This function also calculates the probability density function of each object using Hotelling's T^2 statistic. In Step 5, we use this function to calculate the distances and probabilities of objects in the sulfamethoxazole test set and in the adulterated sulfamethoxazole aliquots in the reject set. The results for the test set are shown in Table 3.3. Each of the test set objects are found to have relatively small distances from the mean and lie inside the 95% confidence interval. For example, test set aliquot 1 has the largest distance from the centroid, 2.1221. Its probability density is 0.398, which indicates it lies at the boundary of the 60.2% confidence interval.

TABLE 3.3
Mahalanobis Distances and Probability Densities from Hotelling's T^2 Statistic for Test Samples of Sulfamethoxazole Compared with the Sulfamethoxazole Training Set

Test Object	Mahalanobis Distance	Probability Density (Hotelling's T^2)
1	2.1221	0.398
2	1.5805	0.680
3	1.2746	0.824
4	2.1322	0.393
5	2.0803	0.418
6	1.2064	0.851
7	1.3914	0.773
8	1.4443	0.748
9	1.8389	0.543
10	2.4675	0.249
11	1.3221	0.804
12	1.2891	0.818
13	1.8642	0.530

Sampling Theory, Distribution Functions 65

MATLAB Example 3.5

```
[d,pr] = m_dist(atrn,arej);
[d,pr]
[d,pr] = m_dist(atrn,atst);
% calculate dist. and prob., test set vs. trn set.
[d pr]
[d,pr] = m_dist(atrn,arej);
% calculate dist. and prob., rej set vs. trn set.
[d pr]
function [d,pr] = m_dist(atrn,atst);
% [d,pr] = m_dist(atrn,atst);
%
% Function to calculate the Mahalanobis Distances of samples, given atrn
% where atrn is a matrix of column-wise vars. The training set, atrn, is
% used to estimate distances for the test set, atst.
%
% The samples' distances from the centroid are returned in d. The
% probability density (chi-squared) for the distance is given in pr.
%
[r,c]=size(atrn);
[rt,ct]=size(atst);
[am,m]=meancorr(atrn);% always use meancorrected data
atm=atst-m(ones(1,rt),:);
s=(am' * am)./(r-1);% calc inv var-covar matrix
d=atm * inv(s) .* atm;% calc distances for test data
d=sqrt(sum(d')');
pr=1-hot_t(c,r,d);% calc prob level
```

3.7.6 STEP 6: FIND "ACCEPTABLE" AND "UNACCEPTABLE" OBJECTS

In Step 6, MATLAB's `find` command is used to find adulterated aliquots of sulfamethoxazole that lie outside the 90 and 95% confidence intervals. Selected examples are shown in Table 3.4. At 3% sulfanilic acid by weight, the Mahalanobis distance is 3.148 and the probability density is 0.077, indicating this sample lies outside the 90% confidence interval. A hypothesis test at the significance level of 0.10 would identify it as an "outlier" or "unacceptable" object. At this point, we conclude the four wavelengths selected for the analysis are not particularly good at detecting sulfanilic acid or sulfanilamide in sulfamethoxazole. Selection of alternative wavelengths can give dramatically better sensitivity for these two contaminants. Additionally, four wavelength variables may not necessarily be an optimal number. Good results might be obtained with just three wavelength variables, or perhaps five wavelength variables are needed. Can you find them? One way to begin approaching this problem would be to plot the derivatives of the training spectra using a green color, and plotting the derivatives of the reject spectra using a red color. The MATLAB `zoom` command can then be used to explore the

TABLE 3.4
Mahalanobis Distances and Probability Densities from Hotelling's T^2 Statistic for Adulterated Samples of Sulfamethoxazole Compared with the Sulfamethoxazole Training Set

Description	Mahalanobis Distance	Probability Density (Hotelling's T^2)
1% SNA	1.773	0.578
2% SNA	2.885	0.126
3% SNA	3.148	0.077
4% SNA	5.093	0.001
5% SNA	5.919	0.000
1% SNM	1.253	0.833
2% SNM	1.541	0.700
3% SNM	3.158	0.075
4% SNM	3.828	0.018
5% SNM	4.494	0.004

Note: SNA = sulfanilic acid; SNM = sulfanilamide.

spectra for regions where there are significant differences between the training and reject spectra.

MATLAB Example 3.6

```
% Search for acceptable samples
t=find(pr>.05)'
nm_rej(t,:)

% Search for samples outside the 99% probability level
t=find(pr<0.01)'
pr(t)'
nm_rej(t,:)

pause
% Replot the spectra, rejects in red, training in green
datrn=diff(a(trn,:)');
darej=diff(a(rej,:)');
plot(w(2:end),darej,'r'); hold on;
plot(w(2:end),datrn,'g'); hold off;
% Use the zoom command to find regions with better discriminating power
```

RECOMMENDED READING

Mark, H. and Workman, J., *Statistics in Spectroscopy*, Academic Press, San Diego, CA, 1991.
Box, G.E.P, *Statistics for Experimenters*, John Wiley & Sons, New York, 1978.

Massart, D.L., Vandeginsted, B.G.M., Buydens, L.M.C., De Jong, S., Lewi, P.J., and Smeyers-Verbeke, J., Eds., *Handbook of Chemometrics and Qualimetrics: Part A*, Elsevier Science B.V., Amsterdam, Netherlands 1997.

Johnson, R.A. and Wichern, D.W., *Applied Multivariate Statistical Analysis*, 3rd ed., Prentice Hall, Upper Saddle River, NJ, 1992.

REFERENCES

1. Rozali bin Othman, M., Hill, J.O., and Magee, R.J., Determination of lead and cadmium in biological samples by potentiometric stripping analysis, *Fresenius' Zeitschrift fuer Analytische Chemie*, 326, 350–353, 1987.
2. Ebdon, L. and Lechotycki, A., The determination of lead in environmental samples by slurry atomization-graphite furnace-atomic absorption spectrophotometry using matrix modification, *Microchemical J.*, 34, 340–348, 1986.
3. Shah, N.K. and Gemperline, P.J., Combination of the Mahalanobis distance and residual variance pattern recognition techniques for classification of near-infrared reflectance spectra, *Anal. Chem.*, 62, 465–470, 1990.
4. De Maesschalck, R., Jouan-Rimbaud, D., and Massart, D.L., The Mahalanobis distance, *Chemom. Intell. Lab. Syst.*, 50, 1–18, 2000.
5. Gemperline, P.J. and Boyer, N.R., Classification of near-infrared spectra using wavelength distances: comparison to the Mahalanobis distance and residual variance methods, *Anal. Chem.*, 67, 160–166, 1995.

4 Principal Component Analysis

Paul J. Gemperline

CONTENTS

4.1 Introduction 70
4.2 Spectroscopic-Chromatographic Data 70
    4.2.1 Basis Vectors 71
4.3 The Principal Component Model 73
    4.3.1 Eigenvectors and Eigenvalues 74
    4.3.2 The Singular-Value Decomposition 76
    4.3.3 Alternative Formulations of the Principal Component Model 77
4.4 Preprocessing Options 77
    4.4.1 Mean Centering 78
    4.4.2 Variance Scaling 78
    4.4.3 Baseline Correction 80
    4.4.4 Smoothing and Filtering 81
    4.4.5 First and Second Derivatives 82
    4.4.6 Normalization 83
    4.4.7 Multiplicative Scatter Correction (MSC) and Standard Normal Variate (SNV) Transforms 83
4.5 PCA Data Exploration Procedure 86
4.6 Influencing Factors 87
    4.6.1 Variance and Residual Variance 89
    4.6.2 Distribution of Error in Eigenvalues 93
    4.6.3 F-Test for Determining the Number of Factors 93
4.7 Basis Vectors 96
    4.7.1 Clustering and Classification with PCA Score Plots 98
4.8 Residual Spectra 98
    4.8.1 Residual Variance Analysis 100
4.9 Conclusions 102
Recommended Reading 103
References 103

4.1 INTRODUCTION

The term principal component analysis (PCA) refers to a method of data analysis for building linear multivariate models of complex data sets [1]. The linear multivariate PCA models are developed using orthogonal basis vectors (eigenvectors), which are usually called principal components. The principal components model the statistically significant variation in the data set as well as the random measurement error. One of the significant goals of PCA is to eliminate the principal components associated with noise, thereby reducing the dimensionality of complex problems and minimizing the effects of measurement error.

Before formally introducing the PCA model in this chapter, the stage for understanding it is set by introducing a hypothetical data set such as high-performance liquid chromatograph with one that might be obtained from UV/visible diode-array detector (HPLC-UV/visible diode array). The mathematical model for this hypothetical data set is easily understood and thus serves as a convenient starting point for introducing the PCA model. The formal introduction of the PCA model immediately follows the descriptions of the HPLC-UV/visible data set and includes a discussion of eigenvectors, eigenvalues, and their relation to the singular-value decomposition (SVD). After the PCA model is introduced, Section 4.4 discusses (a) common preprocessing options and mathematical transformations and (b) their effects on the PCA model. In Section 4.6, the discussion of factors that influence the results of a PCA analysis is meant to provide practical guidance on the selection of the appropriate number of factors to include in PCA models. The chapter finishes up with a section on residuals in PCA models.

4.2 SPECTROSCOPIC-CHROMATOGRAPHIC DATA

Principal component analysis is most easily explained by showing its application on a familiar type of data. In this chapter we show the application of PCA to chromatographic-spectroscopic data. These data sets are the kind produced by so-called hyphenated methods such as gas chromatography (GC) or high-performance liquid chromatography (HPLC) coupled to a multivariate detector such as a mass spectrometer (MS), Fourier transform infrared spectrometer (FTIR), or UV/visible spectrometer. Examples of some common hyphenated methods include GC-MS, GC-FTIR, HPLC-UV/Vis, and HLPC-MS. In all these types of data sets, a response in one dimension (e.g., chromatographic separation) modulates the response of a detector (e.g., a spectrum) in a second dimension.

Consider, as an example, an HPLC-UV/Vis chromatographic data matrix $\mathbf{A}$, where two overlapped peaks elute. Let $\mathbf{A}$ be an $n \times m$ matrix (n rows of spectra recorded at m wavelengths) of mixture spectra. The data matrix $\mathbf{A}$ can be expressed as a product of k vectors representing digitized pure-component chromatograms and k vectors representing digitized pure-component spectra, as shown in Equation 4.1.

Principal Component Analysis

The data matrix **A** expressed by the linear model in Equation 4.1 is often called a *bilinear* data matrix.

$$\mathbf{A} = \mathbf{CP}^T + \boldsymbol{\varepsilon} \qquad (4.1)$$

$$\begin{bmatrix} & & & & \\ & & \mathbf{A} & & \\ & & & & \\ & & & & \end{bmatrix} = \begin{bmatrix} & & \\ & \mathbf{C} & \\ & & \\ & & \end{bmatrix} \begin{bmatrix} & & & & \\ & & \mathbf{P}^T & & \\ & & & & \end{bmatrix}$$

In Equation 4.1, **C** is the $n \times k$ matrix of pure chromatograms (k independently varying components), **P** is the $m \times k$ matrix of pure-component spectra, and the matrix $\boldsymbol{\varepsilon}$ contains unexplained variance, e.g., measurement error. Figure 4.1 shows an example of such a data matrix having two overlapped peaks.

Another way of expressing the model in Equation 4.1 is by breaking **A** up into the sum of the data matrices for each pure component. Notice that one term in the summation is required for each independently varying component in the mixture.

$$\mathbf{A} = \mathbf{c}_1 \mathbf{p}_1^T + \mathbf{c}_2 \mathbf{p}_2^T + \ldots + \mathbf{c}_k \mathbf{p}_k^T + \boldsymbol{\varepsilon} \qquad (4.2)$$

In Equation 4.2, the product $\mathbf{c}_1 \mathbf{p}_1^T$ is the outer product of the concentration profile of the first pure component times its pure spectrum.

$$\mathbf{cp}^T = \begin{bmatrix} c_1 p_1 & c_1 p_2 & \cdots & c_1 p_m \\ c_2 p_1 & c_2 p_2 & & c_2 p_m \\ \vdots & & \ddots & \vdots \\ c_n p_1 & c_n p_2 & \cdots & c_n p_m \end{bmatrix} \qquad (4.3)$$

4.2.1 Basis Vectors

The vectors representing the pure-component spectra in **P** shown in Figure 4.1 can be thought of as *row basis vectors*, since each row in the data matrix **A** can be expressed as a linear combination (mixture) of the pure-component spectra. Similarly, the vectors representing the pure-component chromatograms in **C** can be thought of as *column basis vectors*. Each column in the data matrix **A** can be expressed as a linear combination (mixture) of the pure chromatograms.

Because the example in Figure 4.1 is simulated, there is no random error present and the matrix of residual errors, $\boldsymbol{\varepsilon}$, is zero. In this case we say that *these two sets of basis vectors span the row space and column space* of **A**. The *dimensionality* of the space is 2 because only two basis vectors are required to span it. For example, this means that all of the spectra in **A** lie on the plane defined by the two row basis

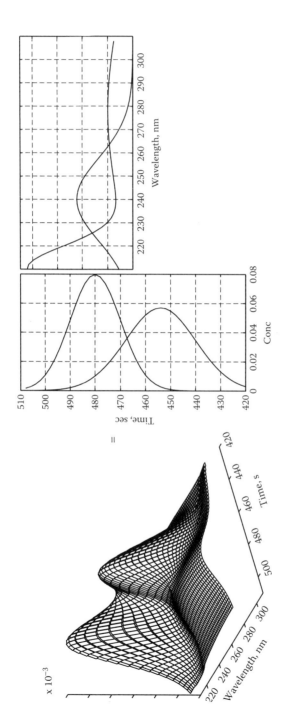

FIGURE 4.1 Simulated HPLC-UV/visible chromatographic data set showing two overlapped peaks with different UV/visible spectra.

vectors. We say that *the rank of* **A** *is two*. As a point of information, we recognize that there exists an infinite number vectors that lie in the same plane, and therefore they will span the same space of **A**. Each pair will be a linear combination of the two sets of basis vectors we have described so far.

Of course, random measurement error is unavoidable when real data are used. If we now suppose **A** is a 50 × 50 data matrix (50 spectra digitized at 50 points) with some random error, the exact solution for Equation 4.2 would require 50 pairs or dyads of basis vectors, one row basis vector and one column basis vector for each pair. The additional 48 pairs of row and column vectors would be required to account for the random variation in **A**. Usually, we are not interested in building a model that includes the random errors. Fortunately, by using the appropriate mathematical operations, we can use our original two basis vectors to reduce the rank or dimensionality of **A** from 50 to 2 without any significant loss of information. This allows us to "ignore" the basis vectors that explain random error. This data compression capability of the PCA model is exploited frequently and is one of its most important features.

4.3 THE PRINCIPAL COMPONENT MODEL

You undoubtedly recognize that we seldom start out knowing the pure spectra and chromatographic profiles of a data set like the one described in Figure 4.1. If we knew them, there would not be any point in applying the technique of PCA. Fortunately, with PCA it is possible to compute unique sets of basis vectors that span the significant space of a data matrix like **A** without prior knowledge.

With PCA, it is possible to build an empirical mathematical model of the data as described by Equation 4.4 where $\mathbf{T}_k$ is the $n \times k$ matrix of principal component scores and $\mathbf{V}_k$ is the $m \times k$ matrix of eigenvectors.

$$\mathbf{A} = \mathbf{T}_k \mathbf{V}_k^T + \varepsilon \tag{4.4}$$

The eigenvectors in $\mathbf{V}_k$ can be used to form a set of orthonormal row basis vectors for **A**. The eigenvectors are called "loadings" or sometimes "abstract factors" or "eigenspectra," indicating that while the vectors form a basis set for the row space of **A**, physical interpretation of the vectors is not always possible (see Figure 4.2).

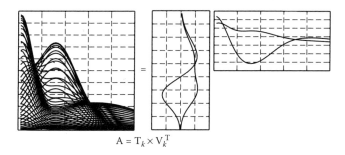

$$A = T_k \times V_k^T$$

FIGURE 4.2 Diagram of a principal component model for the chromatographic-spectroscopic data set shown in Figure 4.1.

The columns of $\mathbf{T}_k$ are called "scores" and are mutually orthogonal but not normalized. They can be used to form a set of column basis vectors for $\mathbf{A}$.

For each independent source of variation in the data, a single principal component (eigenvector) is expected in the model. The first column of scores and the first eigenvector (column in $\mathbf{V}_k$) correspond to the first "factor." The first eigenvector corresponds to the one with the largest eigenvalue. It can be shown that the first factor explains the "maximum" amount of variance possible in the original data (maximum in a least-squares sense). The second factor is the next-most-important factor and corresponds to the second column of scores and the eigenvector associated with the second-largest eigenvalue. It "explains" the maximum amount of variance left in the original data matrix. In short, the data matrix can be partitioned in a sum of k "rank one" $n \times m$ data matrices:

$$\mathbf{A} = \mathbf{t}_1 \mathbf{v}_1^T + \mathbf{t}_2 \mathbf{v}_2^T + \ldots + \mathbf{t}_k \mathbf{v}_k^T + \varepsilon \quad (4.5)$$

The outer vector product, $\mathbf{t}_1 \mathbf{v}_1^T$, is the variance "explained" by the first factor. For the HPLC-UV/Vis data set in Figure 4.1, exactly two terms or "abstract factors" would be required in Equation 4.4 or Equation 4.5 to explain the data, one for each chemical component. The term "abstract factors" is stressed here because the factors do not necessarily correspond to the two chemical components.

4.3.1 Eigenvectors and Eigenvalues

To perform PCA, we need the *complete* set of eigenvectors, $\mathbf{V}$, and eigenvalues, $\mathbf{D}$, that diagonalize the square, symmetric variance–covariance matrix, $\mathbf{Z}$, where $\mathbf{D}$ is the diagonal matrix of eigenvalues.

$$\mathbf{V}^T \mathbf{Z} \mathbf{V} = \mathbf{D} \quad (4.6)$$

$$\mathbf{Z}_{(m \times m)} = \mathbf{A}^T \mathbf{A} \quad (4.7)$$

To construct the principal component model described by Equation 4.4, we define $\mathbf{V}_k$ and $\mathbf{T}_k$ according to Equation 4.8 and Equation 4.9, where $\mathbf{V}_k$ contains the selected k columns from $\mathbf{V}$.

$$\mathbf{V}_k = [\mathbf{v}_1 | \mathbf{v}_2 | \ldots | \mathbf{v}_k] \quad (4.8)$$

$$\mathbf{T}_k = \mathbf{A} \, \mathbf{V}_k \quad (4.9)$$

The two MATLAB commands, `Z=A'*A;` and `[V,D]=eig(Z);`, are all that is needed to produce the matrix of eigenvectors $\mathbf{V}_k$ and eigenvalues $\mathbf{D}$ for $\mathbf{Z}$. The MATLAB command `T=A*V;` will compute the principal component scores. For the noise-free two-component data set in Figure 4.1, only two nonzero eigenvalues and eigenvectors will emerge. In general, for k components, there will be k nonzero eigenvalues. In the absence of random measurement error and machine round-off error, the secondary eigenvalues would be exactly zero (MATLAB's round-off error is about 3.0×10^{-16}).

$$\lambda_1 > \lambda_2 > \ldots \lambda_k > 0 \quad (4.10)$$
$$\text{primary eigenvalues}$$

Principal Component Analysis

$$\lambda_{k+1} = \lambda_{k+2} \ldots \lambda_n = 0 \quad (4.11)$$
$$\text{secondary eigenvalues}$$

For **A** ($n \times m$) with random experimental error and $m < n$, there will always be m nonzero eigenvalues of **Z**. In this case, it is necessary to delete the unwanted eigenvectors and eigenvalues (the ones with very small eigenvalues). The very difficult task of deciding which eigenvalues and eigenvectors should be deleted will be discussed later.

$$\lambda_{k+1} \approx \lambda_{k+2} \ldots \lambda_m \approx 0 \quad (4.12)$$
$$\text{secondary eigenvalues}$$

The MATLAB `eig()` function produces the full set of m eigenvectors and eigenvalues for the $m \times m$ matrix **Z**, while we are only interested in retaining the set of primary eigenvalues and eigenvectors. The MATLAB `eig()` function does not sort the eigenvalues and eigenvectors according to the magnitude of the eigenvalues, so the task of deleting the unwanted ones becomes a little bit harder to do. The following function sorts the eigenvalues and eigenvectors:

MATLAB EXAMPLE 4.1: FUNCTION TO GIVE SORTED EIGENVECTORS AND EIGENVALUES

```
function    [v,d]=sort_eig(z);
% [v,d]=sort_eig(z)
% subroutine to calculate the eigenvectors, v, and
% eigenvalues, d, of the matrix, z. The eigenvalues
% and      eigenvectors are sorted in descending order.
[v,d]=eig(z);
eval=diag   (d);       % get e'vals into a vector
[y,index]=sort(-eval);% sort e'vals in descending order
v=v(:,index            % sort e'vects in descending order
d=diag(-y));           % build diagonal e'vals matrix
```

It is easy to obtain the desired submatrix of primary eigenvalues and eigenvectors using MATLAB's colon notation once the eigenvalues and eigenvectors are sorted. The program in Example 4.2 shows how to put together all of the bits and pieces of code described so far to into one program that performs PCA of a spectroscopic data matrix **A**.

MATLAB EXAMPLE 4.2: PROGRAM TO PERFORM PRINCIPAL COMPONENT ANALYSIS OF A SPECTROSCOPIC DATA SET

```
z=a'*a;             % compute covariance matrix
[V,d]=sort_eig(z);% compute e'vects and e'vals
k=2;                % select the number of factors
```

```
V=V(:,1:k);         % retain the first k columns of V
d=d(1:k,1:k);       % retain the kxk submatrix of e'vals
sc=a*V;             % compute the scores
plot(V);            % plot the e'vects and
format short e
disp(diag(d));      % display the e'vals
```

4.3.2 THE SINGULAR-VALUE DECOMPOSITION

The singular-value decomposition (SVD) is a computational method for simultaneously calculating the *complete* set of column-mode eigenvectors, row-mode eigenvectors, and singular values of any real data matrix. These eigenvectors and singular values can be used to build a principal component model of a data set.

$$\mathbf{A} = \mathbf{U} \mathbf{S} \mathbf{V}^T \tag{4.13}$$

$$\begin{bmatrix} \cdot & \cdot & \cdot & \cdot & \cdot \\ \cdot & \cdot & \cdot & \cdot & \cdot \\ \cdot & \cdot & \mathbf{A} & \cdot & \cdot \\ \cdot & \cdot & \cdot & \cdot & \cdot \\ \cdot & \cdot & \cdot & \cdot & \cdot \end{bmatrix} = \begin{bmatrix} \cdot & \cdot & \cdot \\ \cdot & \cdot & \cdot \\ \cdot & \mathbf{U} & \cdot \\ \cdot & \cdot & \cdot \\ \cdot & \cdot & \cdot \end{bmatrix} \begin{bmatrix} \lambda_1^{1/2} & 0 & 0 \\ 0 & \lambda_2^{1/2} & 0 \\ 0 & 0 & \lambda_3^{1/2} \end{bmatrix} \begin{bmatrix} \cdot & \cdot & \cdot & \cdot & \cdot \\ \cdot & \cdot & \mathbf{V}^T & \cdot & \cdot \\ \cdot & \cdot & \cdot & \cdot & \cdot \end{bmatrix}$$

In Equation 4.13 we seek the k columns of $\mathbf{U}$ that are the column-mode eigenvectors of $\mathbf{A}$. These k columns are the columns with the k largest diagonal elements of $\mathbf{S}$, which are the square root of the eigenvalues of $\mathbf{Z} = \mathbf{A}^T\mathbf{A}$. The k rows of $\mathbf{V}^T$ are the row-mode eigenvectors of $\mathbf{A}$. The following equations describe the relationship between the singular-value decomposition model and the principal component model.

$$\mathbf{T} = \mathbf{U} \mathbf{S} \tag{4.14}$$

$$\mathbf{D}^{1/2} = \mathbf{S} \tag{4.15}$$

The SVD is generally accepted to be the most numerically accurate and stable technique for calculating the principal components of a data matrix. MATLAB has an implementation of the SVD that gives the singular values and the row and column eigenvectors sorted in order from largest to smallest. Its use is shown in Example 4.3. We will use the SVD from now on whenever we need to compute a principal component model of a data set.

MATLAB EXAMPLE 4.3: PRINCIPAL COMPONENT ANALYSIS USING THE SVD

```
[u,s,v]=svd(a);
k=2;                % Trim the unwanted factors from the model
u=u(:,1:k);
s=s(1:k,1:k);
v=v(:,1:k);
```

```
plot(wv,v);      % plot the e'vects and
format short e
disp(diag(s.^2));% display the e'vals
```

4.3.3 ALTERNATIVE FORMULATIONS OF THE PRINCIPAL COMPONENT MODEL

In section 4.3.1, we assumed that the $m \times m$ variance–covariance matrix **Z** was used as the starting point for the analysis. It is also possible to use the $n \times n$ covariance matrix **Z** as the starting point. There is a clear mathematical relationship between the results of the two analyses. First, exactly identical eigenvalues, **D**, will emerge from the diagonalization of either $\mathbf{Z}_{(n \times n)}$ or $\mathbf{Z}_{(m \times m)}$. When $n < m$, the extra $n + 1$ through m eigenvalues from the diagonalization of $\mathbf{Z}_{(m \times m)}$ would be exactly zero if it were not for a very small amount of floating-point round-off error.

$$\mathbf{V}_b^T \mathbf{Z}_{(n \times n)} \mathbf{V}_b = \mathbf{D}_{(n \times n)} \tag{4.16}$$

For this alternative formulation, we define the matrix of eigenvectors $\mathbf{V}_b$ according to Equation 4.17. The corresponding principal component model is given by Equation 4.18.

$$\mathbf{V}_{b(n \times k)} = [\mathbf{v}_1 | \mathbf{v}_2 | \ldots | \mathbf{v}_k]_b \tag{4.17}$$

$$\mathbf{A}^T_{(m \times n)} = \mathbf{T}_{b(m \times k)} \mathbf{V}_b^T{}_{(k \times n)} \tag{4.18}$$

The scores from analysis of $\mathbf{Z}_{(m \times m)}$ are related to the eigenvectors of $\mathbf{Z}_{(n \times n)}$, and vice versa, by their corresponding normalization constants, which are simply the reciprocals of the square roots of their eigenvalues. In other words, all one has to do is normalize the columns of $\mathbf{T}_b$ to obtain $\mathbf{V}_b$.

$$\mathbf{T}\,\mathbf{D}^{-1/2} = \mathbf{V}_b \tag{4.19}$$

The eigenvectors from the analysis of $\mathbf{Z}_{(m \times m)}$ are sometimes referred to as the "row-mode" eigenvectors because they form an orthogonal basis set that spans the row space of **A**. The eigenvectors from the analysis of $\mathbf{Z}_{(n \times n)}$ are sometimes referred to as the "column-mode" eigenvectors because they form an orthogonal basis set that spans the column space of **A**.

4.4 PREPROCESSING OPTIONS

Two data-preprocessing options, called mean centering and variance scaling, are often used in PCA; however, it is sometimes inconvenient to use them when processing chromatographic-spectroscopic data. In this section we will describe these two preprocessing options and explain when their use is appropriate. After introducing two elementary preprocessing options, a hands-on PCA exercise is provided. Additional preprocessing options will be discussed after these two elementary transformations.

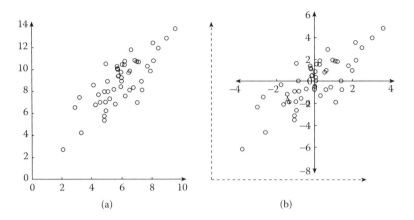

FIGURE 4.3 Graphical illustration showing the effect of mean centering on a bivariate distribution of data points. (a) Original data. (b) Mean-centered data. The original x–y-axes are shown as dashed lines.

4.4.1 MEAN CENTERING

The mean-centering data-preprocessing option is performed by calculating the average data vector or "spectrum" of all n rows in a data set and subtracting it point by point from each vector in the data set. It is slightly inconvenient to use when processing chromatographic-spectroscopic data because it changes the origin of the model. Despite the inconvenience, it is advisable to use mean centering under many circumstances prior to PCA. Graphically, mean centering corresponds to a shift in the origin of a plot, as shown in Figure 4.3.

To use mean centering, it is necessary to substitute the mean-centered data matrix $\mathbf{A}^\dagger$ into the SVD and in all subsequent calculations where $\mathbf{A}$ would normally be used in conjunction with the $\mathbf{U}$, $\mathbf{S}$, or $\mathbf{V}$ from the principal component model.

$$a_{ij}^\dagger = a_{ij} - \frac{1}{n}\sum_{i=1}^{n} a_{ij} \qquad (4.20)$$

The new model based on $\mathbf{A}^\dagger$ can be transformed back to the original matrix, $\mathbf{A}$, by simply adding the mean back into the model as shown in Equation 4.21.

$$\mathbf{A} - \overline{\mathbf{A}} = \mathbf{A}^\dagger = \mathbf{U}\mathbf{S}\mathbf{V}^\mathsf{T} \qquad (4.21)$$

Mean centering changes the number of degrees of freedom in a principal component model from k to $k + 1$. This affects the number of degrees of freedom used in some statistical equations that are described later.

4.4.2 VARIANCE SCALING

Prior to using the variance-scaling preprocessing option, mean centering must first be used. The combination of these two preprocessing options is often called

Principal Component Analysis

autoscaling. It is used to give equal weighting to all portions of the experimentally measured data vectors by normalizing each of the m columns of variables so that the sum of the squared values for each column equals 1. The resulting columns of variables are said to be "scaled to unit variance." Variance scaling is accomplished by simply subtracting the mean and then dividing each column in $\mathbf{A}$ by the standard deviation for that column. The pretreated data matrix $\mathbf{A}^\dagger$ is then used in the SVD and in all subsequent calculations where $\mathbf{A}$ would normally be used in conjunction with the $\mathbf{U}$, $\mathbf{S}$, or $\mathbf{V}$ from the principal component model.

$$a_{ij}^\dagger = \frac{a_{ij} - \bar{a}_j}{s_j} \tag{4.22}$$

$$\bar{a}_j = \frac{1}{n}\sum_{i=1}^{n} a_{ij} \tag{4.23}$$

$$s_j = \left[\frac{1}{n-1}\sum_{i=1}^{n}(a_{ij}-\bar{a}_j)^2\right]^{1/2} \tag{4.24}$$

Variance scaling is most useful when the magnitude of signals or the signal-to-noise ratio varies considerably from variable to variable. When the measurement error is nearly uniform from variable to variable, the use of variance scaling may be unwise. Absorption spectra often meet this requirement (e.g., they have nearly uniform measurement error over the wavelength range under study). Other kinds of data sets may frequently not meet this requirement. For example, consider the case where data vectors consist of trace element concentrations (in ppm) determined by inductively coupled plasma spectroscopy (ICP) in diseased crab tissue samples. It is quite possible that the variability in one element (for example, ppm Ca) could dominate the other variables in the data set, such as ppm Sr or ppm Pb. In this example, variance scaling could be used to reduce the significance of the Ca variable, thereby allowing PCA to give a more balanced representation of the other variables in the data set.

The function in Example 4.4 can be used to autoscale a data matrix. The function determines the size of the argument, its mean vector, and its standard deviation vector. On the last line, a MATLAB programming "trick" is used to extend the mean vector and standard deviation vector into matrices having the same number of rows as the original argument prior to subtraction and division. The expression `ones(r,1)` creates an $r \times 1$ column vector of ones. When used as an index in the statement `mn(ones(r,1),:)`, it instructs MATLAB to replicate the mean vector r times to give a matrix having the dimensions $r \times c$.

MATLAB EXAMPLE 4.4: FUNCTION TO AUTOSCALE A MATRIX

```
function [y,mn,s]=autoscal(x);
% AUTOSCAL  -  Mean center and standardize columns of a matrix
% [y,mn,s]=autoscal(x);
```

```
% or
% [y]=autoscal(x);
[r,c]=size(x);
mn=mean(x);
s=std(x);
y=(x-mn(ones(1,r),:)) ./ s(ones(1,r),:);
```

4.4.3 BASELINE CORRECTION

In many spectroscopic techniques, it is not unusual to encounter baseline offsets from spectrum to spectrum. If present, these kinds of effects can have a profound effect on a PCA model by causing extra factors to appear. In some cases, the baseline effect may consist of a simple offset; however, it is not uncommon to encounter other kinds of baselines with a structure such as a gentle upward or downward sloping line caused by instrument drift, or even a broad curved shape. For example, in Raman emission spectroscopy a small amount of fluorescence background signals can sometimes appear as broad, weak curves.

In the simplest kind of baseline correction, the spectra to be corrected must have a region where there is zero signal. For example, in Figure 4.4, Raman emission spectra are shown with an apparent valley at about 350 cm^{-1}. Assuming there is no Raman emission intensity in this region, it is possible to calculate the average signal over this frequency region for each spectrum and subtract it from each frequency in the respective spectrum, giving the corresponding baseline-corrected spectra on the right-hand side of the figure.

Alternative background correction schemes can be incorporated for more complicated situations. For example, if the background signal is curved and multiple valleys are available in the spectrum, it may be possible to fit a polynomial function

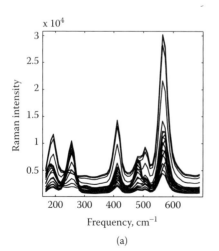
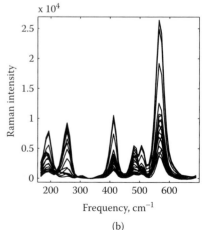

FIGURE 4.4 Illustration of baseline correction of Raman emission spectra. (a) Original spectra. (b) Baseline-corrected spectra.

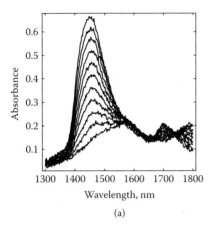 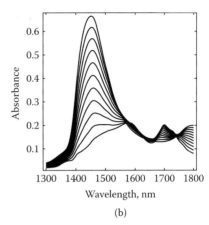

FIGURE 4.5 Illustration of polynomial smoothing on near-infrared spectra of water-methanol mixtures. (a) Original spectra. (b) Smoothed spectra.

through multiple valleys. The resulting curved polynomial line is then subtracted from the corresponding spectrum to be corrected. More sophisticated schemes for background correction have been published [2–5].

4.4.4 SMOOTHING AND FILTERING

With smoothing, it is possible to improve the signal-to-noise ratio of a signal recorded, for example, as a function of time or wavelength. Figure 4.5 shows a graphical illustration of smoothing applied to noisy near-infrared spectra. A detailed discussion of filtering and smoothing is presented in Chapter 10 of this book. Caution must be used when smoothing data. Strong smoothing gives better signal-to-noise ratios than weak smoothing, but strong smoothing may adversely reduce the resolution of the signal. For example, if a method that gives strong smoothing is used on a spectrum with sharp peaks or shoulders, these will be smoothed in a manner similar to noise.

The simplest method of smoothing is to calculate a running average for a narrow window of points. The smoothed spectrum is generated by using the average value from the window. This causes problems at the endpoints of the curve, and numerous authors have described different methods for treating them. The most commonly used type of smoothing is polynomial smoothing, also called Savitzky-Golay smoothing, after the names of two authors of a paper describing the technique published in 1964 [6].

Polynomial smoothing works by least-squares fitting of a smooth polynomial function to the data in a sliding window of width w, where w is usually an odd number. Smoothed points are generated by evaluating the polynomial function at its midpoint. After the polynomial is evaluated to determine a smoothed point, the window is moved to the right by dropping the oldest point from the window and adding the newest point to the window. Another polynomial is fitted to the new window, and its midpoint is estimated. This process is continued, one point at a time, until the entire curve has been smoothed. The degree of smoothing is controlled by varying the width of the window, w, and by changing the degree of the fitted polynomial function. Increasing the width of the window gives stronger smoothing. Increasing the degree of the polynomial, say from a quadratic to a quartic, allows more complex curves to be fitted to the data.

Polynomial smoothing does not possess an ideal frequency-response function and can potentially introduce distortions and artifacts in smoothed signals [7]. Other methods of smoothing do not possess these shortcomings. A detailed discussion of these important points is presented in Chapter 10.

4.4.5 First and Second Derivatives

Taking the derivative of a continuous function can be used to remove baseline offsets, because the derivative of a constant is zero [8]. In practice, the derivative of a digitized curve can be closely approximated by numerical methods to effectively remove baseline offsets. The derivative transformation is linear, and curves produced by taking the derivative retain the quantitative aspects of the original signal. The most commonly used method is based on polynomial smoothing. As in polynomial smoothing, a sliding window is used; however, the coefficients for the smoothing operation produce the derivative of the polynomial function fitted to the data. As in polynomial smoothing, the frequency-response function of these types of filters is not ideal, and it is possible to introduce distortions and artifacts if the technique is misused.

Figure 4.6 shows a graphical illustration of the effect of taking the first derivative on the near-infrared spectra of water-methanol mixtures. In addition to removing baseline offsets, the derivative also functions as a high-pass filter, narrowing and sharpening peaks within the spectrum. Zero crossing points can be used to identify the location of peaks in the original spectra. This process also removes a significant

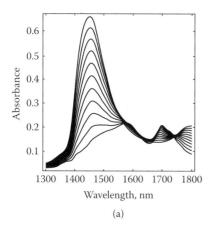
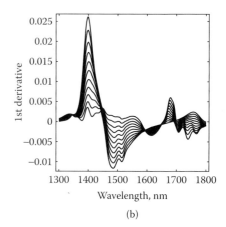

FIGURE 4.6 Illustration of a numerical approximation of the first derivative on near-infrared spectra of water-methanol mixtures. (a) Original spectra. (b) Derivative spectra.

Principal Component Analysis

amount of signal, resulting in a lower signal-to-noise ratio in the derivative curve (note the difference in plotting scales in Figure 4.6).

4.4.6 NORMALIZATION

In some circumstances is it useful to normalize a series of signals such as spectra or chromatograms prior to data analysis. For example, the intensity of Raman emission spectra depends on the intensity of the laser light source used to measure the spectra, and if there are any fluctuations in the intensity of the source during an experiment, these will show up in the spectra as confounding factors in any attempts to perform quantitative analysis. In cases such as these, each spectrum in the experiment can be normalized to constant area, thus removing the effect of the fluctuating signal. The simplest normalization technique is to simply set the sum of squares for each spectrum (a row in **A**) to 1, i.e., each spectrum has unit length. This is exactly the same operation described in Section 4.4.2, Variance Scaling, except the method is applied to rows in the data matrix rather than columns. Many other normalization schemes can be employed, depending on the needs dictated by the application. For example, if a Raman emission band due to solvent alone can be found, then it may be advantageous to normalize the height or area of this band instead of normalizing the total area, thereby avoiding sensitivity to changes in concentration (see Figure 4.7). Another common form of normalization is to normalize a mass spectrum by dividing by the largest peak.

4.4.7 MULTIPLICATIVE SCATTER CORRECTION (MSC) AND STANDARD NORMAL VARIATE (SNV) TRANSFORMS

Two closely related methods — multiplicative scatter correction (MSC) [9] and standard normal variate (SNV) transforms [10] — are discussed in this section. MSC

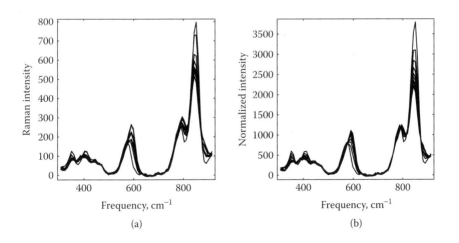

FIGURE 4.7 Illustration of normalization applied to Raman spectra. (a) Original spectra. (b) Normalized spectra.

was first reported in 1989 by Martens and Naes [9] as a method to correct differences in baseline offsets and path length due to differences in particle-size distributions in near-infrared reflectance spectra of powdered samples. A brief discussion of the source of these two effects is presented, followed by a more detailed description of MSC and SNV and an explanation of how they help compensate these kinds of effects.

In NIR reflectance measurements, there are two components of reflected light that reach the detector: specular reflectance and diffuse reflectance. Specular reflectance is light that is reflected from the surface of particles without being absorbed or interacting with the sample. Diffuse reflectance is light that is reflected by the sample after penetrating the sample particles, where some of the light is absorbed by the chemical components present in the particles. Powdered samples with very small uniform particles tend to pack very efficiently compared to samples with large, irregularly shaped particles. Samples with small, efficiently packed particles give a greater intensity of specular reflectance, and after transformation as log(1/reflectance), the higher levels of specular reflectance appear as increased baseline offsets; thus samples with smaller particle-size distributions tend to have larger baseline offsets. Beam penetration is shallow in samples with small, efficiently packed particles; thus these kinds of samples tend to have shorter effective path lengths compared to samples with larger irregularly shaped particles.

MSC attempts to compensate these two measurement artifacts by making a simple linear regression of each spectrum, $\mathbf{x}_i$, against a reference spectrum, $\mathbf{x}_r$. The mean spectrum of a set of training spectra or calibration spectra is usually used as the reference.

$$\mathbf{x}_r \approx \beta_0 + \beta_1 \mathbf{x}_i \qquad (4.25)$$

The least-squares coefficients, β_0 and β_1 (shown in Equation 4.25) are first estimated and then used to calculate the MSC-corrected spectrum, $\mathbf{x}_i^*$.

$$\mathbf{x}_i^* = \beta_0 + \beta_1 \mathbf{x}_i \qquad (4.26)$$

The MSC has been shown to work well in several empirical studies [9, 10], which showed an improvement in the performance of multivariate calibrations and a reduction in the number of factors in PCA. For example, NIR reflectance spectra of 20 powder samples of microcrystalline cellulose are shown in Figure 4.8a. Due to differences in particles size from sample to sample, there are significantly different baseline offsets. The same spectra are shown in Figure 4.8b after multiplicative scatter correction. The different baseline offsets observed in Figure 4.8a are so large that they mask important differences in the water content of these samples. These differences are revealed in the water absorption band at 1940 nm after the baseline offsets have been removed by MSC.

In the SNV transform, the mean of each spectrum is subtracted and the length is normalized to 1. The mathematical similarity to MSC is shown in Equation 4.27, with $\beta_0 = -\bar{\mathbf{x}}_i$ and $\beta_1 = 1/\|\mathbf{x}_i\|$, where the notation $\|\mathbf{x}\|$ represents the norm of $\mathbf{x}$.

$$\mathbf{x}_i^* = \beta_0 + \beta_1 \mathbf{x}_i \qquad (4.27)$$

Principal Component Analysis

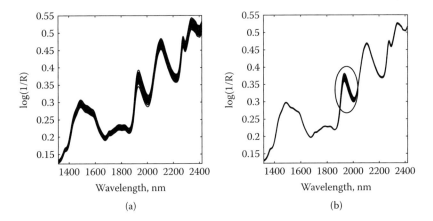

FIGURE 4.8 Illustration of multiplicative scatter correction (MSC). (a) NIR reflectance spectra of 20 powdered samples of microcrystalline cellulose. (b) Same NIR reflectance spectra after multiplicative scatter correction, revealing differences in moisture content.

The SNV transformation produces results similar to MSC in many cases, which sometimes makes it difficult to choose between the two methods. In practice, it is best to try both methods and select the preprocessing method that gives superior performance. Notice, for example, the similarity of the results shown in Figure 4.8 and Figure 4.9. Notice how the abscissa (y-axis) changes after SNV processing. The spectra are centered about the zero axis, which is a result of the mean subtraction. Additionally, the magnitude of the scale is significantly different.

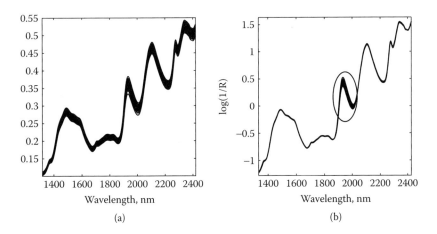

FIGURE 4.9 Illustration of standard normal variate (SNV) preprocessing. (a) NIR reflectance spectra of 20 powdered samples of microcrystalline cellulose. (b) Same NIR reflectance spectra after SNV preprocessing, revealing differences in moisture content.

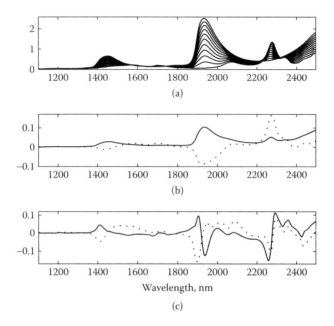

FIGURE 4.10 (a) Plots of NIR spectra of water-methanol mixtures. (b) Eigenvectors 1 (solid line) and 2 (dashed line). (c) Eigenvectors 3 (solid line) and 4 (dashed line).

4.5 PCA DATA EXPLORATION PROCEDURE

In the following example, MATLAB commands are used to perform PCA of NIR spectra of water-methanol mixtures. Plots of the spectra and eigenvectors (loadings) are shown in Figure 4.10. A total of 11 spectra are plotted in the top of the figure. The upper spectrum at 1940 nm is pure water, the bottom spectrum is pure methanol, and the nine spectra in between are mixtures of water and methanol in increments of 10% v/v, e.g., 90, 80, 70%, and so on. The first eigenvector shown in Figure 4.10 has an appearance similar to the average of all 11 spectra. The second eigenvector shows a peak going down at about 1940 nm (water) and a peak going up at about 2280 nm (methanol). This eigenvector is highly correlated with the methanol concentration and inversely correlated with the water concentration. The third and fourth eigenvectors are much more difficult to analyze, but in general, they show derivative-like features in locations where absorption bands give apparent band shifts.

MATLAB EXAMPLE 4.5: PCA PROCEDURE

```
% Load the data set into memory
load meohwat.mat
a=a';
c=c';
% Compute the PCA model & save four factors
[u,s,v]=svd(a);
u=u(:,1:4);
```

Principal Component Analysis

```
s=s(1:4,1:4);
v=v(:,1:4);
% Make plots of the eigenvectors
figure(2); plot(w,a);
figure(1); plot(w,v);
figure(1); plot(w,v(:,1:2));
figure(1); plot(w,v(:,3:4));
% Make scatter plots of scores
figure(1);
lab_plot(u(:,1),u(:,2));
xlabel('Scores for PC 1'); ylabel('Scores for PC 2'); pause;
lab_plot(u(:,2),u(:,3));
xlabel('Scores for PC 2'); ylabel('Scores for PC 3'); pause;
lab_plot(u(:,3),u(:,4));
xlabel('Scores for PC 3'); ylabel('Scores for PC 4'); pause;
```

PCA score plots of the water-methanol mixture spectra are shown in Figure 4.11. Each circle represents the location of a spectrum projected into the plane defined by the corresponding pairs of principle component axes (a detailed discussion of projection into subspaces is given in Section 4.7). The points are labeled consecutively in order of increasing methanol concentration, where the label 1 represents pure water, 2 represents 10% methanol, 3 represents 20% methanol, and so on up to the point labeled 11, which represents pure methanol. In the plane defined by PC1 and PC2, the points tend to lie on a slightly curved line. In the plane defined by PC2 and PC3, the points lie on a curve that is approximately parabolic in shape. In the plane defined by PC3 and PC4, the points lie on a curve having the shape of "α"

The curvature observed in the water-methanol score plots can be described well by simple polynomial functions such as quadratic or cubic functions. The reason for this behavior is the sensitivity of the NIR spectral region to hydrogen bonding. The presence of hydrogen bonding increases the length of O-H bonds, thereby perturbing the frequency of O-H vibration to shorter frequencies or longer wavelengths. Because it is possible for water and methanol molecules to participate in multiple hydrogen bonds — both as proton donors and proton acceptors, as shown in Figure 4.12 [11–13] — these solutions can be considered to consist of equilibrium mixtures of different hydrogen-bonded species, so that the underlying absorption bands can be considered to be a composite of many different kinds of hydrogen-bonded species, as shown in Figure 4.12 [13, 14]. Apparent band shifts in these peaks are the result of changing equilibrium concentrations of the different species. These shifting equilibrium mixtures are described by polynomial functions, which are manifest in the score plots shown in Figure 4.11.

4.6 INFLUENCING FACTORS

Until now we have not said much about how to select the proper number of principal components for a model. Recall that in the presence of random measurement error, there will be l nonzero eigenvalues and eigenvectors for an $n \times m$ data matrix, where l is the smaller of n and m, i.e., $l = min(n,m)$, some of which must be deleted. We

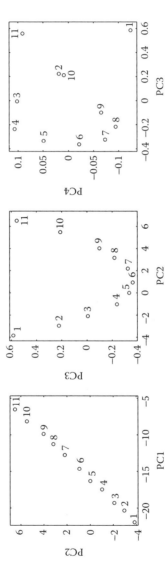

FIGURE 4.11 PCA score plots of the water-methanol mixtures shown in Figure 4.10. Each circle represents the location of a spectrum in a two-dimensional plane defined by the corresponding pair of PC axes. Points are consecutively labeled in order of increasing methanol concentration, 1 = 0%, 2 = 10%, 3 = 20%, ...

Principal Component Analysis

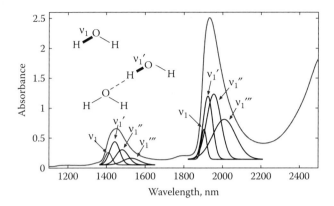

FIGURE 4.12 Illustration of multiple overlapping absorption bands in the NIR spectrum of water, v_1, represents the absorption band of nonhydrogen-bonded water (free O-H), whereas v_1', v_1'', and v_1''' represent water in various hydrogen-bonded states.

also mentioned in Section 4.3.4 that we expect our principal component models to have nonzero eigenvalues and corresponding eigenvectors for each component represented in a data matrix. We must now qualify this statement with several "except when" clauses. These include clauses like "except when the spectra of the overlapping peaks are almost identical," "except when the peaks are almost completely overlapped," "except when a component's signal has almost the same magnitude as the measurement error," and "except when matrix effects are occurring, such as a chemical interaction." Will it be possible to define what we mean by "except when" in these clauses? The answer is yes; however, all of these effects work together in a complicated way to determine whether or not the signal due to a component can be detected in a data matrix.

Before we can begin a discussion of our "except when" clauses and their complicated interrelationships, it is important to have a thorough understanding of measurement error and how it affects principal component models. This will enable us to turn our attention to several statistical tests for determining the number of significant principal components needed to model a data set. Having discussed how measurement error effects principal component models, we will finally return to a discussion of our "except when" clauses. We shall begin our discussion of measurement error by discussing the meaning of variance and residual variance as it applies to principal component models.

4.6.1 Variance and Residual Variance

The total variance in a data matrix $\mathbf{A}$ is the sum of the diagonal elements in $\mathbf{A}^T\mathbf{A}$ or $\mathbf{A}\mathbf{A}^T$ (also called the trace of $\mathbf{A}^T\mathbf{A}$ or the trace of $\mathbf{Z}$). This total sum of squares represents the total amount of variability in the original data. The magnitude of the eigenvalues is directly proportional to the amount of variation explained by a corresponding principal component. In fact, the sum of all of the eigenvalues is equal to the trace of $\mathbf{Z}$.

$$\text{trace}(\mathbf{Z}) = \lambda_1 + \lambda_2 + \ldots + \lambda_l \tag{4.28}$$

It can be shown that the scores for the first eigenvector and principal component extract the maximum possible amount of variance from the original data matrix using a linear factor [15]. In other words, the first principal component is a least-squares result that minimizes the residual matrix. The second principal component extracts the maximum amount of variance from whatever is left in the first residual matrix.

$$\mathbf{R}_1 = \mathbf{A} - \mathbf{u}_1 s_1 \mathbf{v}_1^T \tag{4.29}$$

$$\mathbf{R}_2 = \mathbf{R}_1 - \mathbf{u}_2 s_2 \mathbf{v}_2^T \tag{4.30}$$

$$\mathbf{R}_k = \mathbf{R}_{k-1} - \mathbf{u}_k s_k \mathbf{v}_k^T \tag{4.31}$$

The variance explained by the jth principal component is simply the ratio of the jth eigenvalue and the total variance in the original data matrix, e.g., the trace of $\mathbf{Z}$.

$$\%\mathrm{Var}_j = \frac{\lambda_j}{\sum_{i=1}^{l} \lambda_i} \times 100\% \tag{4.32}$$

The cumulative variance is the variance explained by a principal component model constructed using factors 1 through j.

$$\%\mathrm{CumVar}_j = \frac{\sum_{i=1}^{j} \lambda_i}{\sum_{i=1}^{l} \lambda_i} \times 100\% \tag{4.33}$$

When random experimental error is present in a data set, the total variance can be partitioned into two parts: the part due to statistically significant variation and the part due to random fluctuations.

$$\mathrm{trace}(\mathbf{Z}) = \underbrace{\sum_{i=1}^{k} \lambda_t}_{\text{significant variace}} + \underbrace{\sum_{i=k+1}^{l} \lambda_t}_{\text{residual variance}} \tag{4.34}$$

When the true number of factors, k, is known, the residual matrix, $\mathbf{R}_k$, is a good approximation of the random measurement errors, ε. Using the residual variance, it is possible to calculate an estimate of the experimental error according to Malinowski's *RE* function [15].

$$RE = \left[\sum_{i=k+1}^{l} \lambda_t \Big/ (n-k)(m-k) \right]^{1/2} \tag{4.35}$$

where n and m are the numbers of rows and columns in $\mathbf{A}$, respectively. If mean centering is used, then $(n - k - 1)(m - k - 1)$ should be used in the denominator

Principal Component Analysis

of Equation 4.35 or Equation 4.36. It is possible for computer round-off errors to accumulate in the smallest eigenvalues in these equations; therefore, it is more accurate to calculate *RE* using the alternative formula given in Equation 4.36.

$$RE = \left[trace(\mathbf{Z}) - \sum_{i=1}^{k} \lambda_i \Big/ (n-k)(m-k) \right]^{1/2} \quad (4.36)$$

The plot shown in Figure 4.13 shows the results from analysis of the simulated data set in Figure 4.1 with normal random error added ($\sigma = 0.0005$, $\bar{x} = 0$). The $\bar{x} = 0$ function in the following Example 4.6 was used to calculate and plot Malinowski's *RE*.

Inspecting plots of *RE* as a function of the number of principal components is a good method for determining the number of principal components in a data set. Usually we observe a large decrease in *RE* as significant factors are added to the principal component model. Once all of the statistically significant variance is modeled, *RE* levels off to a nearly constant value and thereafter continues to decrease only slightly (see Figure 4.13). Additional principal components model purely random error. Including these factors in the principal component model reduces the estimated error slightly. In Figure 4.13, we can see a substantial decrease in *RE* when we go from one to two principal components. This is a strong indication that the first two principal components are important. When additional principal components are added to the model, we only see a slight decrease in *RE*. This provides further evidence that only two principal components are needed to model statistically significant variance. For the example shown in Figure 4.13, we correctly conclude that two principal components are significant.

Another function for determining the number of significant principal components is Malinowski's empirical indicator function (*IND*) [15] is shown in Equation 4.37.

$$IND = \frac{RE}{(l-k)^2} \quad (4.37)$$

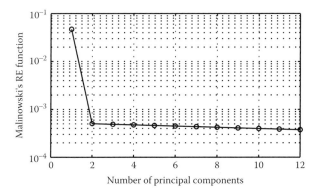

FIGURE 4.13 Malinowski's RE function vs. the number of principal components included in the principal component model for the simulated data set shown in Figure 4.1 with random noise added ($\sigma = 0.0005$).

Malinowski and others have observed that the indicator function often reaches a minimum value when the correct number of factors is used in a principal component model. We finish this section by giving a MATLAB function in Example 4.6 for calculating eigenvalues, variance, cumulative variance, Malinowski's *RE*, and Malinowski's *REV* and *F* (described in Section 4.6.2). Note that the function uses the SVD to determine the eigenvalues.

MATLAB EXAMPLE 4.6: FUNCTION FOR CALCULATING MALINOWSKI'S RE, IND, AND REV FUNCTIONS

```
function [lambda,var,cum_var,err,rv,f]=re_anal(a);
% [lambda,var,cum_var,err,rv,f]=re_anal(a);
% Function re_anal calculates Malinowski's RE, REV, and other stats.
% for determining the number of factors to use for matrix A
% lambda: eigenvalues of a'*a
% var: variance described by the eigenvalues of a'*a
% cum_var: cumulative variance described by eigenvalues of a'*a
% err: Malinowski's RE
% rv: Malinowski's reduced error eigenvalues
% f: Malinowski's f-test
nfac=min(size(a)) - 1; % We'll determine stats for n-1 factors

% Allocate vectors for results
lambda = zeros(nfac,1);
var    = zeros(nfac,1);
cum_var = zeros(nfac,1);
err    = zeros(nfac,1);
rv     = zeros(nfac,1);
f      = zeros(nfac,1);
pr     = zeros(nfac,1);

% calculate degrees of freedom
[r,c]=size(a);
y=min(r,c);
x=max   (r,c);

s=svd   (a',0).^2;           % get singular values
Trace_of_a=sum(s);           % calc total sum of squares
lambda=s(1:nfac);            % get e'vals
var    =100.0*lambda/Trace_of_a; % calc pct. variance

ssq=0.0;
resid_ssq=Trace_of_a;
for i=1:nfac                 % loop to calc RE, cum_var, rv
ssq=ssq+s(i);
resid_ssq=resid_ssq-s(i);
err(i)=sqrt(resid_ssq/((r-i)*(c-i)));
cum_var(i)=100.0*ssq/Trace_of_a;
end    ;
```

Principal Component Analysis

```
nf=min(r,c);
for i=1:nf                    % loop to calc rev
rv(i)=s(i)/((r-i+1)*(c-i+1)); % calculate the vector of reduced eigenvalues
end   ;

% calculate F
for i=1:nfac
den=sum(rv(i+1:nf));
f(i)=(nf-i)*(rv(i)/den);
end;
rv(nf)=[];
```

4.6.2 Distribution of Error in Eigenvalues

In 1989, Malinowski observed that the magnitude of secondary eigenvalues (called "error eigenvalues") with pure random error are proportional to the degrees of freedom used to determine the eigenvalue [16].

$$\lambda_j^\circ = N\,(m - j + 1)(n - j + 1)\,\sigma^2 \tag{4.38}$$

In Equation 4.38, n and m are the numbers of rows and columns in the original data matrix, N is a proportionality constant, and σ is the standard deviation of the error in the original data matrix. Malinowski proposed calculation of so-called "reduced error eigenvalues," which are directly proportional to the square of the measurement error, σ:

$$REV_j = \frac{\lambda_j}{(m - j + 1)(n - j + 1)} \tag{4.39}$$

4.6.3 F-Test for Determining the Number of Factors

A simple hypothesis test can be devised using the reduced eigenvalues to test for the significance of a factor, j [17]

$$F_{(1,n-j)} = \frac{REV_j}{\sum_{i=j+1}^{n} REV_i}(n - j) \tag{4.40}$$

The F-test is used to determine the number of real factors in a data matrix by starting with the next-to-smallest eigenvalue. The next-to-smallest eigenvalue is tested for significance by comparing its variance to the variance of the remaining eigenvalue. If the calculated F is less than the tabulated F at the desired significance level (usually $\alpha = .05$ or $.01$), then the eigenvalue is judged not significant. The next-smallest eigenvalue is tested by comparing its variance to the variance of the pool of nonsignificant eigenvalues. The process of adding eigenvalues to the set of nonsignificant factors is repeated until the variance ratio of the jth eigenvalue exceeds the tabulated F-value. This marks the division between the set of real and error vectors.

In Example 4.6, the statistics in Table 4.1 have been determined for the simulated chromatographic data set shown in Figure 4.1. Random noise ($\sigma = 0.0005$ absorbance

TABLE 4.1
Results from Factor Analysis of Simulated Chromatographic Data

Trace = 47.890421

Factor	Eigenvalue	% Variance	% Cum. Variance	RE	IND ($\times 10^{-7}$)	REV	F-Ratio	Prob. Level
1	43.015138	89.8199	89.8199	0.04755	245.62	1.912×10^{-2}	371	0.000
2	4.874733	10.1789	99.9989	0.00052	2.79	2.261×10^{-3}	141987	0.000
3	0.000047	0.0001	99.9989	0.00050	2.86	2.260×10^{-8}	1.43	0.238
4	0.000039	0.0001	99.9990	0.00050	2.95	1.955×10^{-8}	1.25	0.271
5	0.000037	0.0001	99.9991	0.00049	3.05	1.936×10^{-8}	1.24	0.272
6	0.000033	0.0001	99.9992	0.00048	3.15	1.831×10^{-8}	1.18	0.284
7	0.000032	0.0001	99.9992	0.00047	3.27	1.844×10^{-8}	1.19	0.281
8	0.000029	0.0001	99.9993	0.00046	3.39	1.779×10^{-8}	1.16	0.289
9	0.000028	0.0001	99.9994	0.00046	3.51	1.810×10^{-8}	1.18	0.284
10	0.000026	0.0001	99.9994	0.00045	3.65	1.735×10^{-8}	1.14	0.293
11	0.000023	0.0000	99.9995	0.00044	3.81	1.676×10^{-8}	1.10	0.301
12	0.000021	0.0000	99.9995	0.00043	3.98	1.608×10^{-8}	1.06	0.311

Principal Component Analysis

units, $\bar{x} = 0$) was added to the simulated data. Working backwards from the bottom to the top of the column labeled *F*-ratio in Table 4.1, we see that the first significant principal component occurs at $j = 2$. The probability that the difference between REV_2 and the sum of the remaining eigenvalues is due to random error is given in the column labeled "Prob. Level" in Table 4.1. The actual probability level at $j = 2$ was so small (ca. 1×10^{-7}) that it was rounded to zero. This very low probability level indicates that the difference between REV_2 and the sum of the remaining eigenvalues is highly significant. Selecting two principal components, we find the estimated residual error is about 0.0005 absorbance units (AU) in very good agreement with the actual random error added to this data matrix. The MATLAB code shown in Example 4.7 was MATLAB used to calculate the *REV* values and *F*-ratios.

EXAMPLE 4.7: DETERMINING THE NUMBER OF SIGNIFICANT PRINCIPAL COMPONENTS IN A DATA MATRIX

1. Using the MATLAB function called `re_anal.m`, calculate values for each column shown in Table 4.1. Use the sample data file called "pca _dat". Use the results to make and interpret plots of the eigenvalues and Malinowski's *RE* and *REV* functions.

```
load pca_dat
[lm,vr,cu,er,rv,f]=re_anal(an);
format short e
[lm,vr,cu,er,rv,f]
% Plot e'vals
semilogy(lm,'o'); hold on; semilogy(lm); hold off;
title('Plot of eigenvalues');
% Plot REV
semilogy(rv,'o'); hold on; semilogy(rv); hold off;
title('Plot of Malinowski''s reduced eigenvalues');
% Plot RE
semilogy(er,'o'); hold on; semilogy(er); hold off;
title('Plot of Malinowski''s RE function');
```

2. Using the MATLAB `svd` function, calculate the row-mode and column-mode eigenvectors, U and V, for the sample data file called "pca_dat". Make plots of the first four row-mode and column-mode eigenvectors.

```
% Do svd, plot row-mode and column-mode eigenvectors
[u,s,v]=svd(an);
plot(u(:,1:4)); title('Column-mode eigenvectors');
xlabel('Retention time (s)'); ylabel('Absorbance');
pause % Hit return to continue
plot(wv,v(:,1:4)); title('Row-mode eigenvectors');
xlabel('Wavelength (nm)'); ylabel('Absorbance');
pause % Hit return to continue
```

3. From the table and plots obtained in parts 1 and 2, the number of significant factors in the data set appears to be two, and the experimental error in the data set is estimated to be 0.00052 using RE.

4.7 BASIS VECTORS

When the true intrinsic rank of a data matrix (the number of factors) is properly determined, the corresponding eigenvectors form an orthonormal set of basis vectors that span the space of the original data set. The coordinates of a vector **a** in an m-dimensional space (for example, a $1 \times m$ mixture spectrum measured at m wavelengths) can be expressed in a new coordinate system defined by a set of orthonormal basis vectors (eigenvectors) in the lower-dimensional space. Figure 4.14 illustrates this concept. The projection of **a** onto the plane defined by the basis vectors **x** and **y** is given by $\mathbf{a}^{\ddagger}$. To find the coordinates of any vector on a normalized basis vector, we simply form the inner product. The new vector $\mathbf{a}^{\ddagger}$, therefore, has the coordinates $a_1 = \mathbf{a}^T\mathbf{x}$ and $a_2 = \mathbf{a}^T\mathbf{y}$ in the two-dimensional plane defined by **x** and **y**.

To find a new data vector's coordinates in the subspace defined by the k basis vectors in **V**, we simply take the inner product of that vector with the basis vectors.

$$\mathbf{t} = \mathbf{aV} \qquad (4.41)$$

In Equation 4.41 the row vector of scores, **t**, contains the coordinates of a new spectrum **a** in the subspace defined by the k columns of **V**. Note that the pretreated spectrum **a** must be used in Equation 4.41 if any preprocessing options were used when the principal component model was computed.

Because experimental error is always present in a measured data matrix, the corresponding row-mode eigenvectors (or eigenspectra) form an orthonormal set of basis vectors that *approximately* span the row space of the original data set. Figure 4.14 illustrates this concept. The distance between the endpoints of **a** and $\mathbf{a}^{\ddagger}$ is equal to the variance in **a** not explained by **x** and **y**, that is, the residual variance.

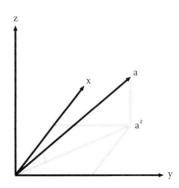

FIGURE 4.14 Projection of a three-dimensional vector **a** onto a two-dimensional subspace formed by the basis vectors **x** and **y** to form $\mathbf{a}^{\ddagger}$.

Principal Component Analysis

The percent cumulative variance explained by the principal component model can be used to judge the quality of the approximation. For example, recall the data matrix of two overlapped peaks in Figure 4.1 consisting of 50 spectra measured at 50 wavelengths. From Table 4.1, it can be seen that over 99.99% of the variance is explained by a two-component model; therefore we judge the approximation to be a very good one for this example. This means that each spectrum in 50-dimensional space can be expressed as a point in a two-dimensional space while still preserving over 99.99% of the information in the original data. This is one of the primary advantages of using PCA. Complex multivariate measurements can be expressed in low-dimensional spaces that are easier to interpret, often without any significant loss of information.

Clearly we cannot imagine a 50-dimensional space. It is possible, however, to view the position of points relative to each other in a 50-dimensional space by plotting them in the new coordinate system defined by the first two basis vectors in **V**. All we have to do is use the first column of **t** as the *x*-axis plotting coordinate and the second column of **t** as the *y*-axis plotting coordinate. An example of such a plot is shown in Figure 4.15. The MATLAB `plot(t(:,1),t(:,2),'o');` statement was used to create the scatter plot in Figure 4.15. The elements of the column vector `t(:,1)` are used as *x*-axis plotting coordinates, and the elements of the column vector `t(:,2)` are used as the *y*-axis plotting coordinates.

As the first pure component begins to elute, the principal component scores increase in Figure 4.15 along the axis labeled "pure component 1." As the second component begins to elute, the points shift way from the component 1 axis and toward the component 2 axis. As the concentration of the second component begins to decrease, the principal component scores decrease along the axis labeled "pure component 2." Points that lie between the two pure-component

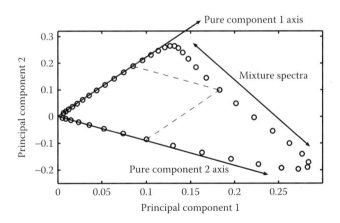

FIGURE 4.15 Scatter plot of the principal component scores from the analysis of the HPLC-UV/visible data set shown in Figure 4.1. The principal component axes are orthogonal, whereas the pure-component axes are not. Distances from the origin along the pure-component axes are proportional to concentration. Pure spectra lie on the pure-component axes. Mixture spectra lie between the two pure-component axes. Dashed lines show the coordinates (e.g., concentrations) of one point on the pure-component axes.

axes represent mixture spectra obtained at moments in time when the two peaks are overlapped.

4.7.1 CLUSTERING AND CLASSIFICATION WITH PCA SCORE PLOTS

An important application of PCA is classification and pattern recognition. This particular application of PCA is described in detail in Chapter 9. The fundamental idea behind this approach is that data vectors representing objects in a high-dimensional space can be efficiently projected into a low-dimensional space by PCA and viewed graphically as scatter plots of PC scores. Objects that are similar to each other will tend to cluster in the score plots, whereas objects that are dissimilar will tend to be far apart. By "efficient," we mean the PCA model must capture a large fraction of the variance in the data set, say 70% or more, in the first few principal components.

Examples illustrating the use of PCA for identification and classification are given in Chapter 9, including classification of American Indian obsidian artifacts by trace element analysis, identification of fuel spills by gas chromatography, identification of recyclable plastics by Raman spectroscopy, and classification of bees by gas chromatography of wax samples.

4.8 RESIDUAL SPECTRA

We have already hinted that the change of bases described above only works when the new basis vectors span the space of the data matrix. For example, suppose the overlapping peaks in Figure 4.15 were actually acetophenone and benzophenone. So long as Beer's law holds, the eigenvectors from the analysis of such a data set should span the space of any mixture of acetophenone and benzophenone. This means that over 99.9% of the variance in an unknown spectrum, a_u, should be explained by the basis vectors.

There may be times when an orthogonal basis may not span the space of a mixture spectrum. For example, suppose a mixture was contaminated with a substance like benzyl alcohol that has a UV/visible spectrum different from acetophenone and benzophenone. In this case, the absorption signal due to all three components will not be modeled by the two eigenvectors from the factor analysis of acetophenone and benzophenone mixtures.

If the contamination is great enough, the total variance in an unknown spectrum explained by the eigenvectors will be significantly less than 99.9%. This is easily demonstrated by calculating and plotting an unknown sample's residual spectrum using k factors.

$$\mathbf{t}_u = \mathbf{a}_u \mathbf{V} \quad (4.42)$$

$$\mathbf{r}_u = \mathbf{a}_u - \mathbf{t}_u \mathbf{V}^T \quad (4.43)$$

In Equation 4.43, $\mathbf{r}_u$ is the sample's residual vector or residual spectrum, $\mathbf{a}_u$ is the pretreated spectrum, and the quantity $\mathbf{t}_u \mathbf{V}^T$ is the unknown sample's reproduced spectrum. The scores for the spectrum must be determined from Equation 4.42 using the basis vectors $\mathbf{V}$ determined from a data set that does not contain the contamination.

Principal Component Analysis

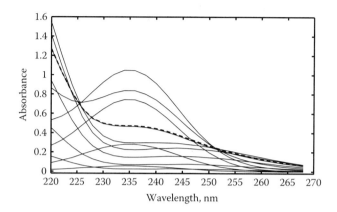

FIGURE 4.16 Plot of simulated two-component mixture spectra (solid line) and a spectrum contaminated with a third unknown component (dashed line).

As an example, the mixture spectra shown as solid lines in Figure 4.16 were used as a "training set" to determine **V**. The residual spectra from the training set are plotted in Figure 4.17 along with the contaminated mixture's residual spectrum. The residual spectra from the training set all have small random deviations. They appear as a solid black line about absorbance = 0 in Figure 4.17. The contaminated spectrum has a larger residual spectrum, indicating that it contains a source of variation not explained by the principal component model.

You can try this above analysis yourself using the simulated data shown in Figure 4.16 and Figure 4.17 and the MATLAB program in Example 4.8. The contaminated spectra are stored in the variable called au in the data file called "residvar.mat". The training spectra are saved in the variable called a.

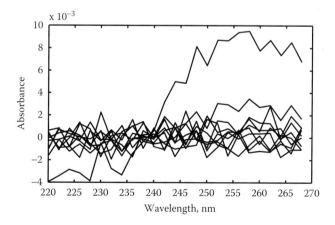

FIGURE 4.17 Plot of residual spectra from the training set (solid line) and the residual spectrum for the unknown spectrum (dashed line) shown in Figure 4.16.

MATLAB EXAMPLE 4.8: COMPUTATION OF RESIDUAL SPECTRA

```
% Load residvar data set
load residvar.mat
whos
plot   (x,a,'-'); title('Training spectra');

% Calc PC model for training set
[u,s,v]=svd(a);
k=2; % use 2 princ. components
[u1  ,s1,v1]=trim(k,u,s,v);

% Calc and plot training set residual spectra
r=a-u1*s1*v1';
plot   (x,r,'-'); title('Training residaul spectra');
% Project unknowns into space of training set and calc residual spectra
unk_scores=au*v1; % Calc    unknown scores
r_unk=au-unk_scores*v1'; % Calc unknown residual spectra
plot(x,r_unk,'-'); title('Unknown residaul spectra');
```

4.8.1 RESIDUAL VARIANCE ANALYSIS

A sample can be classified by calculating the sum of the squares of the difference between its measured spectrum vector and the same spectrum reproduced using a principal component model. The residual variance, s_i^2, of a data vector i fitted to the training set for class q indicates how similar the spectrum is to class q. For data vectors from the training set, the residual variance of a sample is given by Equation 4.44.

$$s_i^2 = \frac{1}{m-k} \sum_{j=1}^{m} r_{ij}^2 \qquad (4.44)$$

In Equation 4.44, r_{ij} is the residual absorbance of the ith sample at the jth variable, m is the number of wavelengths, and k is the number of principal components used in constructing the principal component model. If mean correction is used, then the denominator in Equation 4.44 should be changed to $m - k - 1$. For unknown data vectors (vectors not used in the training set), Equation 4.45 is used to calculate the residual variance, where r_{ij} is a residual absorbance datum for the ith sample's spectrum when fit to class q.

$$s_i^2 = \frac{1}{m} \sum_{j=1}^{m} r_{ij}^2 \qquad (4.45)$$

The notation, "class q" is used in case there happens to be more than one class. The degrees of freedom in Equation 4.45 are unaffected by mean correction.

The expected residual class variance for class q is calculated by using the residual data vectors for all samples in the training set. The resulting residual matrix is used to calculate the residual variance within class q. This value is an indication of how "tight" a class cluster is in multidimensional space. It is calculated according to Equation 4.46, where s_o^2 is the residual variance in class q and n is the number of samples in class q.

$$s_o^2 = \frac{1}{(m-k)(n-k)} \sum_{i=1}^{n} \sum_{j=1}^{m} r_{ij}^2 \qquad (4.46)$$

The summations in Equation 4.46 are carried over all samples in class q and all wavelengths in the residual spectra. Notice the definition for s_o^2 is the same as Malinowski's *RE*. The degrees of freedom in the denominator of Equation 4.46 should be changed to $(n-k-1)(m-k-1)$ when mean correction is used.

If one assumes that the original data are normally distributed and the principal component model is sufficient to describe the original data, then it can be shown that the residuals will be normally distributed. In this case, the variance ratio in Equation 4.47 can be calculated to test the null hypothesis H_0: $s_i^2 = s_o^2$ against H_1: $s_i^2 \neq s_o^2$. The null hypothesis is rejected at probability level α when the calculated ratio is greater than the critical value of F. In our work using NIR spectra, we have found that the degrees of freedom for the F-test shown in Equation 4.47 give satisfactory performance. In other applications, different authors have suggested different degrees of freedom for this test.

$$\frac{s_i^2}{s_o^2} \geq F_{1,n-k}(\alpha) \qquad (4.47)$$

The terms 1 and $n-k$ gives the degrees of freedom used for comparing the calculated F-value of a single unknown spectrum with a tabulated F-value. The quantity $n-k$ is used when no mean correction is used. If mean correction is used, then the quantity $n-k-1$ should be substituted in Equation 4.47. The data vectors are then classified according to the probability levels from the F-test.

In Example 4.9, the results from Example 4.8 are used to compute the residual variance and F-ratios for the data set described in Figure 4.16. The F-values for the 10 "unknown" spectra are shown in Table 4.2. The unknown spectrum contaminated with a minor level of an impurity is shown in the first row. All samples in the training set have small residual variances and F-ratios less than the critical value of $F = 4.105$. The "unacceptable" unknown spectrum has a very large F-value, indicating with a high degree of confidence that it is not a member of the parent population represented by the training set.

TABLE 4.2
Residual Variance Analysis Example

Sample	F-Ratio
1	34.9456
2	0.8979
3	0.6670
4	0.9222
5	0.9133
6	0.8089
7	0.8397
8	0.7660
9	0.9602
10	3.4342

Note: The F-ratios have df = 1, 37. The tabulated value of $F(\alpha = 0.05, 1, 37) = 4.105$. The null hypothesis is rejected at the $1.00 - 0.05 = 0.95$ probability level for all F-ratios > 4.105.

MATLAB EXAMPLE 4.9: RESIDUAL VARIANCE ANALYSIS

```
% Calc residual class variance
[n,m]=size(a);
class_so=sum(sum(r.^2))/((m-k)*(n-k))
r1=1/(m-k)*sum(r'.^2);
F1=r1/class_so;
[(1:n)'  F1']
[nunks,nvars]=size(r_unk);
r2=(1/m)*sum(r_unk'.^2);
F2=r2/class_so;
[(1:nunks)'  F2']
```

4.9 CONCLUSIONS

Principal component analysis is ideally suited for the analysis of bilinear data matrices produced by hyphenated chromatographic-spectroscopic techniques. The principle component models are easy to construct, even when large or complicated data sets are analyzed. The basis vectors so produced provide the fundamental starting point for subsequent computations. Additionally, PCA is well suited for determining the number of chromatographic and spectroscopically unique components in bilinear data matrices. For this task, it offers superior sensitivity because it makes use of all available data points in a data matrix.

RECOMMENDED READING

Malinowski, E.R., *Factor Analysis in Chemistry*, 2nd ed., John Wiley & Sons, New York, 1991.
Sharaf, M.A., Illman, D.L., and Kowalski, B.R., *Chemometrics*, John Wiley & Sons, New York, 1986.
Massart, D.L., Vandeginste, B.G.M., Deming, S.N., Michotte, Y., and Kaufman, L., *Chemometrics: a Textbook*, Elsevier, Amsterdam, 1988.
Jolliffe, I.T., *Principal Component Analysis*, Springer-Verlag, New York, 1986.

REFERENCES

1. Thielemans, A. and Massart, D.L., The use of principal component analysis as a display method in the interpretation of analytical chemical, biochemical, environmental, and epidemiological data, *Chimia*, 39, 236–242, 1985.
2. Maeder, M., Neuhold, Y.-M., Olsen, A., Puxty, G., Dyson, R., and Zilian, A., Rank annihilation correction for the amendment of instrumental inconsistencies, *Analytica Chimica Acta*, 464, 249–259, 2002.
3. Karstang, T.V. and Kvalheim, O., Multivariate prediction and background correction using local modeling and derivative spectroscopy, *Anal. Chem.*, 63, 767–772, 1991.
4. Liang, Y.Z., Kvalheim, O.M., Rahmani, A., and Brereton, R.G., A two-way procedure for background correction of chromatographic/spectroscopic data by congruence analysis and least-squares fit of the zero-component regions: comparison with double-centering, *Chemom. Intell. Lab. Syst.*, 18, 265–279, 1993.
5. Vogt, F., Rebstock, K., and Tacke, M., Correction of background drifts in optical spectra by means of "pseudo principal components," *Chemom. Intell. Lab. Syst.*, 50, 175–178, 2000.
6. Savitzky, A. and Golay, M.J.E., Smoothing and differentiation of data by simplified least squares procedures, *Anal. Chem.*, 36, 1627–1639, 1964.
7. Marchand, P. and Marmet, L., Binomial smoothing filter: a way to avoid some pitfalls of least squares polynomial smoothing, *Rev. Sci. Instrum.*, 54, 1034–1041, 1983.
8. Brown, C.D., Vega-Montoto, L., and Wentzell, P.D., Derivative preprocessing and optimal corrections for baseline drift in multivariate calibration, *Appl. Spectrosc.*, 54, 1055–1068, 2000.
9. Geladi, P., MacDougall, D., and Martens, H., Linearization and scatter-correction for near-infrared reflectance spectra of meat, *Appl. Spectrosc.*, 39, 491–500, 1985.
10. Barnes, R.J., Dhanoa, M.S., and Lister, S.J., Standard normal variate transformation and de-trending of near-infrared diffuse reflectance spectra, *Appl. Spectrosc.*, 43, 772–777, 1989.
11. Katz, E.D., Lochmuller, C.H., and Scott, R.P.W., Methanol-water association and its effect on solute retention in liquid chromatography, *Anal. Chem.*, 61, 349–355, 1989.
12. Alam, M.K. and Callis, J.B., Elucidation of species in alcohol-water mixtures using near-IR spectroscopy and multivariate statistics, *Anal. Chem.*, 66, 2293–2301, 1994.
13. Zhao, Z. and Malinowski, E.R., Detection and identification of a methanol-water complex by factor analysis of infrared spectra, *Anal. Chem.*, 71, 602–608, 1999.
14. Adachi, D., Katsumoto, Y., Sato, H., and Ozaki, Y., Near-infrared spectroscopic study of interaction between methyl group and water in water-methanol mixtures, *Appl. Spectrosc.*, 56, 357–361, 2002.

15. Malinowski, E.R., *Factor Analysis of Chemistry*, 2nd ed., John Wiley & Sons, New York, 1991.
16. Malinowski, E.R., Theory of the distribution of error eigenvalues resulting from principal component analysis with applications to spectroscopic data, *J. Chemom.*, 1, 33–40, 1987.
17. Malinowski, E.R., Statistical F-tests for abstract factor analysis and target testing, *J. Chemom.*, 3, 49–60, 1988.

5 Calibration

John H. Kalivas and Paul J. Gemperline

CONTENTS

5.1	Data Sets	107
	5.1.1 Near Infrared Spectroscopy	107
	5.1.2 Fundamental Modes of Vibration, Overtones, and Combinations	108
	5.1.3 Water–Methanol Mixtures	108
	5.1.4 Solvent Interactions	108
5.2	Introduction to Calibration	109
	5.2.1 Univariate Calibration	109
	5.2.2 Nonzero Intercepts	110
	5.2.3 Multivariate Calibration	111
	5.2.4 Curvilinear Calibration	112
	5.2.5 Selection of Calibration and Validation Samples	113
	5.2.6 Measurement Error and Measures of Prediction Error	114
5.3	A Practical Calibration Example	116
	5.3.1 Graphical Survey of NIR Water-Methanol Data	116
	5.3.2 Univariate Calibration	118
	5.3.2.1 Without an Intercept Term	118
	5.3.2.2 With an Intercept Term	119
	5.3.3 Multivariate Calibration	119
5.4	Statistical Evaluation of Calibration Models Obtained by Least Squares	121
	5.4.1 Hypothesis Testing	122
	5.4.2 Partitioning of Variance in Least-Squares Solutions	123
	5.4.3 Interpreting Regression ANOVA Tables	125
	5.4.4 Confidence Interval and Hypothesis Tests for Regression Coefficients	126
	5.4.5 Prediction Confidence Intervals	127
	5.4.6 Leverage and Influence	128
	5.4.7 Model Departures and Outliers	129
	5.4.8 Coefficient of Determination and Multiple Correlation Coefficient	130
	5.4.9 Sensitivity and Limit of Detection	131
	5.4.9.1 Sensitivity	131

		5.4.9.2 Limit of Detection 132

5.4.9.2 Limit of Detection 132
5.4.10 Interference Effects and Selectivity 134
5.5 Variable Selection 135
    5.5.1 Forward Selection 136
    5.5.2 Efroymson's Stepwise Regression Algorithm 136
        5.5.2.1 Variable-Addition Step 136
        5.5.2.2 Variable-Deletion Step 137
        5.5.2.3 Convergence of Algorithm 137
    5.5.3 Backward Elimination 137
    5.5.4 Sequential-Replacement Algorithms 138
    5.5.5 All Possible Subsets 138
    5.5.6 Simulated Annealing and Genetic Algorithm 138
    5.5.7 Recommendations and Precautions 138
5.6 Biased Methods of Calibration 139
    5.6.1 Principal Component Regression 140
        5.6.1.1 Basis Vectors 141
        5.6.1.2 Mathematical Procedures 142
        5.6.1.3 Number of Basis Vectors 144
        5.6.1.4 Example PCR Results 145
    5.6.2 Partial Least Squares 147
        5.6.2.1 Mathematical Procedure 148
        5.6.2.2 Number of Basis Vectors Selection 149
        5.6.2.3 Comparison with PCR 149
    5.6.3 A Few Other Calibration Methods 150
        5.6.3.1 Common Basis Vectors and a Generic Model 150
    5.6.4 Regularization 151
    5.6.5 Example Regularization Results 153
5.7 Standard Addition Method 153
    5.7.1 Univariate Standard Addition Method 154
    5.7.2 Multivariate Standard Addition Method 155
5.8 Internal Standards 156
5.9 Preprocessing Techniques 156
5.10 Calibration Standardization 157
    5.10.1 Standardization of Predicted Values 157
    5.10.2 Standardization of Instrument Response 158
    5.10.3 Standardization with Preprocessing Techniques 159
5.11 Software 159
Recommended Reading 160
References 160

When one is provided with quantitative information for the target analyte, e.g., concentration, in a series of calibration samples, and when the respective instrumental responses have been measured, there are two central approaches to stating the calibration model. These methods are often referred to as classical least squares

Calibration

and inverse least squares. Classical least squares implies that the spectral response is the dependent variable, while the quantitative information for the target analyte denotes the independent variable for a linear model. If spectral absorbencies are measured, this would be Beer's law. Inverse least squares involves the reverse. In either approach, least squares has nothing to do with the form of the model. Least squares is a method to determine model parameters for a specified model relationship. Thus, one should say that the model (model parameters) was obtained using the method of least squares. This chapter will be based only on the inverse least-squares representation of a calibration model, and the phrase inverse least squares will not be used. It is interesting to note that other phrases have been used to designate the model. For example, expressions such as "inverse regression" and "reverse calibration" have been used to imply a classical least-squares model description, and phrases like "ordinary least squares" and "forward calibration" have been utilized to communicate an inverse least-squares-type model [1].

5.1 DATA SETS

Near-infrared (NIR) spectra of water-methanol mixtures are examined to demonstrate the fundamental aspects of calibration. These spectra are used because they present unique challenges to calibration. Another reference NIR data set is also briefly evaluated. The reader should remember that the information presented is generic and applies to all calibration situations, not just spectroscopic data. Additionally, for discussion purposes, the quantitative information for the target analyte will be concentration. However, other chemical or physical properties can also be modeled. Throughout this chapter, unless noted otherwise as in Sections 5.3 and 5.4, it will be assumed that the described models have had the intercept term eliminated. The easiest way to accomplish this is to mean-center the data.

5.1.1 NEAR INFRARED SPECTROSCOPY

NIR spectroscopy is a popular method for qualitative and quantitative analysis. It is finding widespread use in many different industries for monitoring the identity and quality of raw materials and finished products in the food, agricultural, polymer, pharmaceutical, and organic chemical manufacturing industries.

Prior to the widespread availability of desktop computers and multivariate calibration software, the near-infrared spectral region (700 to 3000 nm) was considered useless for most routine analytical analysis tasks because so many chemical compounds give broad overlapping absorption bands in this region. Now NIR spectroscopy is rapidly replacing many time-consuming conventional methods of analysis such as the Karl Fisher moisture titration, the Kjeldahl nitrogen titration method for determining total protein, and the American Society for Testing Materials (ASTM) gasoline engine method for determining motor octane ratings of gasoline. These applications would be impossible without chemometric methods like multivariate calibration that can be used to "unmix" complicated patterns of broad overlapping absorption bands observed in the information-rich NIR spectral region.

5.1.2 Fundamental Modes of Vibration, Overtones, and Combinations

Absorption bands in the near infrared spectral region (700 to 3000 nm) are the result of overtones or combinations of fundamental modes of vibration in the mid-infrared range (4000 to 600 cm^{-1}). Correlation charts are available that show where certain functional groups can be expected to give absorption in the near-infrared spectral region.

Consider as an example the fundamental stretching frequency of an OH bond that occurs at a frequency of about 3600 cm^{-1}. The first, second, and third overtones of this fundamental mode of vibration can be observed in the near-infrared spectral region at about 7,200, 9,800, and 13,800 cm^{-1}, respectively. Stringed musical instruments like a guitar offer a useful analogy. The first overtone of a guitar string vibrating at its fundamental tone will produce a tone one octave higher, e.g., twice the frequency. The fundamental mode of vibration of a molecule corresponds to a transition from the ground-state energy level $v = 0$, $E_v = \frac{1}{2} h\upsilon$, to the first excited state $v = 1$, $E_v = (1 + \frac{1}{2})h\upsilon$. The first, second, and third overtones correspond to forbidden transitions from the ground state to $v = 2$, $v = 3$, and $v = 4$, respectively. If the vibrating molecular bonds behaved like perfect harmonic oscillators, then these energy levels would be equally spaced. In fact, molecules are anharmonic oscillators, and the energy levels are not perfectly spaced. Because of anharmonicity, the forbidden transitions can be observed, although these transitions are 10 to 100 times weaker than the fundamental transition. Each overtone band becomes successively weaker. The third overtone can only be observed for very strong fundamental bands, and the fourth overtone is usually too weak to be observed.

Combination bands correspond to simultaneous transitions in two modes. For example, a molecule that possesses a carbonyl ($\upsilon_1 = 1750$ cm^{-1}) and hydroxyl ($\upsilon_2 = 3600$ cm^{-1}) functional group in close proximity to each other can show a combination band at $\upsilon_1 + \upsilon_2 = 5350$ cm^{-1}.

5.1.3 Water–Methanol Mixtures

Nine mixtures of methanol and water were prepared having concentrations of 10, 20, 30, ..., 90% methanol by volume. The spectra of the nine mixtures plus the spectrum of pure water and pure methanol were measured in a 0.5-mm flow cell using an NIRSystems model 6500 NIR spectrophotometer. Spectra were recorded from 1100 to 2500 nm in 2-nm increments, giving a total of 700 points per spectrum. No attempt was made to thermostat the sample cell during the 1-hour measurement process. The spectra are plotted in Figure 5.1.

5.1.4 Solvent Interactions

Water and methanol can form strong hydrogen bonds in solutions. These kinds of solvent–solvent interactions have a pronounced effect in the NIR spectral region. For example, in pure methanol solutions it is possible to have dimers, trimers, and other intermolecular hydrogen-bonded species in equilibrium. Equilibrium concentrations of these species are very sensitive to impurity concentrations and temperature changes.

Calibration

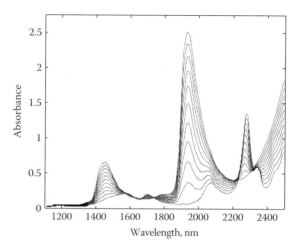

FIGURE 5.1 NIR absorbance spectra of water–methanol mixtures.

The addition of water creates a larger range of possible intermolecular hydrogen-bonded species. Calibration of such complicated two-component systems is difficult.

5.2 INTRODUCTION TO CALIBRATION

5.2.1 UNIVARIATE CALIBRATION

The simplest form of a linear calibration model is $y_i = b_1 x_i + e_i$, where y_i represents the concentration of the ith calibration sample, x_i denotes the corresponding instrument reading, b_1 symbolizes the calibration coefficient (slope of the fitted line), and e_i signifies the error associated with the ith calibration sample, assumed to be normal distributed random, $N(0,1)$. A single instrument response, e.g., absorbance at a single wavelength, is measured for each calibration sample. In matrix algebra notation, the model is depicted on the left in Figure 5.2 and is expressed as

$$\mathbf{y} = \mathbf{x}b_1 + \mathbf{e} \tag{5.1}$$

$$\begin{bmatrix} y_1 \\ y_2 \\ \cdot \\ \cdot \\ \cdot \\ y_n \end{bmatrix} = \begin{bmatrix} x_1 \\ x_2 \\ \cdot \\ \cdot \\ \cdot \\ x_n \end{bmatrix} \begin{bmatrix} b_1 \end{bmatrix} \qquad \begin{bmatrix} y_1 \\ y_2 \\ \cdot \\ \cdot \\ \cdot \\ y_n \end{bmatrix} = \begin{bmatrix} 1 & x_1 \\ 1 & x_2 \\ \cdot & \cdot \\ \cdot & \cdot \\ \cdot & \cdot \\ 1 & x_n \end{bmatrix} \begin{bmatrix} b_0 \\ b_1 \end{bmatrix} \qquad \begin{bmatrix} y_1 \\ y_2 \\ \cdot \\ \cdot \\ \cdot \\ y_n \end{bmatrix} = \begin{bmatrix} 1 & x_{1,1} & x_{1,2} \\ 1 & x_{2,1} & x_{2,2} \\ \cdot & \cdot & \cdot \\ \cdot & \cdot & \cdot \\ \cdot & \cdot & \cdot \\ 1 & x_{n,1} & x_{n,2} \end{bmatrix} \begin{bmatrix} b_0 \\ b_1 \\ b_2 \end{bmatrix}$$

FIGURE 5.2 Diagram of three different types of linear models with n standards. Left: the simplest model has a slope and no intercept. The center model adds a nonzero intercept. The right model is typically noted in the literature as the multiple linear regression (MLR) model because it uses more than one response variable, and $n \geq (m + 1)$ with an intercept term and $n \geq m$ without an intercept term. This model is shown with a nonzero intercept.

where **y**, **x**, and **e** are $n \times 1$ vectors for n calibration samples. It should be noted that while other constituents can be present in the calibration samples, the selected wavelength must be spectrally pure for the analyte, i.e., other constituents do not respond at the wavelength. Additionally, matrix effects must be absent at the selected wavelength, i.e., inter- and intramolecular interactions are not present.

Values in **y** and **x** are used to estimate the model parameter b_1 by the least-squares procedure. This least-squares estimate, $\hat{b}_1$, is computed by

$$\hat{b}_1 = (\mathbf{x}^T\mathbf{x})^{-1}\mathbf{x}^T\mathbf{y} \tag{5.2}$$

In Equation 5.2, the symbol, $\hat{b}_1$, called "b-hat," is used to emphasize its role as an estimate of b_1. The resulting calibration model is used to predict the analyte concentration for an unknown sample, $\hat{y}_{unk}$, by

$$\hat{y}_{unk} = x_{unk}\hat{b}_1 \tag{5.3}$$

where x_{unk} represents the response for the unknown sample measured at the calibrated wavelength. This kind of calibration is called univariate calibration because only one response variable is used.

5.2.2 NONZERO INTERCEPTS

Equation 5.1 and Equation 5.3 assume that the instrument response provides a value of zero when the analyte concentration is zero. In this respect, the above calibration model forces the calibration line through the origin, i.e., when the instrument response is zero, the estimated concentration must likewise equal zero. In such circumstances, the instrument response is frequently set to zero by subtracting the blank sample response from the calibration sample readings. The instrument response for the blank is subject to errors, as are all the calibration measurements. Repeated measures of the blank would give small, normally distributed, random fluctuations about zero. However, for many samples it is difficult if not impossible to obtain a blank sample that matrix-matches the samples and does not contain the analyte.

An intercept of zero for a model can be obtained if **y** and **x** are mean-centered to respective means before using Equation 5.1 through Equation 5.3. The concentration estimate obtained from Equation 5.3 must then be unmean-centered. While the calibration line for mean-centered **y** and **x** has an intercept of zero, inherently, a nonzero intercept is generally involved. The nonzero intercept is removed by the mean-centering process. Thus, mean-centering **y** and **x** to generate a zero intercept is not the same as using the original data and constraining the model to have an intercept of zero.

In the absence of mean centering, it is possible to include a nonzero intercept, b_0, in a calibration model by expressing the model as

$$y_i = b_0 + x_i b_1 + e_i \tag{5.4}$$

In matrix notation, the model is written by augmenting the instrument response vector, **x**, with a column of ones, producing the response matrix as shown in the middle of Figure 5.2 ($\mathbf{y} = \mathbf{Xb} + \mathbf{e}$). Least-squares estimates of the model parameters b_0 and b_1 are computed by

$$\hat{\mathbf{b}} = (\mathbf{X}^T\mathbf{X})^{-1}\mathbf{X}^T\mathbf{y} \tag{5.5}$$

where $\hat{\mathbf{b}}$ symbolizes the 2×1 vector of estimated regression coefficients. As with the univariate model without an intercept, matrix effects must be absent, and the selected wavelength must be spectrally pure for the analyte.

5.2.3 MULTIVARIATE CALIBRATION

Univariate calibration is specific to situations where the instrument response depends only on the target analyte concentration. With multivariate calibration, model parameters can be estimated where responses depend on the target analyte in addition to other chemical or physical variables and, hence, multivariate calibration corrects for these interfering effects. For the *i*th calibration sample, the model with a nonzero intercept can be written as

$$y_i = b_0 + x_{i1}b_1 + x_{i2}b_2 + \ldots + x_{ij}b_j + e_i \tag{5.6}$$

where x_{ij} denotes the response measured at the *j*th instrument response (wavelength). In matrix notation, Equation 5.6 is illustrated on the right in Figure 5.2 for two wavelengths and becomes

$$\mathbf{y} = \mathbf{Xb} + \mathbf{e} \tag{5.7}$$

where **y** and **e** are as before, **X** now has dimensions $n \times (m + 1)$ for *m* wavelengths and a column of ones if an intercept term is to be used, and **b** increases dimensions to $(m + 1) \times 1$. If the **y** and **X** are mean centered, the intercept term is removed from Equation 5.6 and Equation 5.7.

With multivariate calibration, wavelengths no longer have to be selective for only the analyte, but can now respond to other chemical species in the samples. However, the spectrum for the target analyte must be partially different from the spectra of all other responding species. Additionally, a set of calibration standards must be selected that are representative of the samples containing any interfering species. In other words, interfering species must be present in the calibration set in variable amounts. Under the above two conditions, it is possible to build a calibration model that compensates for the interfering species in a least-squares sense. It should be noted that if the roles of spectral responses and concentrations are reversed in Equation 5.7, as is often done in introductory quantitative analysis courses with Beer's law, then quantitative information of *all* chemical and physical effects, i.e., anything causing a response at the measured wavelengths, must be known and included in the model [1, 2]. Thus, there are distinct advantages to expressing the calibration model as in Equation 5.7, with concentration and spectral responses as the dependent and independent variables, respectively.

To obtain an estimate of the regression vector **b** by use of Equation 5.5, i.e., to ensure that the inverse $(\mathbf{X}^T\mathbf{X})^{-1}$ exists, the determinant of $(\mathbf{X}^T\mathbf{X})$ must not be zero. At a minimum, this means that it is necessary for $n \geq (m + 1)$ with an intercept term and $n \geq m$ without an intercept term. Thus, complete spectra cannot be used, and the user must select the wavelengths to be modeled (Section 5.5 discusses this further). This type of model, requiring selected wavelengths to keep $\mathbf{X}^T\mathbf{X}$ nonsingular, is often referred to as the multiple linear regression (MLR) model in the literature. Even though wavelengths have been selected such that $n \geq (m + 1)$ with an intercept term or $n \geq m$ without an intercept term, $\mathbf{X}^T\mathbf{X}$ may still be singular or nearly singular, with the second situation being more common due to spectroscopic noise. This is the spectral collinearity problem (spectral overlap or selectivity), and concentration estimates can be seriously degraded. Thus, selection of specific wavelengths to be included in the model is critical to the performance of the model. In-depth discussions on collinearity (spectral orthogonality) as well as methods for diagnosing collinearity and the extent of involvement by each chemical species are available in the literature [1, 3]. Sections 5.2.6 and 5.4 discuss some of these model performance diagnostics and figure of merits. Methods to select proper wavelengths are described in Section 5.5.

Generally, collinearity (near singularity) is not a problem with biased regression techniques such as principal component regression (PCR), partial least squares (PLS), ridge regression (RR), etc. Section 5.6 describes some of these biased methods that do not require wavelengths to be selected in order to estimate the regression vector in Equation 5.7. However, formation of models by these methods requires determination of at least one metaparameter (regularization parameter), where the metaparameter(s) is used to avoid the near singularity of **X**. Wavelength selection techniques can also be used with these biased methods, but the requirement $n \geq (m + 1)$ or $n \geq m$ is not applicable.

As a final note, the model in Equation 5.7 can be expressed to include other target analytes besides just one. In this situation, the model expands to $\mathbf{Y} = \mathbf{XB} + \mathbf{E}$, where **Y** is $n \times a$ for a analytes and **B** increases to $m \times a$ with a column of regression coefficients for each analyte. A solution for the regression matrix is still obtained by Equation 5.5, with **Y** and $\hat{\mathbf{B}}$ replacing **y** and $\hat{\mathbf{b}}$. When a model is built for multiple analytes, compromise wavelengths are selected, in contrast to the analyte-specific models expressed by Equation 5.7, which are based on selecting wavelengths pertinent to each target analyte.

5.2.4 CURVILINEAR CALIBRATION

When simple univariate or multivariate linear models are inadequate, higher-order models can be pursued. For example, in the case of only one instrument response (wavelength), Equation 5.8

$$y_i = b_0 + x_i b_1 + x_i^2 b_2 + e_i \tag{5.8}$$

describes a linear second-order model for a single instrument response. A second-order curvilinear model can be handled as before, with **b** and **X** dimensionally modified to account for the x_i^2 term. Least squares is used to obtain estimates of the model parameters, where b_1 is typically designated the *linear effect* and b_2 the *curvature effect*. Higher-order models for the single instrument response can be utilized. However, powers higher than three are not generally used because interpretation of model parameters becomes difficult. A model of sufficiently high degree can always be established to fit the data exactly. Hence, a chemist should always be suspicious of high-order curvilinear models used to obtain a good fit. Such a model will generally not be robust to future samples.

Models similar to Equation 5.8 can be defined for multiple instrument responses (wavelengths). Model parameters for linear effects of each wavelength and respective curvature effects would be incorporated. Additionally, model parameters for wavelength combinations can be included.

Curvilinear regression should not be confused with the nonlinear regression methods used to estimate model parameters expressed in a nonlinear form. For example, the model parameters a and b in $y = ax^b$ cannot be estimated by a linear least-squares algorithm. Information in Chapter 7 describes nonlinear approaches to use in this case. Alternatively, a transformation to a linear model can sometimes be used. Implementing a logarithmic transformation on $y_i = ax_i^b$ produces the model $\log y_i = \log a + b \log x_i$, which can now be utilized with a linear least-squares algorithm. The literature [4, 5] should be consulted for additional information on linear transformations.

5.2.5 SELECTION OF CALIBRATION AND VALIDATION SAMPLES

Calibration samples must include representation for every responding chemical species in a system under study. Spectral shifts and changes in instrument readings for mixtures due to interactions between components, changes in pH, temperature, ionic strength, and index of refraction are well known. The use of mixtures instead of pure standards during calibration enables multivariate calibration methods to form approximate linear models for such interactions over narrow assay working ranges, thereby providing more precise results.

The calibration samples must cover a sufficiently broad range of composition that a suitable change in measured response is instrumentally detectable. For simple systems, it is usually possible to prepare mixtures according to the principles of experimental design, where concentrations for all ingredients are varied over a suitable range. This is necessary to ensure that the measured set of mixtures has exemplars where different interactions between ingredients are present.

Often, calibration of natural products and materials is a desirable goal. In these kinds of assays, it is usually not feasible to control the composition of calibration and validation standards. Some well-known examples include the determination of protein, starch, and moisture in whole-wheat kernels and the determination of gasoline octane number by NIR spectroscopy. In cases such as these, sets of randomly selected samples must be obtained and analyzed by reference methods.

Because it is more desirable to make interpolations rather than extrapolations when making predictions from a calibration model, the range of concentrations in the calibration standards should exceed the expected working range of the assay. Calibration sample compositions should give a fairly uniform coverage across the range of interest. The ASTM recommends the minimum calibration concentration range of variation to be five times greater than the reference method of analysis uncertainty. A wider range, say ten times or more, is highly advisable, especially in light of the fact that the Association of Official Analytical Chemists (AOAC) recommends that the minimum signal for the limit of quantitation (LOQ) in univariate assays should be at least ten times the signal of the blank. However, if the range is too large, deviations from linearity could begin to appear. The recommended minimum number of calibration samples is 30 to 50, although this depends on the complexity of the problem. A lengthy discussion regarding the repeatability of the reference values and the use of multiple reference measurements can be found in the literature [6].

Validation of a multivariate calibration model is a critical step that must take place prior to widespread adoption and use of the calibration model for routine assays or in production environments. Standards describing acceptable practices for multivariate spectroscopic assays are beginning to emerge, most notably a standard recently released by the ASTM [6]. The purpose of model validation is to determine the reproducibility of a multivariate calibration, its bias against a known method or accepted values, and its long-term ruggedness. In general, the properties described above for the ideal calibration data set apply to validation standards as well, with the following additional considerations. It is very important that validation sets do not contain aliquots of samples used for calibration. The validation sample set must form a truly independent set of samples. For samples having controlled composition, these should be prepared separately from the calibration samples. Another equally important point is that the composition of validation samples should be designed to lie at points in between calibration points, so as to exercise the interpolating ability of the calibration model. For randomly selected samples of complex materials or natural products, this may not be possible.

Different validation data sets should be prepared to investigate every source of expected variation in the response. For example, validation sets might be designed to study short-term or long-term variation in instrument response, variation from instrument to instrument, variation due to small changes in sample temperature, and so on.

5.2.6 MEASUREMENT ERROR AND MEASURES OF PREDICTION ERROR

Because of measurement errors, the estimated parameters for calibration models always show some small, random deviations, e_i, from the "true values." For the calibration models presented in this chapter, it is assumed that the errors in y_i are small, random, uncorrelated, follow the normal distribution, and are greater than the errors in x_i. Note that this may not always be the case.

Practitioners of multivariate calibration typically use different strategies for determining the level of prediction error for a model. Three figures of merit for

Calibration

estimating errors in y_i are discussed in this section. They are (1) the root mean square error of calibration (RMSEC), (2) the root mean square error of prediction (RMSEP), also known as RMSEV for validation, and (3) the root mean square error of cross-validation (RMSECV).

The RMSEC describes the degree of agreement between the calibration model estimated concentration values for the calibration samples and the accepted true values for the calibration samples used to obtain the model parameters in Equation 5.7 according to

$$\text{RMSEC} = \left[\frac{1}{n-m-1} \sum_{i=1}^{n} (y_i - \hat{y}_i)^2 \right]^{1/2} \quad (5.9)$$

Because estimation of model parameters, $b_0, b_1, \ldots, b_m$ uses $m + 1$ degrees of freedom, the remaining $n - m - 1$ degrees of freedom are used to estimate RMSEC. If the intercept b_0 is omitted from the calibration model, then the number of degrees of freedom for RMSEC is $n - m$. If the data has been mean-centered, the degrees of freedom remain $n - m - 1$. Typically, RMSEC provides overly optimistic estimates of a calibration model's predictive ability for samples measured in the future. This is because a portion of the noise in the standards is inadvertently modeled by the estimated parameters. A better estimate of the calibration model's predictive ability may be obtained by the method of cross-validation with the calibration samples or from a separate set of validation samples.

To obtain the RMSEP, the validation samples prepared and measured independently from the calibration samples are used. The number of validation samples, p, should be large, so that the estimated prediction error accurately reflects all sources of variability in the calibration method. The RMSEP is computed by

$$\text{RMSEP} = \left[\frac{1}{p} \sum_{i=1}^{p} (y_i - \hat{y}_i)^2 \right]^{1/2} \quad (5.10)$$

The cross-validation approach can also be used to estimate the predictive ability of a calibration model. One method of cross-validation is leave-one-out cross-validation (LOOCV). Leave-one-out cross-validation is performed by estimating n calibration models, where each of the n calibration samples is left out one at a time in turn. The resulting calibration models are then used to estimate the sample left out, which acts as an independent validation sample and provides an independent prediction of each y_i value, $\hat{y}_{(i)}$, where the notation i indicates that the ith sample was left out during model estimation. This process of leaving a sample out is repeated until all of the calibration samples have been left out. The predictions $\hat{y}_{(i)}$ can be used in Equation 5.10 to estimate the RMSECV. However, LOOCV has been shown to determine models that are overfitting (too many parameters are included) [7, 8]. The same is true for v-fold cross-validation, where the calibrations set is split into

v disjunct groups of approximately the same size and a group is left out on each cycle to serve as an independent validation sample set. This deficiency can be overcome if a Monte Carlo CV (MCCV) [7–9], also called leave-multiple-out CV (LMOCV) [8] is used. With MCCV, the calibration set is split such that the number of validation samples is greater than the number of calibration samples. An average MCCV value is obtained from a large number of random splits. A variation of this approach is to use repeated v-fold CV, where B cycles of v-fold CV are used with different random splits into the v disjoint groups [10]. In summary, while many authors prefer the LOOCV approach when small numbers of calibration samples are used, the resulting RMSECV also tends to give an overly optimistic estimate of a calibration model's predictive ability.

5.3 A PRACTICAL CALIBRATION EXAMPLE

5.3.1 Graphical Survey of NIR Water–Methanol Data

Before any model is constructed, the spectra should be plotted. Since this is the data used to build the models, a graphical survey of the spectra allows determination of spectral quality. Example items to investigate include the signal-to-noise ratio across the wavelengths, collinearity, background shifts, and any obvious abnormalities such as a spectrum significantly different than the rest, suggesting the spectrum to be an outlier. Pictured in Figure 5.1 are the spectra for the 11 water–methanol samples. Plotted in Figure 5.3, Figure 5.4a, and Figure 5.4b are the water and methanol pure-component spectra and the first- and second-derivative spectra for the 11 samples, respectively. From Figure 5.1, several observations can be made. There does not appear to be any obvious abnormality. However, there is a definite trend in the baseline with increasing wavelength. The first derivative appears to achieve some correction at the lower wavelengths, and the second derivative presents visual correction for a complete spectrum. Thus, the best

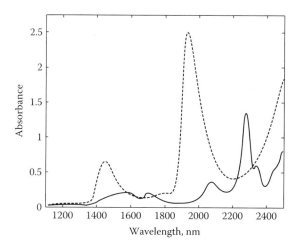

FIGURE 5.3 Pure-component NIR spectra for water (---) and methanol (—).

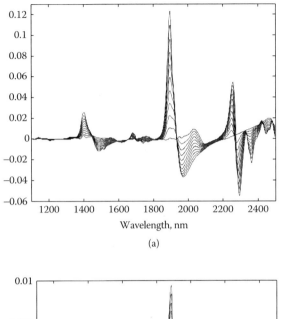

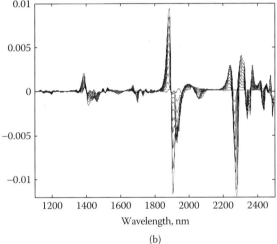

FIGURE 5.4 (a) First-derivative and (b) second-derivative spectra of the water–methanol mixtures in Figure 5.1.

results may be gained from using the second-derivative spectra in the calibration model. However, the signal-to-noise ratio degrades with successive derivatives.

Regardless of whether or not a derivative is used, proper wavelengths must be determined. A graphical survey of the spectra can sometimes assist with this. Selected wavelengths should offer good signal-to-noise ratios, be linear, and exhibit a large amount of variability with respect to changes in composition. From Figure 5.1 and Figure 5.3, wavelengths 1452 and 1932 nm seem appropriate for water, and wavelengths 2072 and 2274 nm appear satisfactory for methanol.

TABLE 5.1
Water Results from Univariate Calibration of the Water–Methanol Mixture

Wavelength (intercept model)	RMSEC (% water)[a]	RMSEV (% water)[a]
1452 nm (no intercept)	10.85	8.59
1452 nm (with intercept)	0.53	0.45
1932 nm (no intercept)	4.98	4.17
1932 nm (with intercept)	3.95	2.62

[a] Values are for six calibration and five validation samples.

Now that some potentially reasonable wavelengths have been identified, several models will be built and compared. These include assorted combinations of univariate and multivariate models with and without intercept terms. For this section, derivative spectra are not considered. It should be noted that results equal to the inclusion of an intercept term could be obtained by mean centering.

For the remaining subsections of Section 5.3, the water–methanol data will be considered split such that the 6 odd-numbered samples of the 11 denote the calibration set and the remaining 5 even-numbered samples compose the validation set. In some situations as noted, the calibration set consists of all 11 samples.

5.3.2 UNIVARIATE CALIBRATION

5.3.2.1 Without an Intercept Term

Listed in Table 5.1 and Table 5.2 are RMSEC and RMSEV values using only one wavelength suggested for water and methanol, respectively. When an intercept term is not included, prediction errors are clearly unacceptable. To uncover problems

TABLE 5.2
Methanol Results from Univariate and Multivariate Calibration of the Water–Methanol Mixture

Wavelength (intercept model)	RMSEC (% methanol)[a]	RMSEV (% methanol)[a]
2072 nm (no intercept)	49.35	37.85
2072 nm (with intercept)	2.82	1.73
2274 nm (no intercept)	18.16	13.46
2274 nm (with intercept)	3.03	1.86
1452, 2274 nm (no intercept)	2.07	1.29
1452, 2274 nm (with intercept)	0.48	0.36
1452, 1932, 2072, 2274 nm (no intercept)	0.45	0.37
1452, 1932, 2072, 2274 nm (with intercept)	0.24	0.18

[a] Values are for six calibration and five validation samples.

Calibration

occurring with these univariate models, some graphical diagnostics should be performed. Because there is only one wavelength being modeled, the first graphic is to plot the calibration concentrations used in **y** against the measured responses in **x**. Placed in this plot should also be the actual model calibration line. Such a plot is shown in Figure 5.5a for methanol using the 2274-nm wavelength and all 11 calibration samples. From this graphic, it is determined that the model does not fit the data at all, and the pattern indicates that an offset is involved. This result is seen again in the calibration residual plot displayed in Figure 5.5b, where the estimated residuals $\hat{e}_i = y_i - \hat{y}_i$ are plotted against corresponding $\hat{y}_i$ values for the 11 calibration samples. The distinguishing pattern indicates that the intercept term has been omitted from the model. While not obvious, there is an acute curvature in the residual plot, denoting that some nonlinearity is involved due to chemical interactions. Further discussion about trends in residual plots is presented in Section 5.4.7. Another useful graphic is the plot of y_i against $\hat{y}_i$ where the ideal result consists of having all the plotted points fall on the line of equality ($y_i = \hat{y}_i$). This plot is presented in Figure 5.5c for the calibration samples and reveals similar problems noted for Figure 5.5a and Figure 5.5b. The graphical description and problems discussed for Figure 5.5 are also applicable to methanol at 2072 nm and water at 1452 and 1932 nm. Similar plots to those shown in Figures 5.5b and Figure 5.5c were also generated for the validation samples. The conclusions are the same, but because the number of validation samples is small for this data set, trends observed in the plots are not as easily discerned.

These simple one-wavelength calibration models with no intercept term are severely limited. Spectral data is used from only one wavelength, which means a lot of useful data points recorded by the instrument are thrown away. Nonzero baseline offsets cannot be accommodated. Worst of all, because spectral data from only one wavelength is used, absorbance signals from other constituents in the mixtures can interfere with analysis. Some of the problems revealed for models without an intercept term can be reduced when an intercept term is incorporated.

5.3.2.2 With an Intercept Term

Tabulated in Table 5.1 and Table 5.2 are the RMSEC and RMSEV values for the four models when a nonzero intercept is allowed in the model. Results significantly improve. Plots provided in Figure 5.6 for methanol at 2274 nm further document the improvement. However, the residual plot in Figure 5.6b discloses that nonlinearity is now a dominating feature. Additionally, the large absorbance at zero percent methanol shown in Figure 5.6a suggests that other constituents are present in the mixture and have not been accounted for. Supplementary wavelengths are needed to correct for the chemical interactions and spectral overlaps.

5.3.3 MULTIVARIATE CALIBRATION

As a first attempt to compensate for the presence of interfering substances in the mixtures, the two wavelengths 1452 and 2274 nm were used to quantitate methanol. Table 5.2 contains the methanol RMSEC and RMSEV results. A substantial improvement has occurred even without an intercept term compared with the one-wavelength model without an intercept. The residual plots again show the nonlinear pattern observed previously.

Models were formulated using the four wavelengths 1452, 1932, 2072, and 2274 nm. Results listed in Table 5.2 disclose further improvements from the two-wavelength situations. Using the plots shown in Figure 5.7, it is observed that more of the

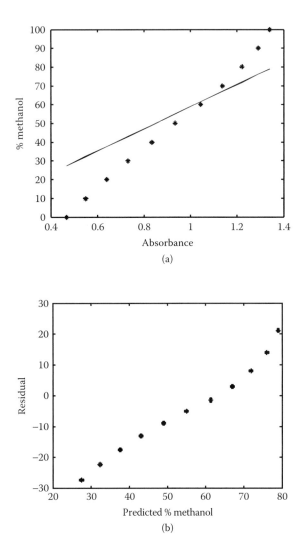

FIGURE 5.5 Graphical displays for the methanol model at 2274 nm without an intercept term (the model is constrained to go through the origin) using all 11 calibration samples. The RMSEC is 15.96% methanol. (a) Actual calibration model (-) and measured values (*). (b) Calibration residual plot. (c) A plot of estimated values against the actual values for the calibration samples; the drawn line is the line of equality.

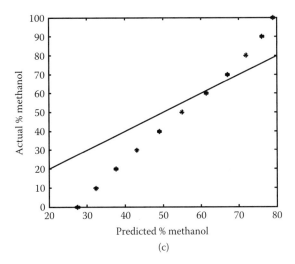

FIGURE 5.5 (Continued).

nonlinearity has been modeled with the extra wavelengths. Additionally, spectral overlap has been corrected. By using more than one wavelength, the presence of interfering constituents can be compensated. The difficult question is then deciding which wavelengths are important. More information on this concern is presented in Section 5.5.

5.4 STATISTICAL EVALUATION OF CALIBRATION MODELS OBTAINED BY LEAST SQUARES

Least squares is used to determine the model parameters for concentration prediction of unknown samples. This is achieved by minimizing the usual sum of the squared errors, $(\mathbf{y} - \hat{\mathbf{y}})^T (\mathbf{y} - \hat{\mathbf{y}})$. As stated before, the errors in $\mathbf{y}$ are assumed to be much larger than the errors in $\mathbf{X}$ for these models. Because the regression parameters are determined from measured data, measurement errors propagate into the estimated coefficients of the regression vector $\hat{\mathbf{b}}$ and the estimated values in $\hat{\mathbf{y}}$. In fact, we can only estimate the residuals, $\hat{\mathbf{e}}$, in the $\mathbf{y}$ measurements, as shown in Equation 5.12 through Equation 5.14. Summarizing previous discussions and equations, the model is defined in Equation 5.11 as

$$\mathbf{y} = \mathbf{Xb} + \mathbf{e} \tag{5.11}$$

The following equations are then used for computing the least-squares estimates, $\hat{\mathbf{b}}$, $\hat{\mathbf{y}}$, and $\hat{\mathbf{e}}$.

$$\hat{\mathbf{b}} = (\mathbf{X}^T\mathbf{X})^{-1}\mathbf{X}^T\mathbf{y} \tag{5.12}$$

$$\hat{\mathbf{y}} = \mathbf{X}\hat{\mathbf{b}} \tag{5.13}$$

$$\hat{\mathbf{e}} = \mathbf{y} - \hat{\mathbf{y}} \tag{5.14}$$

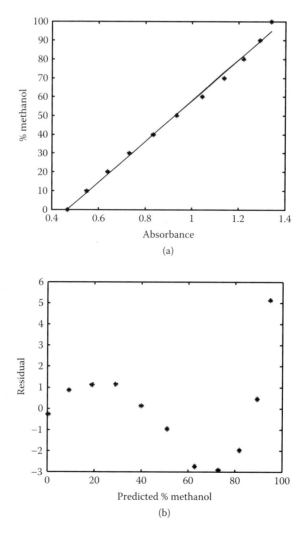

FIGURE 5.6 Graphical displays for the methanol model at 2274 nm with a nonzero intercept using all 11 calibration samples. The RMSEC is 2.37% methanol. (a) Actual calibration model (-) and measured values (*). (b) Calibration residual plot. (c) A plot of estimated values against the actual values for the calibration samples; the drawn line is the line of equality.

5.4.1 Hypothesis Testing

A statistical hypothesis denotes a statement about one or more parameters of a population distribution requiring verification. The *null hypothesis*, H_0, designates the hypothesis being tested. If the tested H_0 is rejected, the alternative hypothesis, H_1, must be accepted. When testing the null hypothesis, acceptance or rejection errors are possible. Rejecting the H_0 when it is actually true results in a type I error. Likewise, accepting

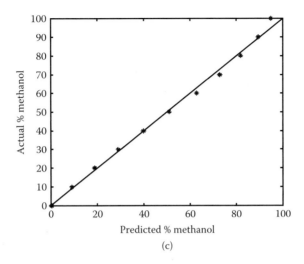

FIGURE 5.6 (Continued).

H_0 when it is false results in a type II error. The probability of making a type I error is fixed by specifying the level of confidence (or significance), α. If $\alpha = 0.05$, the probability of making a type I error translates to 0.05 (5%), and the probability of correct acceptance of H_0 becomes $1 - \alpha$, or 0.95 (95%). The probability of making a type II error is β, and $1 - \beta$ denotes the probability of making a correct rejection. Keeping α small helps reduce the type I error. However, as the probability of producing a type I error becomes smaller, the probability of making a type II error increases, and vice versa.

5.4.2 Partitioning of Variance in Least-Squares Solutions

All of the statistical figures of merit used for judging the quality of least-squares fits are based upon the fundamental relationship shown in Equation 5.15, which describes how the total sum of squares is partitioned into two parts: (1) the sum of squares explained by the regression and (2) the residual sum of squares, where $\bar{y}$ is the mean concentration value for the calibration samples.

$$\underbrace{\sum_{i=1}^{n}(y_i - \bar{y})^2}_{\substack{\text{total sum of squares} \\ \text{about the mean} \\ SS_{tot} \\ \text{degrees of freedom} = n-1}} = \underbrace{\sum_{i=1}^{n}(\hat{y}_i - \bar{y})^2}_{\substack{\text{sum of squares explained} \\ \text{by the regression model} \\ SS_{regr} \\ \text{degrees of freedom} = m}} + \underbrace{\sum_{i=1}^{n}(\hat{y}_i - y_i)^2}_{\substack{\text{residual (error)} \\ \text{sum of squares} \\ SS_{resid} \\ \text{degrees of freedom} = n-m-1}} \quad (5.15)$$

Each term in Equation 5.15 has associated with it a certain number of degrees of freedom. The total sum of squares has $n - 1$ degrees of freedom because the

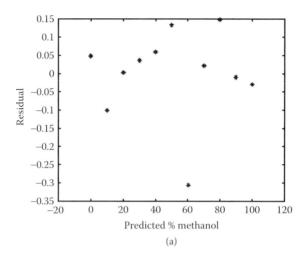

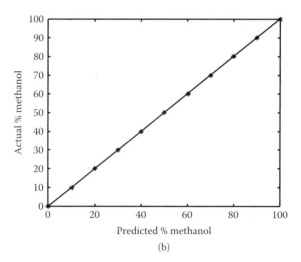

FIGURE 5.7 Graphical displays for the methanol model at 1452, 1932, 2072, and 2274 nm with a nonzero intercept using all 11 calibration samples. The RMSEC is 0.16% methanol. (a) Calibration residual plot. (b) A plot of estimated values against the actual values for the calibration samples; the drawn line is the line of equality. (c) Validation residual plot after the 11 samples were split to 6 calibration (odd-numbered samples) and 5 validation (even-numbered samples).

mean $\bar{y}$ is subtracted. Estimation of the calibration model, excluding the intercept, uses m degrees of freedom, one for every estimated parameter. The number of degrees of freedom in the residual sum of squares is simply the number of degrees of freedom remaining, $n - m - 1$. The residual sum of squares is used to compute RMSEC, as shown previously in Equation 5.9, by dividing by the degrees of freedom and taking the square root.

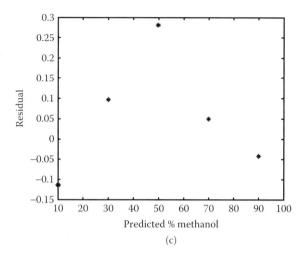

FIGURE 5.7 (Continued).

5.4.3 INTERPRETING REGRESSION ANOVA TABLES

Standard statistical packages for computing models by least-squares regression typically perform an analysis of variance (ANOVA) based upon the relationship shown in Equation 5.15 and report these results in a table. An example of a table is shown in Table 5.3 for the water model computed by least squares at 1932 nm.

The two sums of squares on the right-hand side of Equation 5.15 are shown in the table along with their degrees of freedom. The "mean square" is obtained by

TABLE 5.3
Summary Statistics for NIR Calibration of Water in Water-Methanol Mixtures Using One Wavelength and a Nonzero Intercept

Source	Sum of Squares	df	Mean Square	F-Ratio
Regression	6937.44	1	6937.440	444
Residual	62.56	4	15.641	—

Variable	Coefficient	s.e. of Coeff.	t-Ratio	Probability
Constant	−6.1712	3.12	−1.98	0.1189
1932 nm	40.3222	1.92	21.10	≤ 0.0001

Note: $y_{water} = b_0 + A_{1932\,nm}b_1$; $R^2 = 99.1\%$, R^2 (adjusted) = 98.9%; $s_y = 3.955$ with $6 - 2 = 4$ degrees of freedom.

dividing the sum-of-squares term by its respective degrees of freedom. The estimate of the error, s_y, in the y-measurements (standard error of y) is estimated by

$$s_y = \left[\frac{1}{n-m-1} \sum_{i=1}^{n} (\hat{y}_i - y_i)^2 \right]^{1/2}$$

The null hypothesis tested with the F-ratio is a general hypothesis stating that the true coefficients are all zero (note that b_0 is not included). The F-ratio has an F-distribution with $df_{regr} = m$ and $df_{resid} = n - m - 1$ degrees of freedom in the numerator and denominator, respectively,

$$H_0 : b_1 = b_2 = ... = b_m = 0$$

H_1 : one or more b_i are not zero

If $\dfrac{SS_{regr}/df_{regr}}{SS_{resid}/df_{resid}} \geq F_{m,n-m-1}(\alpha)$ then reject H_0

and SS_{regr} is the sum of squares explained by the regression model, and SS_{resid} is the residual sum of squares (see Equation 5.15). For sufficiently large F-ratios, we reject the null hypothesis at confidence level α. This means that the variance explained by the regression is too large for it to have happened by chance alone.

5.4.4 CONFIDENCE INTERVAL AND HYPOTHESIS TESTS FOR REGRESSION COEFFICIENTS

After calculating calibration coefficients, it is worthwhile to examine the errors existing in **b** and establish confidence intervals. The standard error of each regression coefficient is computed according to

$$s_{\hat{b}_i} = \sqrt{\text{Var}(\hat{b}_i)}$$

where $\text{Var}(\hat{b}_i)$ is the estimated variance of the least-squares regression coefficient $\hat{b}_i$ provided by the ith diagonal element of $s_y^2(\mathbf{X}^T\mathbf{X})^{-1}$. Interpretation of standard errors in the coefficients can be facilitated by calculating t-ratios and confidence intervals for the regression coefficients where

$$t_{b_i} = \frac{b_i}{s_{b_i}}$$

$$b_i = \hat{b}_i \pm \sqrt{\text{Var}(\hat{b}_i)} \sqrt{(m+1)F_{m+1,n-m-1}(\alpha)} \quad (5.16)$$

Calibration

However, many computer programs and practitioners often ignore the "simultaneous" confidence intervals computed in Equation 5.16 and use instead a "one-at-a-time" t-value for F as shown in the following equation

$$b_i = \hat{b}_i \pm t_{n-m-1}\left(\frac{\alpha}{2}\right)\sqrt{Var(\hat{b}_i)}$$

Generally, a regression coefficient is important when its standard error is small compared to its magnitude. The t-ratio can be used to help judge the significance of individual regression coefficients. Typically, when a t-value is less than 2.0 to 2.5, the coefficient is not especially useful for prediction of y-values. Specifically, if a regression coefficient's t-value is less than the critical value at t_{n-m-1}, then we should accept the null hypothesis, H_0: $b_i = 0$. This condition indicates that the coefficient's confidence interval includes a value of zero. The t-ratio can also be thought of as the signal-to-noise ratio.

As a reminder from Section 5.4, the above discussion pertains to those situations where errors in the variables (**X**) are not included. If errors in the variables are to be integrated, then the literature [11–17] should be consulted.

5.4.5 PREDICTION CONFIDENCE INTERVALS

The $100\%(1 - \alpha)$ confidence interval for the model at $\mathbf{x}_0$ can be computed from

$$y_0 = \hat{y}_0 \pm t_{n-m-1}\left(\frac{\alpha}{2}\right)\sqrt{s_y^2(\mathbf{x}_0(\mathbf{X}^T\mathbf{X})^{-1}\mathbf{x}_0^T)}$$

where $\mathbf{x}_0$ represents the estimated average of all possible sample aliquots with value $\mathbf{x}_0$ for the predictors, i.e., let $\mathbf{x}_0$ be a selected value of $\mathbf{x}$ with the predicted mean value of $\hat{y}_0 = \mathbf{x}_0^T\mathbf{b}$. The probability density, α, for the t-value is divided by two because the confidence interval is a two-sided distribution. For example, the 95% confidence interval would be obtained by selecting a critical value of t at $\alpha = 0.10$. In this case we can say, "There is a 95% probability that the true calibration line lies within this interval." The confidence interval for the model has a parabolic shape with a minimum at the mean values of x and y, as shown in Figure 5.8.

Prediction of an unknown sample is the primary motivation for developing a calibration model and is easily accomplished by use of Equation 5.13. Often, statisticians refer to this as *forecasting*. The $100\%(1 - \alpha)$ confidence interval for prediction at $\mathbf{x}_0$ is given by

$$y_0 = \hat{y}_0 \pm t_{n-m-1}\left(\frac{\alpha}{2}\right)\sqrt{s_y^2(1 + \mathbf{x}_0(\mathbf{X}^T\mathbf{X})^{-1}\mathbf{x}_0^T)}$$

where $\mathbf{x}_0$ denotes a new set of observed measurements for which the response y_0 is yet unobserved. Note that the prediction interval is wider than the confidence interval

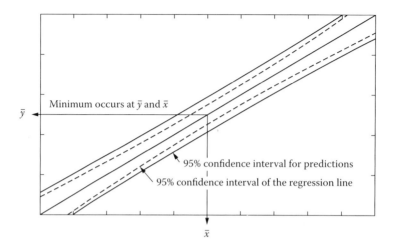

FIGURE 5.8 Illustration of confidence intervals for the regression line and for predictions.

for the regression model of the fitted values because the measurement error, e_0, at x_0 is unknown. For predictions of this type we can say, "There is a 95% probability that the true value, y_0, lies within this interval." For discussion of variance expressions when errors in the variables are to be included, see the literature [11–17].

5.4.6 LEVERAGE AND INFLUENCE

The effect of individual cases (calibration samples) on a calibration model can be large in certain circumstances. For example, there might be calibration samples that are outliers, either in the y-value or in one or more x-values. Several statistical figures of merit are presented in this section to identify influential cases.

The leverage, h_i, of the ith calibration sample is the ith diagonal of the hat matrix, **H**. The leverage is a measure of how far the ith calibration sample lies from the other $n-1$ calibration samples in **X**-space. The matrix **H** is called the hat matrix because it is a projection matrix that projects the vector **y** into the space spanned by the **X** matrix, thus producing **y**-hat. Notice the similarity between leverage and the Mahalanobis distance described in Chapter 4.

$$\mathbf{H} = \mathbf{X}(\mathbf{X}^T\mathbf{X})^{-1}\mathbf{X}^T$$

$$\hat{\mathbf{y}} = \mathbf{H}\mathbf{y}$$

The leverage h_i, signifying the ith diagonal element of **H**, takes on values from 0 to 1. Samples far from center of the x-values, $\bar{\mathbf{x}}$, generally having higher leverage values and are potentially the most *influential*.

The concept of leverage in statistics is comparable to the physical model of a lever. The fulcrum for the calibration line lies at $\bar{x}$, the center of the x-values. Calibration samples close to the mean of the x-values tend to exert little force on the slope of the calibration curve. Calibration samples farthest from the mean of the x-values can put forth a greater force on the slope of the calibration curve, so that their residuals are made as small as possible. Some authors recommend points with a leverage exceeding $2m/n$ or $3m/n$ should be carefully scrutinized as possible influential outliers.

One method for identifying influential cases is to examine plots of the residuals $\hat{e}_i = y_i - \hat{y}_i$. Here, the problem is that residuals for calibration samples near the mean of the x-values have greater variance than residuals for cases at the extreme x-values. A common method for solving this scaling problem is to standardize the residuals to give the so-called studentized residuals, r_i, defined as

$$r_i = \frac{\hat{e}_i}{s_y \sqrt{1-h_i}}$$

Calibration samples having large studentized residuals should be carefully scrutinized as possible outliers.

Distance measures, such as Cook's distance, combine the concept of leverage and residuals to compute an overall measure of a calibration sample's influence on the calibration model. Cook's distance is computed as

$$D_i = \left(\frac{1}{m+1}\right)\left(\frac{h_i}{1-h_i}\right)\left(\frac{\hat{e}_i^2}{s_y \sqrt{1-h_i}}\right)$$

and follows approximately the F-distribution with $m + 1$ and $n - m - 1$ degrees of freedom in the numerator and denominator, respectively. A large value for Cook's distance, e.g., greater than the appropriate critical value of F, indicates that the corresponding calibration point exerts a large influence on the least-squares parameters and should be carefully scrutinized as a possible outlier.

5.4.7 MODEL DEPARTURES AND OUTLIERS

Assessment of model departures from model assumptions can be interpreted from the residual plot. Additionally, as noted in the previous section, some outliers can be identified in the residual plot. Sections 5.3.2 and 5.3.3 provided some brief discussions on using the residual plot to diagnose model departures.

Example residual plots are depicted in Figure 5.9. If all assumptions about the model are correct, a plot of residuals (computed by Equation 5.14) against the estimated $\hat{y}_i$ values should show a horizontal band, as illustrated in Figure 5.9a. A plot similar to Figure 5.9b indicates a dependence on the predicted value, suggesting that numerical calculations are incorrect or an intercept term has been omitted from the

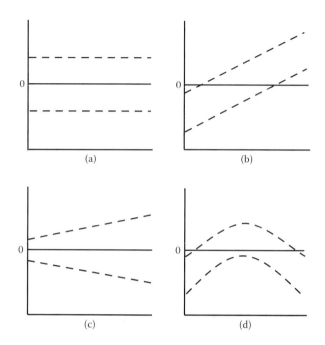

FIGURE 5.9 Possible residual plots with the estimated residual on the y-axis ($\hat{e}_i = y_i - \hat{y}_i$) and the estimated concentration on the x-axis ($\hat{y}_i$). Residuals can be located anywhere between the dashed lines. See Section 5.4.7 for discussion on patterns and implications.

model. The pattern of residuals illustrated in Figure 5.9c implies that the variance is not constant and increases with each increment of the predicted value. Thus, instead of variance being homoscedastic, as assumed, the variance is instead, heteroscedastic. Transformations or a weighted least-squares approach (or both) are required. Figure 5.9d characterizes nonlinear trends existing in the data, indicating that transformations or curvilinear calibration with inclusion of extra terms are needed.

5.4.8 Coefficient of Determination and Multiple Correlation Coefficient

The R^2 statistic computed by

$$R^2 = \frac{\sum_{i=1}^{n}(\hat{y}_i - \bar{y})^2}{\sum_{i=1}^{n}(y_i - \bar{y})^2} = \frac{SS_{tot} - SS_{resid}}{SS_{tot}}$$

Calibration

is the called the *coefficient of determination* and takes on values in the range from 0 to 1 (SS_{tot} and SS_{resid} are defined in Equation 5.15). The magnitude of R^2 provides the proportion of total variation in **y** explained by the calibration model. When R^2 is exactly 1, there is perfect correlation, and all residual errors are zero. When R^2 is exactly 0, the regression coefficients in $\hat{\mathbf{b}}$ have no ability to predict **y**. The square root is the *multiple correlation coefficient*.

The *adjusted* R^2 calculated by the following equation

$$R^2_{adj} = \frac{(SS_{tot}/df_{tot}) - (SS_{resid}/df_{resid})}{SS_{tot}/df_{tot}} = \frac{MS_{tot} - MS_{resid}}{MS_{tot}}$$

is more appropriate for multivariate calibration, where R^2 is expected to increase as new terms are added to the model, even when the new terms are random variables and have no useful predictive ability. The adjusted R^2 accounts for this effect to more accurately indicate the effect of adding new variables to the regression. In the above equation, SS_x represents sum of squares, and df_x represents degrees of freedom as defined in Equation 5.15. MS_x represents the mean square, which is obtained by dividing the sum of squares by the corresponding degrees of freedom.

5.4.9 SENSITIVITY AND LIMIT OF DETECTION

5.4.9.1 Sensitivity

For univariate calibration, the International Union of Pure and Applied Chemistry (IUPAC) defines sensitivity as the slope of the calibration curve when the instrument response is the dependent variable, i.e., **y** in Equation 5.4, and the independent variable is concentration. This is also known as the calibration sensitivity, contrasted with the analytical sensitivity, which is the calibration sensitivity divided by the standard deviation of an instrumental response at a specified concentration [18]. Changing concentration to act as the dependent variable, as in Equation 5.4, shows that the slope of this calibration curve, $\hat{b}_1$, is related to the inverse of the calibration sensitivity. In either case, confidence intervals for concentration estimates are linked to sensitivity [1, 19–22].

In the multivariate situation, the sensitivity figure of merit is a function of all wavelengths involved in the regression model. It is commonly presented as equal to $1/\|\hat{\mathbf{b}}\|$ when the dependent variable is defined as concentration [22], where $\|\cdot\|$ defines the Euclidean norm. Note that $\|\hat{\mathbf{b}}\|$ denotes the length of $\hat{\mathbf{b}}$, thus models with high sensitivity are characterized by regression vectors having short lengths. When instrumental responses are used as the dependent variables, sensitivity has been defined as $\|\mathbf{k}_i\|$, where $\mathbf{k}_i$ is the pure-component spectrum for the *i*th analyte at unit concentration. A result of this is that the sensitivity value $1/\|\hat{\mathbf{b}}\|$ can be expressed as the product of $\|\mathbf{k}_i\|$ and the selectivity for the analyte (see Section 5.4.10 for information on selectivity). Thus, $1/\|\hat{\mathbf{b}}\|$ is representative of the *effective* sensitivity, i.e., the pure-component spectrum sensitivity $\|\mathbf{k}_i\|$ scaled by the degree of spectral interferences from the other sample constituents. If no interferences exist, the selectivity is 1, and $1/\|\hat{\mathbf{b}}\| = \|\mathbf{k}_i\|$.

5.4.9.2 Limit of Detection

Often, trace analysis must be preformed. Prior to transforming a measured signal to concentration, it must be discerned whether or not the signal is significantly above the background. There is some disagreement in the literature on how to define "significantly above the background." The terminology introduced by Currie [23] will be used here.

5.4.9.2.1 Univariate Decision Limit

The decision limit corresponds to the critical level for a signal, x_c, at which an observed signal can be reliably distinguished from the background. If interferences are absent and measurement errors for the blank and sample containing the analyte follow normal distributions, then the distributions can be viewed as in Figure 5.10, where $\bar{x}_b$ and $\bar{x}_s$ symbolize the bank and sample measurement means, respectively, and s_b and s_s represent corresponding standard deviations. Distributions drawn in Figure 5.10 are when $s_b = s_s$, which is usually true at trace levels. If x_b and x_s specify

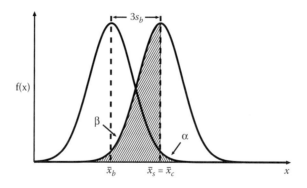

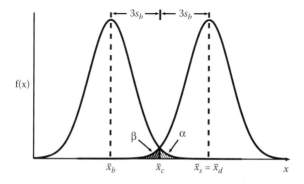

FIGURE 5.10 Graphical representation of (a) decision limits with $\alpha = 0.0013$, $z_\alpha = 3.0$, and $\beta = 0.5$ and (b) detection limits with $\alpha = 0.0013$, $z_\alpha = 3.0$, and $\beta = 0.0013$. (Reprinted from Haswell, S.J., Ed., *Practical Guide to Chemometrics*, Marcel Dekker, New York, 1992. With permission.)

signals measured for the blank and sample, respectively, then in terms of hypothesis testing, the null hypothesis becomes: "The analyte is not present" or "The measured signal does not significantly differ from the blank" (H_0: $x_s = x_b$). The alternative hypothesis is: "The analyte is present" or "The measured signal is significantly different than the blank" (H_1: $x_s > x_b$). Accepting the alternative hypothesis when the analyte is not present invokes a type I error (false positive) with probability α. Accepting the null hypothesis when the analyte is present renders a type II error (false negative) with probability β.

Acceptance or rejection of the null hypothesis is based on a set critical level for the measured signal, y_c, commonly expressed as

$$x_c = \bar{x}_b + k_c s_b$$

where k_c specifies a numerical value governed by the risk accepted for a type I error. Determining $\bar{x}_b$ and s_b from many measurements implies suitable estimates of the corresponding population values μ_b and σ_b. Therefore, z_α can be used for k_c (see Chapter 3 for a discussion on z values). If the risk of a type I error is set to $\alpha = 0.0013$, the critical z values corresponds to 3.00 for the $1 - \alpha = 99.87\%$ confidence level. The decision limit becomes

$$x_c = \mu_b + z_b \sigma_b = \mu_b + 3\sigma_b \qquad (5.17)$$

With this risk level, a 0.13% chance exists that a sample without the analyte would be interpreted as having the analyte present. Unfortunately, the chance of making a type II error is then $\beta = 0.50$, expressing a risk of failing to detect the analyte 50% of the time. Figure 5.10a graphically shows the problem.

In practice, only a limited number of measurements are made to compute $\bar{x}_b$ and s_b. The value for k_c is then determined from the appropriate t value with the proper degrees of freedom. Once x_c has been estimated, it can be used in the calibration model to obtain the corresponding concentration value y_c.

5.4.9.2.2 Univariate Detection Limit

The detection limit, x_d, represents the signal level that can be relied upon to imply detection. To avoid the large β value observed with decision limits in the previous section, a larger critical signal becomes necessary. The blank and sample signals would have analogous statistical distributions, as noted in the previous section, but the sample signal would be centered around a greater value for x_d. Choosing x_d such that $\alpha = \beta = 0.0013$ substantially reduces the probability of obtaining a measurement below x_c, as defined in Equation 5.17. Figure 5.10b illustrates the situation. The signal level at which this occurs identifies the detection limit expressed by

$$x_d = x_c + 3\sigma_b = \mu_b + 6\sigma_b \qquad (5.18)$$

Thus, requiring a larger critical signal level considerably diminishes the chance of making a type II error. Similar to the decision limit, appropriate t-values for the proper degrees of freedom should be used if a small number of measurements is used. Substitution of x_d into the calibration model will provide the detection limit concentration y_d.

An alternative definition for detection limit prevalent in the literature substitutes 3.00 for the 6.00 in Equation 5.18. This formulates a decision limit of $x_c = \mu_b + 1.5\sigma_b$. For this other definition, the probabilities of making type I or II detection limit errors are modified to $\alpha = \beta = 0.067$ for a $1 - \alpha = 93.30\%$ confidence level. Thus, this definition runs a greater risk of making errors. A figure analogous to Figure 5.10 can be made for this situation with replacement of $3s_b$ by $1.5s_b$. Apparently, numerous definitions are possible for detection limit, with each definition depending on the designated level of confidence. For example, at the 95% confidence level, $\alpha = 0.05$, $z_\alpha = 1.645$, $x_c = \mu_b + 1.645\sigma_b$, and $x_d = \mu_b + 3.29\sigma_b$. Therefore, reported detection limits should be accompanied by the level of significance selected. In general, the greater the confidence level, the larger the detection limit.

5.4.9.2.3 Determination Limit

The determination limit, x_q, designates the signal level at which an acceptable quantitative analysis can be made. A value of

$$x_q = \mu_b + 10\sigma_b$$

is typically used. This is also known as the limit of quantitation (LOQ).

5.4.9.2.4 Multivariate Detection Limit

Various approaches have been used to define detection limit for the multivariate situation [24]. The first definition was developed by Lorber [19]. This multivariate definition is of limited use because it requires concentration knowledge of all analytes and interferences present in calibration samples or spectra of all pure components in the calibration samples. However, the work does introduce the important concept of net analyte signal (NAS) vector for multivariate systems. The NAS representation has been extended to the more usual multivariate situations described in this chapter [25–27], where the NAS is related to the regression vector **b** in Equation 5.11. Mathematically, $\mathbf{b} = \text{NAS}/\|\text{NAS}\|$ and $\|\text{NAS}\| = 1/\|\mathbf{b}\|$. Thus, the norm of the NAS vector is the same as the effective sensitivity discussed in Section 5.4.9.1 A simple form of the concentration multivariate limit of detection (LOD) can be expressed as $\text{LOD} = 3\|\varepsilon\|\|\mathbf{b}\|$, where ε denotes the vector of instrumental noise values for the m wavelengths. The many proposed practical approaches to multivariate detection limits are succinctly described in the literature [24].

5.4.10 INTERFERENCE EFFECTS AND SELECTIVITY

Interferences are common in chemical analysis. In general, interferences are classified as physical, chemical, or spectral. Physical interferences are caused by the effects from physical properties of the sample on the physical process involved in the analytical measurements. Viscosity, surface tension, and vapor pressure of a sample solution are physical properties that commonly cause interferences in atomic absorption and atomic emission. Chemical interferences influence the analytical signal, and they result from chemical interactions between the analyte and other substances present in the sample as well as analyte interactions with the analyte (intermolecular and intramolecular interactions are present). Spectral interferences are those that arise when a wavelength is not completely selective for the analyte, and these are quite common in most spectroscopic methods of analysis.

Physical and chemical effects can be combined for identification as sample matrix effects. Matrix effects alter the slope of calibration curves, while spectral interferences cause parallel shifts in the calibration curve. The water-methanol data set contains matrix effects stemming from chemical interferences. As already noted in Section 5.2, using the univariate calibration defined in Equation 5.4 requires an interference-free wavelength. Going to multivariate models can correct for spectral interferences and some matrix effects. The standard addition method described in Section 5.7 can be used in some cases to correct for matrix effects. Severe matrix effects can cause nonlinear responses requiring a nonlinear modeling method.

Selectivity describes the degree of spectral interferences, and several measures have been proposed. Most definitions refer to situations where pure-component spectra of the analyte and interferences are accessible [19–21, 28]. In these situations, the selectivity is defined as the sine of the angle between the pure-component spectrum for the analyte and the space spanned by the pure-component spectra for all the interfering species. Recently, equations have been presented to calculate selectivity for an analyte in the absence of spectral knowledge of the analyte or interferences [25–27]. These approaches depend on computing the NAS, defined as the signal due only to the analyte. Methods have been presented to compute selectivity values for N-way data sets (see Section 5.6.4 for the definition of N-way) [29, 30].

5.5 VARIABLE SELECTION

As noted previously, the single most important question to be answered when using least squares to form the multivariate regression model is: Which variables (wavelengths) should be included? It is tempting to include all variables known to affect or are believed to affect the prediction properties; however, this may lead to suboptimal models or, even worse, inclusion of highly correlated variables in the model. When highly correlated variables are included in the model, computation of the inverse $(\mathbf{X}^T\mathbf{X})^{-1}$ becomes unstable, i.e., $\mathbf{X}^T\mathbf{X}$ is singular or nearly singular. Additionally, the reader is reminded that for $\mathbf{X}^T\mathbf{X}$ to be nonsingular, $n \geq m + 1$ for models with an intercept and $n \geq m$ for models without an intercept, where m is the number of wavelengths used in the model, i.e., full spectra have been measured at w wavelengths, and m is the number of wavelengths in the model subset. Unless the true form of the relationship between $\mathbf{X}$ and $\mathbf{y}$ is known, it is necessary to select appropriate variables to develop a calibration model that gives an adequate and representative statistical description for use in prediction.

Most approaches to variable selection are based on minimizing a prediction-error criterion. In this case, it is important to provide a data set for validating (testing) the model with the final selected variables. For example, if the RMSEV (RMSEP) or RMSECV is used as a criterion for evaluating selected variables and choosing the final model, then an additional data set independent of the data sets used in evaluating the selected variables is needed for a concluding test of the final model.

5.5.1 FORWARD SELECTION

In forward selection, the first variable (wavelength) selected is that variable x_j that minimizes the residual sum of squares, RSS, according to

$$RSS = \sum_{i=1}^{n} (y_i - \hat{b}_j x_{ij})^2$$

where $\hat{b}_j$ is the corresponding least-squares regression coefficient. The variable selected first, x_1, is forced into all further subsets. New variables x_2, x_3, ..., x_m are progressively added to the model, each variable being chosen because it minimizes the residual sum of squares when added to those already selected. Various rules can be used as stopping criteria [3, 5].

5.5.2 EFROYMSON'S STEPWISE REGRESSION ALGORITHM

There are two important problems with the simple forward-selection procedure described above.

1. In general, the subset of m variables providing the smallest residual sum of squares does not necessarily contain the subset of $(m-1)$ variables that gives the smallest residual sum of squares for $(m-1)$ variables.
2. There is no guarantee that forward selection will find the best-fitting subsets of any size except for $m = 1$ and $m = w$.

In order to address these two problems, a test is made to see if any of the previously selected variables can be deleted without appreciably increasing the residual sum of squares. The test is performed after each variable other than when the first is added to the set of selected variables. Before introducing the complete algorithm, two different types of steps are described, the variable-addition step and the variable-deletion step.

5.5.2.1 Variable-Addition Step

Let RSS_m denote the residual sum of squares for a model with m variables and an intercept term, b_0. Suppose the smallest RSS that can be obtained by adding another variable to the present set is RSS_{m+1}. The calculated ratio R according to

$$R = \frac{RSS_m - RSS_{m+1}}{RSS_{n+1}/(n-m-2)}$$

Calibration

is compared with an "F-to-enter" value, say F_e. If R is greater than F_e, the variable is added to the selected set.

5.5.2.2 Variable-Deletion Step

With m variables and a constant in the selected subset, let RSS_{m-1} be the smallest RSS that can be obtained after deleting any variable from the previously selected variables. The ratio computed by

$$R = \frac{RSS_{m-1} - RSS_m}{RSS_m/(n-m-2)}$$

is compared with an "F-to-delete (or drop)" value, say F_d. If R is less than F_d, the variable is deleted from the selected variables set.

5.5.2.3 Convergence of Algorithm

The above two steps can be combined to form a complete algorithm. It can be proved that when a successful addition step is followed by a successful deletion step, the new RSS^* will be less than the previous RSS and

$$RSS_m^* \leq RSS_m \cdot \frac{1 + F_d/(n-m-2)}{1 + F_e/(n-m-2)}$$

The procedure stops when no further additions or deletions are possible that satisfy the criteria. As each step is bounded below by the smallest RSS for any subset of m variables, by ensuring that the RSS is reduced each time that a new subset of m variables is found, convergence is guaranteed. A sufficient condition for convergence is that $F_d < F_e$. As with forward selection, there is no guarantee that this algorithm will locate the best-fitting subset, although it often performs better than forward selection when some of the predictors are highly correlated.

5.5.3 BACKWARD ELIMINATION

In this procedure, we start with all w variables, including a constant if there is one, in the selected set. Let RSS_w be the corresponding residual sum of squares. A variable is chosen for deletion that yields the smallest value of RSS_{w-1} after deletion. The process continues until there is only one variable left, or until some stopping criterion is satisfied. Note that:

- In some cases, the first variable deleted in backward elimination is the first one inserted in forward selection.
- A backward-elimination analogue of the Efroymson procedure is possible.
- Both forward selection and backward elimination can fare arbitrarily poorly in finding the best-fitting subsets.

5.5.4 Sequential-Replacement Algorithms

Once two or more variables have been selected, it is determined whether any of those variables can be replaced with another variable to generate a smaller RSS. With some of these attempts, there will be no variable that yields a reduction in the RSS, in which case the process moves on to identifying the next variable. Sometimes, variables that have been replaced will return. The process continues until no further reduction is possible by replacing any variable. Note that:

- The sequential-replacement algorithm can be obtained by taking the forward-selection algorithm and applying a replacement procedure after each new variable is added.
- Replacing two variables at a time substantially reduces the maximum number of stationary subsets and means that there is a greater chance of finding good subsets.
- Even if the best-fitting subset of a certain size is located, there is no way of knowing whether it is indeed the best one.

5.5.5 All Possible Subsets

It is sometimes feasible to generate all possible subsets of variables, provided that the number of predictor variables is not too large. After the complete search has been carried out, a small number of the more promising subsets can be examined in greater detail. The obvious disadvantage of generating all subsets is computation time. The number of possible subsets of one or more variables out of w is $(2^w - 1)$. For example, when $w = 10$, the total number of subsets is about 1000; however, when $w = 20$, the total number of possible subsets is more than 1,000,000.

5.5.6 Simulated Annealing and Genetic Algorithm

Except for testing all possible combinations, the above methods of variable selection primarily suffer from the fact that suboptimal subsets can result. Said another way, the above methods can easily converge to a locally optimal combination of variables and not result in the global subset. The methods of simulated annealing (SA) and genetic algorithm (GA) are known to be global optimization methods and are applicable to variable selection [31–33]. Both methods are stochastic-search heuristic approaches and have been shown to perform equally well for wavelength selection [34, 35]. For SA, the user is required to specify how many wavelengths are desired, while with GA this is generally not necessary. The method of GA does mandate more algorithm operational parameters to be set than does SA, and generalized SA (GSA) needs even fewer [36].

5.5.7 Recommendations and Precautions

In general, if it is feasible to carry out an exhaustive search, then that is to be recommended. As the sequential-replacement algorithm is fairly fast, it can always be used first to provide an indication of the maximum size of the subset that is likely

Calibration

to be of interest for the exhaustive search, or it can be used as a starting point for SA or GA. When it is not feasible to carry out the exhaustive search, the use of random starts followed by sequential replacement, or two-at-a-time replacement, can be used, though there can be no guarantee of finding the best-fitting subsets. The methods of SA and GA are applicable as well.

In all cases, graphical or other methods should be used to access the adequacy of the fit obtained. These examinations often uncover residual patterns that may indicate the suitability of using a transformation, or some kind of weighting, or adding extra variables such as quadratic or interaction terms. Unfortunately, inference becomes almost impossible if the total subset of available predictors is augmented subjectively in this way.

A number of derogatory names have been used in the past to describe the practices of subset selection, such as data grubbing, data mining, and even "torturing the data until they confess." Given a sufficiently exhaustive search, some apparent pattern can always be found, even if all of the predictions have come from a random number generator. The best subset for prediction may not be the one that gives the best fit to the sample data. In general, a number of the better-fitting subsets should be retained and examined in detail. If possible, an independent sample should be obtained to test the adequacy of the prediction equation. Alternatively, the data set can be divided into two parts; one part is used for model selection and calibration of parameters, and the second part for testing the adequacy of the predictions.

5.6 BIASED METHODS OF CALIBRATION

Biased approaches to calibration do not mandate wavelength selection prior to determining the calibration regression vector. Thus, these methods permit using more wavelengths than calibration samples and offer a form of signal-averaging advantage that can help cancel random errors in measured responses. Basically, the estimated model coefficients are obtained by

$$\hat{\mathbf{b}} = \mathbf{X}^+ \mathbf{y} \qquad (5.19)$$

where $\mathbf{X}^+$ designates a generalized inverse of $\mathbf{X}$. The biased approaches essentially differ in the computation of $\mathbf{X}^+$.

Diagnostic information can be obtained to determine whether the calibration model provides an adequate fit to the standards, e.g., nonlinearity or other kinds of model errors can be detected, or whether an unknown sample is adequately fitted by the calibration model. A large lack of fit is usually due to background signals different from those present in the calibration standards. This is what some people have called the "false sample" problem. For example, suppose a calibration model was developed for the spectroscopic determination of iron in dissolved carbon steel samples. This model might be expected to provide a poor performance in the determination of iron in stainless steel samples. In this case, a figure-of-merit calculated from the biased model would detect the "false sample."

Principal component regression (PCR), partial least squares (PLS), and ridge regression (RR) are three of the most popular biased-calibration methods. These methods have gained widespread acceptance. Industry implements routine analytical methods employing multivariate calibration methods because enhanced speed and accuracy over other methods are typically obtained. Frequently, the methods can be applied to mixtures without resorting to time-consuming chemical separation using chromatography. While PCR, PLS, RR, and other methods do not require wavelength selection, other metaparameters must be established. With PCR and PLS, the number of basis vectors to be used in generating the model is the metaparameter to be determined. Other terms for this are the number of factors, latent vectors, principal components, or basis vectors. The role of the metaparameter in the case of PCR and PLS is to reduce the dimensionality of the regression space and shrink the regression vector. The method of RR necessitates settling on a ridge-parameter value for the metaparameter and also forces the model to use less of the complete calibration space. As with variable selection (wavelength selection), it is important to perform a validation (test) of the final optimized metaparameter utilizing an independent data set not used in determining the final metaparameter value.

For the subsections of this section, variances and confidence intervals formulas are not furnished. The literature [11–17] provides excellent discussions on this subject. However, if only a rough estimate is needed, the equations previously presented in Sections 5.4.4 and 5.4.5 are often adequate.

5.6.1 PRINCIPAL COMPONENT REGRESSION

Recall from Chapter 4, Principal Component Analysis, that a mean-centered data matrix with n rows of mixture spectra recorded at m wavelengths, where each mixture contains up to k constituents, can be expressed as a product of k vectors representing concentrations and k vectors representing spectra for the pure constituents in the mixtures, as shown in Equation 5.20.

$$\mathbf{X} = \mathbf{YK}^T + \mathbf{E} \qquad (5.20)$$

The concentration of the ith component in the mixture is specified by the ith column of $\mathbf{Y}$, and the ith row of $\mathbf{K}$ contains the pure-component spectrum for the ith component.

With principal component analysis, it is possible to build an empirical mathematical model for the mean-centered data matrix $\mathbf{X}$, as shown by

$$\mathbf{X}_d = \mathbf{U}_d \mathbf{S}_d \mathbf{V}_d^T + \mathbf{E} \qquad (5.21)$$

or $\hat{\mathbf{X}}_d = \mathbf{U}_d \mathbf{S}_d \mathbf{V}_d^T$, where the product $\mathbf{U}_d \mathbf{S}_d$ represents the $n \times d$ matrix of principal component scores, $\mathbf{V}_d$ denotes the $m \times d$ matrix of eigenvectors, and d symbolizes the number of respective vectors used from the complete set available obtained by the singular value decomposition (SVD) of $\mathbf{X} = \mathbf{USV}^T$. In simple chemical systems, the value of d is often equal to k, the number of constituents. See Chapter 4 for additional information on the SVD.

Calibration

The eigenvectors in $\mathbf{V}_d$ are also referred to as abstract factors, eigenspectra, basis vectors, loading vectors, or latent vectors, indicating that while the vectors form a basis set for the row space of $\mathbf{X}$, physical interpretation of the vectors is not very useful. The columns of $\mathbf{V}_d$ are mutually orthogonal and normalized. Often the first eigenvector looks like the average spectrum for the calibration set. In spectral analysis, sometimes positive or negative peaks can be observed in the eigenvectors corresponding to overlapped or hidden bands in the calibration spectra. The columns of $\mathbf{U}_d$ are also mutually orthogonal and normalized. They can be used to form a set of column basis vectors for $\mathbf{X}$.

For each independent source of variation in the data, a single principal component (eigenvector) is expected in the model. For the NIR water-methanol data set, one factor for each chemical species in the mixture is expected, including intermolecular hydrogen-bonded species. The first column of scores (column in the product $\mathbf{US}$) and the first eigenvector (row in $\mathbf{V}^T$) denote the first factor. The first eigenvector corresponds to the one with the largest eigenvalue. It can be shown that the first factor explains the maximum amount of variation possible in the original data (maximum in a least-squares sense). The second factor is the next-most-important factor and corresponds to the second column of scores and the eigenvector associated with the second-largest eigenvalue. It explains the maximum amount of variation left in the original data matrix. Figure 5.11 shows a three-component principal component model for the NIR spectra of the water-methanol mixtures characterizing the similarities between Equations 5.20 and 5.21.

5.6.1.1 Basis Vectors

As described in Section 4.2.1, the $\mathbf{V}$ eigenvectors in Figure 5.11 can be thought of as row basis vectors, since each row in the data matrix $\mathbf{X}$ can be expressed as a linear combination (mixture) of the three eigenvectors. Similarly, the columns in $\mathbf{U}$ can be thought of as column basis vectors. Each column in the data matrix $\mathbf{X}$ can be expressed as a linear combination (mixture) of the columns in $\mathbf{U}$.

The coordinates of a vector $\mathbf{x}$ in an m dimensional space, e.g., an $m \times 1$ mixture spectrum measured at $m = 700$ wavelengths, can be expressed in a new coordinate system defined by a set of orthonormal basis vectors (eigenvectors) in the lower-dimensional space. Clearly, we cannot imagine a 700-dimensional space. It is

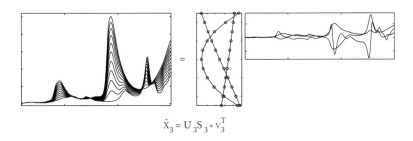

$$\hat{X}_3 = U_3 S_3 * V_3^T$$

FIGURE 5.11 Diagram of a three-factor principal component model for NIR spectra of water–methanol mixtures.

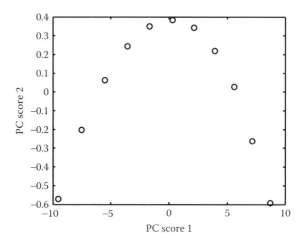

FIGURE 5.12 Scatter plot of the principal component (PC) scores from the SVD analysis of the water–methanol NIR mixture data.

possible, however, to view the position of points relative to each other in a 700-dimensional space by plotting them in the new coordinate system defined by two basis vectors from **V**. Using the first column of scores as x-axis plotting coordinates and the second column of scores as the y-axis plotting coordinates is a good initial plot. An example of such a plot using the NIR water–methanol data set is shown in Figure 5.12, where the elements of column vector 2 from **US** are used as the y-axis plotting coordinates and the elements of column vector 1 from **US** are used as the x-axis plotting coordinates. Curvature in the plot arises because the concentration of intermolecular hydrogen-bonded species is a nonlinear function of concentration.

5.6.1.2 Mathematical Procedures

Principal component regression is accomplished in two steps, a calibration step and an unknown prediction step. In the calibration step, concentrations of the constituent(s) to be quantitated in each calibration standard sample are assembled into a matrix, **y**, and mean-centered. Spectra of standards are measured, assembled into a matrix **X**, mean-centered, and then an SVD is performed. Calibration spectra are projected onto the d principal components (basis vectors) retained and are used to determine a vector of regression coefficients that can be then used to estimate the concentration of the calibrated constituent(s).

5.6.1.2.1 Calibration Steps

1. Compute the projected calibration spectra:

$$\hat{\mathbf{X}}_d = \mathbf{U}_d \mathbf{S}_d \mathbf{V}_d^T$$

2. Compute a regression vector using the calibration samples:

$$\hat{\mathbf{b}} = \hat{\mathbf{X}}_d^+ \mathbf{y} = \mathbf{V}_d \mathbf{S}_d^{-1} \mathbf{U}_d^T \mathbf{y}$$

3. Calibration prediction step:

$$\hat{\mathbf{y}} = \hat{\mathbf{X}}_d \hat{\mathbf{b}}$$

Note that $\mathbf{X}$ can be used instead of $\hat{\mathbf{X}}_d$, as $\hat{\mathbf{b}}$ is based on only d basis vectors.

4. Estimate the RMSEC, where n is the number of calibration samples used:

$$\text{RMSEC} = \left[\frac{1}{n-d-1} \sum_{i=1}^{n} (y_i - \hat{y}_i)^2 \right]^{1/2}$$

Some users of PCR do not mean-center the $\mathbf{X}$ and $\mathbf{y}$ matrices first, in which case the degrees of freedom become $n - d$ not $n - d - 1$. Other users of PCR choose not to subtract the number of factors, as this is an arbitrary constraint, and use n for no mean centering or $n - 1$ with mean centering. The literature [37–39] should be consulted on using effective rank instead of d. Unless noted otherwise, mean centering is used, and the degrees of freedom are $n - d - 1$ in this chapter.

5.6.1.2.2 Unknown Prediction Steps

1. Prediction step: Note that because the calibration mean-centered data is used, concentration predictions obtained by Equation 5.22 must be un-mean-centered if actual prediction values are to be reported.

$$\hat{\mathbf{y}}_{unk} = \mathbf{X}_{unk} \hat{\mathbf{b}} \tag{5.22}$$

2. Validation step: If concentrations of some of the unknowns are actually known, i.e., pseudo-unknowns, they can be used to determine the RMSEP (RMSEV):

$$\text{RMSEP} = \left[\frac{1}{p} \sum_{i=1}^{p} (y_i - \hat{y}_i)^2 \right]^{1/2}.$$

where p is the number of pseudo-unknowns used in the validation.

If more than one analyte is to be modeled simultaneously, then $\mathbf{y}$ is expanded to an $n \times a$ matrix $\mathbf{Y}$, resulting in an $m \times a$ matrix of regression coefficients $\mathbf{B}$, where each column is the regression vector for the *a*th analyte. Unless individual models are generated, the number of basis vectors used to form $\hat{\mathbf{B}}$ is now a compromise for all analytes.

5.6.1.3 Number of Basis Vectors

So far, the number of basis vectors that should be used in the calibration model has not been discussed. It is standard practice during PCR calibration modeling to use one principal component (PC), two PCs, three PCs, and so on. The error from this prediction is used to calculate the RMSEP figure of merit. Plots of RMSEC and RMSEP against the number of PCs used in the calibration model are used to determine the optimum number of factors. Usually, a continuous decrease in RMSEC is observed as more PCs are added into the calibration model; however, the predictive performance of the calibration model often reaches a minimum RMSEP at the optimum number of factors and begins to increase thereafter.

An alternative criterion that can be used is the RMSECV, as described in Section 5.2.6. A plot of RMSECV vs. the number of factors frequently shows a minimum or levels off at the optimum number of factors. As a reminder from Section 5.2.6, LOOCV commonly overfits, and MCCV is a better choice.

Common practice is to use only RMSEC, RMSEP, or RMSECV to assess the optimum number of basis vectors. However, these diagnostics only evaluate the bias of the model with respect to prediction error. As Figure 5.13 shows, there is a trade-off of variance for prediction estimates with respect to bias. As more basis vectors are utilized to generate the regression vector, the bias decreases at a sacrifice of a variance increase.

A graphic that can be produced to better describe the actual situation is the plot of $\|\hat{\mathbf{b}}\|$ against $\|\mathbf{y}-\hat{\mathbf{y}}\|$, where $\|\cdot\|$ symbolizes the Euclidean vector norm [40–46]. As Figure 5.14 discloses, an *L*-shaped curve results with the best model occurring at the bend, which reflects a harmonious model with the least amount of compromise in the trade-off between minimization of the residual and regression vector norms. The regression vector norm acts as an indicator of variance for the

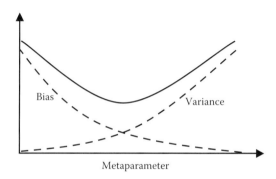

FIGURE 5.13 A generic situation for model determination showing the bias/variance trade-off with selection of the metaparameter.

Calibration

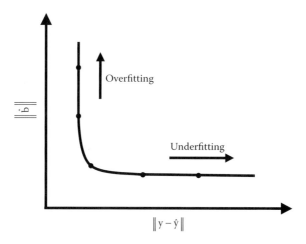

FIGURE 5.14 A generic plot of a variance indicator ($\|\hat{\mathbf{b}}\|$) against a bias measure ($\|\mathbf{y} - \hat{\mathbf{y}}\|$). The dots denote models with different metaparameters.

concentration estimates, and other measures of bias and variance can be utilized [11–17, 45, 46]. Because of the L shape, the harmonious plot is sometimes referred to as the L-curve. The foundation of using such a plot stems from Tikhonov regularization, as described in Section 5.6.4.

Overfitting the PCR calibration model is easily accomplished by including too many factors. For this reason, it is very important to use test data to judge the performance of the calibration model. The test data set should be obtained from standards or samples prepared independently from the calibration data set. These test standards are treated as pseudo-unknown samples. In other words, the final PCR calibration model is used to estimate the concentration of these test samples. Using the harmonious approach noted in Figure 5.14 significantly reduces the chance of obtaining an overfitted model.

It should be noted that other approaches to selecting basis vectors for PCR have been proposed [47 and references therein]. The most popular approach includes those basis vectors that are maximally correlated to $\mathbf{y}$ [48 and references therein].

5.6.1.4 Example PCR Results

Using the water–methanol data, PCR was performed with results graphically presented in Figure 5.15 and Figure 5.16. From the plot of only the bias criteria RMSEC and RMSEP in Figure 5.15a, it is not obvious as to the proper number of basis vectors. While the RMSEC increase from the three- to the four-factor model, respective RMSEP values decrease. Using the calibration residual plot presented in Figure 5.15b does not really assist in the decision. Regardless of the model, there still appears to be some nonlinearity not modeled. This becomes especially obvious when the validation residuals are inspected in Figure 5.15c. While the nonlinearity is not clearly

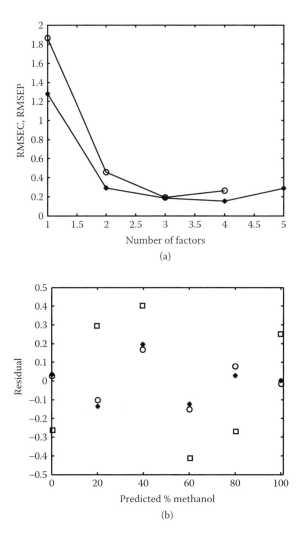

FIGURE 5.15 PCR methanol: (a) RMSEC (o) and RMSEP(*); (b) and (c) are calibration and validation residuals, respectively, for (□) two, (*) three, and (o) four PCs.

observable with calibration residuals based only on six calibration samples, the pattern becomes apparent when all 11 samples are used as the calibration set, as in Figure 5.7 with the four-wavelength data set.

The harmonious plot shown in Figure 5.16 aids in deciding on the number of basis vectors for the PCR model. The actual trade-off between improving the bias by including another basis vector and the degradation to variance can be assessed. Assisting the decision are the R^2 values for the validation set, which are 0.99899, 0.99997, 0.99998, 0.99999, and 0.99995 for one-, two-, three-, four-, and, five-factor models,

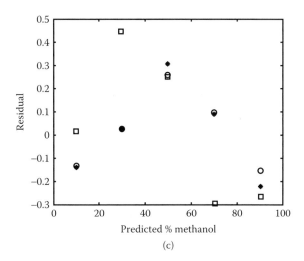

FIGURE 5.15 (Continued)

respectively. Corresponding calibration values are 0.99901, 0.99996, 0.99999, 1.00000, and 1.00000. Even though the improvements are all small, the small increase from the three- to the four-factor models, coupled with the other information discussed, results in a conclusion of three basis vectors as optimal. However, it is possible to argue that a model based only on two basis vectors is better because the gain in bias from proceeding to the three-factor model indicated by the RMSEC, RMSEP, and R^2 validation values may not be worth the corresponding increase in the variance indicator $\|\hat{\mathbf{b}}\|$.

The reader is reminded that PCR with wavelength selection could provide better results and is worth exploring. Similarly, using only a small, select set of wavelengths such that MLR can be implemented may also prove to be better and should likewise be investigated.

5.6.2 PARTIAL LEAST SQUARES

Partial least squares (PLS) was first developed by H. Wold in the field of econometrics in the late 1960s. During the late 1970s, groups led by S. Wold and H. Martens popularized use of the method for chemical applications. It should be noted that the well-known conjugate gradient method reviewed by Hansen [42] is equivalent to PLS [49, 50]. Two different methods are available, called PLS1 and PLS2. In PLS1, separate calibration models are built for each column in $\mathbf{Y}$. With PLS2, one calibration model is built for all columns of $\mathbf{Y}$ simultaneously.

The statistical properties of PLS2 are still not well understood and may not even be optimal for many calibration problems. The solution produced by PLS2 is dependent on how its iterative computations are initialized. A usual practice is to initialize PLS2 with the column from $\mathbf{Y}$ with the greatest correlation to $\mathbf{X}$. Initialization with other columns of $\mathbf{Y}$ produces different results.

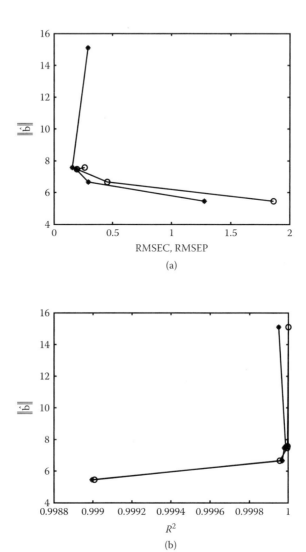

FIGURE 5.16 Harmonious PCR plot for methanol: (a) $\|\hat{\mathbf{b}}\|$ against RMSEC (o) and RMSEP(*); (b) $\|\hat{\mathbf{b}}\|$ against R2 for calibration (o) and validation (*).

5.6.2.1 Mathematical Procedure

In PLS, the response matrix $\mathbf{X}$ is decomposed in a fashion similar to principal component analysis, generating a matrix of scores, $\mathbf{T}$, and loadings or factors, $\mathbf{P}$. (These vectors can also be referred to as basis vectors.) A similar analysis is performed for $\mathbf{Y}$, producing a matrix of scores, $\mathbf{U}$, and loadings, $\mathbf{Q}$.

$$\mathbf{X} = \mathbf{TP}^\mathrm{T} + \mathbf{E}$$

$$\mathbf{Y} = \mathbf{UQ}^\mathrm{T} + \mathbf{F}$$

Calibration

The goal of PLS is to model all the constituents forming **X** and **Y** so that the residuals for the **X** block, **E**, and the residuals for the **Y** block, **F**, are approximately equal to zero. An inner relationship is also constructed that relates the scores of the **X** block to the scores of the **Y** block.

$$U = TW$$

The above model is improved by developing the so-called inner relationship. Because latent (basis) vectors are calculated for both blocks independently, they may have only a weak relation to each other. The inner relation is improved by exchanging the scores, **T** and **U**, in an iterative calculation. This allows information from one block to be used to adjust the orientation of the latent vectors in the other block, and vice versa. An explanation of the iterative method is available in the literature [42, 51, 52]. Once the complete model is calculated, the above equations can be combined to give a matrix of regression vectors, one for each component in **Y**:

$$\hat{B} = P(P^T P)^{-1} W Q^T \tag{5.23}$$

$$\hat{Y} = X\hat{B}$$

Various descriptions of the PLS algorithm exist in the literature. Some of the differences arise from the way normalization is used. In some descriptions, neither the scores nor the loadings are normalized. In other descriptions, either the loadings or scores may be normalized. These differences result in different expressions for the PLS calculations; however, the estimated regression vectors for **b** should be the same, except for differences in round-off error.

5.6.2.2 Number of Basis Vectors Selection

Similar to PCR, the number of basis vectors to use in Equation 5.23 must be discerned. The same methods described in Section 5.6.1.3 are used with PLS too.

5.6.2.3 Comparison with PCR

The simultaneous use of information from **X** and **Y** makes PLS more complex than PCR. However, it can allow PLS to develop better regression vectors, i.e., more harmonious with respect to the bias/variance trade-off. Some authors also report that PLS can sometime provide acceptable solutions for low-precision data where PCR cannot. Other authors have reported that PLS has a greater tendency to overfit noisy **Y** data compared to PCR. It is often reported in the literature that PLS is preferred because it uses fewer factors than PCR and, hence, forms a more parsimonious model. This is not the case, and the literature [38, 39, 43, 45, 53] should be consulted.

Even though problems exist, there may be situations where PLS2 is useful, particularly when extra variables with a strong correlation to **Y** are available that can be included in **Y**. For example, design variables or variables describing experimental conditions can be included in **Y**. Inclusion of these design variables may make it easier to interpret the final regression vectors, **b**.

Personal experience has shown that PLS often provides lower RMSEC values than PCR. The improvement in calibration performance must also manifest itself in predictions for independent samples. Therefore, a thorough evaluation of PCR versus PLS in any calibration application must involve using a large external validation data set with a comparison of $RMSEP_{PCR}$ and $RMSEP_{PLS}$ in conjunction with respective regression vector norms or other variance expressions.

5.6.3 A Few Other Calibration Methods

Besides PCR and PLS, other approaches to obtaining an estimate for the model coefficients in Equation 5.7 exist, and they are briefly mentioned here. Some of these methods are ridge regression (RR) [54], generalized RR (GRR) [54, 55], continuum regression (CR) [56], cyclic subspace regression (CSR) [57], and ridge variations of PCR, PLS, etc. [43, 58, and references therein]. The methods of GRR, CR, and CSR can generate the least-squares, PCR, and PLS models. Geometrical interrelationships of CR and CSR have been expressed as well as describing modifications to GRR to form PLS models [59].

These mentioned approaches in addition to PCR and PLS result in a regression vector having a smaller length (smaller $\|\mathbf{b}\|$) relative to the least-squares solution. Each of these methods requires determination of a metaparameter(s). In the case of RR and GRR, appropriate ridge parameters are necessary. An exponential value is needed with CR. The method of CSR first projects $\mathbf{X}$ based on a set of basis eigenvectors from $\mathbf{V}$ obtained through the SVD of $\mathbf{X}$, and then a PLS1 algorithm is used, which necessitates determining the number of PLS basis vectors to use from the eigenvector-projected $\mathbf{X}$. Recently, an approach was developed that first projects $\mathbf{X}$ using a subset of PLS basis vector from the original $\mathbf{X}$, and then an SVD is performed on the PLS-projected $\mathbf{X}$, requiring selection of how many basis eigenvectors to then use for the final model [60]. Other variations of projections with $\mathbf{V}$ and PLS basis vectors combined with RR have been described [61–63], as well as variations combining variable selection with RR [64].

While beyond the scope of this chapter, N-way modeling methods are being used more widely in the literature [65]. The idea here is to use other dimensions of information. For example, first-order data consists of only the spectroscopic order for a spectrum or the chromatographic order for a chromatogram. Second-order data is that formed by combing data from two first-order instruments. Variance expressions for N-way modeling have been derived [66, 67]. See Chapter 12 for more information.

Using artificial neural networks to develop calibration models is also possible. The reader is referred to the literature [68–70] for further information. Neural networks are commonly utilized when the data set maintains a large degree of nonlinearity. Additional multivariate approaches for nonlinear data are described in the literature [71, 72].

5.6.3.1 Common Basis Vectors and a Generic Model

The approaches described or mentioned to obtain model coefficients in Equation 5.7 can be expressed using a common basis set. For example, the literature commonly describes PCR and PLS as using different basis sets to span respective calibration spaces. In reality, PCR and PLS regression vectors can be written as linear combinations of a specified basis set. Using the **V** eigenvectors from the SVD of **X** results in

$$\hat{\mathbf{b}} = \mathbf{V}\beta \tag{5.24}$$

where β represent a vector of weights [43, 45, 59 and references therein], and $\hat{\mathbf{b}}$ can be the PCR, PLS, RR, GRR, CR, CSR, etc. regression vector. An analogous equation can be formed using the PLS basis set, as well as other basis sets. Because values in β identify the importance of a basis vector direction, it is useful to compare values obtained from different modeling procedures. That is, once regression vectors have been estimated by various modeling methods, the corresponding weights for a specific basis set can be computed, thus allowing intermodel comparisons in that basis set. Because of Equation 5.24 and other equations for respective basis sets, the concept of the most parsimonious model with respect to models compared in different basis sets is not practical. A generic expression, as in Equation 5.24, can be written based on filter values, further demonstrating the interrelationships of different modeling methods [39, 42]. Summarizing, a goal of multivariate calibration is then to find weight values for β, using a given basis set, that are optimal with respect to specified criteria. The next section further discusses this concept.

5.6.4 REGULARIZATION

Regularization is a term coined to describe processes that replace $(\mathbf{X}^T\mathbf{X})^{-1}$ in Equation 5.12 or $\mathbf{X}^+$ in Equation 5.19 by a family of approximate inverses [73]. In the case of multivariate calibration, the goal is to balance variance with bias, much like the H-principle [74]. An analogy in image restoration is to seek a balance between noise suppression and the loss of details in the restored image. Thus, the purpose of regularization is to single out a useful and stable solution. The methods of PCR, PLS, and those listed in Section 5.6.4 can all be classified as methods of regularization. The most well-known form of regularization is Phillips-Tikhonov regularization, usually referred to as Tikhonov regularization [75–78]. The approach is to use a modified least-squares problem by defining a regularized solution **b** as the minimizer of the following weighted combination of the residual norm model and coefficient norm

$$\mathbf{b}_\lambda = \operatorname*{argmin}\left(\|\mathbf{X}\mathbf{b} - \mathbf{y}\|^2 + \lambda\|\mathbf{L}\mathbf{b}\|^2\right) \tag{5.25}$$

for some matrix **L** and regularization parameter that controls the weight. A large value for λ, and hence a large amount of regularization, favors a small-solution norm at the cost of a large-residual norm; conversely, a small λ, and hence very little regularization, has the opposite effect. When **L** is the identity matrix, the regularization problem is said to be in standard form and RR results, the statisticians' name for Tikhonov regularization. However, it should be noted that in determining the optimal RR value for λ, only prediction-error criteria, such as RMSEP or RMSECV, are commonly used. A recent comparison study documents the importance of using a variance indicator such as $\|\mathbf{b}\|$ in addition to a prediction-error criterion [46]. Other diagnostic measures could be included in the minimization problem of Equation 5.25.

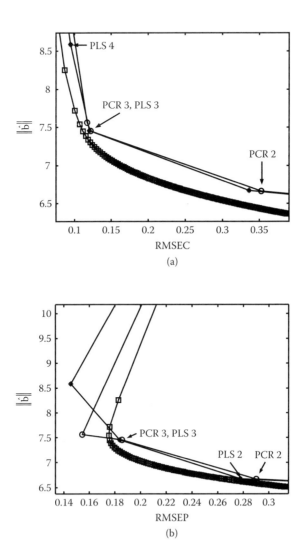

FIGURE 5.17 PCR (o), PLS (*), and RR (□) harmonious plots for methanol: (a) RMSEC and (b) RMSEP. The PCR and PLS2 factor models are in the lower right corner, with the RR ridge value beginning at 0.0011 in the upper left corner for (a) and (b) and ending at 0.4731 for (a) and 0.1131 for (b) in the lower right corner in increments of 0.001. The RMSEC values are with $n-1$ degrees of freedom.

Calibration

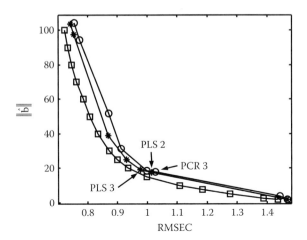

FIGURE 5.18 PCR (o), PLS (*), and simplex ( ) harmonious plots for NIR analysis of moisture in soy samples. The PCR and PLS1 factor models are in the lower right corner, and the respective models are 8 and 7 factors in the upper left corner. The simplex models converged to RR and GRR models. The RMSEC values are with $n - 1$ degrees of freedom.

Besides Tikhonov regularization, there are numerous other regularization methods with properties appropriate to distinct problems [42, 53, 73]. For example, an iterated form of Tikhonov regularization was proposed in 1955 [77]. Other situations include using different norms instead of the Euclidean norm in Equation 5.25 to obtain variable-selected models [53, 79, 80] and different basis sets such as wavelets [81].

5.6.5 Example Regularization Results

A regularization approach has been used to compare PLS, PCR, CR, CSR, RR, and GRR [45]. Simplex optimization [82] was used for GRR and minimization of Equation 5.25 in conjunction with Equation 5.24 using different basis sets. Plotted in Figure 5.17 and Figure 5.18 are the harmonious plots for two data sets. The curve identified as simplex in Figure 5.18 is also the curve obtained by RR and GRR. Thus, from Figure 5.18, the PLS and PCR models are passed over in the simplex optimization, and the models are converged to those of RR within round-off error. Other harmonious graphics using different variance indicators from prediction-variance equations (Equation 5.11 through Equation 5.17) are provided in the literature [44, 46].

5.7 STANDARD ADDITION METHOD

In Sections 5.2.1 and 5.2.2, it was stated that the samples must be matrix-effect-free for univariate models, e.g., inter- and intramolecular interactions must not be present. The standard addition method can be used to correct sample matrix effects. It should be noted that most descriptions of the standard addition method in the literature use a model form, where the instrument response signifies the dependent variable, and concentration represents the independent variable. For consistency with the discussion in this chapter, the reverse shall be used.

5.7.1 Univariate Standard Addition Method

Procedural steps encompass dividing the sample up into several equal-volume aliquots, adding increasing amounts of an analyte standard to all aliquots except one, diluting each aliquot to the same volume, and measuring instrument responses. The model implied by the standard addition method is

$$y_o = x_o b_1 \tag{5.26}$$

$$y_i = x_i b_1 \tag{5.27}$$

where y_o denotes the analyte concentration in the aliquot with no standard addition, y_i symbolizes the total analyte concentration after the ith standard addition of the analyte, x_o and x_i signify the corresponding instrument responses, and b_1 represents the model coefficient. As with the univariate model of Section 5.2.1, zero concentration of the analyte (and matrix) should evoke a zero response. Subtracting Equation 5.26 from Equation 5.27 results in $\Delta y_i = \Delta x_i b_1$, where Δy_i is the concentration of the standard added on the ith addition, and similar meaning is given to Δx_i. In matrix algebra, the model is expressed as $\Delta \mathbf{y} = \Delta \mathbf{x} b_1$, which can be solved for the regression coefficient as in Equation 5.2. Thus, plotting concentration of the standard added against the change in signal will provide a calibration curve with a slope equal to $\hat{b}_1$. The estimated slope is then used in Equation 5.26 to obtain an estimate of the analyte concentration in the unknown sample.

An alternative approach is to write the model as

$$\frac{v_i y_s}{v_T} + \frac{v_o y_o}{v_T} = x_i b_1 \tag{5.28}$$

where v_i expresses the volume of standard added in the ith addition, y_s denotes the corresponding analyte standard stock solution concentration, v_T symbolizes the total volume that all aliquots are diluted to, v_0 and y_0 represent the volume and analyte concentration of the unknown sample, and x_i designates the corresponding measured response. Equation 5.28 can be rearranged to

$$v_i = -\frac{v_o y_o}{y_s} + \frac{v_T b_1}{y_s} x_i \tag{5.29}$$

revealing that a plot of the volume of standard added against the respective measured response will produce a calibration curve with an intercept of $-v_0 y_0/y_s$ that can be solved for the concentration of the analyte in the unknown sample. Using the notation of Equation 5.29, Equation 5.30 is derived for an analyte concentration estimate based on only one standard addition ($i = 1$).

$$y_o = \frac{x_o v_1 y_s}{v_o (x_1 - x_o)} \qquad (5.30)$$

At best, this estimate is semiquantitative.

Approaches have been described for using standard additions without diluting to a constant volume. Multiplying measured responses by a ratio of total to initial volume accomplishes a correction for dilution. However, the matrix concentration is also diluted by the additions, creating a nonlinear matrix effect that may or may not be transformed into a linear effect by the volume correction. Kalivas [83] demonstrated the critical importance of maintaining constant volume.

In summary, requirements for the univariate standard addition methods are that (1) the response for the analyte should be zero when the concentration equals zero (as well as for the matrix), (2) the response is a linear function of the analyte concentration, (3) sample matrix effects are independent of the ratio of the analyte and matrix, and perhaps most importantly, (4) a standard solution of only the analyte is available for the additions.

5.7.2 Multivariate Standard Addition Method

The standard addition method has been generalized to correct for both spectral interferences and matrix effects simultaneously [84, 85] as well as compensate for instrument drift [86]. Original derivations of the generalization were with instrument responses signifying the dependent variables and with constituent concentrations representing the independent variables. For consistency with the discussion in this chapter, the reverse shall be used. For an analyte in a multianalyte mixture, Equation 5.26 becomes $y_0 = \mathbf{x}_0^T \mathbf{b}$, where y_0 denotes the analyte concentration in the aliquot with no standard addition and $\mathbf{b}$ symbolizes the respective column of $\mathbf{B}$ from $\mathbf{y}_0^T = \mathbf{x}_0^T \mathbf{B}$ and $\mathbf{Y} = \mathbf{XB}$, where $\mathbf{y}_0$ represents the concentration vector of all responding constituents that form $\mathbf{x}_0$, $\mathbf{Y}$ symbolizes total concentrations after respective standard additions, and $\mathbf{X}$ designates the respective measurements. The $\mathbf{B}$ matrix (or respective column $\mathbf{b}$) is obtained from the corresponding difference equation $\Delta \mathbf{Y} = \Delta \mathbf{XB}$ ($\Delta \mathbf{y} = \Delta \mathbf{Xb}$). Note that to obtain estimates of $\mathbf{B}$ or $\mathbf{b}$, standard additions must be made for all responding constituents that form $\mathbf{x}_0$. Additionally, wavelengths must be selected, or a biased approach such as RR, PCR, PLS, etc. needs to be used (refer to Sections 5.2.3 and 5.6). Alternatively, as expressed in the original derivations [84], writing the model as $\mathbf{x}_0^T = \mathbf{y}_0^T \mathbf{K}$ and $\mathbf{X} = \mathbf{YK}$ — with the difference being $\Delta \mathbf{X} = \Delta \mathbf{YK}$, where $\mathbf{K}$ denotes the $a \times m$ matrix of regression coefficients for a analytes (responding constituents) at m wavelengths — does not require wavelengths to be selected.

A standard addition method has been studied for use with second-order data [87]. The specific application investigated was analysis of trichloroethylene in samples that have matrix effects caused by an interaction with chloroform.

5.8 INTERNAL STANDARDS

In some instances, uncontrolled experimental factors during sample preparation, measurement, or prediction procedures can cause undesired systematic changes in the instrument response. Examples of such factors might include variation in extraction efficiency or variation in the effective optical path length. In cases such as these, an internal standard can be used to improve the calibration precision. An internal standard is a substance added in constant amount to all samples. Calibration is obtained by using the ratio of instrument readings to an instrument reading specific only to the internal standard added. If the internal standard and analyte (as well as all other responding constituents) respond proportionally to the random fluctuations, compensation is possible because the ratios of the instrument readings are independent of the fluctuations. If the two readings are influenced the same way by matrix effects, compensation for these effects also transpires. The process is the same whether univariate or multivariate calibration is being used. For either calibration approach, a measurement variable (e.g., wavelength) must exist that is selective to only the internal standard.

An example of the use of an internal standard is the precision enhancement for univariate quantitative chromatography based on peak area. With manual sample injection, the reproducibility of volume size is not consistent. For this approach to be successful, the internal standard peak must be well separated from the peaks of any other sample constituent. Use of an internal standard can be avoided with an autosampler.

In the case of multivariate calibration, an example consists of using KSCN as an internal standard for analysis of serum with mid-IR spectra of dry films [88]. In this study, KSCN was added to serum samples, and a small volume of a serum sample was then spread on a glass slide and allowed to dry. The accuracy of this approach suffers from the variation of sample volume and placement, causing nonreproducibility of spectra. Thus, by using a ratio of the spectral measurements to an isolated band for KSCN, the precision of the analysis improved.

5.9 PREPROCESSING TECHNIQUES

Preprocessing of instrument response data can be a critical step in the development of successful multivariate calibration models. Oftentimes, selection of an appropriate preprocessing technique can remove unwanted artifacts such as variable path lengths or different amounts of scatter from optical reflectance measurements. Preprocessing techniques can be applied to rows of the data matrix (by object) or columns (by variable).

One such pretreatment, mean centering, has already been introduced (see Section 5.2.2). Other preprocessing methods consist of using derivatives (first and second are common) to remove baseline offsets and scatter. The method of multiplicative scatter correction (MSC) works by regressing the spectra to be corrected against a reference spectrum [89]. A simple linear regression model is used, giving a baseline-offset correction and a multiplicative path-length correction. Often the mean spectrum of the calibration data set is used as the reference. Orthogonal signal correction (OSC) is a preprocessing technique designed to remove variance from the spectral data in $\mathbf{X}$ that is unrelated (orthogonal) to the chemical information in **y** [90]. The standard normal variate (SNV) method is another preprocessing approach sometimes used [91] and has an effect much like that of MSC. Using score plots, it was shown that with spectroscopic data contaminated with scatter, using the second derivative and MSC preprocessing provided the best spectral reproducibility [92].

Deciding on the type of preprocessing to include is not always straightforward and often requires comparison of modeling diagnostics between different preprocessing steps. Chapter 3 in [93] and Chapter 10 in [94] provide explicit discussions of the many preprocessing forms.

5.10 CALIBRATION STANDARDIZATION

After a calibration model has been built, situations can arise that may cause it to become invalid. For example, instrumental drift, spectral shifts, and intensity changes can invalidate a multivariate calibration model. These disturbances could be induced by uncontrolled experimental factors such as dirt on fiber-optic probes, or even maintenance events as simple as replacing a lamp in a spectrometer. Additionally, it is often advantageous to develop calibration models on a single master instrument and distribute it to many instruments in the field. Small differences in response from instrument to instrument could also invalidate the multivariate calibration model in such applications. In all of these situations, if the change in instrument response is large enough, it may be necessary to recalibrate the model with fresh calibration standards. Because recalibration can be lengthy and costly, alternative calibration standardization (transfer) methods have been developed. There are three main categories of calibration-transfer methods: (1) standardization of predicted values (e.g., slope and bias correction), (2) standardization of instrument (spectral) response, and (3) methods based on preprocessing techniques mentioned in Section 5.9. Overviews of these standardization methods are available in the literature [95–97 and references therein].

Most calibration-transfer methods require a small set of standards that must be measured on the pair of instruments to be standardized. These calibration-transfer standards may be a subset of calibration samples or other reference materials whose spectra adequately span the spectral domain of the calibration model. Calibration samples with high leverage or large influence are recommended by some authors as good candidates for calibration transfer [98]. Alternatively, other authors recommend selecting a few samples with the largest distance from each other [99]. A very small number of calibration-transfer standards (three to five) can be used; however, a larger number provides a better matching between the pair of instruments to be standardized.

5.10.1 STANDARDIZATION OF PREDICTED VALUES

To employ the simple slope-and-bias-correction method, a subset of calibration standards is measured on both instruments. The calibration model from the primary instrument is then used to predict the sample concentrations or properties of the measurements from the primary and the secondary instruments, giving predicted values $\hat{\mathbf{y}}_P$ and $\hat{\mathbf{y}}_S$. The secondary instrument is the primary instrument at a later time than when the original calibration model was built on another instrument. The bias and slope correction factors are determined by simple linear regression of $\hat{\mathbf{y}}_S$ on $\hat{\mathbf{y}}_P$, which is subsequently used to calculate the corrected estimates for the secondary instrument, $\hat{\mathbf{y}}_S(corr)$.

$$\hat{\mathbf{y}}_S(corr) = bias + slope \times \hat{\mathbf{y}}_S$$

This method assumes that the differences between the primary and secondary instruments follow a simple linear relationship. In fact, the differences may be much more complex, in which case more-advanced methods like piecewise direct standardization (PDS) may be more useful.

5.10.2 Standardization of Instrument Response

The goal of methods that standardize instrument response is to find a function that maps the response of the secondary instrument to match the response of the primary instrument. This concept is used in the statistical analysis procedure known as Procrustes analysis [97]. One such method for standardizing instrument response is the piecewise direct standardization (PDS) method, first described in 1991 [98, 100]. PDS was designed to compensate for mismatches between spectroscopic instruments due to small differences in optical alignment, gratings, light sources, detectors, etc. The method has been demonstrated to work well in many NIR assays where PCR or PLS calibration models are used with a small number of factors.

In the PDS algorithm, a small set of calibration-transfer samples are measured on a primary instrument and a secondary instrument, producing spectral response matrices $\overline{\mathbf{X}}_1$ and $\overline{\mathbf{X}}_2$. A permutation matrix $\mathbf{F}$ (Procrustes transfer matrix) is used to map spectra measured on the secondary instrument so that they match the spectra measured on the primary instrument.

$$\overline{\mathbf{X}}_1 = \overline{\mathbf{X}}_2 \mathbf{F}$$

The procedure for computing $\mathbf{F}$ employs numerous local regression models to map narrow windows of responses at wavelengths, $i - j$ to $i + j$, from the secondary instrument, giving an estimate of the corrected secondary response at wavelength i. At each wavelength i, a least-squares regression vector $\mathbf{b}_i$ is computed for the window of responses that bracket the point of interest, x_i.

$$\overline{x}_{1,i} = \overline{\mathbf{x}}_2 \mathbf{b}_i$$

These regression vectors are then assembled to form the banded diagonal transformation matrix, $\mathbf{F}$, where p is the number of response values to be converted.

$$\mathbf{F} = \text{diag}(\mathbf{b}_1^T, \mathbf{b}_2^T, \ldots, \mathbf{b}_i^T, \ldots, \mathbf{b}_p^T)$$

Either PLS or PCR can be used to compute $\mathbf{b}_i$ at less than full rank by discarding factors associated with noise. Because of the banded diagonal structure of the transformation matrix used by PDS, localized multivariate differences in spectral response between the primary and secondary instrument can be accommodated, including intensity differences, wavelength shifts, and changes in spectral bandwidth. The flexibility and power of the PDS method has made it one of the most popular instrument standardization methods.

5.10.3 STANDARDIZATION WITH PREPROCESSING TECHNIQUES

The previously discussed standardization methods require that calibration-transfer standards be measured on both instruments. There may be situations where transfer standards are not available, or where it is impractical to measure them on both instruments. In such cases, if the difference between the two instruments can be approximated by simple baseline offsets and path-length differences, preprocessing techniques such as baseline correction, first derivatives, or MSC can be used to remove one or more of these effects. In this approach, the desired preprocessing technique is applied to the calibration data from the primary instrument before the calibration model is developed. Prediction of samples from the primary or secondary instrument is accomplished simply by applying the identical preprocessing technique prior to prediction. See Section 5.9 for a brief overview of preprocessing methods and Chapter 4 for a more detailed discussion. A few methods are briefly discussed next.

The preprocessing approach of mean centering has been shown to correct for much for the spectral differences between the primary and secondary instruments [97, 101]. The mean-centering process can correct for baseline offsets, wavelength shifts, and intensity changes [97]. Finite impulse response (FIR) filters can be used to achieve a similar correction to that of MSC. However, because a moving window is used, greater flexibility is offered, allowing for locally different baseline offsets and path-length corrections [102, 103]. With FIR and MSC, there is a possibility that some chemical information may be lost. In instrument-standardization applications of OSC, it is assumed that baseline offsets, drift, and variation between different instruments is unrelated to $\mathbf{y}$ and therefore is completely removed by OSC prior to calibration or prediction [104, 105]. Several preprocessing methods for calibration transfer, including derivatives, MSC, and OSC, are also compared in the literature [102, 103].

5.11 SOFTWARE

Almost all of the approaches described in this chapter are readily available from commercial software packages. A short list includes the PLS_Toolbox from Eigenvector Research (note that PLS_Toolbox is not restricted to PLS, but also includes all aspects of multivariate calibration as well as numerous calibration approaches), Unscrambler from Camo, SIMCA from Umetrics, Infometrix maintains Pirouette, Thermo Galactic has GRAMS, and DeLight is available from DSquared Development. All of these software packages are considered to be user friendly. It should be noted that there are also other good and user-friendly calibration software packages. The PLS_Toolbox is MATLAB-based, allowing easy adaptation to user-specific problems. The Unscrambler package also maintains a MATLAB interface.

An abundance of MATLAB routines are available from many independent Web sites as well as tutorial Web sites, all of which are too numerous to mention here. Additionally, most instrument companies supply software that performs many of the calibration topics discussed in this chapter.

RECOMMENDED READING

Beebe, K.R., Pell, R.J., and Seasholtz, M.B., *Chemometrics: A Practical Guide*, John Wiley & Sons, New York, 1998.

Kramer, R., *Chemometric Techniques for Quantitative Analysis*, Marcel Dekker, New York, 1998.

Naes, T., Isaksson, T., Fearn, T., and Davies, T., *A User-Friendly Guide to Multivariate Calibration and Classification*, NIR Publications, Chichester, U.K., 2002.

Wickens, T.D., *The Geometry of Multivariate Statistics*, Lawrence Erlbaum Associates, Hillsdale, NJ, 1995.

Johnson, R.A. and Wichern, D.W., *Applied Multivariate Statistical Analysis*, 2nd ed., Prentice Hall, UpperSaddle River, NJ, 1988.

Weisberg, S., *Applied Linear Regression*, 2nd ed., John Wiley & Sons, New York, 1985.

Neter, J., Wasserman, W., and Kutner, M.H., *Applied Linear Statistical Models*, 3rd ed., Irwin, Boston, 1990.

Green, P.E., *Mathematical Tools for Applied Multivariate Analysis*, Academic Press, New York, 1978.

REFERENCES

1. Kalivas, J.H. and Lang, P.M., Mathematical Analysis of Spectral Orthogonality, Marcel Dekker, New York, 1994.
2. Mark, H., Principles and Practices of Spectroscopic Calibration, John Wiley & Sons, New York, 1991.
3. Belsley, D.A., Kuh, E., and Welsch, R.E., *Regression Diagnostics: Identifying Influential Data and Sources of Collinearity*, John Wiley & Sons, New York, 1980.
4. Neter, J., Wasserman, W., and Kutner, M.H., Applied Linear Statistical Models, 3rd ed., Irwin, Boston, 1990.
5. Weisberg, S., *Applied Linear Regression*, 2nd ed., John Wiley & Sons, New York, 1985, pp. 140–156.
6. ASTM E1655-97: Standard Practices for Infrared, Multivariate, Quantitative Analysis, ASTM, West Conshohocken, PA, 1999; available on-line at http://www.astm.org.
7. Shao, J., Linear model selection by cross-validation, *J. Am. Stat. Assoc.*, 88, 486–494, 1993.
8. Baumann, K., Cross-validation as the objective function for variable-selection techniques, *Trends Anal. Chem.*, 22, 395–406, 2003.
9. Xu, Q.S. and Liang, Y.Z., Monte Carlo cross validation, *Chemom. Intell. Lab. Syst.*, 56, 1–11, 2001.
10. Cruciani, G., Baroni, M., Clementi, S., Costantino, G., and Riganelli, D., Predictive ability of regression models, part II: selection of the best predictive PLS model, *J. Chemom.*, 6, 347–356, 1992.
11. Faber, K. and Kowalski, B.R., Propagation of measurement errors for the validation of predictions obtained by principal component regression and partial least squares, *J. Chemom.*, 11, 181–238, 1997.

12. Faber, K. and Kowalski, B.R., Prediction error in least squares regression: further critique on the deviation used in the unscrambler, *Chemom. Intell. Lab. Syst.*, 34, 283–292, 1996.
13. Faber, N.M., Song, X.H., and Hopke, P.K., Sample-specific standard error of prediction for partial least squares regression, *Trends Anal. Chem.*, 22, 330–334, 2003.
14. Fernández Pierna, J.A., Jin, L., Wahl, F., Faber, N.M., and Massart, D.L., Estimation of partial least squares regression prediction uncertainty when the reference values carry a sizable measurement error, *Chemom. Intell. Lab. Syst.*, 65, 281–291, 2003.
15. Lorber, A. and Kowalski, B.R., Estimation of prediction error for multivariate calibration, *J. Chemom.*, 2, 93–109, 1988.
16. Faber, N.M., Uncertainty estimation for multivariate regression coefficients, *Chemom. Intell. Lab. Syst.*, 64, 169–179, 2002.
17. Olivieri, A.C., A simple approach to uncertainty propagation in preprocessed multivariate calibration, *J. Chemom.*, 16, 207–217, 2002.
18. Skoog, D.A., Holler, F.J., and Nieman, T.A., *Principles of Instrumental Analysis*, Saunders College Publishing, Philadelphia, 1998, pp. 12–13.
19. Lorber, A., Error propagation and figures of merit for quantification by solving matrix equations, *Anal. Chem.*, 58, 1167–1172, 1986.
20. Kalivas, J.H. and Lang, P.M., Interrelationships between sensitivity and selectivity measures for spectroscopic analysis, *Chemom. Intell. Lab. Syst.*, 32, 135–149, 1996.
21. Kalivas, J.H. and Lang, P.M., Response to "Comments on interrelationships between sensitivity and selectivity measures for spectroscopic analysis," K. Faber et al., *Chemom. Intell. Lab. Syst.*, 38, 95–100, 1997.
22. Faber, K., Notes on two competing definitions of multivariate sensitivity, *Anal. Chim. Acta.*, 381, 103–109, 1999.
23. Currie, L.A., Limits for qualitative detection and quantitative determination, *Anal. Chem.*, 40, 586–593, 1968.
24. Boqué, R. and Rius, F.X., Multivariate detection limits estimators, *Chemom. Intell. Lab. Syst.*, 32, 11–23, 1996.
25. Ferré, J., Brown, S.D., and Rius, F.X., Improved calculation of the net analyte signal in the inverse calibration, *J. Chemom.*, 15, 537–553, 2001.
26. Lorber, A., Faber, K., and Kowalski, B.R., Net analyte signal calculation in multivariate calibration, *Anal. Chem.*, 69, 1620–1626, 1997.
27. Faber, N.M., Efficient computation of net analyte signal vector in inverse multivariate calibration models, *Anal. Chem.*, 70, 5108–5110, 1998.
28. Faber, N.M., Ferré, J., Boqué, R., and Kalivas, J.H., Quantifying selectivity in spectrometric multicomponent analysis, *Trends Anal. Chem.*, 22, 352–361, 2003.
29. Messick, N.J., Kalivas, J.H., and Lang, P.M., Selectivity and related measures for nth-order data, *Anal. Chem.*, 68, 1572–1579, 1996.
30. Faber, K., Notes on analytical figures of merit for calibration of nth-order data, *Anal. Lett.*, 31, 2269–2278, 1998.
31. Kalivas, J.H., Ed., *Adaption of Simulated Annealing to Chemical Optimization Problems,* Elsevier, Amsterdam, 1995.
32. Goldberg, D.E., *Genetic Algorithms in Search, Optimization, and Machine Learning,* Addison-Wesley, Reading, MA, 1989.
33. Leardi, R., Genetic algorithms in chemometrics and chemistry: a review, *J. Chemom.*, 15, 559–570, 2001.
34. Lucasius, C.B., Beckers, M.L.M., and Kateman, G., Genetic algorithm in wavelength selection: a comparative study, *Anal. Chim. Acta,* 286, 135–153, 1994.

35. Hörchner, U. and Kalivas, J.H., Further investigation on a comparative study of simulated annealing and genetic algorithm for wavelength selection, *Anal. Chim. Acta,* 311, 1–13, 1995.
36. Bohachevsky, I.O., Johnson, M.E., and Stein, M.L., Simulated annealing and generalizations, in *Adaption of Simulated Annealing to Chemical Optimization Problems,* Kalivas, J.H., Ed., Elsevier, Amsterdam, 1995, pp. 3–24.
37. Gilliam, D.S., Lund, J.R., and Vogel, C.R., Quantifying information content for ill-posed problems, *Inv. Prob.,* 6, 725–736, 1990.
38. van der Voet, H., Pseudo-degrees of freedom for complex predictive models: the example of partial least squares, *J. Chemom.,* 13, 195–208, 1999.
39. Seipel, H.A. and Kalivas, J.H., Effective rank for multivariate calibration methods, *J. Chemom.,* 18, 306–311, 2004.
40. Lawson, C.L. and Hanson, R.J., *Solving Least Squares Problems,* Prentice Hall, Upper Saddle River, NJ, 1974, pp. 200–206.
41. Hansen, P.C., Truncated singular value decomposition solutions to discrete ill-posed problems with ill-determined numerical rank, *SIAM J. Sci. Stat. Comput.,* 11, 503–519, 1990.
42. Hansen, P.C., *Rank-Deficient and Discrete Ill-Posed Problems: Numerical Aspects of Linear Inversion,* SIAM, Philadelphia, 1998.
43. Kalivas, J.H., Basis sets for multivariate regression, *Anal. Chim. Acta,* 428, 31–40, 2001.
44. Green, R.L. and Kalivas, J.H., Graphical diagnostics for regression model determinations with consideration of the bias/variance tradeoff, *Chemom. Intell. Lab. Syst.,* 60, 173–188, 2002.
45. Kalivas, J.H. and Green, R.L., Pareto optimal multivariate calibration for spectroscopic data, *Appl. Spectrosc.,* 55, 1645–1652, 2001.
46. Forrester, J.B. and Kalivas, J.H., Ridge regression optimization using a harmonious approach, *J. Chemom.,* 18, 372–384, 2004.
47. Joliffe, I.T., *Principal Component Analysis,* Springer Verlag, New York, 1986, pp. 135–138.
48. Fairchild, S.Z. and Kalivas, J.H., PCR eigenvector selection based on the correlation relative standard deviations, *J. Chemom.,* 15, 615–625, 2001.
49. Manne, R., Analysis of two partial-least-squares algorithms for multivariate calibration, *Chemom. Intell. Lab. Syst.,* 2, 187–197, 1987.
50. Phatak, A. and De Hoog, F., Exploiting the connection between PLS, Lanczos methods and conjugate gradients: alternative proofs of some properties of PLS, *J. Chemom.,* 16, 361–367, 2002.
51. Geladi, P. and Kowalski, B.R., Partial least-squares regression: a tutorial, *Anal. Chim. Acta,* 185, 1–17, 1986.
52. Geladi, P. and Kowalski, B.R., An example 2-block predictive partial least-squares regression with simulated data, *Anal. Chim. Acta,* 185, 19–32, 1986.
53. Frank, I.E. and Friedman, J.H., A statistical view of some chemometrics regression tools, *Technometrics,* 35, 109–148, 1993.
54. Hoerl, A.E. and Kennard, R.W., Ridge regression: biased estimation for nonorthogonal problems, *Technometrics,* 12, 55–67, 1970.
55. Hocking, R.R., Speed, F.M., and Lynn, M.J., A class of biased estimators in linear regression, *Technometrics,* 18, 425–437, 1976.
56. Stone, M. and Brooks, R.J., Continuum regression: cross-validated sequentially constructed prediction embracing ordinary least squares, partial least squares and principal components regression, *J. R. Stat. Soc. B,* 52, 237–269, 1990.

57. Lang, P.M., Brenchley, J.M., Nieves, R.G., and Kalivas, J.H., Cyclic subspace regression, *J. Multivariate Anal.*, 65, 58–70, 1998.
58. Xu, Q.-S., Liang, Y.-Z., and Shen, H.-L., Generalized PLS regression, *J. Chemom.*, 15, 135–148, 2001.
59. Kalivas, J.H., Interrelationships of multivariate regression methods using the eigenvector basis sets, *J. Chemom.*, 13, 111–132, 1999.
60. Wu, W. and Manne, R., Fast regression methods in a Lanczos (PLS-1) basis: theory and applications, *Chemom. Intell. Lab. Syst.*, 51, 145, 2000.
61. O'Leary, D.P. and Simmons, J.A., A bidiagonalization — regularization procedure for large scale discretizations of ill-posed problems, *SIAM J. Sci. Stat. Comput.*, 2, 474–489, 1981.
62. Kilmer, M.E. and O'Leary, D.P., Choosing regularization parameters in iterative methods for ill-posed problems, *SIAM J. Matrix Anal. Appl.*, 22, 1204–1221, 2001.
63. Engl, H.W., Hanke, M., and Neubauer, A., *Regularization of Inverse Problems*, Kluwer Academic, Boston, 1996.
64. Hoerl, R.W., Schuenemeyer, J.H., and Hoerl, A.E., A simulation of biased estimation and subset selection regression techniques, *Technometrics*, 28, 369–380, 1986.
65. Anderson, C. and Bro, R., Eds., Special issue: multiway analysis, *J. Chemom.*, 14, 103–331, 2000.
66. Faber, N.M. and Bro, R., Standard error of prediction for multiway PLS, 1: Background and a simulation study, *Chemom. Intell. Lab. Syst.*, 61, 133–149, 2002.
67. Olivieri, A.C. and Faber, N.M., Standard error of prediction in parallel factor analysis of three-way data, *Chemom. Intell. Lab. Syst.*, 70, 75–82, 2004.
68. Lon, J.R., Gregoriou, V.G., and Gemperline, P.J., Spectroscopic calibration and quantitation using artificial neural networks, *Anal. Chem.*, 62, 1791–1797, 1990.
69. Gemperline, P.J., Long, J.R., and Gregoriou, V.G., Nonlinear multivariate calibration using principal components and artificial neural networks, *Anal. Chem.*, 63, 2313–2323, 1991.
70. Naes, T., Kvaal, K., Isaksson, T., and Miller, C., Artificial neural networks in multivariate calibration, *J. Near Infrared Spectrosc.*, 1, 1–11, 1993.
71. Naes, T., Isaksson, T., Fearn, T., and Davies, T., *A User-Friendly Guide to Multivariate Calibration and Classification*, NIR Publications, Chichester, U.K., 2002, pp. 93–104, 137–153.
72. Sekulic, S., Seasholtz, M.B., Wang, Z., Kowalski, B.R., Lee, S.E., and Holt, B.R., Nonlinear multivariate calibration methods in analytical chemistry, *Anal. Chem.*, 65, 835A–845A, 1993.
73. Neumaier, A., Solving ill-conditioned and singular linear systems: a tutorial on regularization, *SIAM Rev.*, 40, 636–666, 1998.
74. Höskuldsson, A., Dimension of linear models, *Chemom. Intell. Lab. Syst.*, 32, 37–55, 1996.
75. Tikhonov, A.N., Solution of incorrectly formulated problems and the regularization method, *Soviet Math. Dokl.*, 4, 1035–1038, 1963.
76. Tikhonov, A.N. and Goncharsky, A.V., *Solutions of Ill-Posed Problems*, Winston & Sons, Washington, D.C., 1977.
77. Riley, J.D., Solving systems of linear equations with a positive definite symmetric but possibly ill-conditioned matrix, *Math. Table Aids Comput.*, 9, 96–101, 1955.
78. Phillips, D.L., A technique for the numerical solution of certain integral equations of the first kind, *J. Assoc. Comput. Mach.*, 9, 84–97, 1962.
79. Tibshirani, R., Regression shrinkage and selection via the lasso, *J. R. Stat. Soc. B*, 58, 267–288, 1996.

80. Öjelund, H., Madsen, H., and Thyregod, P., Calibration with absolute shrinkage, *J. Chemom.*, 15, 497–509, 2001.
81. Tenorio, L., Statistical regularization of inverse problems, *SIAM Rev.*, 43, 347–366, 2001.
82. Nelder, J.A. and Mead, R., A simplex method for function minimization, *Comp. J.*, 7, 308–313, 1965.
83. Kalivas, J.H., Evaluation of volume and matrix effects for the generalized standard addition method, *Talanta*, 34, 899–903, 1987.
84. Saxberg, B.E.H. and Kowalski, B.R., Generalized standard addition method, *Anal. Chem.*, 51, 1031–1038, 1979.
85. Frank, I.E., Kalivas, J.H., and Kowalski, B.R., Partial least squares solutions for multicomponent analysis, *Anal. Chem.*, 55, 1800–1804, 1983.
86. Kalivas, J.H. and Kowalski, B.R., Compensation for drift and interferences in multicomponent analysis, *Anal. Chem.*, 54, 560–565, 1982.
87. Booksh, K., Henshaw, J.M., Burgess, L.W., and Kowalski, B.R., A second-order standard addition method with application to calibration of a kinetics-spectroscopic sensor for quantitation of trichloroethylene, *J. Chemom.*, 9, 263–282, 1995.
88. Shaw, R.A. and Mantsch, H.H., Multianalyte serum assays from mid-IR spectra of dry films on glass slides, *Appl. Spectrosc.*, 54, 885–889, 2000.
89. Geladi, P., McDougall, D., and Martens, H., Linearization and scatter correction for near infrared reflectance spectra of meat, *Appl. Spectrosc.*, 39, 491–500, 1985.
90. Wold, S., Antii, S.H., Lindgren, F., and Öhman, J., Orthogonal signal correction of near-infrared spectra, *Chemom. Intell. Lab. Syst.*, 44, 175–185, 1998.
91. Barnes, R.J., Dhanoa, M.S., and Lister, S.J., Standard normal variate transformation and detrending of near infrared diffuse reflectance, *Appl. Spectrosc.*, 43, 772–777, 1989.
92. de Noord, O.E., The influence of data preprocessing on the robustness of parsimony of multivariate calibration models, *Chemom. Intell. Lab. Syst.*, 23, 65–70, 1994.
93. Beebe, K.R., Pell, R.J., and Seasholtz, M.B., *Chemometrics: A Practical Guide*, John Wiley & sons, New York, 1998, pp. 26–55.
94. Naes, T., Isaksson, T., Fearn, T., and Davies, T., *A User-Friendly Guide to Multivariate Calibration and Classification*, NIR Publications, Chichester, U.K., 2002.
95. Fearn, T., Standardization and calibration transfer for near infrared instruments: a review, *J. Near Infrared Spectrosc.*, 9, 229–244, 2001.
96. Feudale, R.N., Woody, N.A., Tan, H., Myles, A.J., Brown, S.D., and Ferré, J., Transfer of multivariate calibration models: a review, *Chemom. Intell. Lab. Syst.*, 64, 181–192, 2002.
97. Anderson, C.E. and Kalivas, J.H., Fundamentals of calibration transfer through Procrustes analysis, *Appl. Spectrosc.*, 53, 1268–1276, 1999.
98. Wang, Y., Veltkamp, D.J., and Kowalski, B.R., Multivariate instrument standardization, *Anal. Chem.*, 63, 2750–2756, 1991.
99. Kennard, R.W. and Stone, L.A., Computer aided design of experiments, *Technometrics*, 11, 137–148, 1969.
100. Wang, Y. and Kowalski, B.R., Calibration transfer and measurement stability of near-infrared spectrometers, *Appl. Spectrosc.*, 46, 764–771, 1992.
101. Swierenga, H., Haanstra, W.G., de Weijer, A.P., and Buydens, L.M.C., Comparison of two different approaches toward model transferability in NIR spectroscopy, *Appl. Spectrosc.*, 52, 7–16, 1998.
102. Blank, T.B., Sum, S.T., Brown, S.D., and Monfre, S.L., Transfer of near-infrared multivariate calibrations without standards, *Anal. Chem.*, 68, 2987–2995, 1996.

103. Tan, H., Sum, S.T., and Brown, S.D., Improvement of a standard-free method for near-infrared calibration transfer, *Appl. Spectrosc.,* 56, 1098–1106, 2002.
104. Sjoblom, J., Svensson, O., Josefson, M., Kullberg, H., and Wold, S., An evaluation of orthogonal signal correction applied to calibration transfer of near infrared spectra, *Chemom. Intell. Lab. Syst.,* 44, 229–244, 1998.
105. Greensill, C.V., Wolfs, P.J., Spiegelman, C.H., and Walsh, K.B., Calibration transfer between PDA-based NIR spectrometers in the NIR assessment of melon soluble solids content, *Appl. Spectrosc.,* 55, 647–653, 2001.

6 Robust Calibration

Mia Hubert

CONTENTS

- 6.1 Introduction 168
- 6.2 Location and Scale Estimation 169
  - 6.2.1 The Mean and the Standard Deviation 169
  - 6.2.2 The Median and the Median Absolute Deviation .. 171
  - 6.2.3 Other Robust Estimators of Location and Scale 171
- 6.3 Location and Covariance Estimation in Low Dimensions .. 173
  - 6.3.1 The Empirical Mean and Covariance Matrix 173
  - 6.3.2 The Robust MCD Estimator 174
  - 6.3.3 Other Robust Estimators of Location and Covariance .. 176
- 6.4 Linear Regression in Low Dimensions 176
  - 6.4.1 Linear Regression with One Response Variable ... 176
    - 6.4.1.1 The Multiple Linear Regression Model ... 176
    - 6.4.1.2 The Classical Least-Squares Estimator ... 177
    - 6.4.1.3 The Robust LTS Estimator 178
    - 6.4.1.4 An Outlier Map 180
    - 6.4.1.5 Other Robust Regression Estimators 182
  - 6.4.2 Linear Regression with Several Response Variables .. 183
    - 6.4.2.1 The Multivariate Linear Regression Model .. 183
    - 6.4.2.2 The Robust MCD-Regression Estimator .. 184
    - 6.4.2.3 An Example 185
- 6.5 Principal Components Analysis 185
  - 6.5.1 Classical PCA 185
  - 6.5.2 Robust PCA Based on a Robust Covariance Estimator .. 187
  - 6.5.3 Robust PCA Based on Projection Pursuit 188
  - 6.5.4 Robust PCA Based on Projection Pursuit and the MCD .. 189
  - 6.5.5 An Outlier Map 191
  - 6.5.6 Selecting the Number of Principal Components .. 193
  - 6.5.7 An Example 194
- 6.6 Principal Component Regression 194
  - 6.6.1 Classical PCR 194
  - 6.6.2 Robust PCR 197

		6.6.3	Model Calibration and Validation	198
		6.6.4	An Example	199
6.7	Partial Least-Squares Regression			202
		6.7.1	Classical PLSR	202
		6.7.2	Robust PLSR	203
		6.7.3	An Example	204
6.8	Classification			207
		6.8.1	Classification in Low Dimensions	207
			6.8.1.1 Classical and Robust Discriminant Rules	207
			6.8.1.2 Evaluating the Discriminant Rules	208
			6.8.1.3 An Example	209
		6.8.2	Classification in High Dimensions	211
6.9	Software Availability			211
References				212

6.1 INTRODUCTION

When collecting and analyzing real data, it often occurs that some observations are different from the majority of the samples. More precisely, they deviate from the model that is suggested by the major part of the data, or they do not satisfy the usual assumptions. Such observations are called outliers. Sometimes they are simply the result of transcription errors (e.g., a misplaced decimal point or the permutation of two digits). Often the outlying observations are not incorrect but were made under exceptional circumstances, or they might belong to another population (e.g., it may have been the concentration of a different compound) and consequently they do not fit the model well. It is very important to be able to detect these outliers. They can then be used, for example, to pinpoint a change in the production process or in the experimental conditions.

To find the outlying observations, two strategies can be followed. The first approach is to apply a classical method, followed by the computation of several diagnostics that are based on the resulting residuals. Consider, for example, the Cook's distance in regression. For each observation $i = 1, \ldots, n$, it is defined as

$$D_i = \frac{\sum_{j=1}^{n}(\hat{y}_j - \hat{y}_{j,-i})^2}{ps^2} \tag{6.1}$$

where $\hat{y}_{j,-i}$ is the fitted value for observation j obtained by deleting the ith observation from the data set, p is the number of regression parameters, and s^2 is the estimate of the residual variance. See Section 6.4 for more details about the regression setting. This leave-one-out diagnostic thus measures the influence on all fitted values when the ith sample is removed. It explicitly uses a property of the classical least-squares method for multiple linear regression (MLR), namely that it is very sensitive to the presence of outliers. If the ith sample is outlying, the parameter estimates and the fitted values can change a lot if we remove it, hence D_i will become large. This approach can work appropriately, but it has a very important disadvantage. When outliers occur in groups (even small groups with only two samples), the fit will not necessarily modify drastically when only one observation at a time is removed. In this case, one should rely on diagnostics that measure the influence on the fit when several items are deleted simultaneously. But this becomes very time consuming, as we cannot know in advance how many outliers are grouped.

In general, classical methods can be so strongly affected by outliers that the resulting fitted model does not allow the detection of the deviating observations. This is called the masking effect. Additionally, some good data points might even show up as outliers, which is known as swamping.

A second strategy to detect outliers is to apply robust methods. The goal of robust statistics is to find a fit that is similar to the fit we would have found without the outliers. That solution then allows us to identify the outliers by their residuals from that robust fit. From Frank Hampel [1], one of the founders of robust statistics, we cite:

> Outliers are a topic of constant concern in statistics.... The main aim (of robust statistics) is to accomodate the outliers, that is, to play safe against their potential dangers and to render their effects in the overall result harmless.... A second aim is to identify outliers in order to learn from them (e.g., about their sources, or about a better model). Identification can be achieved by looking at the residuals from robust fits. In this context, it is much more important not to miss any potential outlier (which may give rise to interesting discoveries) than to avoid casting any doubt on "good" observations.... On the third and highest level, outliers are discussed and interpreted in the full context of data analysis, making use not only of formal statistical procedures but also of the background knowledge and general experience from applied statistics and the subject-matter field, as well as the background of the particular data set at hand.

In this chapter we describe robust procedures for the following problems:

Location and scale estimation (Section 6.2)
Location and covariance estimation in low dimensions (Section 6.3)
Linear regression in low dimensions (Section 6.4)
Location and covariance estimation in high dimensions: PCA (Section 6.5)
Linear regression in high dimensions: PCR and PLS (Sections 6.6 and 6.7)
Classification in low and high dimensions (Section 6.8)

Finally, Section 6.9 discusses software availability.

6.2 LOCATION AND SCALE ESTIMATION

6.2.1 THE MEAN AND THE STANDARD DEVIATION

The location-and-scale model states that the n univariate observations x_i are independent and identically distributed (i.i.d.) with distribution function $F[(x - \theta)/\sigma]$, where F is known. Typically F is the standard Gaussian distribution function Φ.

We then want to find estimates for the center θ and the scale parameter σ (or for the variance σ^2). The classical estimates are the sample mean

$$\hat{\theta} = \bar{x} = \frac{1}{n}\sum_{i=1}^{n} x_i$$

and the standard deviation

$$\hat{\sigma} = s = \sqrt{\frac{1}{n-1}\sum_{i=1}^{n}(x_i - \bar{x})^2}$$

The mean and the standard deviation are, however, very sensitive to aberrant values. Consider the following example data [2], listed in Table 6.1, which depicts the viscosity of an aircraft primer paint in 15 batches. From the raw data, it can be seen that an upward shift in viscosity has occurred at batch 13, resulting in a higher viscosity for batches 13 to 15. The mean of all 15 batches is $\bar{x} = 34.07$, and the standard deviation is $s = 1.16$. However, if we only consider the first 12 batches,

TABLE 6.1
Viscose Data Set and Standardized Residuals Obtained with Different Estimators of Location and Scale

		Standardized Residual Based on		
Batch Number	Viscosity	Mean Stand. Dev.	Median MAD[a]	Huber MAD[a]
1	33.75	−0.27	0.14	0.04
2	33.05	−0.88	−1.25	−1.35
3	34.00	−0.06	0.63	0.54
4	33.81	−0.22	0.26	0.16
5	33.46	−0.53	−0.44	−0.53
6	34.02	−0.04	0.67	0.58
7	33.68	−0.33	0	−0.10
8	33.27	−0.69	−0.81	−0.91
9	33.49	−0.50	−0.38	−0.47
10	33.20	−0.75	−0.95	−1.05
11	33.62	−0.39	−0.12	−0.22
12	33.00	−0.92	−1.35	−1.44
13	36.11	1.77	4.82	4.72
14	35.99	1.66	4.58	4.49
15	36.53	2.13	5.65	5.56

Note: Outlying batches with absolute standardized residual larger than 2.5 are underlined.

[a] Median Absolute Deviation

Robust Calibration

the mean is 33.53 and $s = 0.35$. We thus see that the outlying batches 13 to 15 have caused an upward shift of the mean and a serious increase of the scale estimate.

To detect outliers, we could use the rule that all observations outside the interval $\bar{x} \pm 2.5s$ are suspicious under the normal assumption. Equivalently, we pinpoint outliers as the batches whose absolute standardized residual $|(x_i - \bar{x})/s|$ exceeds 2.5. However, with $\bar{x} = 34.07$ and $s = 1.16$, none of the batches has such a large standardized residual (see Table 6.1), hence no outlying batches are detected. We also notice that the residual of each of the regular observations is negative. This is because the mean $\bar{x} = 34.07$ is larger than any of the first 12 samples.

6.2.2 THE MEDIAN AND THE MEDIAN ABSOLUTE DEVIATION

When robust estimates of θ and σ are used, the situation is much different. The most popular robust estimator of location is the sample median, defined as the middle of the ordered observations. If n is even, the median is defined as the average of the two middlemost points. For the viscose data, the median is 33.68 which corresponds to the viscosity of batch 7.

It is clear that the median can resist up to 50% outliers. More formally, it is said that the median has a breakdown value of 50%. This is the minimum proportion of observations that need to be replaced in the original data set to make the location estimate (here, the median) arbitrarily large or small. The sample mean on the other hand has a zero breakdown value, as one observation can pull the average toward $+\infty$ or $-\infty$.

A simple robust estimator of σ is the median absolute deviation (MAD) given by the median of all absolute distances from the sample median:

$$\text{MAD} = 1.483 \underset{j=1,\ldots,n}{\text{median}} |x_j - \underset{i=1,\ldots,n}{\text{median}}(x_i)|$$

The constant 1.483 is a correction factor that makes the MAD unbiased at the normal distribution. The MAD also has a 50% breakdown value and can be computed explicitly. For the viscose data, we find $\text{MAD} = 0.50$. If we compute the standardized residuals based on the median and the MAD, we obtain for batches 13 to 15, respectively, 4.82, 4.58, and 5.65 (see Table 6.1). Thus, all three are correctly identified as being different from the other batches.

6.2.3 OTHER ROBUST ESTIMATORS OF LOCATION AND SCALE

Although the median is very robust, it is not a very efficient estimator for the Gaussian model as it is primarily based on the ranks of the observations. A more efficient location estimator is the (k/n)-trimmed average, which is the average of the data set except for the k smallest and the k largest observations. Its breakdown value is k/n and thus can be chosen as any value between 0% and 50% by an appropriate choice of k. A disadvantage of the trimmed average is that it rejects a fixed percentage of observations at each side of the distribution. This might be too large if they are not all outlying, but even worse, this might be too small if k is chosen too low.

More-adaptive procedures can be obtained using M-estimators [3]. They are defined implicitly as the solution of the equation

$$\sum_{i=1}^{n} \psi\left(\frac{x_i - \hat{\theta}}{\hat{\sigma}}\right) = 0 \qquad (6.2)$$

with ψ an odd, continuous, and monotone function. The denominator $\hat{\sigma}$ is an initial robust scale estimate such as the MAD. A solution to Equation 6.2 can be found by the Newton-Raphson algorithm, starting from the initial location estimate $\hat{\theta}^{(0)} = \text{median}(x_i)$. From each $\hat{\theta}^{(k-1)}$, the next $\hat{\theta}^{(k)}$ is then computed by

$$\hat{\theta}^{(k)} = \hat{\theta}^{(k-1)} + \hat{\sigma}\frac{\sum_i \psi((x_i - \hat{\theta}^{(k-1)})/\hat{\sigma})}{\sum_i \psi'((x_i - \hat{\theta}^{(k-1)})/\hat{\sigma})}$$

Often, a single iteration step is sufficient, which yields the one-step M-estimator. Both the fully iterated and the one-step M-estimator have a 50% breakdown value if the ψ function is bounded. Huber proposed the function $\psi(x) = \min\{b, \max\{-b, x\}\}$, which is now named after him, where typically $b = 1.5$. For the viscose data, the one-step Huber estimator yields $\hat{\theta} = 33.725$, whereas the fully iterated one is hardly distinguishable with $\hat{\theta} = 33.729$. Again, the standardized residuals based on the (fully iterated) Huber estimator and the MAD detect the correct outliers (last column of Table 6.1).

An alternative to the MAD is the Q_n estimator [4], which attains a breakdown value of 50%. The Q_n estimator is defined as

$$Q_n = 2.2219 c_n \{|x_i - x_j|; i < j\}_{(k)} \qquad (6.3)$$

with $k = \binom{h}{2} \approx \binom{n}{2}/4$ and $h = [\frac{n}{2}]+1$. The notation (k) stands for the kth-order statistic out of the $\binom{n}{2} = \frac{n(n-1)}{2}$ possible differences $|x_i - x_j|$, and $[z]$ stands for the largest integer smaller or equal to z. This scale estimator is essentially the first quartile of all pair-wise differences between two data points. The constant c_n is a small-sample correction factor, which makes Q_n an unbiased estimator (note that c_n only depends on the sample size n, and that $c_n \to 1$ for increasing n). As with the MAD, the Q_n can be computed explicitly, but it does not require an initial estimate of location. A fast algorithm of $O(n \log n)$ time has been developed for its computation.

As for location, M-estimators of scale can be defined as the solution of an implicit equation [3]. Then, again, an initial scale estimate is needed, for which the MAD is usually taken. Simultaneous M-estimators of location and scale can also be considered, but they have a smaller breakdown value, even in small samples [5].

Note that all of the mentioned estimators are location and scale equivariant. That is, if we replace our data set $\mathbf{X} = \{x_1, ..., x_n\}$ by $a\mathbf{X} + b = \{ax_1 + b, ..., ax_n + b\}$, then a location estimator $\hat{\theta}$ and a scale estimator $\hat{\sigma}$ must satisfy

$$\hat{\theta}(a\mathbf{X}+b) = a\hat{\theta}(\mathbf{X}) + b$$

$$\hat{\sigma}(a\mathbf{X}+b) = |a|\hat{\sigma}(\mathbf{X}).$$

6.3 LOCATION AND COVARIANCE ESTIMATION IN LOW DIMENSIONS

6.3.1 THE EMPIRICAL MEAN AND COVARIANCE MATRIX

In the multivariate location and scatter setting, we assume that the data are stored in an $n \times p$ data matrix $\mathbf{X} = (\mathbf{x}_1, ..., \mathbf{x}_n)^T$, with $\mathbf{x}_i = (x_{i1}, ..., x_{ip})^T$ being the ith observation. Hence n stands for the number of objects and p for the number of variables. In this section we assume, in addition, that the data are low-dimensional. Here, this means that p should at least be smaller than $n/2$ (or equivalently that $n > 2p$). Based on the measurements $\mathbf{X}$, we try to find good estimates for their center μ and their scatter matrix Σ.

To illustrate the effect of outliers, consider the following simple example presented in Figure 6.1, which depicts the concentration of inorganic phosphorus and organic phosphorus in the soil [6]. On this plot the classical tolerance ellipse is superimposed, defined as the set of p-dimensional points $\mathbf{x}$ whose Mahalanobis distance

$$\mathrm{MD}(\mathbf{x}) = \sqrt{(\mathbf{x} - \bar{\mathbf{x}})^T \mathbf{S}_x^{-1} (\mathbf{x} - \bar{\mathbf{x}})} \tag{6.4}$$

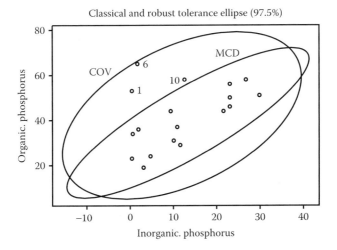

FIGURE 6.1 Classical and robust tolerance ellipse of the phosphorus data set.

equals $\sqrt{\chi^2_{2,0.975}}$, the square root of the 0.975 quantile of the chi-square distribution with $p = 2$ degrees of freedom. In Equation 6.4, we use the classical estimates for the location and shape of the data, which are the mean

$$\bar{\mathbf{x}} = \frac{1}{n}\sum_{i=1}^{n} \mathbf{x}_i$$

and the empirical covariance matrix

$$\mathbf{S}_x = \frac{1}{n-1}\sum_{i=1}^{n} (\mathbf{x}_i - \bar{\mathbf{x}})(\mathbf{x}_i - \bar{\mathbf{x}})^T$$

of the $\mathbf{x}_i$. In general, the cutoff value $c = \sqrt{\chi^2_{p,0.975}}$ stems from the fact that the squared Mahalanobis distances of normally distributed data are asymptotically χ^2_p distributed. The distance $MD(\mathbf{x}_i)$ should tell us how far away $\mathbf{x}_i$ is from the center of the cloud, relative to the size of the cloud. It is well known that this approach suffers from the masking effect, as multiple outliers do not necessarily have a large $MD(\mathbf{x}_i)$. Indeed, in Figure 6.1 we see that this tolerance ellipse is highly inflated and even includes all the outliers.

Note that any p-dimensional vector μ and $p \times p$ positive-definite matrix Σ defines a statistical distance:

$$D(\mathbf{x}, \mu, \Sigma) = \sqrt{(\mathbf{x} - \mu)^T \Sigma^{-1}(\mathbf{x} - \mu)} \quad (6.5)$$

From Equation 6.4 and Equation 6.5, it follows that $MD(\mathbf{x}) = D(\mathbf{x}, \bar{\mathbf{x}}, \mathbf{S}_x)$.

6.3.2 THE ROBUST MCD ESTIMATOR

Contrary to the classical mean and covariance matrix, a robust method yields a tolerance ellipse that captures the covariance structure of the majority of the data points. In Figure 6.1, this robust tolerance ellipse is obtained by applying the highly robust minimum covariance determinant (MCD) estimator of location and scatter [7] to the data, yielding $\hat{\mu}_{MCD}$ and $\hat{\Sigma}_{MCD}$, and by plotting the points $\mathbf{x}$ whose robust distance

$$RD(\mathbf{x}) = D(\mathbf{x}, \hat{\mu}_{MCD}, \hat{\Sigma}_{MCD}) = \sqrt{(\mathbf{x} - \hat{\mu}_{MCD})^T \hat{\Sigma}_{MCD}^{-1}(\mathbf{x} - \hat{\mu}_{MCD})} \quad (6.6)$$

is equal to $\sqrt{\chi^2_{2,0.975}}$. This robust tolerance ellipse is much narrower than the classical one. Consequently, the outliers have a much larger robust distance and are recognized as deviating from the majority.

The MCD method looks for the $h > n/2$ observations (out of n) whose classical covariance matrix has the lowest possible determinant. The MCD estimate of location $\hat{\mu}_0$ is then the average of these h points, whereas the MCD estimate of scatter $\hat{\Sigma}_0$ is their covariance matrix, multiplied with a consistency factor.

Based on the raw MCD estimates, a reweighing step can be added that increases the finite-sample efficiency considerably. In general, we can weigh each $\mathbf{x}_i$ by $w_i = w(D(\mathbf{x}_i, \hat{\mu}_0, \hat{\Sigma}_0))$, for instance by putting

$$w_i = \begin{cases} 1 & \text{if } D(\mathbf{x}_i, \hat{\mu}_0, \hat{\Sigma}_0) \leq \sqrt{\chi^2_{p,0.975}} \\ 0 & \text{otherwise.} \end{cases}$$

The resulting one-step reweighed mean and covariance matrix are then defined as

$$\hat{\mu}_{\text{MCD}}(\mathbf{X}) = \left(\sum_{i=1}^{n} w_i \mathbf{x}_i \right) \bigg/ \left(\sum_{i=1}^{n} w_i \right)$$

$$\hat{\Sigma}_{\text{MCD}}(\mathbf{X}) = \left(\sum_{i=1}^{n} w_i (\mathbf{x}_i - \hat{\mu}_{\text{MCD}})(\mathbf{x}_i - \hat{\mu}_{\text{MCD}})^T \right) \bigg/ \left(\sum_{i=1}^{n} w_i - 1 \right).$$

The final robust distances $RD(\mathbf{x}_i) = D(\mathbf{x}_i, \hat{\mu}_{\text{MCD}}(\mathbf{X}), \hat{\Sigma}_{\text{MCD}}(\mathbf{X}))$ are then obtained by inserting $\hat{\mu}_{\text{MCD}}(\mathbf{X})$ and $\hat{\Sigma}_{\text{MCD}}(\mathbf{X})$ into Equation 6.6.

The MCD estimates have a breakdown value of $(n - h + 1)/n$, hence the number h determines the robustness of the estimator. Note that for a scatter matrix, breakdown means that its largest eigenvalue becomes arbitrarily large, or that its smallest eigenvalue becomes arbitrarily close to zero. The MCD has its highest possible breakdown value when $h = [(n + p + 1)/2]$. When a large proportion of contamination is presumed, h should thus be chosen close to $0.5n$. Otherwise, an intermediate value for h, such as $0.75n$, is recommended to obtain a higher finite-sample efficiency.

The robustness of a procedure can also be measured by means of its influence function [8]. Robust estimators ideally have a bounded influence function, which means that a small contamination at a certain point can only have a small effect on the estimator. This is satisfied by the MCD estimator [9].

The MCD location and scatter estimates are affine equivariant, which means that they behave properly under affine transformations of the data. That is, for a data set $\mathbf{X}$ in IR^p, the MCD estimates $(\hat{\mu}, \hat{\Sigma})$ satisfy

$$\hat{\mu}(\mathbf{X}\mathbf{A} + 1_n \mathbf{v}^T) = \hat{\mu}(\mathbf{X})\mathbf{A} + \mathbf{v}$$

$$\hat{\Sigma}(\mathbf{X}\mathbf{A} + 1_n \mathbf{v}^T) = \mathbf{A}^T \hat{\Sigma}(\mathbf{X})\mathbf{A}$$

for all nonsingular $p \times p$ matrices $\mathbf{A}$ and vectors $\mathbf{v} \in IR^p$. The vector $1_n = (1,1,\ldots,1)^T$ of length n.

The computation of the MCD estimator is nontrivial and naively requires an exhaustive investigation of all h-subsets out of n. In Rousseeuw and Van Driessen [10], a fast algorithm is presented (FAST-MCD) that avoids such a complete enumeration. The results of this algorithm are approximate estimates of the MCD minimization problem. This means that although the h-subset with the lowest covariance determinant may not be found, another h-subset whose covariance determinant is close to the minimal one will be. In small dimensions p, the FAST-MCD algorithm often yields the exact solution, but the approximation is rougher in higher dimensions. Note that the MCD can only be computed if $p < h$; otherwise, the covariance matrix of any h-subset has zero determinant. Since $n/2 < h$, we thus require that $p < n/2$. However, detecting several outliers (or, equivalently, fitting the majority of the data) becomes intrinsically tenuous when n/p is small. This is an instance of the "curse of dimensionality." To apply any method with 50% breakdown, it is recommended that $n/p > 5$. For small n/p, it is preferable to use a method with lower breakdown value such as the MCD with $h \approx 0.75n$, for which the breakdown value is 25%.

Note that the univariate MCD estimator of location and scale reduces to the mean and the standard deviation of the h-subset with lowest variance. It can be computed exactly and swiftly by ordering the data points and considering all contiguous h-subsets [6].

6.3.3 Other Robust Estimators of Location and Covariance

Many other affine equivariant and robust estimators of location and scatter have been presented in the literature. The first such estimator was proposed independently by Stahel [11] and Donoho [12] and investigated by Tyler [13] and Maronna and Yohai [14]. Multivariate M-estimators [15] have a relatively low breakdown value due to possible implosion of the estimated scatter matrix. Together with the MCD estimator, Rousseeuw [16] introduced the minimum-volume ellipsoid. Davies [17] also studied one-step M-estimators. Other classes of robust estimators of multivariate location and scatter include S-estimators [6, 18], CM-estimators [19], τ-estimators [20], MM-estimators [21], estimators based on multivariate ranks or signs [22], depth-based estimators [23–26], methods based on projection pursuit [27], and many others.

6.4 LINEAR REGRESSION IN LOW DIMENSIONS

6.4.1 Linear Regression with One Response Variable

6.4.1.1 The Multiple Linear Regression Model

The multiple linear regression model assumes that in addition to the p independent x-variables, a response variable y is measured, which can be explained as an affine combination of the x-variables (also called the regressors). More precisely, the model says that for all observations $(\mathbf{x}_i, y_i)$ with $i = 1, \ldots, n$, it holds that

$$y_i = \beta_0 + \beta_1 x_{i1} + \cdots + \beta_p x_{ip} + \varepsilon_i = \beta_0 + \beta^T \mathbf{x}_i + \varepsilon_i \qquad i = 1,\ldots,n \qquad (6.7)$$

Robust Calibration

where the errors ε_i are assumed to be independent and identically distributed with zero mean and constant variance σ^2. The vector $\beta = (\beta_1, \ldots, \beta_p)^T$ is called the slope, and β_0 the intercept. For regression without intercept, we require that $\beta_0 = 0$. We denote $\mathbf{x}_i = (x_{i1}, \ldots, x_{ip})^T$ and $\theta = (\beta_0, \beta^T)^T = (\beta_0, \beta_1, \ldots, \beta_p)$. Applying a regression estimator to the data yields $p + 1$ regression coefficients. The residual r_i of case i is defined as the difference between the observed response y_i and its estimated value:

$$r_i(\hat{\theta}) = y_i - \hat{y}_i = y_i - (\hat{\beta}_0 + \hat{\beta}_1 x_{i1} + \cdots + \hat{\beta}_p x_{ip})$$

6.4.1.2 The Classical Least-Squares Estimator

The classical least-squares method for multiple linear regression (MLR) to estimate θ minimizes the sum of the squared residuals. Formally, this can be written as

$$\text{minimize} \sum_{i=1}^{n} r_i^2$$

This is a very popular method because it allows us to compute the regression estimates explicitly as $\hat{\theta} = (\mathbf{X}^T\mathbf{X})^{-1}\mathbf{X}^T\mathbf{y}$ (where the design matrix $\mathbf{X}$ is enlarged with a column of ones for the intercept term and $\mathbf{y} = (y_1, \ldots, y_n)^T$ and, moreover, the least-squares method is optimal if the errors are normally distributed.

However, MLR is extremely sensitive to regression outliers, which are the observations that do not obey the linear pattern formed by the majority of the data. This is illustrated in Figure 6.2 for simple regression (where there is only one regressor x, or $p = 1$), which illustrates a Hertzsprung-Russell diagram of 47 stars. The diagram plots the logarithm of the stars' light intensity vs. the logarithm of their surface temperature [6]. The four outlying observations are giant stars, and they

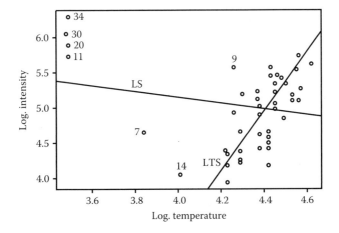

FIGURE 6.2 Stars regression data set with classical and robust fit.

clearly deviate from the main sequence of stars. Also, the stars with labels 7, 9, and 14 seem to be outlying. The least-squares fit is added to this plot and clearly is highly attracted by the giant stars.

In regression, we can distinguish between different types of outliers. Leverage points are observations ($\mathbf{x}_i$, y_i) whose $\mathbf{x}_i$ are outlying, i.e., $\mathbf{x}_i$ deviates from the majority in x-space. We call such an observation ($\mathbf{x}_i$, y_i) a good leverage point if ($\mathbf{x}_i$, y_i) follows the linear pattern of the majority. If, on the other hand, ($\mathbf{x}_i$, y_i) does not follow this linear pattern, we call it a bad leverage point, like the four giant stars and case 7 in Figure 6.2. An observation whose $\mathbf{x}_i$ belongs to the majority in x-space, but where ($\mathbf{x}_i$, y_i) deviates from the linear pattern is called a vertical outlier, like observation 9. A regression data set can thus have up to four types of points: regular observations, vertical outliers, good leverage points, and bad leverage points. Leverage points attract the least-squares solution toward them, so bad leverage points are often masked in a classical regression analysis.

To detect regression outliers, we could look at the standardized residuals r_i/s, where s is an estimate of the scale of the error distribution σ. For MLR, an unbiased estimate of σ^2 is given by $s^2 = \frac{1}{n-p-1} \sum_{i=1}^{n} r_i^2$. One often considers observations for which $|r_i/s|$ exceeds the cutoff 2.5 to be regression outliers (because values generated by a Gaussian distribution are rarely larger than 2.5 σ), whereas the other observations are thought to obey the model. In Figure 6.3a, this strategy fails: the standardized MLR (or LS) residuals of all 47 points lie inside the tolerance band between -2.5 and 2.5. There are two reasons why this plot hides (masks) the outliers: the four leverage points in Figure 6.2 have attracted the MLR line so much that they have small residuals r_i from it; and the MLR scale estimate s computed from all 47 points has become larger than the scale of the 43 points in the main sequence.

In general, the MLR method tends to produce normal-looking residuals, even when the data themselves behave badly.

6.4.1.3 The Robust LTS Estimator

In Figure 6.2, a robust regression fit is superimposed. The least-trimmed squares estimator (LTS) proposed by Rousseeuw [7] is given by

$$\text{minimize} \sum_{i=1}^{h} (r^2)_{i:n} \tag{6.8}$$

where $(r^2)_{1:n} \leq (r^2)_{2:n} \leq \ldots \leq (r^2)_{n:n}$ are the ordered squared residuals (note that the residuals are first squared and then ordered). Because the criterion of Equation 6.8 does not count the largest squared residuals, it allows the LTS fit to steer clear of outliers. The value h plays the same role as in the definition of the MCD estimator. For $h \approx n/2$, we find a breakdown value of 50%, whereas for larger h, we obtain a breakdown value of $(n - h + 1)/n$ [79]. A fast algorithm for the LTS estimator (FAST-LTS) has been developed [28].

Robust Calibration

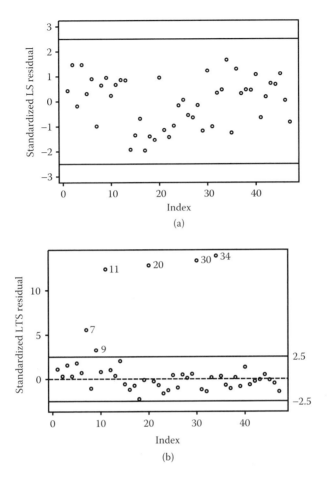

FIGURE 6.3 Standardized residuals of the stars data set, based on (a) classical MLR and (b) robust LTS estimator.

The LTS estimator is regression, scale, and affine equivariant. That is, for any $\mathbf{X} = (\mathbf{x}_1, \ldots, \mathbf{x}_n)^T$ and $\mathbf{y} = (y_1, \ldots, y_n)^T$, it holds that

$$\hat{\theta}(\mathbf{X}, \mathbf{y} + \mathbf{X}\mathbf{v} + 1_n c) = \hat{\theta}(\mathbf{X}, \mathbf{y}) + (\mathbf{v}^T, c)^T$$

$$\hat{\theta}(\mathbf{X}, c\mathbf{y}) = c\hat{\theta}(\mathbf{X}, \mathbf{y})$$

$$\hat{\theta}(\mathbf{X}\mathbf{A}^T + 1_n \mathbf{v}^T, \mathbf{y}) = (\hat{\beta}^T(\mathbf{X}, \mathbf{y})\mathbf{A}^{-1}, \beta_0(\mathbf{X}, \mathbf{y}) - \hat{\beta}^T(\mathbf{X}, \mathbf{y})\mathbf{A}^{-1}\mathbf{v})^T$$

for any vector $\mathbf{v} \in IR^p$, any constant c, and any nonsingular $p \times p$ matrix $\mathbf{A}$. Again $1_n = (1,1,\ldots,1)^T \in IR^n$. It implies that the estimate transforms correctly under affine transformations of the x-variables and of the response variable y.

When using LTS regression, the scale of the errors σ can be estimated by

$$\hat{\sigma}_{LTS} = c_{h,n}\sqrt{\frac{1}{h}\sum_{i=1}^{h}(r^2)_{i:n}}$$

where r_i are the residuals from the LTS fit, and $c_{h,n}$ makes $\hat{\sigma}$ consistent and unbiased at Gaussian error distributions [79]. We can then identify regression outliers by their standardized LTS residuals $r_i/\hat{\sigma}_{LTS}$. This yields Figure 6.3b, from which we clearly see the different outliers.

It should be stressed that LTS regression does not throw away a certain percentage of the data. Instead, it finds a majority fit, which can then be used to detect the actual outliers. The purpose is not to delete and forget the points outside the tolerance band, but to study the residual plot in order to find out more about the data. For instance, we notice the star 7 intermediate between the main sequence and the giants, which might indicate that this star is evolving to its final stage.

In regression analysis, inference is very important. The LTS by itself is not suited for inference because of its relatively low finite-sample efficiency. This can be resolved by carrying out a reweighed least-squares step. To each observation i, one assigns a weight w_i based on its standardized LTS residual $r_i/\hat{\sigma}_{LTS}$, e.g., by putting $w_i = w(|r_i/\hat{\sigma}_{LTS}|)$ where w is a decreasing continuous function. A simpler way, but still effective, is to put

$$w_i = \begin{cases} 1 & \text{if } |r_i/\hat{\sigma}_{LTS}| \leq 2.5 \\ 0 & \text{otherwise} \end{cases}$$

Either way, the reweighed LTS fit is then defined by

$$\text{minimize} \sum_{i=1}^{n} w_i r_i^2 \qquad (6.9)$$

which can be computed quickly. The result inherits the breakdown value, but is more efficient and yields all the usual inferential output such as t-statistics, F-statistics, an R^2 statistic, and the corresponding p-values.

6.4.1.4 An Outlier Map

Residuals plots such as those in Figure 6.3 become even more important in multiple regression with more than one regressor, as then we can no longer rely on a scatter plot of the data. A diagnostic display can be constructed that does not solely expose the regression outliers, i.e., the observations with large standardized residual, but that also classifies the observations according their leverage [29]. Remember that leverage points are those that are outlying in the space of the independent x-variables. Hence, they can be detected by computing robust distances (Equation 6.6) based on, for example, the MCD estimator that is applied on the x-variables.

For the artificial data of Figure 6.4a, the corresponding diagnostic plot is shown in Figure 6.4b. It exposes the robust residuals $r_i/\hat{\sigma}_{LTS}$ vs. the robust

Robust Calibration 181

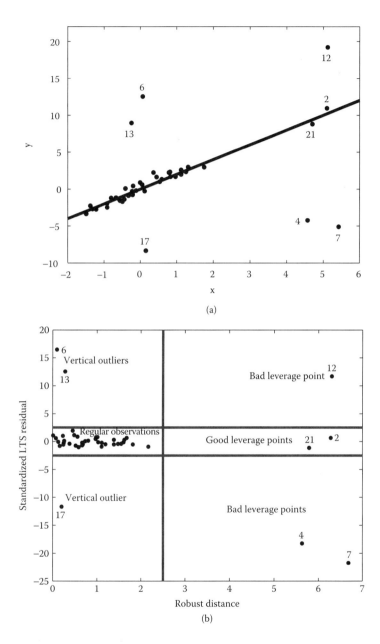

FIGURE 6.4 (a) Artificial regression data and (b) their corresponding outlier map. (Adapted from [82].)

distances $D(\mathbf{x}_i, \hat{\mu}_{MCD}, \hat{\Sigma}_{MCD})$. Because this figure classifies the observations into several types of points, it is also called an outlier map.

Figure 6.5 illustrates this outlier map on the stars data. We see that star 9 is a vertical outlier because it only has an outlying residual. Observation 14 is a good leverage point; it has an outlying surface temperature, but it still follows the linear

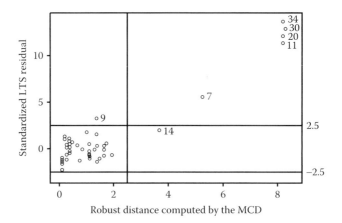

FIGURE 6.5 Outlier map for the stars data set.

trend of the main sequence. Finally, the giant stars and star 7 are bad leverage points, with both a large residual and a large robust distance.

Note that the most commonly used diagnostics to flag leverage points have traditionally been the diagonal elements h_{ii} of the hat matrix $\mathbf{H} = \mathbf{X}(\mathbf{X}^T\mathbf{X})^{-1}\mathbf{X}^T$. These are equivalent to the Mahalanobis distances $MD(\mathbf{x}_i)$ because of the monotone relation

$$h_{ii} = \frac{(MD(\mathbf{x}_i))^2}{n-1} + \frac{1}{n}$$

Therefore, the h_{ii} are masked whenever the $MD(\mathbf{x}_i)$ are. In particular, Cook's distance in Equation 6.1 can fail, as it can be rewritten as

$$D_i = \frac{r_i^2}{ps^2} \left[\frac{h_{ii}}{(1-h_{ii})^2} \right]$$

6.4.1.5 Other Robust Regression Estimators

The earliest systematic theory of robust regression was based on M-estimators [3, 30], given by

$$\text{minimize} \sum_{i=1}^{n} \rho(r_i/\sigma)$$

where $\rho(t) = |t|$ yields least absolute values (L^1) regression as a special case. For general ρ, one needs a robust $\hat{\sigma}$ to make the M-estimator scale equivariant. This $\hat{\sigma}$ either needs to be estimated in advance or estimated jointly with the regression parameters. Unlike M-estimators, scale equivariance holds automatically for R-estimators [31] and L-estimators [32] of regression.

The breakdown value of all regression M-, L-, and R-estimators is 0% because of their vulnerability to bad leverage points. If leverage points cannot occur, as in fixed-design studies, a positive breakdown value can be attained [33].

The next step was the development of generalized M-estimators (GM-estimators), with the purpose of bounding the influence of outlying $\mathbf{x}_i$ by giving them a small weight. This is why GM-estimators are often called bounded-influence methods. A survey is given in Hampel et al. [8]. Both M- and GM-estimators can be computed by iteratively reweighed least squares or by the Newton-Raphson algorithm [34]. Unfortunately, the breakdown value of all GM-estimators goes down to zero for increasing p, when there are more opportunities for outliers to occur.

In the special case of simple regression ($p = 1$) several earlier methods exist, such as the Brown-Mood line, Tukey's resistant line, and the Theil-Sen slope. These methods are reviewed in Rousseeuw and Leroy [6] together with their breakdown values.

For multiple regression, the least median of squares (LMS) of Rousseeuw [7] and the LTS described previously were the first equivariant methods to attain a 50% breakdown value. Their low finite-sample efficiency can be improved by carrying out a one-step reweighed least-squares fit (Equation 6.9) afterward. Another approach is to compute a one-step M-estimator starting from LMS or LTS, which also maintains the breakdown value and yields the same efficiency as the corresponding M-estimator. In order to combine these advantages with those of the bounded-influence approach, it was later proposed to follow the LMS or LTS by a one-step GM-estimator [35].

A different approach to improving on the efficiency of the LMS and the LTS is to replace their objective functions by a more efficient scale estimator applied to the residuals r_i. This direction has led to the introduction of efficient positive-breakdown regression methods, such as S-estimators [36], MM-estimators [37], CM-estimators [38], and many others.

To extend the good properties of the median to regression, the notion of regression depth [39] and deepest regression [40, 41] was introduced and applied to several problems in chemistry [42].

6.4.2 LINEAR REGRESSION WITH SEVERAL RESPONSE VARIABLES

6.4.2.1 The Multivariate Linear Regression Model

The regression model can be extended to the case where we have more than one response variable. For p-variate predictors $\mathbf{x}_i = (x_{i1}, \ldots, x_{ip})^T$ and q-variate responses $\mathbf{y}_i = (y_{i1}, \ldots, y_{iq})^T$, the multivariate (multiple) regression model is given by

$$\mathbf{y}_i = \beta_0 + \mathbf{B}^T \mathbf{x}_i + \varepsilon_i \qquad (6.10)$$

where $\mathbf{B}$ is the $p \times q$ slope matrix, β_0 is the q-dimensional intercept vector, and the errors are i.i.d. with zero mean and with $Cov(\varepsilon) = \Sigma_\varepsilon$, a positive definite matrix of size q. Note that for $q = 1$, we obtain the multiple regression model (Equation 6.7).

On the other hand, putting $p = 1$ and $x_i = 1$ yields the multivariate location and scatter model. The least-squares solution can be written as

$$\hat{\mathbf{B}} = \hat{\Sigma}_x^{-1} \hat{\Sigma}_{xy} \tag{6.11}$$

$$\hat{\beta}_0 = \hat{\mu}_y - \hat{\mathbf{B}}^T \hat{\mu}_x \tag{6.12}$$

$$\hat{\Sigma}_\varepsilon = \hat{\Sigma}_y - \hat{\mathbf{B}}^T \hat{\Sigma}_x \hat{\mathbf{B}} \tag{6.13}$$

where

$$\hat{\mu} = \begin{pmatrix} \hat{\mu}_x \\ \hat{\mu}_y \end{pmatrix} \quad \text{and} \quad \hat{\Sigma} = \begin{pmatrix} \hat{\Sigma}_x & \hat{\Sigma}_{xy} \\ \hat{\Sigma}_{yx} & \hat{\Sigma}_y \end{pmatrix}$$

are the empirical mean and covariance matrix of the joint (x, y) variables.

6.4.2.2 The Robust MCD-Regression Estimator

Vertical outliers and bad leverage points highly influence the least-squares estimates in multivariate regression, and they can make the results completely unreliable. Therefore, robust alternatives have been developed.

In Rousseeuw et al. [43], it is proposed to use the MCD estimates for the center μ and the scatter matrix Σ of the joint (x, y) variables in Equation 6.11 to Equation 6.13. The resulting estimates are called MCD-regression estimates. They inherit the breakdown value of the MCD estimator. To obtain a better efficiency, the reweighed MCD estimates are used in Equation 6.11 to Equation 6.13 and followed by a regression reweighing step. For any fit $\hat{\theta} = (\hat{\beta}_0^T, \hat{\mathbf{B}}^T)^T$, denote the corresponding q-dimensional residuals by $\mathbf{r}_i(\hat{\theta}) = \mathbf{y}_i - \hat{\mathbf{B}}^T \mathbf{x}_i - \hat{\beta}_0$. Then the residual distance of the ith case is defined as

$$\text{ResD}_i = D(\mathbf{r}_i, 0, \hat{\Sigma}_\varepsilon) = \sqrt{\mathbf{r}_i^T \hat{\Sigma}_\varepsilon^{-1} \mathbf{r}_i}. \tag{6.14}$$

Next, a weight can be assigned to every observation according to its residual distance, e.g.,

$$w_i = \begin{cases} 1 & \text{if } ResD_i \leq \sqrt{\chi_{q,0.975}^2} \\ 0 & \text{otherwise} \end{cases} \tag{6.15}$$

The reweighed regression estimates are then obtained as the weighted least-squares fit with weights w_i. If the hard rejection rule is used as in Equation 6.15, this means that the multivariate least-squares method is applied to the observations with weight 1. The final residual distances are then given by Equation 6.14, where the residuals and $\hat{\Sigma}_\varepsilon$ are based on the reweighed regression estimates.

6.4.2.3 An Example

To illustrate MCD-regression, we analyze a data set obtained from Shell's polymer laboratory in Ottignies, Belgium, by courtesy of Prof. Christian Ritter. The data set consists of $n = 217$ observations, with $p = 4$ predictor variables and $q = 3$ response variables. The predictor variables describe the chemical characteristics of a piece of foam, whereas the response variables measure its physical properties such as tensile strength. The physical properties of foam are determined by the chemical composition used in the production process. Therefore, multivariate regression is used to establish a relationship between the chemical inputs and the resulting physical properties of foam. After an initial exploratory study of the variables, a robust multivariate MCD-regression was used. The breakdown value was set equal to 25%.

To detect leverage points and vertical outliers, the outlier map can be extended to multivariate regression. Then the final robust distances of the residuals, ResD_i, (Equation 6.14) are plotted vs. the robust distances $\text{RD}(\mathbf{x}_i)$ of the $\mathbf{x}_i$ (Equation 6.6). This yields the classification as given in Table 6.2.

Figure 6.6 shows the outlier map of the Shell foam data. Observations 215 and 110 lie far from both the horizontal cutoff line at $\sqrt{\chi^2_{3,0.975}} = 3.06$ and the vertical cutoff line at $\sqrt{\chi^2_{4,0.975}} = 3.34$. These two observations can be classified as bad leverage points. Several observations lie substantially above the horizontal cutoff but not to the right of the vertical cutoff, which means that they are vertical outliers (their residuals are outlying but their x-values are not).

6.5 PRINCIPAL COMPONENTS ANALYSIS

6.5.1 CLASSICAL PCA

Principal component analysis is a popular statistical method that tries to explain the covariance structure of data by means of a small number of components. These components are linear combinations of the original variables, and often allow for an interpretation and a better understanding of the different sources of variation. Because PCA is concerned with data reduction, it is widely used for the analysis of high-dimensional data, which are frequently encountered in chemometrics. PCA is then often the first step of the data analysis, followed by classification, cluster analysis, or other multivariate techniques [44]. It is thus important to find those principal components that contain most of the information.

TABLE 6.2
Overview of the Different Types of Observations Based on Their Robust Distance (RD) and Their Residual Distance (ResD)

Distances	Small RD	Large RD
Large ResD	Vertical outlier	Bad leverage point
Small ResD	Regular observation	Good leverage point

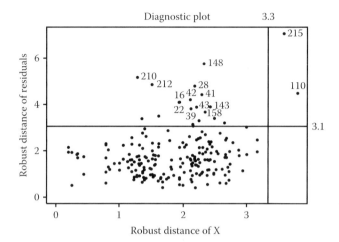

FIGURE 6.6 Outlier map of robust residuals vs. robust distances of the carriers for the foam data set [43].

In the classical approach, the first principal component corresponds to the direction in which the projected observations have the largest variance. The second component is then orthogonal to the first and again maximizes the variance of the data points projected on it. Continuing in this way produces all the principal components, which correspond to the eigenvectors of the empirical covariance matrix. Unfortunately, both the classical variance (which is being maximized) and the classical covariance matrix (which is being decomposed) are very sensitive to anomalous observations. Consequently, the first components are often attracted toward outlying points and thus may not capture the variation of the regular observations. Therefore, data reduction based on classical PCA (CPCA) becomes unreliable if outliers are present in the data.

To illustrate this, let us consider a small artificial data set in $p = 4$ dimensions. The Hawkins-Bradu-Kass data set [6] consist of $n = 75$ observations in which two groups of outliers were created, labeled 1–10 and 11–14. The first two eigenvalues explain already 98% of the total variation, so we select $k = 2$. If we project the data on the plane spanned by the first two principal components, we obtain the CPCA scores plot depicted in Figure 6.7a. In this figure we can clearly distinguish the two groups of outliers, but we see several other undesirable effects. We first observe that, although the scores have zero mean, the regular data points lie far from zero. This stems from the fact that the mean of the data points is a poor estimate of the true center of the data in the presence of outliers. It is clearly shifted toward the outlying group, and consequently the origin even falls outside the cloud of the regular data points. On the plot we have also superimposed the 97.5% tolerance ellipse. We see that the outliers 1–10 are within the tolerance ellipse, and thus do not stand out based on their Mahalanobis distance. The ellipse has stretched itself to accommodate these outliers.

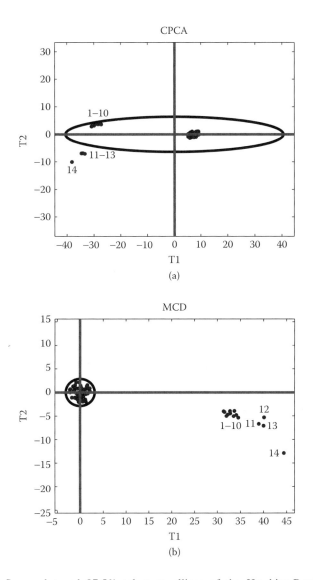

FIGURE 6.7 Score plot and 97.5% tolerance ellipse of the Hawkins-Bradu-Kass data obtained with (a) CPCA and (b) MCD [44].

6.5.2 ROBUST PCA BASED ON A ROBUST COVARIANCE ESTIMATOR

The goal of robust PCA methods is to obtain principal components that are not influenced much by outliers. A first group of methods is obtained by replacing the classical covariance matrix with a robust covariance estimator, such as the reweighed MCD estimator [45] (Section 6.3.2). Let us reconsider the Hawkins-Bradu-Kass data in $p = 4$ dimensions. Robust PCA using the reweighed MCD estimator yields the score plot in Figure 6.7b. We now see that the center is correctly estimated in the middle of the regular observations. The 97.5% tolerance ellipse nicely encloses these points and excludes all of the 14 outliers.

Unfortunately, the use of these affine equivariant covariance estimators is limited to small to moderate dimensions. To see why, let us again consider the MCD estimator. As explained in Section 6.3.2, if p denotes the number of variables in our data set, the MCD estimator can only be computed if $p < h$. Otherwise the covariance matrix of any h-subset has zero determinant. Because $h < n$, the number of variables p may never be larger than n. A second problem is the computation of these robust estimators in high dimensions. Today's fastest algorithms can handle up to about 100 dimensions, whereas there are fields like chemometrics that need to analyze data with dimensions in the thousands. Moreover the accuracy of the algorithms decreases with the dimension p, so it is recommended that small data sets not use the MCD in more than 10 dimensions.

Note that classical PCA is not affine equivariant because it is sensitive to a rescaling of the variables. But it is still orthogonally equivariant, which means that the center and the principal components transform appropriately under rotations, reflections, and translations of the data. More formally, it allows transformations **XA** for any orthogonal matrix **A** (that satisfies $\mathbf{A}^{-1} = \mathbf{A}^T$). Any robust PCA method only has to be orthogonally equivariant.

6.5.3 ROBUST PCA BASED ON PROJECTION PURSUIT

A second and orthogonally equivariant approach to robust PCA uses projection pursuit (PP) techniques. These methods maximize a robust measure of spread to obtain consecutive directions on which the data points are projected. In Hubert et al. [46], a projection pursuit (PP) algorithm is presented, based on the ideas of Li and Chen [47] and Croux and Ruiz-Gazen [48]. The algorithm is called RAPCA, which stands for reflection algorithm for principal components analysis.

If $p \geq n$, the RAPCA method starts by reducing the data space to the affine subspace spanned by the n observations. This is done quickly and accurately by a singular-value decomposition (SVD) of $\mathbf{X}_{n,p}$. From here on, the subscripts to a matrix serve to recall its size, e.g., $\mathbf{X}_{n,p}$ is an $n \times p$ matrix. Let $\tilde{\mathbf{X}}$ denote the mean-centered data matrix. Standard SVD computes the eigenvectors of $\tilde{\mathbf{X}}^T\tilde{\mathbf{X}}$, which is a matrix of size $p \times p$. As the dimension p can be in the hundreds or thousands, this is computationally expensive. The computational speed can be increased by computing the eigenvectors **v** of $\tilde{\mathbf{X}}\tilde{\mathbf{X}}^T$, which is an $n \times n$ matrix. The transformed $\tilde{\mathbf{X}}^T\mathbf{v}$ vectors then yield the eigenvectors of $\tilde{\mathbf{X}}^T\tilde{\mathbf{X}}$, whereas the eigenvalues remain the same. This is known as the kernel version of the eigenvalue decomposition [49]. Note that this singular-value decomposition is just an affine transformation of the data. It is not used to retain only the first eigenvectors of the covariance matrix of **X**. This would imply that classical PCA is performed, which is of course not robust. Here, the data are merely represented in their own dimensionality $r = \text{rank}(\tilde{\mathbf{X}}) \leq n-1$. This step is useful as soon as $p > r$. When $p \gg n$ we obtain a huge reduction. For spectral data, e.g., $n = 50$, $p = 1000$, this reduces the 1000-dimensional original data set to one in only 49 dimensions.

The main step of the RAPCA algorithm is then to search for the direction in which the projected observations have the largest robust scale. To measure the univariate scale, the Q_n estimator as defined in Equation 6.3 is used. Comparisons using other scale estimators are presented in Croux and Ruiz-Gazen [50] and Cui et al. [51]. To make the algorithm computationally feasible, the collection of directions to be investigated is restricted to all directions that pass through the L^1-median and a data point. The L^1-median is a highly robust (50% breakdown value) and orthogonally equivariant location estimator, also known as the spatial median. It is defined as the point θ, which minimizes the sum of the distances to all observations, i.e.,

$$\text{minimize} \sum_{i=1}^{n} \|\mathbf{x}_i - \theta\|$$

When the first direction, $\mathbf{v}_1$, has been found, the data are reflected such that the first eigenvector is mapped onto the first basis vector. Then the data are projected onto the orthogonal complement of the first eigenvector. This is simply done by omitting the first component of each (reflected) point. Doing so, the dimension of the projected data points can be reduced by 1 and, consequently, all the computations do not need to be done in the full r-dimensional space.

The method can then be applied in the orthogonal complement to search for the second eigenvector, and so on. It is not necessary to compute all eigenvectors, which would be very time consuming for high p, and the computations can be stopped as soon as the required number of components has been found.

Note that a PCA analysis often starts by prestandardizing the data to obtain variables that all have the same spread. Otherwise, the variables with a large variance compared with the others will dominate the first principal components. Standardizing by the mean and the standard deviation of each variable yields a PCA analysis based on the correlation matrix instead of the covariance matrix. We can also standardize each variable j in a robust way, e.g., by first subtracting its median, $\text{med}(x_{1j}, \ldots, x_{nj})$, and then dividing by its robust scale estimate, $Q_n(x_{1j}, \ldots, x_{nj})$.

6.5.4 ROBUST PCA BASED ON PROJECTION PURSUIT AND THE MCD

Another approach to robust PCA has been proposed by Hubert et al. [52] and is called ROBPCA. This method combines ideas of both projection pursuit and robust covariance estimation. The projection pursuit part is used for the initial dimension reduction. Some ideas based on the MCD estimator are then applied to this lower-dimensional data space. Simulations have shown that this combined approach yields more accurate estimates than the raw projection pursuit algorithm RAPCA. The complete description of the ROBPCA method is quite involved, so here we will only outline the main stages of the algorithm.

First, as in RAPCA, the data are preprocessed by reducing their data space to the affine subspace spanned by the n observations. As a result, the data are represented using at most $n - 1 = \text{rank}(\tilde{\mathbf{X}}_{n,p})$ variables without loss of information.

In the second step of the ROBPCA algorithm, a measure of outlyingness is computed for each data point [11, 12]. This is obtained by projecting the high-dimensional data points on many univariate directions **d** through two data points. On every direction, the univariate MCD estimator of location $\hat{\mu}_{MCD}$ and scale $\hat{\sigma}_{MCD}$ is computed on the projected points $\mathbf{x}_j^T \mathbf{d}$ ($j = 1,\ldots,n$), and for every data point its standardized distance to that center is measured. Finally, for each data point, its largest distance over all of the directions is considered. This yields the outlyingness

$$\text{outl}(\mathbf{x}_i) = \max_{\mathbf{d}} \frac{|\mathbf{x}_i^T \mathbf{d} - \hat{\mu}_{MCD}(\mathbf{x}_j^T \mathbf{d})|}{\hat{\sigma}_{MCD}(\mathbf{x}_j^T \mathbf{d})}$$

Next, the covariance matrix $\hat{\Sigma}_h$ of the h data points with smallest outlyingness is computed. The last stage of ROBPCA consist of projecting all of the data points onto the k-dimensional subspace spanned by the k dominant eigenvectors of $\hat{\Sigma}_h$ and then of computing their center and shape by means of the reweighed MCD estimator. The eigenvectors of this scatter matrix then determine the robust principal components, which can be collected in a loading matrix $\mathbf{P}_{p,k}$ with orthogonal columns. The MCD location estimate $\hat{\mu}_x$ serves as a robust center.

Because the loadings are orthogonal, they determine a new coordinate system in the k-dimensional subspace that they span. The k-dimensional scores of each data point $\mathbf{t}_i$ are computed as the coordinates of the projections of the robustly centered $\mathbf{x}_i$ onto this subspace, or equivalently

$$\mathbf{t}_i = \mathbf{P}_{k,p}^T (\mathbf{x}_i - \hat{\mu}_x)$$

The orthogonal distance measures the distance between an observation $\mathbf{x}_i$ and its projection $\hat{\mathbf{x}}_i$ in the k-dimensional PCA subspace:

$$\hat{\mathbf{x}}_i = \hat{\mu}_x + \mathbf{P}_{p,k} \mathbf{t}_i \tag{6.16}$$

$$OD_i = \|\mathbf{x}_i - \hat{\mathbf{x}}_i\|. \tag{6.17}$$

Let $\mathbf{L}_{k,k}$ denote the diagonal matrix that contains the k eigenvalues l_j of the MCD scatter matrix, sorted from largest to smallest. Thus $l_1 \geq l_2 \geq \ldots \geq l_k$. The score distance of the ith sample measures the robust distance of its projection to the center of all the projected observations. Hence, it is measured within the PCA subspace, where due to the knowledge of the eigenvalues, we have information about the covariance structure of the scores. Consequently, the score distance is defined as in Equation 6.6:

$$SD_i = \sqrt{\mathbf{t}_i^T \mathbf{L}^{-1} \mathbf{t}_i} = \sqrt{\sum_{j=1}^{k} (t_{ij}^2 / l_j)} \tag{6.18}$$

Robust Calibration

Moreover, the k robust principal components generate a $p \times p$ robust scatter matrix $\hat{\boldsymbol{\Sigma}}_x$ of rank k given by

$$\hat{\boldsymbol{\Sigma}}_x = \mathbf{P}_{p,k}\mathbf{L}_{k,k}\mathbf{P}_{k,p}^T \qquad (6.19)$$

Note that all results (the scores $\mathbf{t}_i$, the scores distances, the orthogonal distances, and the scatter matrix $\hat{\boldsymbol{\Sigma}}_x$) depend on the number of components k. But to simplify the notations, we do not explicitly add a subscript k.

We also mention the robust LTS-subspace estimator and its generalizations [6, 53]. The idea behind these approaches consists in minimizing a robust scale of the orthogonal distances, similar to the LTS estimator and S-estimators in regression. Also, the orthogonalized Gnanadesikan-Kettenring estimator [54] is fast and robust, but it is not orthogonally equivariant.

6.5.5 AN OUTLIER MAP

The result of the PCA analysis can be represented by means of a diagnostic plot or outlier map [52]. As in regression, this figure highlights the outliers and classifies them into several types. In general, an outlier is defined as an observation that does not obey the pattern of the majority of the data. In the context of PCA, this means that an outlier either lies far from the subspace spanned by the k eigenvectors, or that the projected observation lies far from the bulk of the data within this subspace. This outlyingness can be expressed by means of the orthogonal and the score distances. These two distances define four types of observations, as illustrated in Figure 6.8a.

Regular observations have a small orthogonal and a small score distance. When samples have a large score distance but a small orthogonal distance, we call them good leverage points. Observations 1 and 4 in Figure 6.8a can be classified into this category. These observations lie close to the space spanned by the principal components but far from the regular data. This implies that they are different from the majority, but there is only a little loss of information when we replace them by their fitted values in the PCA-subspace.

Orthogonal outliers have a large orthogonal distance, but a small score distance, as, for example, case 5. They cannot be distinguished from the regular observations once they are projected onto the PCA subspace, but they lie far from this subspace. Consequently, it would be dangerous to replace that sample with its projected value, as its outlyingness would not be visible anymore.

Bad leverage points, such as observations 2 and 3, have a large orthogonal distance and a large score distance. They lie far outside the space spanned by the principal components, and after projection they are far from the regular data points. Their degree of outlyingness is high in both directions, and typically they have a large influence on classical PCA, as the eigenvectors will be tilted toward them.

The outlier map displays the OD_i vs. the SD_i and, hence, classifies the observations according to Table 6.3 and Figure 6.8b. On this plot, lines are drawn to distinguish the observations with a small and a large OD, and with a small and a large SD. For the latter distances, we use the property that normally distributed data

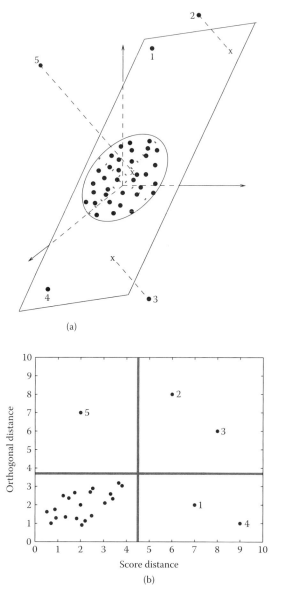

FIGURE 6.8 (a) Different types of outliers when a three-dimensional data set is projected on a robust two-dimensional PCA subspace, with (b) the corresponding outlier map [44].

have normally distributed scores and, consequently, their squared Mahalanobis distances have asymptotically a χ_k^2 distribution. Hence, we use as cutoff value $c = \sqrt{\chi_{k,0.975}^2}$. For the orthogonal distances, the approach of Box [55] is followed. The squared orthogonal distances can be approximated by a scaled χ^2 distribution, which in its turn can be approximated by a normal distribution using the Wilson-Hilferty transformation.

**TABLE 6.3
Overview of the Different Types of Observations Based on Their Score Distance (SD) and Their Orthogonal Distance (OD)**

Distances	Small SD	Large SD
Large OD	Orthogonal outlier	Bad PCA-leverage point
Small OD	Regular observation	Good PCA-leverage point

The mean and variance of this normal distribution are then estimated by applying the univariate MCD to the $OD_i^{2/3}$. Observations that exceed the horizontal or the vertical cutoff value are then classified as the PCA outliers.

6.5.6 Selecting the Number of Principal Components

To choose the optimal number of loadings k_{opt}, there are many criteria. For a detailed overview, see Joliffe [56]. A very popular graphical one is based on the scree plot, which exposes the eigenvalues in decreasing order. The index of the last component before the plot flattens is then selected.

A more formal criterion considers the total variation that is explained by the first k loadings and requires, for example, that

$$\left(\sum_{j=1}^{k_{opt}} l_j\right) \bigg/ \left(\sum_{j=1}^{p} l_j\right) \geq 80\% \qquad (6.20)$$

Note that this criterion cannot be used with ROBPCA, as the method does not yield all of the p eigenvalues (as then it would become impossible to compute the MCD estimator in the final stage of the algorithm). But we can apply it on the eigenvalues of the covariance matrix of $\hat{\Sigma}_h$ that was constructed in the second stage of the algorithm.

One can also choose k_{opt} as the smallest value for which

$$l_k / l_1 \geq 10^{-3}$$

Another criterion that is based on the predictive ability of PCA is the predicted sum of squares (PRESS) statistic. To compute the (cross validated) PRESS value at a certain k, we remove the ith observation from the original data set (for $i = 1, \ldots, n$), estimate the center and the k loadings of the reduced data set, and then compute the fitted value of the ith observation following Equation 6.16, now denoted as $\hat{\mathbf{x}}_{-i}$. Finally, we set

$$\text{PRESS}_k = \sum_{i=1}^{n} \left\| \mathbf{x}_i - \hat{\mathbf{x}}_{-i} \right\|^2 \qquad (6.21)$$

The value k for which the PRESS$_k$ is small enough is then considered as the optimal number of components k_{opt}. One could also apply formal F-type tests based on successive PRESS values [57, 58].

However, the PRESS$_k$ statistic is not suitable for use with contaminated data sets because it also includes the prediction error of the outliers. Even if the fitted values are based on a robust PCA algorithm, their prediction error might increase the PRESS$_k$ because they fit the model poorly. Consequently, the decision about the optimal number of components k_{opt} could be wrong.

To obtain a robust PRESS value, we can apply the following procedure. For each PCA model under investigation ($k = 1, \ldots, k_{max}$), the outliers are marked. As discussed in Section 6.5.5, these are the observations that exceed the horizontal or vertical cutoff value on the outlier map. Next, all the outliers are collected (over all k) and removed from the sum in Equation 6.21. By doing this, the robust PRESS$_k$ value is based on the same set of observations for each k. Moreover, fast methods to compute x_{-i} have been developed [62].

6.5.7 AN EXAMPLE

We illustrate ROBPCA and the outlier map on a data set that consists of spectra of 180 ancient glass pieces over $p = 750$ wavelengths [59]. The measurements were performed using a Jeol JSM 6300 scanning electron microscope equipped with an energy-dispersive Si(Li) x-ray detection system. We first performed ROBPCA with default value $h = 0.75$, $n = 135$. However, the outlier maps then revealed a large amount of outliers. Therefore, we analyzed the data set a second time with $h = 0.70$, $n = 126$. Three components are retained for CPCA and ROBPCA yielding a classical explanation percentage of 99% and a robust explanation percentage (see Equation 6.20) of 96%.

The resulting outlier maps are shown in Figure 6.9. From the classical diagnostic plot in Figure 6.9a, we see that CPCA does not find large outliers. On the other hand, the ROBPCA plot of Figure 6.9b clearly distinguishes two major groups in the data, a smaller group of bad leverage points, a few orthogonal outliers, and the isolated case 180 in between the two major groups. A high-breakdown method, such as ROBPCA, treats the smaller group with cases 143–179 as one set of outliers. Later, it turned out that the window of the detector system had been cleaned before the last 38 spectra were measured. As a result of this, less radiation (x-rays) was absorbed and more could be detected, resulting in higher x-ray intensities. The other bad leverage points, 57–63 and 74–76, are samples with a large concentration of calcic. The orthogonal outliers (22, 23, and 30) are borderline cases, although it turned out that they have larger measurements at the wavelengths 215–245. This might indicate a larger concentration of phosphor.

6.6 PRINCIPAL COMPONENT REGRESSION

6.6.1 CLASSICAL PCR

Principal component regression is typically used for linear regression models (Equation 6.7 or Equation 6.10), where the number of independent variables p is very large or where the regressors are highly correlated (this is known as multicollinearity).

Robust Calibration

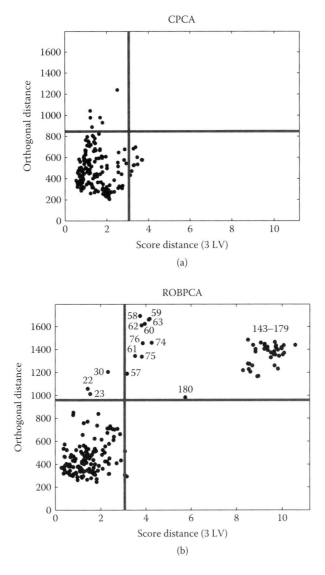

FIGURE 6.9 Outlier map of the glass data set based on three principal components computed with (a) CPCA and (b) ROBPCA [52].

An important application of PCR is multivariate calibration, whose goal is to predict constituent concentrations of a material based on its spectrum. This spectrum can be obtained via several techniques, including fluorescence spectrometry, near-infrared spectrometry (NIR), nuclear magnetic resonance (NMR), ultraviolet spectrometry (UV), energy-dispersive x-ray fluorescence spectrometry (ED-XRF), etc. Because a spectrum typically ranges over a large number of wavelengths, it is a high-dimensional vector with hundreds of components. The number of concentrations, on the other hand, is usually limited to about five, at most.

In the univariate approach, only one concentration at a time is modeled and analyzed. The more general problem assumes that the number of response variables q is larger than 1, which means that several concentrations are to be estimated together. This model has the advantage that the covariance structure between the concentrations is also taken into account, which is appropriate when the concentrations are known to be strongly intercorrelated with each other. As argued in Martens and Naes [60], the multivariate approach can also lead to better predictions if the calibration data for one important concentration, say y_1, are imprecise. When this variable is highly correlated with some other constituents that are easier to measure precisely, then a joint calibration may give better understanding of the calibration data and better predictions for y_1 than a separate univariate calibration for this analyte. Moreover, the multivariate calibration can be very important to detect outlying samples that would not be discovered by separate regressions. Here, we will write down the formulas for the general multivariate setting (Equation 6.10) for which $q \geq 1$, but they can, of course, be simplified when $q = 1$.

The PCR method (and PLS, partial least squares, discussed in the Section 6.7) assumes that the linear relation (Equation 6.10) between the x- and y-variables is in fact a bilinear model that depends on scores t:

$$\mathbf{x}_i = \bar{\mathbf{x}} + \mathbf{P}_{p,k} \tilde{\mathbf{t}}_i + \mathbf{f}_i \qquad (6.22)$$

$$\mathbf{y}_i = \bar{\mathbf{y}} + \mathbf{A}_{q,k}^T \tilde{\mathbf{t}}_i + \mathbf{g}_i \qquad (6.23)$$

with $\bar{\mathbf{x}}$ and $\bar{\mathbf{y}}$ the mean of the x- and y-variables.

Consequently, classical PCR (CPCR) starts by mean-centering the data. Then, in order to cope with the multicollinearity in the x-variables, the first k principal components of $\mathbf{X}_{n,p}$ are computed. As outlined in Section 6.5.1, these loading vectors $\tilde{\mathbf{P}}_{p,k} = (\mathbf{p}_1,\ldots,\mathbf{p}_k)$ are the k eigenvectors that correspond to the k dominant eigenvalues of the empirical covariance matrix $\mathbf{S}_x = \frac{1}{n-1}\tilde{\mathbf{X}}^T\tilde{\mathbf{X}}$. Next, the k-dimensional scores of each data point $\tilde{\mathbf{t}}_i$ are computed as $\tilde{\mathbf{t}}_i = \tilde{\mathbf{P}}_{k,p}^T \tilde{\mathbf{x}}_i$. In the final step, the centered response variables $\tilde{\mathbf{y}}_i$ are regressed onto $\tilde{\mathbf{t}}_i$ using MLR. This yields parameter estimates $\hat{\mathbf{A}}_{k,q} = (\mathbf{T}^T\mathbf{T})^{-1}_{k,k}\mathbf{T}^T_{k,n}\tilde{\mathbf{Y}}_{n,q}$ and fitted values $\hat{\mathbf{y}}_i = \bar{\mathbf{y}} + \hat{\mathbf{A}}^T_{q,k}\tilde{\mathbf{t}}_i = \bar{\mathbf{y}} + \hat{\mathbf{A}}^T_{q,k}\tilde{\mathbf{P}}^T_{k,p}(\mathbf{x}_i - \bar{\mathbf{x}})$. The unknown regression parameters in the model presented in Equation 6.10 are then estimated as

$$\hat{\mathbf{B}}_{p,q} = \tilde{\mathbf{P}}_{p,k}\hat{\mathbf{A}}_{k,q}$$

$$\boldsymbol{\beta}_0 = \bar{\mathbf{y}} - \hat{\mathbf{B}}^T_{q,p}\bar{\mathbf{x}}.$$

Finally, the covariance matrix of the errors can be estimated as the empirical covariance matrix of the residuals

$$\mathbf{S}_\varepsilon = \frac{1}{n-1}\sum_{i=1}^{n}\mathbf{r}_i\mathbf{r}_i^T = \frac{1}{n-1}\sum_{i=1}^{n}(\mathbf{y}_i - \hat{\mathbf{y}}_i)(\mathbf{y}_i - \hat{\mathbf{y}}_i)^T$$

$$= \mathbf{S}_y - \hat{\mathbf{A}}^T\mathbf{S}_t\hat{\mathbf{A}} \qquad (6.24)$$

with S_y and S_t being the empirical covariance matrices of the y- and the t-variables. Note that Equation 6.24 follows from the fact that the fitted MLR values are orthogonal to the MLR residuals.

6.6.2 ROBUST PCR

The robust PCR (RPCR) method proposed by Hubert and Verboven [61] combines robust PCA for high-dimensional data (ROBPCA, see Section 6.5.4) with a robust regression technique such as LTS regression (Section 6.4.1.3) or MCD regression (Section 6.4.2.2). In the first stage of the algorithm, robust scores t_i are obtained by applying ROBPCA on the x-variables and retaining k components. In the second stage of the RPCR method, the original response variables y_i are regressed on the t_i using a robust regression method. Note that here a regression model with intercept is fitted:

$$y_i = \alpha_0 + A^T t_i + \breve{\varepsilon}_i \tag{6.25}$$

with $\text{Cov}(\breve{\varepsilon}) = \Sigma_{\breve{\varepsilon}}$. If there is only one response variable ($q = 1$), the parameters in Equation 6.25 can be estimated using the reweighed LTS estimator. If $q > 1$, the MCD regression is performed. As explained in Section 6.4.2.2, it starts by computing the reweighed MCD estimator on the (t_i, y_i) jointly, leading to a $(k + q)$-dimensional location estimate $\hat{\mu} = (\hat{\mu}_t^T, \hat{\mu}_y^T)^T$ and a scatter estimate $\hat{\Sigma}_{k+q,k+q}$, which can be split into a scatter estimate of the t-variables, the y-variables, and of the cross-covariance between the ts and ys:

$$\hat{\Sigma}_{\text{MCD}} = \begin{pmatrix} \hat{\Sigma}_t & \hat{\Sigma}_{ty} \\ \hat{\Sigma}_{yt} & \hat{\Sigma}_y \end{pmatrix}$$

Robust parameter estimates are then obtained following Equation 6.11 to Equation 6.13 as

$$\hat{A}_{k,q} = \hat{\Sigma}_t^{-1} \hat{\Sigma}_{ty} \tag{6.26}$$

$$\hat{\alpha}_0 = \hat{\mu}_y - \hat{A}^T \hat{\mu}_t \tag{6.27}$$

$$\hat{\Sigma}_{\breve{\varepsilon}} = \hat{\Sigma}_y - \hat{A}^T \hat{\Sigma}_t \tag{6.28}$$

Note the correspondence of Equation 6.28 with Equation 6.24. Next, a reweighing step can be added based on the residual distances (Equation 6.14).

The regression parameters in the model depicted by Equation 6.10 are then derived as:

$$\hat{B}_{p,q} = P_{p,k} \hat{A}_{k,q}$$

$$\hat{\beta}_0 = \hat{\alpha}_0 - \hat{B}_{q,p}^T \hat{\mu}_x$$

$$\hat{\Sigma}_\varepsilon = \hat{\Sigma}_{\breve{\varepsilon}}$$

Note that, as for the MCD estimator, the robustness of the RPCR algorithm depends on the value of h, which is chosen in the ROBPCA algorithm and in the LTS and MCD regression. Although it is not really necessary, it is recommended that the same value be used in both steps.

Following the practice in Table 6.2, Table 6.3 and observations can now be classified as regular observations, PCA outliers, or regression outliers. This will be illustrated in Section 6.6.4.

6.6.3 MODEL CALIBRATION AND VALIDATION

An important issue in PCR is the selection of the optimal number of principal components k_{opt}, for which several methods have been proposed. A popular approach consists of minimizing the root mean squared error of cross-validation criterion RMSECV$_k$. For one response variable ($q = 1$), it equals

$$\text{RMSECV}_k = \sqrt{\frac{1}{n}\sum_{i=1}^{n}(y_i - \hat{y}_{-i,k})^2} \tag{6.29}$$

with $\hat{y}_{-i,k}$ the predicted value for observation i, where i was left out of the data set when performing the PCR method with k principal components. For multiple y-variables, it is usually defined as

$$\text{RMSECV}_k = \sqrt{\frac{1}{n}\sum_{i=1}^{n}\sum_{j=1}^{q}(y_{ij} - \hat{y}_{-i,j,k})^2} \tag{6.30}$$

The goal of the RMSECV$_k$ statistic is twofold. It yields an estimate of the root mean squared prediction error $E(y - \hat{y})^2$ when k components are used in the model, whereas the curve of RMSECV$_k$ for $k = 1, \ldots, k_{max}$ is a popular graphical tool to choose the optimal number of components.

As argued for the PRESS statistic (Equation 6.21) in PCA, this RMSECV$_k$ statistic is also not suitable for use with contaminated data sets because it includes the prediction error of the outliers. A robust RMSECV (R-RMSECV) measure can be constructed in analogy with the robust PRESS value [61]. Roughly said, for each PCR model under investigation ($k = 1, \ldots, k_{max}$), the regression outliers are marked and then removed from the sum in Equation 6.29 or Equation 6.30. By doing this, the RMSECV$_k$ statistic is based on the same set of observations for each k. The optimal number of components is then taken as the value k_{opt} for which RMSECV$_k$ is minimal or sufficiently small.

Once the optimal number of components k_{opt} is chosen, the PCR model can be validated by estimating the prediction error. A robust root mean squared error of prediction (R-RMSEP) is obtained as in Equation 6.29 or Equation 6.30 by eliminating the outliers found by applying RPCR with k_{opt} components. It thus includes all the regular observations for the model with k_{opt} components, which is larger than the set used to obtain RMSECV$_k$. Hence, in general R-RMSEP$_{k_{opt}}$ will be different from R-RMSECV$_{k_{opt}}$.

The R-RMSECV$_k$ values are rather time consuming because, for every choice of k, they require the whole RPCR procedure to be performed n times. Faster algorithms for cross validation are described [80]. They avoid the complete recomputation of resampling methods, such as the MCD, when one observation is removed from the data set. Alternatively, one could also compute a robust R^2-value [61]. For $q = 1$ it equals:

$$R_k^2 = 1 - \frac{\sum_i r_{i,k}^2}{\sum_i (y_i - \bar{y})^2}$$

where $r_{i,k}$ is the ith residual obtained with a RPCR model with k components, and the sum is taken over all regular observations for $k = 1, \ldots, k_{\max}$. The optimal number of components k_{opt} is then chosen as the smallest value k for which R_k^2 attains, e.g., 80%, or the R_k^2 curve becomes nearly flat. This approach is fast because it avoids cross validation by measuring the variance of the residuals instead of the prediction error.

6.6.4 AN EXAMPLE

To illustrate RPCR, we analyze the biscuit dough data set [63]. It contains 40 NIR spectra of biscuit dough with measurements every 2 nm, from 1200 nm up to 2400 nm. The data are first scaled using a logarithmic transformation to eliminate drift and background scatter. Originally the data set consisted of 700 variables, but the ends were discarded because of the lower instrumental reliability. Then the first differences were used to remove constants and sudden shifts. After this preprocessing, we ended up with a data set of $n = 40$ observations in $p = 600$ dimensions. The responses are the percentages of four constituents in the biscuit dough: y_1 = fat, y_2 = flour, y_3 = sucrose, and y_4 = water. Because there is a significant correlation among the responses, a multivariate regression is performed. The robust R-RMSECV$_k$ curve plotted in Figure 6.10 suggests the selection of $k = 2$ components.

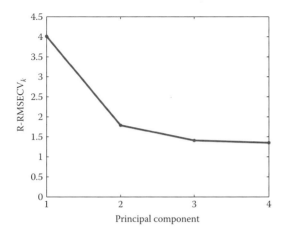

FIGURE 6.10 Robust R-RMSECV$_k$ curve for the biscuit dough data set [61].

Differences between CPCR and RPCR show up in the loading vectors and in the calibration vectors. Figure 6.11 shows the second loading vector and the second calibration vector for y_3 (sucrose). For instance, we notice (between wavelengths 1390 and 1440) a large discrepancy in the C-H bend.

Next, we can construct outlier maps as in Sections 6.5.5 and 6.4.2.3. ROBPCA yields the PCA outlier map displayed in Figure 6.12a. We see that there are no PCA leverage points, but there are some orthogonal outliers, the largest being 23, 7, and 20. The result of the regression step is shown in Figure 6.12b. It exposes the robust distances of the residuals (or the standardized residuals if $q = 1$) vs. the score

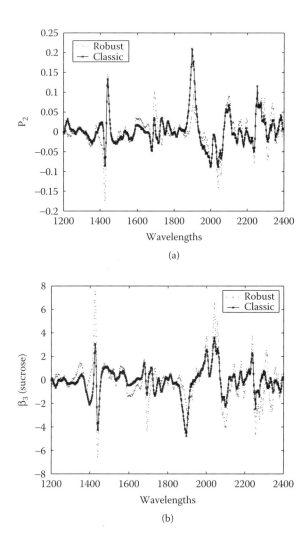

FIGURE 6.11 Second loading vector and calibration vector of sucrose for the biscuit dough data set, computed with (a) second loading of vector and (b) calibration vector of sucrose for the biscuit dough data set, computed with CPCR and RPCR [61].

Robust Calibration

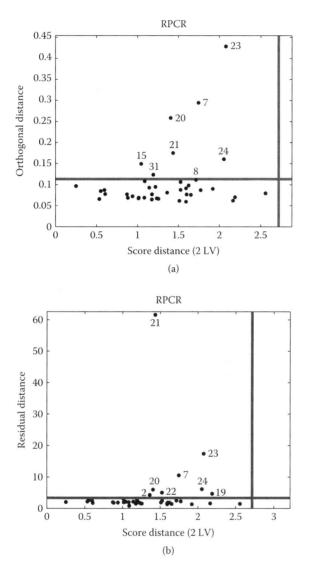

FIGURE 6.12 (a) PCA outlier map when applying RPCR to the biscuit dough data set; (b) corresponding regression outlier map [61].

distances, and thus identifies the outliers with respect to the model depicted in Equation 6.25. RPCR shows that observation 21 has an extremely high residual distance. Other vertical outliers are 23, 7, 20, and 24, whereas there are a few borderline cases. In Hubert and Verboven [61], it is demonstrated that case 21 never showed up as such a large outlier when performing four univariate calibrations. It is only by using the full covariance structure of the residuals in Equation 6.14 that this extreme data point is found.

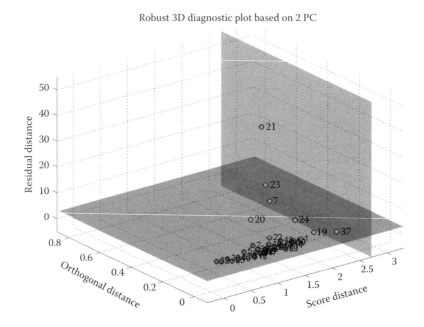

FIGURE 6.13 Three-dimensional outlier map of the biscuit dough data set obtained with RPCR [61].

Finally, three-dimensional outlier maps can be made by combining the PCA and the regression outlier maps, i.e., by plotting for each observation the triple (SD_i, OD_i, $ResD_i$). Figure 6.13 shows the result for the biscuit dough data. It is particularly interesting to create such three-dimensional plots with interactive software packages (such as MATLAB or S-PLUS) that allow you to rotate and spin the whole figure.

6.7 PARTIAL LEAST-SQUARES REGRESSION

6.7.1 CLASSICAL PLSR

Partial least-squares regression (PLSR) is similar to PCR. Its goal is to estimate regression coefficients in a linear model with a large number of x-variables that are highly correlated. In the first step of PCR, the scores were obtained by extracting the main information present in the x-variables by performing a principal component analysis on them without using any information about the y-variables. In contrast, the PLSR scores are computed by maximizing a covariance criterion between the x- and y-variables. Hence, the first stage of this technique already uses the responses.

More precisely, let $\tilde{\mathbf{X}}_{n,p}$ and $\tilde{\mathbf{Y}}_{n,q}$ denote the mean-centered data matrices. The normalized PLS weight vectors $\mathbf{r}_a$ and $\mathbf{q}_a$ (with $\|\mathbf{r}_a\| = \|\mathbf{q}_a\| = 1$) are then defined as the vectors that maximize

$$\text{cov}(\tilde{\mathbf{Y}}\mathbf{q}_a, \tilde{\mathbf{X}}\mathbf{r}_a) = \mathbf{q}_a^T \frac{\tilde{\mathbf{Y}}^T \tilde{\mathbf{X}}}{n-1} \mathbf{r}_a = \mathbf{q}_a^T \mathbf{S}_{yx} \mathbf{r}_a \qquad (6.31)$$

for each $a = 1, \ldots, k$, where $\mathbf{S}_{yx}^T = \mathbf{S}_{xy} = \frac{\tilde{\mathbf{X}}^T \tilde{\mathbf{Y}}}{n-1}$ is the empirical cross-covariance matrix between the x- and the y-variables. The elements of the scores $\tilde{\mathbf{t}}_i$ are then defined as linear combinations of the mean-centered data: $\tilde{t}_{ia} = \tilde{\mathbf{x}}_i^T \mathbf{r}_a$, or equivalently $\tilde{\mathbf{T}}_{n,k} = \tilde{\mathbf{X}}_{n,p} \mathbf{R}_{p,k}$ with $\mathbf{R}_{p,k} = (\mathbf{r}_1, \ldots, \mathbf{r}_k)$.

The computation of the PLS weight vectors can be performed using the SIMPLS algorithm [64]. The solution of the maximization problem in Equation 6.31 is found by taking $\mathbf{r}_1$ and $\mathbf{q}_1$ as the first left and right singular eigenvectors of $\mathbf{S}_{xy}$. The other PLSR weight vectors $\mathbf{r}_a$ and $\mathbf{q}_a$ for $a = 2, \ldots, k$ are obtained by imposing an orthogonality constraint to the elements of the scores. If we require that $\sum_{i=1}^n t_{ia} t_{ib} = 0$ for $a \neq b$, a deflation of the cross-covariance matrix $\mathbf{S}_{xy}$ provides the solutions for the other PLSR weight vectors. This deflation is carried out by first calculating the x-loading

$$\mathbf{p}_a = \mathbf{S}_x \mathbf{r}_a / (\mathbf{r}_a^T \mathbf{S}_x \mathbf{r}_a) \tag{6.32}$$

with $\mathbf{S}_x$ the empirical covariance matrix of the x-variables. Next an orthonormal base $\{\mathbf{v}_1, \ldots, \mathbf{v}_a\}$ of $\{\mathbf{p}_1, \ldots, \mathbf{p}_a\}$ is constructed, and $\mathbf{S}_{xy}$ is deflated as

$$\mathbf{S}_{xy}^a = \mathbf{S}_{xy}^{a-1} - \mathbf{v}_a (\mathbf{v}_a^T \mathbf{S}_{xy}^{a-1})$$

with $\mathbf{S}_{xy}^1 = \mathbf{S}_{xy}$. In general, the PLSR weight vectors $\mathbf{r}_a$ and $\mathbf{q}_a$ are obtained as the left and right singular vector of $\mathbf{S}_{xy}^a$.

6.7.2 ROBUST PLSR

A robust method, RSIMPLS, has been developed by Hubert and Vanden Branden [65]. It starts by applying ROBPCA on the joint x- and y-variables to replace $\mathbf{S}_{xy}$ and $\mathbf{S}_x$ by robust estimates, and then proceeds analogously to the SIMPLS algorithm. More precisely, to obtain robust scores, ROBPCA is first applied on the joint x- and y-variables $\mathbf{Z}_{n,m} = (\mathbf{X}_{n,p}, \mathbf{Y}_{n,q})$ with $m = p + q$. Assume that we select k_0 components. This yields a robust estimate of the center of $\mathbf{Z}$, $\hat{\boldsymbol{\mu}}_z = (\hat{\boldsymbol{\mu}}_x^T, \hat{\boldsymbol{\mu}}_y^T)^T$ and, following Equation 6.19, an estimate of its shape, $\hat{\boldsymbol{\Sigma}}_z$, which can be split into

$$\hat{\boldsymbol{\Sigma}}_z = \begin{pmatrix} \hat{\boldsymbol{\Sigma}}_x & \hat{\boldsymbol{\Sigma}}_{xy} \\ \hat{\boldsymbol{\Sigma}}_{yx} & \hat{\boldsymbol{\Sigma}}_y \end{pmatrix} \tag{6.33}$$

The cross-covariance matrix $\boldsymbol{\Sigma}_{xy}$ is then estimated by $\hat{\boldsymbol{\Sigma}}_{xy}$, and the PLS weight vectors $\mathbf{r}_a$ are computed as in the SIMPLS algorithm, but now starting with $\hat{\boldsymbol{\Sigma}}_{xy}$ instead of $\mathbf{S}_{xy}$. In analogy with Equation 6.32, the x-loadings $\mathbf{p}_j$ are defined as $\mathbf{p}_j = \hat{\boldsymbol{\Sigma}}_x \mathbf{r}_j / (\mathbf{r}_j^T \hat{\boldsymbol{\Sigma}}_x \mathbf{r}_j)$. Then the deflation of the scatter matrix $\hat{\boldsymbol{\Sigma}}_{xy}^a$ is performed as in SIMPLS. In each step, the robust scores are calculated as:

$$t_{ia} = \tilde{\mathbf{x}}_i^T \mathbf{r}_a = (\mathbf{x}_i - \hat{\boldsymbol{\mu}}_x)^T \mathbf{r}_a$$

where $\tilde{\mathbf{x}}_i = \mathbf{x}_i - \hat{\boldsymbol{\mu}}_x$ are the robustly centered observations.

Next, a robust regression has to be applied of the y_i against the t_i. This could again be done using the MCD regression method of Section 6.4.2.2, but a faster approach goes as follows. The MCD regression method starts by applying the reweighed MCD estimator on (t, y) to obtain robust estimates of their center μ and scatter Σ. This reweighed MCD corresponds to the mean and the covariance matrix of those observations that are considered not to be outlying in the $(k + q)$-dimensional (t, y) space. To obtain the robust scores, t_i, ROBPCA was first applied to the (x, y)-variables, and hereby a k_0-dimensional subspace K_0 was obtained that represented these (x, y)-variables well. Because the scores were then constructed to summarize the most important information given in the x-variables, we might expect that outliers with respect to this k_0-dimensional subspace are often also outlying in the (t, y) space. Hence, the center μ and the scatter Σ of the (t, y)-variables can be estimated as the mean and covariance matrix of those (t_i, y_i) whose corresponding (x_i, y_i) are not outlying to K_0. It is those observations whose score distance and orthogonal distance do not exceed the cutoff values on the outlier map, as defined in Section 6.5.4.

Having identified the regular observations (x_i, y_i) with ROBPCA, we thus compute the mean $\hat{\mu}$ and covariance $\hat{\Sigma}$ of the corresponding (t_i, y_i). Then, the method proceeds as in the MCD-regression method. These estimates are plugged into Equation 6.26 to Equation 6.28, residual distances are computed as in Equation 6.14, and a reweighed MLR is performed. This reweighing step has the advantage that it might again include outlying observations from ROBPCA that are not regression outliers.

Note that when performing the ROBPCA method on $Z_{n,m}$, we need to determine k_0, which should be a good approximation of the dimension of the space spanned by the x- and y-variables. If k is known, k_0 can be set as $\min(k, 10) + q$. The number $k + q$ represents the sum of the number of x-loadings that give a good approximation of the dimension of the x-variables and the number of response variables. The maximal value $k_{max} = 10$ is included to ensure a good efficiency of the FAST-MCD method in the last stage of ROBPCA, but it can be increased if enough observations are available. Other ways to select k_0 are discussed in Section 6.5.6. By doing this, one should keep in mind that it is logical that k_0 be larger than the number of components k that will be retained in the regression step.

This RSIMPLS approach yields bounded-influence functions for the weight vectors r_a and q_a and for the regression estimates [66]. Also, the breakdown value is inherited from the MCD estimator. Model calibration and validation is similar to the RPCR method and proceeds as in Section 6.6.3.

6.7.3 AN EXAMPLE

The robustness of RSIMPLS is illustrated on an octane data set [67] consisting of NIR absorbance spectra over $p = 226$ wavelengths ranging from 1102 nm to 1552 nm, with measurements every 2 nm. For each of the $n = 39$ production gasoline samples, the octane number y was measured, so $q = 1$. It is known that the octane data set contains six outliers (25, 26, 36–39) to which alcohol was added. From the R-RMSECV values [68], it follows that $k = 2$ components should be retained.

Robust Calibration

The resulting outlier maps are shown in Figure 6.14. The robust PCA outlier map is displayed in Figure 6.14a. Note that according to the model presented in Equation 6.22, its score distance SD_i is displayed on the horizontal axis for each observation

$$SD_i = D(\mathbf{t}_i, \hat{\mu}_t, \hat{\Sigma}_t) = \sqrt{(\mathbf{t}_i - \hat{\mu}_t)^T \hat{\Sigma}_t^{-1}(\mathbf{t}_i - \hat{\mu}_t)}$$

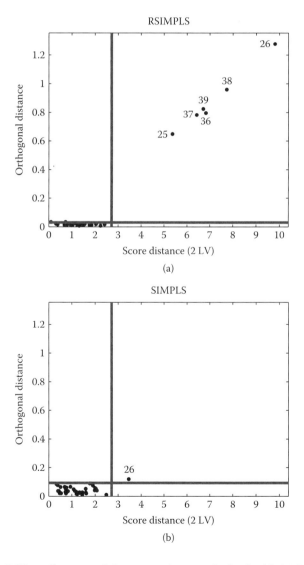

FIGURE 6.14 PCA outlier map of the octane data set obtained with (a) RSIMPLS and (b) SIMPLS. Regression outlier map obtained with (c) RSIMPLS and (d) SIMPLS.

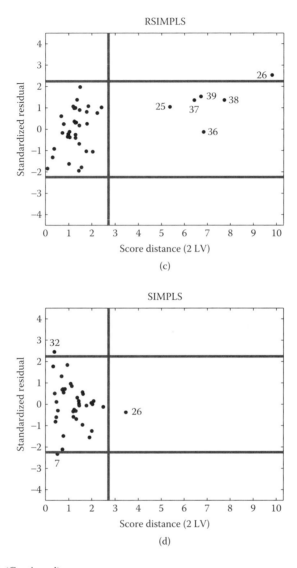

FIGURE 6.14 (Continued)

where $\hat{\mu}_t$ and $\hat{\Sigma}_t$ are derived in the regression step of the RSIMPLS algorithm. The vertical axis of Figure 6.14a shows the orthogonal distance of an observation to the t-space:

$$\text{OD}_i = \left\| \mathbf{x}_i - \hat{\mu}_x - \mathbf{P}_{p,k} \mathbf{t}_i \right\|$$

We immediately spot the six samples with added alcohol. The SIMPLS outlier map is shown in Figure 6.14b. We see that this analysis only detects the outlying spectrum 26, which does not even stick out much above the borderline. The robust regression outlier map in Figure 6.14c shows that the outliers are good leverage points, whereas SIMPLS in Figure 6.14d again reveals only case 26.

6.8 CLASSIFICATION

6.8.1 CLASSIFICATION IN LOW DIMENSIONS

6.8.1.1 Classical and Robust Discriminant Rules

The goal of classification, also known as discriminant analysis or supervised learning, is to obtain rules that describe the separation between known groups of observations. Moreover, it allows the classification of new observations into one of the groups. We denote the number of groups by l and assume that we can describe our experiment in each population π_j by a p-dimensional random variable X_j with distribution function (density) f_j. We write p_j for the membership probability, i.e., the probability for an observation to come from π_j.

The maximum-likelihood rule classifies an observation $\mathbf{x} \in IR^p$ into π_a if $\ln(p_a f_a(\mathbf{x}))$ is the maximum of the set $\{\ln(p_j f_j(\mathbf{x})); j = 1, \ldots, l\}$. If we assume that the density f_j for each group is Gaussian with mean μ_j and covariance matrix Σ_j, then it can be seen that the maximum-likelihood rule is equivalent to maximizing the discriminant scores $d_j^Q(\mathbf{x})$ with

$$d_j^Q(\mathbf{x}) = -\frac{1}{2}\ln|\Sigma_j| - \frac{1}{2}(\mathbf{x}-\mu_j)^T \Sigma_j^{-1}(\mathbf{x}-\mu_j) + \ln(p_j). \qquad (6.34)$$

That is, $\mathbf{x}$ is allocated to π_a if $d_a^Q(\mathbf{x}) \geq d_j^Q(\mathbf{x})$ for all $j = 1, \ldots, l$ (see, e.g., Johnson and Wichern [69]).

In practice, μ_j, Σ_j, and p_j have to be estimated. Classical quadratic discriminant analysis (CQDA) uses the group's mean and empirical covariance matrix to estimate μ_j and Σ_j. The membership probabilities are usually estimated by the relative frequencies of the observations in each group, hence $\hat{p}_j^C = n_j/n$, where n_j is the number of observations in group j.

A robust quadratic discriminant analysis (RQDA) [70] is derived by using robust estimators of μ_j, Σ_j, and p_j. In particular, if the number of observations is sufficiently large with respect to the dimension p, we can apply the reweighed MCD estimator of location and scatter in each group (Section 6.3.2). As a by-product of this robust procedure, outliers (within each group) can be distinguished from the regular observations. Finally, the membership probabilities can be robustly estimated as the relative frequency of the regular observations in each group, yielding $\hat{p}_j^R$.

When all the covariance matrices are assumed to be equal, the quadratic scores (Equation 6.34) can be simplified to

$$d_j^L(\mathbf{x}) = \mu_j^T \Sigma^{-1} \mathbf{x} - \frac{1}{2}\mu_j^T \Sigma^{-1} \mu_j + \ln(p_j) \qquad (6.35)$$

where Σ is the common covariance matrix. The resulting scores (Equation 6.35) are linear in $\mathbf{x}$, hence the maximum-likelihood rule belongs to the class of linear discriminant analysis. It is well known that if we have only two populations ($l = 2$) with a common covariance structure, and if both groups have equal membership probabilities, then this rule coincides with Fisher's linear discriminant rule. Again, the common covariance matrix can be estimated by means of the MCD estimator, e.g., by pooling the MCD estimates in each group. Robust linear discriminant analysis, based on the MCD estimator (or S-estimators), has been studied by several authors [70–73].

6.8.1.2 Evaluating the Discriminant Rules

One also needs a tool to evaluate a discriminant rule, i.e., we need an estimate of the associated probability of misclassification. To do this, we could apply the rule to our observed data and count the (relative) frequencies of misclassified observations. However, it is well known that this yields an overly optimistic misclassification error, as the same observations are used to determine and to evaluate the discriminant rule. Another very popular approach is cross validation [74], which computes the classification rule by leaving out one observation at a time and then looking to see whether each observation is correctly classified or not. Because it makes little sense to evaluate the discriminant rule on outlying observations, one could apply this procedure by leaving out the nonoutliers one by one and counting the percentage of misclassified ones. This approach is rather time consuming, especially with large data sets. For the classical linear and quadratic discriminant rules, updating formulas are available [75] that avoid the recomputation of the discriminant rule if one data point is deleted. Because the computation of the MCD estimator is much more complex and based on resampling, updating formulas can not be obtained exactly, but approximate methods can be used [62].

A faster, well-known alternative for estimating the classification error consists of splitting the observations randomly into (a) a training set that is then used to compose the discriminant rule and (b) a validation set used to estimate the misclassification error. As pointed out by Lachenbruch [76] and others, such an estimate is wasteful of data and does not evaluate the discriminant rule that will be used in practice. With larger data sets, however, there is less loss of efficiency when we use only part of the data set, and if the estimated classification error is acceptable, then the final discriminant rule can still be constructed from the whole data set. Because it can happen that this validation set also contains outlying observations that should not be taken into account, we estimate the misclassification probability MP_j of group j by the proportion of nonoutliers from the validation set that belong to group j and that are misclassified. An estimate of the overall misclassification probability (MP) is then given by the weighted mean of the misclassification probabilities of all the groups, with weights equal to the estimated membership probabilities, i.e.,

$$MP = \sum_{j=1}^{l} \hat{p}_j^R MP_j \qquad (6.36)$$

6.8.1.3 An Example

We obtained a data set containing the spectra of three different cultivars of the same fruit (cantaloupe, *Cucumis melo* L. *cantalupensis*) from Colin Greensill (Faculty of Engineering and Physical Systems, Central Queensland University, Rockhampton, Australia). The cultivars (named D, M, and HA) had sizes 490, 106, and 500, and all spectra were measured in 256 wavelengths. The data set thus contains 1096 observations and 256 variables.

First, we applied a robust principal component analysis (RAPCA, see Section 6.5.3) to reduce the dimension of the data space. From the scree plot (not shown) and based on the ratio of the ordered eigenvalues and the largest one ($\lambda_2/\lambda_1 = 0.045$, $\lambda_3/\lambda_1 = 0.018$, $\lambda_4/\lambda_1 = 0.006$, $\lambda_5/\lambda_1 < 0.0005$), it was decided to retain four principal components. We then randomly divided the data into a training set and a validation set, containing 60% and 40% of the observations, respectively. Because there was no prior knowledge of the covariance structure of the three groups, the quadratic discriminant rule RQDR was applied. The membership probabilities were estimated as the proportion of nonoutliers in each group of the training set, yielding and $\hat{p}_D^R = 54\%$, $\hat{p}_M^R = 10\%$ and $\hat{p}_{HA}^R = 36\%$. The robust misclassification probabilities MP_j were computed by only considering the "good" observations from the validation set. To the training set, the classical quadratic discriminant rule CQDR was also applied and evaluated using the same reduced validation set. The results are presented in Table 6.4. The misclassifications for the three groups are listed separately first. The fourth column MP shows the overall misclassification probability as defined in Equation 6.36. We see that the overall misclassification probability of CQDR is more than three times larger than the misclassification of RQDR. The most remarkable difference is obtained for the cultivar HA, which contains a large group of outlying observations.

This is clearly seen in the plot of the data projected onto the first two principal components. Figure 6.15a shows the training data. In this figure, the cultivar D is marked with crosses, cultivar M with circles, and cultivar HA with diamonds. We see that cultivar HA has a cluster of outliers that are far from the other observations. As it turns out, these outliers were caused by a change in the illumination system.

For illustrative purposes, we have also applied the linear discriminant rule (Equation 6.35) with a common covariance matrix Σ. In Figure 6.15a, we have superimposed the robust tolerance ellipses for each group. Figure 6.15b shows the same data with the corresponding classical tolerance ellipses. Note how strongly the classical

TABLE 6.4
Misclassification Probabilities for RQDR and CQDR Applied to the Fruit Data Set

RQDR				CQDR			
MP_D	MP_M	MP_{HA}	MP	MP_D	MP_M	MP_{HA}	MP
0.03	0.18	0.01	0.04	0.06	0.30	0.21	0.14

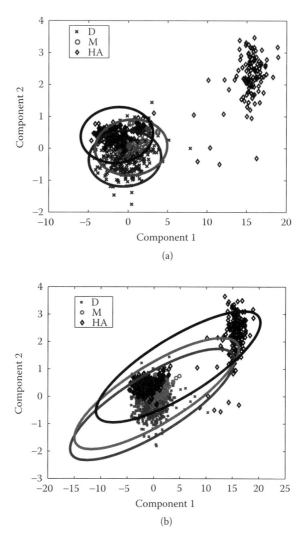

FIGURE 6.15 (a) Robust tolerance ellipses for the fruit data with common covariance matrix; (b) classical tolerance ellipses.

covariance estimator of the common Σ is influenced by the outlying subgroup of cultivar HA. The effect on the resulting classical linear discriminant rules is dramatic for cultivar M. It appears that all of the observations are misclassified because they would have to belong to a region that lies completely outside the boundary of this figure. The robust discriminant analysis does a better job. The tolerance ellipses are not affected by the outliers and the resulting discriminant lines split up the different groups more accurately. The misclassification rates are 17% for cultivar D, 95% for cultivar M, and 6% for cultivar HA, with an overall MP = 23%. The misclassification rate of cultivar M remains very high. This is due to the intrinsic overlap between

Robust Calibration

the three groups and the fact that cultivar M has few data points compared with the others. When we impose the constraint that all three groups are equally important by setting the membership probabilities $\hat{p}_j^R = 1/3$, we obtain a better classification of cultivar M, with 46% of errors. But now the other groups have a worse classification error (MP_D = 30% and MP_{HA} = 17%). The global MP equals 31%, which is higher than with the discriminant analysis based on unequal membership probabilities.

6.8.2 Classification in High Dimensions

When data are high dimensional, the approach of the previous section can no longer be applied because the MCD becomes uncomputable. In the previous example (Section 6.8.1.3), this was solved by applying a dimension-reduction procedure (PCA) on the whole set of observations. Instead, one can also apply a PCA method on each group separately. This is the idea behind the SIMCA method (soft independent modeling of class analogy) [77].

A robust variant of SIMCA can be obtained by applying a robust PCA method, such as ROBPCA (Section 6.5.4), on each group [78]. For example, the number of components in each group can be selected by cross validation, as explained in Section 6.5.6 and, hence, they need not be the same for each population. A classification rule is found by combining the orthogonal distance (Equation 6.17) of a new observation **x** to group π_j, denoted as $OD_j(\mathbf{x})$, and its score distance (Equation 6.18) within that group, yielding $SD_j(\mathbf{x})$. More precisely, let c_j^v be the cutoff value for the orthogonal distances when applying ROBPCA to the j group (see Section 6.5.5), and let c_j^h be the cutoff value for the score distances. Then we define the standardized orthogonal distance as $OD_j(\mathbf{x})/c_j^v$ and the standardized score distance as $SD_j(\mathbf{x})/c_j^h$. Finally, the jth group distance equals

$$GD_j(\mathbf{x}) = \sqrt{\left(\frac{OD_j(\mathbf{x})}{c_j^v}\right)^2 + \left(\frac{SD_j(\mathbf{x})}{c_j^h}\right)^2}$$

Observation **x** could then be allocated to π_a if $GD_a(\mathbf{x}) \leq GD_j(\mathbf{x})$ for all $j = 1, \ldots, l$. Alternative group distances have been considered as well [78].

6.9 SOFTWARE AVAILABILITY

MATLAB™ functions for all of the procedures mentioned in this chapter are part of LIBRA, Library for Robust Analysis [81], which can be downloaded from http://www.wis.kuleuven.be/stat/robust.html.

Stand-alone programs carrying out FAST-MCD and FAST-LTS can be downloaded from the Web site http://www.agoras.ua.ac.be/, as well as MATLAB versions. The MCD is available in the packages S-PLUS and R as the built-in function cov.mcd, and it has also been included in SAS Version 11 and SAS/IML Version 7. These packages all provide the one-step reweighed MCD estimates. The LTS is available in S-PLUS and R as the built-in function ltsreg and has also been incorporated in SAS Version 11 and SAS/IML Version 7.

REFERENCES

1. Hampel, F.R., The breakdown points of the mean combined with some rejection rules, *Technometrics*, 27, 95–107, 1985.
2. Montgomery, D.C., *Introduction to Statistical Quality Control*, John Wiley & Sons, New York, 1985.
3. Huber, P.J., *Robust Statistics*, John Wiley & Sons, New York, 1981.
4. Rousseeuw, P.J. and Croux, C., Alternatives to the median absolute deviation, *J. Am. Stat. Assoc.*, 88, 1273–1283, 1993.
5. Rousseeuw, P.J. and Verboven, S., Robust estimation in very small samples, *Comput. Stat. Data Anal.*, 40, 741–758, 2002.
6. Rousseeuw, P.J. and Leroy, A.M., *Robust Regression and Outlier Detection*, John Wiley & Sons, New York, 1987.
7. Rousseeuw, P.J., Least median of squares regression, *J. Am. Stat. Assoc.*, 79, 871–880, 1984.
8. Hampel, F.R., Ronchetti, E.M., Rousseeuw, P.J., and Stahel, W.A., *Robust Statistics: The Approach Based on Influence Functions*, John Wiley & Sons, New York, 1986.
9. Croux, C. and Haesbroeck, G., Influence function and efficiency of the minimum covariance determinant scatter matrix estimator, *J. Multivar. Anal.*, 71, 161–190, 1999.
10. Rousseeuw, P.J. and Van Driessen, K., A fast algorithm for the minimum covariance determinant estimator, *Technometrics*, 41, 212–223, 1999.
11. Stahel, W.A., Robuste Schätzungen: infinitesimale Optimalität und Schätzungen von Kovarianzmatrizen, Ph.D. thesis, Eidgenössische Technische Hochschule (ETH), Zürich, 1981.
12. Donoho, D.L., Breakdown Properties of Multivariate Location Estimators, Ph.D. thesis, Harvard University, Boston, 1982.
13. Tyler, D.E., Finite-sample breakdown points of projection-based multivariate location and scatter statistics, *Ann. Stat.*, 22, 1024–1044, 1994.
14. Maronna, R.A. and Yohai, V.J., The behavior of the Stahel-Donoho robust multivariate estimator, *J. Am. Stat. Assoc.*, 90, 330–341, 1995.
15. Maronna, R.A., Robust M-estimators of multivariate location and scatter, *Ann. Stat.*, 4, 51–67, 1976.
16. Rousseeuw, P.J., Multivariate estimation with high breakdown point, in *Mathematical Statistics and Applications*, Vol. B, Grossmann, W., Pflug, G., Vincze, I., and Wertz, W., Eds., Reidel Publishing, Dordrecht, The Netherlands, 1985, pp. 283–297.
17. Davies, L., An efficient fréchet differentiable high breakdown multivariate location and dispersion estimator, *J. Multivar. Anal.*, 40, 311–327, 1992.
18. Davies, L., Asymptotic behavior of S-estimators of multivariate location parameters and dispersion matrices, *Ann. Stat.*, 15, 1269–1292, 1987.
19. Kent, J.T. and Tyler, D.E., Constrained M-estimation for multivariate location and scatter, *Ann. Stat.*, 24, 1346–1370, 1996.
20. Lopuhaä, H.P., Multivariate τ-estimators for location and scatter, *Can. J. Stat.*, 19, 307–321, 1991.
21. Tatsuoka, K.S. and Tyler, D.E., On the uniqueness of S-functionals and M-functionals under nonelliptical distributions, *Ann. Stat.*, 28, 1219–1243, 2000.

22. Visuri, S., Koivunen, V., and Oja, H., Sign and rank covariance matrices, *J. Stat. Plan. Infer.*, 91, 557–575, 2000.
23. Donoho, D.L. and Gasko, M., Breakdown properties of location estimates based on halfspace depth and projected outlyingness, *Ann. Stat.*, 20, 1803–1827, 1992.
24. Liu, R.Y., Parelius, J.M., and Singh, K., Multivariate analysis by data depth: descriptive statistics, graphics and inference, *Ann. Stat.*, 27, 783–840, 1999.
25. Rousseeuw, P.J. and Struyf, A., Computing location depth and regression depth in higher dimensions, *Stat. Computing*, 8, 193–203, 1998.
26. Oja, H., Descriptive statistics for multivariate distributions, *Stat. Probab. Lett.*, 1, 327–332, 1983.
27. Huber, P.J., Projection pursuit, *Ann. Stat.*, 13, 435–475, 1985.
28. Rousseeuw, P.J. and Van Driessen, K., An algorithm for positive-breakdown methods based on concentration steps, in *Data Analysis: Scientific Modeling and Practical Application*, Gaul, W., Opitz, O., and Schader, M., Eds., Springer-Verlag, New York, 2000, pp. 335–346.
29. Rousseeuw, P.J. and van Zomeren, B.C., Unmasking multivariate outliers and leverage points, *J. Am. Stat. Assoc.*, 85, 633–651, 1990.
30. Huber, P.J., Robust regression: asymptotics, conjectures and Monte Carlo, *Ann. Stat.*, 1, 799–821, 1973.
31. Jurečková, J., Nonparametric estimate of regression coefficients, *Ann. Math. Stat.*, 42, 1328–1338, 1971.
32. Koenker, R. and Portnoy, S., L-estimation for linear models, *J. Am. Stat. Assoc.*, 82, 851–857, 1987.
33. Mizera, I. and Müller, C.H., Breakdown points and variation exponents of robust M-estimators in linear models, *Ann. Stat.*, 27, 1164–1177, 1999.
34. Marazzi, A., *Algorithms, Routines and S Functions for Robust Statistics*, Wadsworth and Brooks, Belmont, CA., 1993.
35. Simpson, D.G., Ruppert, D., and Carroll, R.J., On one-step GM-estimates and stability of inferences in linear regression, *J. Am. Stat. Assoc.*, 87, 439–450, 1992.
36. Rousseeuw, P.J. and Yohai, V.J., Robust regression by means of S-estimators, in *Robust and Nonlinear Time Series Analysis, Lecture Notes in Statistics No. 26*, Franke, J., Härdle, W., and Martin, R.D., Eds., Springer-Verlag, New York, 1984, pp. 256–272.
37. Yohai, V.J., High breakdown point and high efficiency robust estimates for regression, *Ann. Stat.*, 15, 642–656, 1987.
38. Mendes, B. and Tyler, D.E., Constrained M estimates for regression, in *Robust Statistics; Data Analysis and Computer Intensive Methods, Lecture Notes in Statistics No. 109*, Rieder, H., Ed., Springer-Verlag, New York, 1996, pp. 299–320.
39. Rousseeuw, P.J. and Hubert, M., Regression depth, *J. Am. Stat. Assoc.*, 94, 388–402, 1999.
40. Van Aelst, S. and Rousseeuw, P.J., Robustness of deepest regression, *J. Multivar. Anal.*, 73, 82–106, 2000.
41. Van Aelst, S., Rousseeuw, P.J., Hubert, M., and Struyf, A., The deepest regression method, *J. Multivar. Anal.*, 81, 138–166, 2002.
42. Rousseeuw, P.J., Van Aelst, S., Rambali, B., and Smeyers-Verbeke, J., Deepest regression in analytical chemistry, *Anal. Chim. Acta*, 446, 243–254, 2001.
43. Rousseeuw, P.J., Van Aelst, S., Van Driessen, K., and Agulló, J., Robust multivariate regression, *Technometrics*, 46, 293–305, 2004.
44. Hubert, M. and Engelen, S., Robust PCA and classification in biosciences, *Bioinformatics*, 2004, in *Bioinformatics*, 20, 1728–1736, 2004.

45. Croux, C. and Haesbroeck, G., Principal components analysis based on robust estimators of the covariance or correlation matrix: influence functions and efficiencies, *Biometrika*, 87, 603–618, 2000.
46. Hubert, M., Rousseeuw, P.J., and Verboven, S., A fast robust method for principal components with applications to chemometrics, *Chemom. Intell. Lab. Syst.*, 60, 101–111, 2002.
47. Li, G. and Chen, Z., Projection-pursuit approach to robust dispersion matrices and principal components: primary theory and Monte Carlo, *J. Am. Stat. Assoc.*, 80, 759–766, 1985.
48. Croux, C. and Ruiz-Gazen, A., A fast algorithm for robust principal components based on projection pursuit, in *COMPSTAT 1996* (Barcelona), Physica, Heidelberg, 1996, pp. 211–217.
49. Wu, W., Massart, D.L., and de Jong, S., Kernel-PCA algorithms for wide data, part II: fast cross-validation and application in classification of NIR data, *Chemom. Intell. Lab. Syst.*, 36, 165–172, 1997.
50. Croux, C. and Ruiz-Gazen, A., High breakdown estimators for principal components: the projection-pursuit approach revisited, *J. Multivariate Anal.*, 95, 206–226, 2005.
51. Cui, H., He, X., and Ng, K.W., Asymptotic distributions of principal components based on robust dispersions, *Biometrika*, 90, 953–966, 2003.
52. Hubert, M., Rousseeuw, P.J., and Vanden Branden, K., ROBPCA: a new approach to robust principal components analysis, *Technometrics*, 2004, 47, 64–79, 2005.
53. Maronna, R.A., Principal components and orthogonal regression based on robust scales, 2003, *Technometrics*, 47, 264–273, 2005.
54. Maronna, R. and Zamar, R.H., Robust multivariate estimates for high dimensional data sets, *Technometrics*, 44, 307–317, 2002.
55. Box, G.E.P., Some theorems on quadratic forms applied in the study of analysis of variance problems: effect of inequality of variance in one-way classification, *Ann. Math. Stat.*, 25, 33–51, 1954.
56. Joliffe, I.T., *Principal Component Analysis*, 2nd ed., Springer-Verlag, New York, 2002.
57. Wold, S., Cross-validatory estimation of the number of components in factor and principal components models, *Technometrics*, 20, 397–405, 1978.
58. Eastment, H.T. and Krzanowski, W.J., Cross-validatory choice of the number of components from a principal components analysis, *Technometrics*, 24, 73–77, 1982.
59. Lemberge, P., De Raedt, I., Janssens, K.H., Wei, F., and Van Espen, P.J., Quantitative Z-analysis of 16th–17th century archaeological glass vessels using PLS regression of EPXMA and μ-XRF data, *J. Chemom.*, 14, 751–763, 2000.
60. Martens, H. and Naes, T., *Multivariate Calibration*, John Wiley & Sons, New York, 1998.
61. Hubert, M. and Verboven, S., A robust PCR method for high-dimensional regressors, *J. Chemom.*, 17, 438–452, 2003.
62. Engelen, S. and Hubert, M., Fast cross-validation for robust PCA, *Proc. COMPSTAT 2004*, J. Antoch, Ed., Springer-Verlag, Heidelberg, 989–996, 2004.
63. Osborne, B.G., Fearn, T., Miller, A.R., and Douglas, S., Application of near infrared reflectance spectroscopy to the compositional analysis of biscuits and biscuit dough, *J. Sci. Food Agric.*, 35, 99–105, 1984.
64. de Jong, S., SIMPLS: an alternative approach to partial least squares regression, *Chemom. Intell. Lab. Syst.*, 18, 251–263, 1993.
65. Hubert, M. and Vanden Branden, K., Robust methods for partial least squares regression, *J. Chemom.*, 17, 537–549, 2003.

66. Vanden Branden, K. and Hubert, M., Robustness properties of a robust PLS regression method, *Anal. Chim. Acta*, 515, 229–241, 2004.
67. Esbensen, K.H., Schönkopf, S., and Midtgaard, T., *Multivariate Analysis in Practice*, Camo, Trondheim, Norway, 1994.
68. Engelen, S., Hubert, M., Vanden Branden, K., and Verboven, S., Robust PCR and robust PLS: a comparative study, in *Theory and Applications of Recent Robust Methods: Statistics for Industry and Technology*, Hubert, M., Pison, G., Struyf, G., and Van Aelst, S., Eds., Birkhäuser, Basel, 2004.
69. Johnson, R.A. and Wichern, D.W., *Applied Multivariate Statistical Analysis*, Prentice Hall, Upper Saddle River, NJ, 1998.
70. Hubert, M. and Van Driessen, K., Fast and robust discriminant analysis, *Comput. Stat. Data Anal.*, 45, 301–320, 2004.
71. Hawkins, D.M. and McLachlan, G.J., High-breakdown linear discriminant analysis, *J. Am. Stat. Assoc.*, 92, 136–143, 1997.
72. He, X. and Fung, W.K., High breakdown estimation for multiple populations with applications to discriminant analysis, *J. Multivar. Anal.*, 720, 151–162, 2000.
73. Croux, C. and Dehon, C., Robust linear discriminant analysis using S-estimators, *Can. J. Stat.*, 29, 473–492, 2001.
74. Lachenbruch, P.A. and Mickey, M.R., Estimation of error rates in discriminant analysis, *Technometrics*, 10, 1–11, 1968.
75. McLachlan, G.J., *Discriminant Analysis and Statistical Pattern Recognition*, John Wiley & Sons, New York, 1992.
76. Lachenbruch, P.A., *Discriminant Analysis*, Hafner Press, New York, 1975.
77. Wold, S., Pattern recognition by means of disjoint principal components models, *Patt. Recog.*, 8, 127–139, 1976.
78. Vanden Branden, K. and Hubert, M., Robust classification in high dimensions based on the SIMCA method, *Chemometrics and Intelligent Lab, Syst.*, 79, 10–21, 2005.
79. Rousseeuw, P.J. and Hubert, M., Recent developments in PROGRESS in L_1-*Statistical Procedures and Related Topics*, Y. Dodge, Ed., Institute of Mathematical Statistics Lecture Notes-Monograph Series, Volume 31, Hayward, California, 201–214, 1997.
80. Engelen, S. and Hubert, M., Fast model selection for robust calibration methods, *Analytica Chimica Acta*, 544, 219–228, 2005.
81. Verboven, S. and Hubert, M., LIBRA: A MATLAB Library for Robust Analysis, *Chemometrics and Intelligent Lab. Syst.*, 75, 127–136, 2005.
82. Hubert, M., et al., *Handbook of Statistics*, Ch. 10, 2004, Elsevier.

7 Kinetic Modeling of Multivariate Measurements with Nonlinear Regression

Marcel Maeder and Yorck-Michael Neuhold

CONTENTS

7.1	Introduction	218
7.2	Multivariate Data, Beer-Lambert's Law, Matrix Notation	219
7.3	Calculation of the Concentration Profiles: Case I, Simple Mechanisms	220
7.4	Model-Based Nonlinear Fitting	222
	7.4.1 Direct Methods, Simplex	225
	7.4.2 Nonlinear Fitting Using Excel's Solver	227
	7.4.3 Linear and Nonlinear Parameters	228
	7.4.4 Newton-Gauss-Levenberg/Marquardt (NGL/M)	230
	7.4.5 Nonwhite Noise	237
7.5	Calculation of the Concentration Profiles: Case II, Complex Mechanisms	241
	7.5.1 Fourth-Order Runge-Kutta Method in Excel	242
	7.5.2 Interesting Kinetic Examples	246
	7.5.2.1 Autocatalysis	246
	7.5.2.2 Zeroth-Order Reaction	248
	7.5.2.3 Lotka-Volterra (Sheep and Wolves)	250
	7.5.2.4 The Belousov-Zhabotinsky (BZ) Reaction	251
7.6	Calculation of the Concentration Profiles: Case III, Very Complex Mechanisms	253
7.7	Related Issues	255
	7.7.1 Measurement Techniques	256
	7.7.2 Model Parser	256
	7.7.3 Flow Reactors	256
	7.7.4 Globalization of the Analysis	256

	7.7.5	Soft-Modeling Methods	257
	7.7.6	Other Methods	258
Appendix			258
References			259

7.1 INTRODUCTION

The most prominent technique for investigating the kinetics of chemical processes in solution is light-absorption spectroscopy. This includes IR (infrared), NIR (near-infrared), CD, (circular dichroism) and above all, UV/Vis (ultraviolet/visible) spectroscopy. Absorption spectroscopy is used for slow reactions, where solutions are mixed manually and the measurements are started after introduction of the solution into a cuvette in the instrument. For fast reactions, stopped-flow or temperature-jump instruments are used, and even for very fast reactions (pulse radiolysis, flash photolysis), light absorption is the most useful technique. For these reasons, we develop the methodology presented in this chapter specifically for the analysis of absorption measurements. Many aspects of these methods apply straightforwardly to other techniques. For instance, a series of NMR (nuclear magnetic resonance) spectra can be analyzed in essentially identical ways as long as there are no fast equilibria such as protonation equilibria involved. Similarly, data from emission spectroscopy can be employed. Also, in cases where individual concentrations of some or all reacting components are observed directly (e.g., chromatography), the methods are virtually identical. Generally, the methods can be applied to all measured data as long as the signals are linearly dependent on individual concentrations [1–4].

This chapter deals with multivariate data sets. In the present context, this means that complete spectra are observed as a function of reaction time, e.g., with a diode-array detector. As we will demonstrate, the more commonly performed single-wavelength measurements can be regarded as a special case of multiwavelength measurements.

The chapter begins with a short introduction to the appropriate mathematical handling of multiwavelength absorption data sets. We demonstrate how matrix notation can be used very efficiently to describe the data sets acquired in such investigations. Subsequently, we discuss in detail the two core aspects of the model-based fitting of kinetic data:

1. *Modeling the concentration profiles of the reacting components*. We first discuss simple reaction mechanisms. By this we mean mechanisms for which there are analytical solutions for the sets of differential equations. Later we turn our attention to the modeling of reaction mechanisms of virtually any complexity. In the last section, we look at extensions to the basic modeling methods in an effort to analyze measurements that were recorded under nonideal conditions, such as at varying temperature or pH.
2. *Methods for nonlinear least-squares fitting*, with a demonstration of how these can be applied to the analysis of kinetic data.

We illustrate the theoretical concepts in a few selected computer programs and then apply them to realistic examples. MATLAB™ [5] is the programming language of choice for most chemometricians. The MATLAB code provided in the examples is intended to encourage and guide readers to write their own programs for their specific tasks. Excel is much more readily available than MATLAB, and many quite sophisticated analyses can be performed in Excel. A few examples demonstrate how Excel can be used to tackle problems that are beyond the everyday tasks performed by most scientists. For methods that are clearly beyond the capabilities of Excel, it is possible to write Visual Basic programs of any complexity and link them to a spreadsheet. As an example, routines for the singular-value decomposition (SVD) are readily available on the Internet [6].

In this chapter, we describe the methods required for the model-based analysis of multivariate measurements of chemical reactions. This comprises reactions of essentially any complexity in solution, but it does not include the investigation of gas-phase reactions, for example in flames or in the atmosphere, which involve hundreds or even thousands of steps [7–12].

7.2 MULTIVARIATE DATA, BEER-LAMBERT'S LAW, MATRIX NOTATION

To maximize the readability of mathematical texts, it is helpful to differentiate matrices, vectors, scalars, and indices by typographic conventions. In this chapter, matrices are denoted in boldface capital characters (**M**), vectors in boldface lowercase (**v**), and scalars in lowercase italic characters (s). For indices, lowercase characters are used (j). The symbol "t" indicates matrix and vector transposition (**M**^{t}). Chemical species are given in uppercase italic characters (A).

Beer-Lambert's law states that the total absorption, y_λ, of a solution at one particular wavelength, λ, is the sum over all contributions of dissolved absorbing species, A, B, ..., Z, with molar absorptivities $\varepsilon_{A,\lambda}$, $\varepsilon_{B,\lambda}$, ..., $\varepsilon_{Z,\lambda}$.

$$y_\lambda = [A]\varepsilon_{A,\lambda} + [B]\varepsilon_{B,\lambda} + \ldots + [Z]\varepsilon_{Z,\lambda} \tag{7.1}$$

If complete spectra are measured as a function of time, Equation 7.1 can be written for each spectrum at each wavelength. Such a large collection of equations is very unwieldy, and it is crucial to recognize that the structure of such a system of equations allows the application of the very elegant matrix notation shown in Equation 7.2.

$$nt \begin{array}{c} n\lambda \\ \boxed{\mathbf{Y}} \end{array} = \begin{array}{c} nc \\ \boxed{\mathbf{C}} \end{array} \times \begin{array}{c} n\lambda \\ \boxed{\mathbf{A}} \end{array} nc + nt \begin{array}{c} n\lambda \\ \boxed{\mathbf{R}} \end{array} \tag{7.2}$$

Y is a matrix that consists of all the individual measurements. The absorption spectra, measured at $n\lambda$ wavelengths, form $n\lambda$-dimensional vectors, which are arranged as rows of **Y**. Thus, if nt spectra are measured at nt reaction times, **Y** contains nt rows of $n\lambda$ elements; it is an $nt \times n\lambda$ matrix. As the structures of Beer-Lambert's law and the mathematical law for matrix multiplication are essentially identical, this matrix **Y** can be written as a product of two matrices **C** and **A**, where **C** contains as columns the concentration profiles of the absorbing species. If there are nc absorbing species, **C** has nc columns, each one containing nt elements, the concentrations of the species at the nt reaction times. Similarly, the matrix **A** contains, in nc rows, the molar absorptivities of the absorbing species, measured at $n\lambda$ wavelengths; these are the $\varepsilon_{x,\lambda}$ values of Equation 7.1.

Due to imperfections in any real measurements, the product $\mathbf{C} \times \mathbf{A}$ does not exactly result in $\mathbf{Y}$. The difference is a matrix $\mathbf{R}$ of residuals. Note that $\mathbf{C} \times \mathbf{A}$ and $\mathbf{R}$ have the same dimensions as $\mathbf{Y}$. The task of the analysis is to find the best matrices $\mathbf{C}$ and $\mathbf{A}$ for a given measured $\mathbf{Y}$. We start with the calculation of the matrix $\mathbf{C}$ for simple reaction mechanisms. The computation of $\mathbf{C}$ is the core of any fitting program. We will return to the computation of $\mathbf{C}$ for complex mechanisms toward the end of this chapter.

7.3 CALCULATION OF THE CONCENTRATION PROFILES: CASE I, SIMPLE MECHANISMS

There is a limited number of reaction mechanisms for which there are explicit formulae to calculate the concentrations of the reacting species as a function of time. This set includes all reaction mechanisms that contain only first-order reactions, as well as a very few mechanisms with second-order reactions [1, 3, 13]. A few examples for such mechanisms are given in Equation 7.3.

$$
\begin{aligned}
a) &\quad A \xrightarrow{k} B \\
b) &\quad 2A \xrightarrow{k} B \\
c) &\quad A + B \xrightarrow{k} C \\
d) &\quad A \xrightarrow{k_1} B \xrightarrow{k_2} C
\end{aligned}
\tag{7.3}
$$

Any chemical reaction mechanism is described by a set of ordinary differential equations (ODEs). For the reactions in Equation 7.3, the ODEs are

$$
\begin{aligned}
a) &\quad [\dot{A}] = -[\dot{B}] = -k[A] \\
b) &\quad [\dot{A}] = -2[\dot{B}] = -2k[A]^2 \\
c) &\quad [\dot{A}] = [\dot{B}] = -[\dot{C}] = -k[A][B] \\
d) &\quad [\dot{A}] = -k_1[A], \\
&\quad [\dot{B}] = k_1[A] - k_2[B], \\
&\quad [\dot{C}] = k_2[\dot{B}]
\end{aligned}
\tag{7.4}
$$

where we use the $[\dot{A}]$ notation for the derivative of $[A]$ with respect to time,

$$[\dot{A}] = \frac{d[A]}{dt}$$

Integration of the ODEs results in the concentration profiles for all reacting species as a function of the reaction time. The explicit solutions for the examples shown here are given in Equation 7.5, which lists the equations for each example. Note that in examples (a) to (c), the integrated form of the equation is only given for A. The equations for the concentration(s) of the remaining species can be calculated from the mass balance or closure principle (e.g., in the first example $[B] = [A]_0 - [A]$, where $[A]_0$ is the concentration of A at time zero). In example (d), the integrated form is given for species A and B. Again, the concentration of species C can be determined from the mass balance principle.

a) $\quad [A] = [A]_0 e^{-kt}$

b) $\quad [A] = \dfrac{[A]_0}{1 + 2[A]_0 kt}$

c) $\quad [A] = \dfrac{[A]_0([B]_0 - [A]_0)}{[B]_0 e^{([B]_0 - [A]_0)kt} - [A]_0} \quad ([A]_0 \neq [B]_0) \quad\quad (7.5)$

d) $\quad [A] = [A]_0 e^{-kt},$

$\quad\quad [B] = [A]_0 \dfrac{k_1}{k_2 - k_1}(e^{-k_1 t} - e^{-k_2 t}) \quad ([B]_0 = 0, k_1 \neq k_2)$

Modeling and visualization of a reaction $A \xrightarrow{k} B$ requires only a few lines of MATLAB code (see MATLAB Example 7.1), including a plot of the concentration profiles, as seen in Figure 7.1. Of course this task can equally well be performed in Excel.

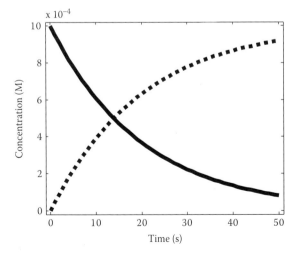

FIGURE 7.1 Concentration profiles for a reaction $A \xrightarrow{k} B$ (— A, ⋯ B) as calculated by MATLAB Example 7.1.

MATLAB Example 7.1

```
% A -> B

t=[0:50]';              % time vector (column vector)
A_0=1e-3;               % initial concentration of A
k=.05;                  % rate constant
C(:,1)=A_0*exp(-k*t);   % [A]
C(:,2)=A_0-C(:,1);      % [B] (Closure)
plot(t,C);              % plotting C vs t
```

Solutions for the integration of ODEs such as those given in Equation 7.5 are not always readily available. For nonspecialists, it is difficult to determine whether there is an explicit solution at all. MATLAB's symbolic toolbox provides a very convenient means of producing the results and also of testing for explicit solutions of ordinary differential equations, e.g., for the reaction $2A \xrightarrow{k} B$, as seen in MATLAB Example 7.2. (Note that MATLAB's symbolic toolbox demands lowercase characters for species names.)

MATLAB Example 7.2

```
% 2A -> B, explicit solution

d=dsolve('Da=-2*k1*a^2','Db=k1*a^2','a(0)=a_0',' b(0)=0');
pretty(simplify(d.a))

                                        a_0
                                   ---------------
                                   2 k1 t a_0 + 1
```

In a section 7.5, we will demonstrate how to deal with more complex mechanisms for which the ODEs cannot be integrated analytically.

7.4 MODEL-BASED NONLINEAR FITTING

Model-based fitting of measured data can be a rather complex process, particularly if there are many parameters to be fitted to many data points. Multivariate measurements can produce very large data matrices, especially if spectra are acquired at many wavelengths. Such data sets may require many parameters for a quantitative description. It is crucial to deal with such large numbers of parameters in efficient ways, and we will describe how this can be done. Large quantities of data are no longer a problem on modern computers, since inexpensive computer memory is easily accessible.

As mentioned previously, the task of model-based data fitting for a given matrix $\mathbf{Y}$ is to determine the best rate constants defining the matrix $\mathbf{C}$, as well as the best molar absorptivities collected in the matrix $\mathbf{A}$. The quality of the fit is represented by the matrix of residuals, $\mathbf{R} = \mathbf{Y} - \mathbf{C} \times \mathbf{A}$. Assuming white noise, i.e., normally distributed noise of constant standard deviation, the sum of the squares, ssq, of all elements $r_{i,j}$ is statistically the "best" measure to be minimized. This is generally called a least-squares fit.

$$ssq = \sum_{i=1}^{nt} \sum_{j=1}^{n\lambda} r_{i,j}^2 \qquad (7.6)$$

(An adaptation using weighted least squares is discussed in a later section for the analysis of data sets with nonwhite noise.) The least-squares fit is obtained by minimizing the sum of squares, *ssq*, as a function of the measurement, **Y**, the chemical model (rate law), and the parameters, i.e., the rate constants, **k**, and the molar absorptivities, **A**.

$$ssq = f(\mathbf{Y}, \text{model}, \text{parameters}) \tag{7.7}$$

It is important to stress here that for the present discussion we do not vary the model; rather, we determine the best parameters for a given model. The determination of the correct model is a task that is significantly more difficult. One possible approach is to fit the complete set of possible models and select the best one defined by statistical criteria and chemical intuition. Because there is usually no obvious limit to the number of potential models, this task is rather daunting. As described in Chapter 11, Multivariate Curve Resolution, model-free analyses can be a very powerful tool to support the process of finding the correct model.

We confidently stated at the very beginning of this chapter that we would deal with multivariate data. The high dimensionality makes graphical representation difficult or impossible, as our minds are restricted to visualization of data in three dimensions. For this reason, we initiate the discussion with monovariate examples, i.e., kinetics measured at only one wavelength. As we will see, the appropriate generalization to many wavelengths is straightforward.

In order to gain a good understanding of the different aspects of the task of parameter fitting, we will start with a simple but illustrative example. We will use the first-order reaction $A \xrightarrow{k} B$, as shown in MATLAB Example 7.1 and also in Figure 7.1. The kinetics is followed at a single wavelength, as shown in Figure 7.2. The measurement is rather noisy. The magnitude of noise is not relevant, but it is easier to graphically discern the difference between original and fitted data.

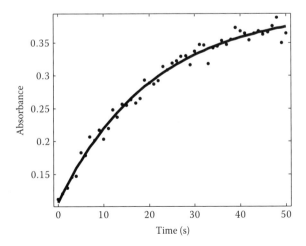

FIGURE 7.2 First-order ($A \xrightarrow{k} B$) kinetic single-wavelength experiment (...) and the result of a least-squares fit (—).

In Appendix 7.1 at the end of this chapter, a MATLAB function (data_ab) is given that generates this absorbance data set.

Because this is a single-wavelength experiment, the matrices **A**, **Y**, and **R** collapse into column vectors **a**, **y**, and **r**, and Equation 7.2 is reduced to Equation 7.8.

$$\underbrace{\begin{vmatrix} \\ nt \\ \\ \end{vmatrix}}_{\mathbf{y}} = \underbrace{\begin{vmatrix} \\ \mathbf{C} \\ \\ \end{vmatrix}}^{nc} \times \underbrace{\begin{vmatrix} nc \\ \mathbf{a} \\ \end{vmatrix}} + \underbrace{\begin{vmatrix} \\ nt \\ \\ \end{vmatrix}}_{\mathbf{r}} \tag{7.8}$$

For this example there are three parameters only, the rate constant k, which defines the matrix **C** of the concentration profiles, and the molar absorptivities $\varepsilon_{A,\lambda}$ and $\varepsilon_{B,\lambda}$ for the components A and B, which form the two elements of the vector **a**.

First, we assume that the molar absorptivities of A and B at the appropriate wavelength λ have been determined independently and are known ($\varepsilon_{A,\lambda} = 100$ M^{-1}cm^{-1}, $\varepsilon_{B,\lambda} = 400$ M^{-1}cm^{-1}); then, the only parameter to be optimized is k. In accordance with Equation 7.6 and Equation 7.8, for any value of k we can calculate a matrix **C** — and subsequently the quality of the fit via the sum of squares, ssq — by multiplying the matrix **C** with the known vector **a**, subtracting the result from **y**, and adding up the squared elements of the vector of differences (residuals), **r**. Figure 7.3 shows a plot of the logarithm of ssq vs. k. The optimal value for the rate constant that minimizes ssq is obviously around $k = 0.05$ s^{-1}.

In a second, more realistic thought experiment, we assume to know the molar absorptivity $\varepsilon_{A,\lambda}$ of species A only, and thus have to fit $\varepsilon_{B,\lambda}$ and k. The equivalent ssq analysis as above leads to a surface in a three-dimensional space when we plot ssq vs. k and $\varepsilon_{B,\lambda}$. This is illustrated in Figure 7.4. Again, the task is to find the minimum of the function defining ssq, or in other words, the bottom of the valley (at $k \cong 0.05$ s^{-1}

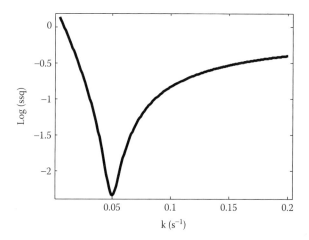

FIGURE 7.3 Logarithm of the square sum ssq of the residual vector **r** as a function of the rate constant k.

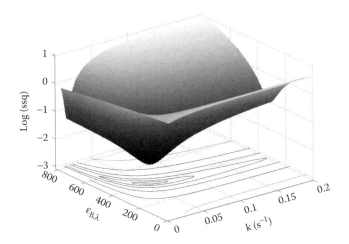

FIGURE 7.4 Square sum *ssq* of the residuals **r** as a function of two parameters k and $\varepsilon_{B,\lambda}$.

and $\varepsilon_{B,\lambda} \cong 400$ M^{-1}cm^{-1}). In the first example, there was only one parameter (k) to be optimized; in the second, there are two (k and $\varepsilon_{A,\lambda}$).

Even more realistically, all three parameters k, $\varepsilon_{A,\lambda}$, and $\varepsilon_{B,\lambda}$ are unknown (e.g., a solution of pure *A* cannot be made, as it immediately starts reacting to form *B*). It is impossible to represent graphically the relationship between *ssq* and the three parameters; it is a hypersurface in a four-dimensional space and beyond our imagination. Nevertheless, as we will see soon, there is a minimum for one particular set of parameters.

It is probably clear by now that highly multivariate measurements need special attention, as there are many parameters that need to be fitted, i.e., the rate constants and all molar absorptivities at all wavelengths. We will come back to this apparently daunting task.

There are many different methods for the task of fitting any number of parameters to a given measurement [14–16]. We can put them into two groups: (a) the direct methods, where the sum of squares is optimized directly, e.g., finding the minimum, similar to the example in Figure 7.4, and (b) the Newton-Gauss methods, where the residuals in **r** or **R** themselves are used to guide the iterative process toward the minimum.

7.4.1 DIRECT METHODS, SIMPLEX

Graphs of the kind shown in Figure 7.3 and Figure 7.4 are simple to produce and the subsequent "manual" location of the optimum is straightforward. However, it requires a great deal of computation time and, more importantly, the direct input of an operator. Additionally, such a method is restricted to only one or two parameters.

Very useful and, thus, heavily used is the simplex algorithm, which is conceptually a very simple method. It is reasonably fast for a modest number of parameters; further, it is very robust and reliable. However, for high-dimensional tasks, i.e., with many parameters, the simplex algorithm becomes extremely slow.

A simplex is a multidimensional geometrical object with $n + 1$ vertices in an n-dimensional space. In two dimensions (two parameters), the simplex is a triangle, in three dimensions (three parameters) it becomes a tetrahedron, etc. At first, the functional values (ssq) at all corners of the simplex have to be determined. Assuming we are searching for the minimum of a function, the highest value of the corners has to be determined. Next, this worst one is discarded and a new simplex is constructed by reflecting the old simplex at the face opposite the worst corner. Importantly, only one new value has to be determined for the new simplex. The new simplex is treated in the same way: the worst vertex is determined and the simplex reflected until there is no more significant change in the functional value.

The process is represented in Figure 7.5. In the initial simplex, the worst value is 14, and the simplex has to be reflected at the opposite face (8,9,11), marked in gray. A new functional value of 7 is determined in the new simplex. The next move would be the reflection at the face (8,9,7), reflecting the corner with value 11. Advanced simplex algorithms include constant adaptation of the size of the simplex [17]. Overly large simplices will not follow the fine structure of the surface and will only result in approximate minima; simplices that are too small will move very slowly. In the example here, we are searching for the minimum, but the process is obviously easily adapted for maximization.

The simplex algorithm works well for a reasonably low number of parameters. Naturally, it is not possible to give a precise and useful maximal number; 10 could be a reasonable estimate. Multivariate data with hundreds of unknown molar absorptivities cannot be fitted without further substantial improvement of the algorithm.

In MATLAB Example 7.3a and 7.3b we give the code for a simplex optimization of the first-order kinetic example discussed above. Refer to the MATLAB manuals for details on the simplex function `fminsearch`. Note that all three parameters k, $\varepsilon_{A,\lambda}$, and $\varepsilon_{B,\lambda}$ are fitted. The minimal ssq is reached at $k = 0.048$ s^{-1}, $\varepsilon_{A,\lambda} = 106.9$ M^{-1}cm^{-1}, and $\varepsilon_{B,\lambda} = 400.6$ M^{-1}cm^{-1}.

MATLAB Example 7.3b employs the function that calculates ssq (and also C). It is repeatedly used by the simplex routine called in MATLAB Example 7.3a. In Figure 7.2 we have already seen a plot of the experimental data together with their fit.

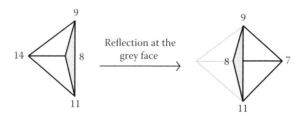

FIGURE 7.5 Principle of the simplex minimization with three parameters.

Kinetic Modeling of Multivariate Measurements with Nonlinear Regression 227

MATLAB Example 7.3a

```
% simplex fitting of k, eps_A and eps_B to the kinetic model A -> B

[t,y]=data_ab;                    % get absorbance data
A_0=1e-3;                         % initial concentration of A
par0=[0.1;200;600];               % start parameter vector
                                  % [k0;eps_A0;eps_B0]
par=fminsearch('rcalc_ab1',par0,[],A_0,t,y)    % simplex call
[ssq,C]=rcalc_ab1(par,A_0,t,y);   % calculate ssq and C with final parameters
y_calc=C*par(2:3);                % determine y_calc from C, eps_A and eps_B
plot(t,y,'.',t,y_calc,'-');       % plot y and y_calc vs t
```

MATLAB Example 7.3b

```
function [ssq,C]=rcalc_ab1(par,A_0,t,y)

C(:,1)=A_0*exp(-par(1)*t);        % concentrations of species A
C(:,2)=A_0-C(:,1);                % concentrations of B
r=y-C*par(2:3);                   % residuals
ssq=sum(r.*r);                    % sum of squares
```

7.4.2 NONLINEAR FITTING USING EXCEL'S SOLVER

Fitting tasks of a modest complexity, like the one just discussed, can straightforwardly be performed in Excel using the Solver tool provided as an Add-In method. The Solver tool does not seem to be very well known, even in the scientific community, and therefore we will briefly discuss its application based on the example above. As with MATLAB, we assume familiarity with the basics of Excel.

Figure 7.6 displays the essential parts of the spreadsheet. The columns A and B (from row 10 downward) contain the given measurements, the vectors **t** and **y**, respectively. Columns C and D contain the concentration profiles [A] and [B], respectively. The equations used to calculate these values in the Excel language are indicated. The rate constant is defined in cell B2, and the molar absorptivities in the cells B3:B4. Next, a vector y_{calc} is calculated in column E. Similarly, the residuals and their squares are given in the next two columns. Finally, the sum over all these squares, *ssq*, is given in cell B6. The task is to modify the parameters, the contents of the cells B2:B4, until *ssq* is minimal. It is a good exercise to try to do this manually. Excel provides the Solver for this task. The operator has to (a) define the Target Cell, in this case, cell B6 containing *ssq*; (b) make sure the Minimize button is chosen; and (c) define the Changing Cells, in this case, the cells containing the variable parameters, B2:B4. Click Solve and in no time the result is found. As with any iterative fitting algorithm, it is important that the initial guesses for the parameters be reasonable, otherwise the minimum might not be found. These initial guesses are entered into the cells B2:B4, and they are subsequently refined by the Solver to yield the result shown in Figure 7.6. For further information on Excel's Solver, we refer the reader to some relevant publications on this topic [18–21].

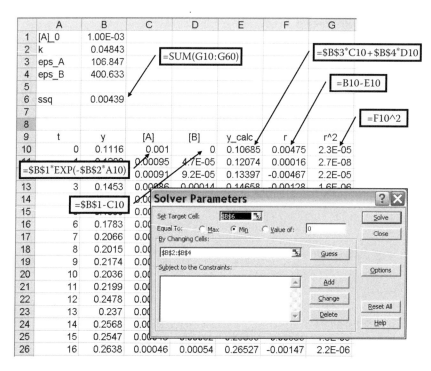

FIGURE 7.6 Using Excel's Solver for nonlinear fitting of a first-order reaction $A \xrightarrow{k} B$.

7.4.3 LINEAR AND NONLINEAR PARAMETERS

As stated in the introduction (Section 7.1), this chapter is about the analysis of multivariate data in kinetics, i.e., measurements at many wavelengths. Compared with univariate data this has two important consequences: (a) there is much more data to be analyzed and (b) there are many more parameters to be fitted.

Consider a reaction scheme with nk reactions (rate constants), involving nc absorbing components. Measurements are done using a diode-array spectrophotometer where nt spectra are taken at $n\lambda$ wavelengths. Thus, we are dealing with $nt \times n\lambda$ individual absorption measurements. The number of parameters to be fitted is $nk + nc \times n\lambda$ (the number of rate constants plus the number of molar absorptivities). Let us look at an example for the reaction scheme $A \rightarrow B \rightarrow C$, with 100 spectra measured at 1024 wavelengths. The number of data points is 1.024×10^5 and, more importantly, the number of parameters is 3074 ($2 + 3 \times 1024$). There is no doubt that "something" needs to be done to reduce this large number, as no fitting method can efficiently deal with that many parameters.

There are two fundamentally different kinds of parameters: a small number of rate constants, which are nonlinear parameters, and the large number of molar absorptivities, which are linear parameters. Fortunately, we can exploit this situation of having to deal with two different sets of parameters.

Kinetic Modeling of Multivariate Measurements with Nonlinear Regression

The rate constants (together with the model and initial concentrations) define the matrix **C** of concentration profiles. Earlier, we have shown how **C** can be computed for simple reactions schemes. For any particular matrix **C** we can calculate the best set of molar absorptivities **A**. Note that, during the fitting, this will not be the correct, final version of **A**, as it is only based on an intermediate matrix **C**, which itself is based on an intermediate set of rate constants (**k**). Note also that the calculation of **A** is a linear least-squares estimate; its calculation is explicit, i.e., noniterative.

$$\mathbf{A} = \mathbf{C}^+ \mathbf{Y} \quad \text{or}$$
$$\mathbf{A} = (\mathbf{C}^t \mathbf{C})^{-1} \mathbf{C}^t \mathbf{Y} \quad \text{or} \quad (7.9)$$
$$\mathbf{A} = \mathbf{C} \backslash \mathbf{Y} \quad \text{(MATLAB notation)}$$

$\mathbf{C}^+$ is the so-called pseudoinverse of **C**. It can be computed as $\mathbf{C}^+ = (\mathbf{C}^t \mathbf{C})^{-1} \mathbf{C}^t$. However, MATLAB provides a numerically superior method for the calculation of **A** by means of the back-slash operator (\). Refer to the MATLAB manuals for details. The important point is that we are now in a position to write the residual matrix **R**, and thus *ssq*, as a function of the rate constants **k** only:

$$\mathbf{R} = \mathbf{Y} - \mathbf{CA} = \mathbf{Y} - \mathbf{CC}^+ \mathbf{Y} = f(\mathbf{Y}, \text{model}, \mathbf{k}) \quad (7.10)$$

The absolutely essential difference between Equation 7.10 and Equation 7.7 is that now there is only a very small number of parameters to be fitted iteratively. To go back to the example above, we have reduced the number of parameters from 3074 to 2 (*nk*). This number is well within the limits of the simplex algorithm. For the example of the consecutive reaction mechanism mentioned above, we give the function that calculates *ssq* in MATLAB Example 7.4b. It is repeatedly used by the simplex routine `fminsearch` called in MATLAB Example 7.4a. A minimum in *ssq* is found for $k_1 = 2.998 \times 10^{-3}$ s^{-1} and $k_2 = 1.501 \times 10^{-3}$ s^{-1}. As before, a MATLAB function (`data_abc`) that generates the absorbance data used for fitting is given in the Appendix at the end of this chapter. It is interesting to note that the calculated best rate constants are very close to the "true" ones used to generate the data. Generally, multivariate data are much better and more robust at defining parameters compared with univariate (one wavelength) measurements.

MATLAB Example 7.4a

```
% simplex fitting to the kinetic model A -> B -> C

[t,Y]=data_abc;                   % get absorbance data
A_0=1e-3;                         % initial concentration of A
k0=[0.005; 0.001];                % start parameter vector
[k,ssq]=fminsearch('rcalc_abc1',k0,[],A_0,t,Y)     % simplex call
```

MATLAB Example 7.4b

```
function ssq=rcalc_abc1(k,A_0,t,Y)

C(:,1)=A_0*exp(-k(1)*t);          % concentrations of species A
C(:,2)=A_0*k(1)/(k(2)-k(1))*(exp(-k(1)*t)-exp(-k(2)*t));   % conc. of B
C(:,3)=A_0-C(:,1)-C(:,2);         % concentrations of C
A=C\Y;                             % elimination of linear parameters
R=Y-C*A;                           % residuals
ssq=sum(sum((R.*R)));              % sum of squares
```

To analyze other mechanisms, all we need to do is to replace the few lines that calculate the matrix **C**. The computation of **A**, **R**, and *ssq* are independent of the chemical model, and generalized software can be written for the fitting task.

In two later sections, we will deal with numerical integration, which is required to solve the differential equations for complex mechanisms. Before that, we will describe nonlinear fitting algorithms that are significantly more powerful and faster than the direct-search simplex algorithm used by the MATLAB function fminsearch. Of course, the principle of separating linear (**A**) and nonlinear parameters (**k**) will still be applied.

7.4.4 Newton-Gauss-Levenberg/Marquardt (NGL/M)

In contrast to methods where the sum of squares, *ssq*, is minimized directly, the NGL/M type of algorithm requires the complete vector or matrix of residuals to drive the iterative refinement toward the minimum. As before, we start from an initial guess for the rate constants, k_0. Now, the parameter vector is continuously improved by the addition of the appropriate ("best") parameter shift vector Δ**k**. The shift vector is calculated in a more sophisticated way that is based on the derivatives of the residuals with respect to the parameters.

We could define the matrix of residuals, **R**, as a function of the measurements, **Y**, and the parameters, **k** and **A**. However, as previously shown, it is highly recommended if not mandatory to define **R** as a function of the nonlinear parameters only. The linear parameters, **A**, are dealt with separately, as shown in Equation 7.9 and Equation 7.10.

At each cycle of the iterative process a new parameter shift vector, δ**k**, is calculated. To derive the formulae for the iterative refinement of **k**, we develop **R** as a function of **k** (starting from **k** = k_0) into a Taylor series expansion. For sufficiently small δ**k**, the residuals, **R**(**k** + δ**k**), can be approximated by a Taylor series expansion.

$$\mathbf{R}(\mathbf{k}+\delta\mathbf{k}) = \mathbf{R}(\mathbf{k}) + \frac{1}{1!} \times \frac{\partial \mathbf{R}(\mathbf{k})}{\partial \mathbf{k}} \times \delta\mathbf{k} + \frac{1}{2!} \times \frac{\partial^2 \mathbf{R}(\mathbf{k})}{\partial \mathbf{k}^2} \times \delta\mathbf{k}^2 + \ldots \quad (7.11)$$

We neglect all but the first two terms in the expansion. This leaves us with an approximation that is not very accurate; however, it is easy to deal with, as it is a linear equation. Algorithms that include additional higher terms in the Taylor expansion often result in fewer iterations but require longer computation times due to the increased complexity. Dropping the higher-order terms from the Taylor series expansion gives the following equation.

$$\mathbf{R}(\mathbf{k} + \delta\mathbf{k}) = \mathbf{R}(\mathbf{k}) + \frac{\partial \mathbf{R}(\mathbf{k})}{\partial \mathbf{k}} \times \delta\mathbf{k} \qquad (7.12)$$

The matrix of partial derivatives, $\partial \mathbf{R}(\mathbf{k})/\partial \mathbf{k}$, is called the Jacobian, $\mathbf{J}$. We can rearrange this equation in the following way:

$$\mathbf{R}(\mathbf{k}) = -\mathbf{J} \times \delta\mathbf{k} + \mathbf{R}(\mathbf{k} + \delta\mathbf{k}) \qquad (7.13)$$

The matrix of residuals, $\mathbf{R}(\mathbf{k})$, is known, and the Jacobian, $\mathbf{J}$, is determined as shown later in this section. The task is to calculate the $\delta\mathbf{k}$ that minimizes the new residuals, $\mathbf{R}(\mathbf{k} + \delta\mathbf{k})$. Note that the structure of Equation 7.13 is identical to that of Equation 7.2, and the minimization problem can be solved explicitly by simple linear regression, equivalent to the calculation of the molar absorptivity spectra $\mathbf{A}$ ($\mathbf{A} = \mathbf{C}^+ \times \mathbf{Y}$) as outlined in Equation 7.9.

$$\delta\mathbf{k} = -\mathbf{J}^+ \times \mathbf{R}(\mathbf{k}) \qquad (7.14)$$

The Taylor series expansion is an approximation, and therefore the shift vector $\delta\mathbf{k}$ is an approximation as well. However, the new parameter vector $\mathbf{k} + \delta\mathbf{k}$ will generally be better than the preceding $\mathbf{k}$. Thus, an iterative process should always move toward the optimal rate constants. As the iterative fitting procedure progresses, the shifts, $\delta\mathbf{k}$, and the residual sum of squares, ssq, usually decrease continuously. The relative change in ssq is often used as a convergence criterion. For example, the iterative procedure can be terminated when the relative change in ssq falls below a preset value μ, typically $\mu = 10^{-4}$.

$$\mathrm{abs}\left(\frac{ssq_{old} - ssq}{ssq_{old}}\right) \leq \mu \qquad (7.15)$$

At this stage, we need to discuss the actual task of calculating the Jacobian matrix $\mathbf{J}$. It is always possible to approximate $\mathbf{J}$ numerically by the method of finite differences. In the limit as Δk_i approaches zero, the derivative of $\mathbf{R}$ with respect to k_i is given by Equation 7.16. For sufficiently small Δk_i, the approximation can be very good.

$$\frac{\partial \mathbf{R}}{\partial k_i} \cong \frac{\mathbf{R}(\mathbf{k} + \Delta k_i) - \mathbf{R}(k)}{\Delta k_i} \qquad (7.16)$$

Here, $(\mathbf{k} + \Delta k_i)$ represents the original parameter vector $\mathbf{k}$ to whose ith element, Δk_1, is added. A separate calculation must be performed for each element of $\mathbf{k}$. In other words, the derivatives with respect to the elements in $\mathbf{k}$ must be calculated one at a time. It is probably most instructive to study the MATLAB code in MATLAB Box 7.5b, where this procedure is defined precisely.

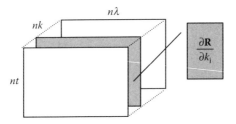

FIGURE 7.7 Three-dimensional representation of the Jacobian **J** as slices of $\partial \mathbf{R}/\partial k_i$.

A few additional remarks with respect to the calculation of the Jacobian matrix **J** are in order here. For reaction mechanisms that have explicit solutions to the set of differential equations, it is always possible to define the derivatives $\partial \mathbf{C}/\partial \mathbf{k}$ explicitly. In such cases, the Jacobian **J** can be calculated in explicit equations, and time-consuming finite-difference approximations are not required. The equations are rather complex, although implementation in MATLAB is straightforward. More information on this topic can be found in the literature [22]. The calculation of numerical derivatives is always possible, and for mechanisms that require numerical integration it is the only option.

The Jacobian matrix, **J**, is the derivative of a matrix with respect to a vector. Further discussion of its structure and the computation of its pseudoinverse are warranted. The most straightforward way to organize **J** is in a three-dimensional array: the derivative of **R** with respect to one particular parameter k_i is a matrix of the same dimensions as **R** itself. The collection of all these nk derivatives with respect to all of the nk parameters can be arranged in a three-dimensional array of dimensions $nt \times n\lambda \times nk$, with the individual matrices $\partial \mathbf{R}/\partial k_i$ written slicewise "behind" each other, as illustrated in Figure 7.7.

Organizing **J** in a three-dimensional array is elegant, but it does not fit well into the standard routines of MATLAB for matrix manipulation. There is no command for the calculation of the pseudoinverse $\mathbf{J}^+$ of such a three-dimensional array. There are several ways around this problem; one of them is discussed in the following. The matrices $\mathbf{R}(\mathbf{k})$ and $\mathbf{R}(\mathbf{k} + \delta \mathbf{k})$ as well as each matrix $\partial \mathbf{R}/\partial k_i$ are vectorized, i.e., unfolded into long column vectors $\mathbf{r}(\mathbf{k})$ and $\mathbf{r}(\mathbf{k} + \delta \mathbf{k})$. The nk vectorized partial derivatives then form the columns of the matricized Jacobian **J**. The structure of the resulting analogue to Equation 7.13 can be represented graphically in Equation 7.17.

$$\mathbf{r}(\mathbf{k}) = -\mathbf{J} \times \delta \mathbf{k} + \mathbf{r}(\mathbf{k} + \delta \mathbf{k})$$

(7.17)

Because **J** now possesses a well-defined matrix structure, the solution to $\delta \mathbf{k}$ can be written without any difficulty.

$$\delta \mathbf{k} = -\mathbf{J}^+ \times \mathbf{r(k)} \tag{7.18}$$

Or, using the MATLAB "\" notation for the pseudoinverse:

$$\delta \mathbf{k} = -\mathbf{J} \backslash \mathbf{r(k)} \tag{7.19}$$

It is important to recall at this point that **k** comprises only the nonlinear parameters, i.e., the rate constants. The linear parameters, i.e., the elements of the matrix **A** containing the molar absorptivities, are solved in a separate linear regression step, as described earlier in Equation 7.9 and Equation 7.10.

The basic structure of the iterative Newton-Gauss method is given in Scheme 7.1. The convergence of the Newton-Gauss algorithm in the vicinity of the minimum is usually excellent (quadratic). However, if starting guesses are poorly chosen, the shift vector, $\delta \mathbf{k}$, as calculated by Equation 7.18, can point in a wrong direction or the step size can be too long. The result is an increased *ssq*, divergence, and a usually quick and dramatic crash of the program. Marquardt [23], based on ideas by Levenberg [24], suggested a very elegant and efficient method to manage the problems associated with divergence.

The pseudoinverse for the calculation of the shift vector has been computed traditionally as $\mathbf{J}^+ = -(\mathbf{J}^t\mathbf{J})^{-1}\mathbf{J}^t$. Adding a certain number, the Marquardt parameter *mp*, to the diagonal elements of the square matrix $\mathbf{J}^t\mathbf{J}$ prior to its inversion has two consequences: (a) it shortens the shift vector $\delta \mathbf{k}$ and (b) it turns its direction toward the direction of steepest descent. The larger the Marquardt parameter, the greater is the effect.

$$\delta \mathbf{k} = -(\mathbf{J}^t\mathbf{J} + mp \times \mathbf{I})^{-1} \mathbf{J}^t \times \mathbf{r(k)} \tag{7.20}$$

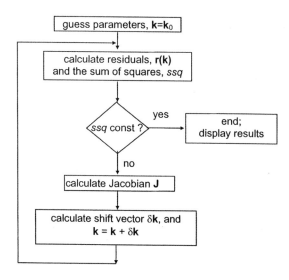

SCHEME 7.1 Flow diagram of a very basic Newton-Gauss algorithm.

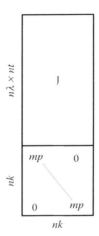

FIGURE 7.8 Appending the Marquardt parameter *mp* to the Jacobian **J**.

where **I** is the identity matrix of the appropriate size. If we want to use the MATLAB backslash notation, $\Delta \mathbf{k} = -\mathbf{J} \setminus \mathbf{r}(\mathbf{k}_0)$, we get the same effect by appending a diagonal matrix containing the Marquardt parameter to the lower end of **J**. This is visualized in Figure 7.8.

As the number of rows in **J** and elements in **r(k)** must be the same, we must also append the same number of *nk* zeros to the end of the vector **r(k)**. It might be a useful exercise for the reader to verify the equivalence of the two approaches.

Depending on the improvement of the sum of squares, *ssq*, the Marquardt parameter, *mp*, is reduced or augmented. There are no general rules on how exactly this should be done in detail; it depends on the specific case. If required, the initial value for the Marquardt parameter has to be chosen sensibly as well; the original suggestion was to use the value of the largest diagonal element of $\mathbf{J}^t\mathbf{J}$. In MATLAB Example 7.5b, if *mp* is required we just set it initially to 1. Usually convergence occurs with no Marquardt parameter at all; in the programming MATLAB Example 7.5b, it is thus initialized as zero.

The simplex and similar algorithms do not deliver standard errors for the parameters. A particularly dangerous feature of the simplex algorithm is the possibility of inadvertently fitting completely irrelevant parameters. The immediate result of the fit gives no indication about the relevance of the fitted parameters (i.e., the kinetic model). This also applies to the Solver algorithm offered by Excel, although appropriate procedures have been suggested as macros in Excel to provide statistical analysis of the results [18]. The NGL/M algorithm allows a direct error analysis on the fitted parameters. In fact, for normally distributed noise, the relevant information is obtained during the calculation of $\delta \mathbf{k}$. According to statistics textbooks [16], the standard error σ_{ki} in the fitted nonlinear parameters k_i can be approximated from the expression

$$\sigma_{k_i} = \sigma_y \sqrt{d_{ii}} \qquad (7.21)$$

where d_{ii} is the ith diagonal element of the inverted Hessian matrix $(\mathbf{J^tJ})^{-1}$ (without the Marquardt parameter added) and σ_y represents the standard deviation of the measurement error in $\mathbf{Y}$.

$$\sigma_y = \sqrt{\frac{ssq}{v}} \quad (7.22)$$

The denominator represents the degree of freedom, v, and is defined as the number of experimental values (elements of $\mathbf{Y}$), minus the number of optimized nonlinear ($\mathbf{k}$) and linear ($\mathbf{A}$) parameters.

$$v = nt \times n\lambda - nk - nc \times n\lambda \quad (7.23)$$

This method of estimating the errors σ_{ki} in the parameters, k_i, is based on ideal behavior, e.g., perfect initial concentrations, disturbed only by white noise in the measurement. Experience shows that the estimated errors tend to be smaller than those determined by statistical analysis of several measurements fitted individually.

We are now in a position to write a MATLAB program based on the Newton-Gauss-Levenberg/Marquardt algorithm. Scheme 7.2 represents a flow diagram. We will apply this NGL/M algorithm to the same data set of a consecutive reaction scheme $A \rightarrow B \rightarrow C$ that was previously subjected to a simplex optimization in Section 7.4.1. Naturally, the results must be the same within error limits. In MATLAB Example 7.5c a function is given that computes the residuals that are repeatedly required by the NGL/M routine, given in MATLAB Example 7.5b, and which in turn is called by the main program shown in MATLAB Example 7.5a.

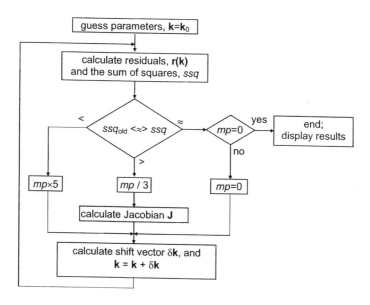

SCHEME 7.2 The Newton-Gauss-Levenberg/Marquardt (NGL/M) algorithm.

Note that the standard errors in the rate constants ($k_1 = 2.996 \pm 0.005 \times 10^{-3}$ s^{-1} and $k_2 = 1.501 \pm 0.002 \times 10^{-3}$ s^{-1}) are delivered in addition to the standard deviation ($\sigma_Y = 9.991 \times 10^{-3}$) in **Y**. The ability to directly estimate errors in the calculated parameters is a distinct advantage of the NGL/M fitting procedure. Furthermore, even for this relatively simple example, the computation times are already faster than using a simplex by a factor of five. This difference dramatically increases with increasing complexity of the kinetic model.

MATLAB Example 7.5a

```
% ngl/m fitting to the kinetic model A -> B -> C

[t,Y]=data_abc;                    % get absorbance data
A_0=1e-3;                          % initial concentration of species A
k0=[0.005;0.001];                  % start parameter vector
[k,ssq,C,A,J]=nglm('rcalc_abc2',k0,A_0,t,Y);       % call ngl/m
k                                                   % display k
ssq                                                 % ssq
sig_y=sqrt(ssq/(prod(size(Y))-length(k)-(prod(size(A)))))  % sigma_y
sig_k=sig_y*sqrt(diag(inv(J'*J)))                          % sigma_par
```

MATLAB Example 7.5b

```
function [k,ssq,C,A,J]=nglm(fname,k0,A_0,t,Y)

ssq_old=1e50;
mp=0;
mu=1e-4;                           % convergence limit
delta=1e-6;                        % step size for numerical diff
k=k0;
while 1
    [r0,C,A]=feval(fname,k,A_0,t,Y);   % call calculation of
                                       % residuals
    ssq=sum(r0.*r0);
    conv_crit=(ssq_old-ssq)/ssq_old;
    if abs(conv_crit) <= mu            % ssq_old=ssq, minimum
                                       % reached !
        if mp==0
            break                      % if Marquardt par zero, stop
        else                           % otherwise
            mp=0;                      % set it to 0, another iteration
            r0_old=r0;
        end
    elseif conv_crit > mu              % convergence !
        mp=mp/3;
        ssq_old=ssq;
        r0_old=r0;
        for i=1:length(k)
            k(i)=(1+delta)*k(i);
            r=feval(fname,k,A_0,t,Y);  % slice wise numerical
            J(:,i)=(r-r0)/(delta*k(i));% differentiation to
            k(i)=k(i)/(1+delta);       % form the Jacobian
        end
```

Kinetic Modeling of Multivariate Measurements with Nonlinear Regression

```
        elseif conv_crit < -mu          % divergence !
            if mp==0
                mp=1;                    % use Marquardt parameter
            else
                mp=mp*5;
            end
            k=k-delta_k;                 % and take shifts back
        end
        J_mp=[J;mp*eye(length(k))];      % augment Jacobian matrix
        r0_mp=[r0_old;zeros(size(k))];   % augment residual vector
        delta_k=-J_mp\r0_mp;             % calculate parameter shifts
        k=k+delta_k;                     % add parameter shifts
end
```

MATLAB Example 7.5c

```
function [r,C,A]=rcalc_abc2(k,A_0,t,Y)

C(:,1)=A_0*exp(-k(1)*t);                 % concentrations of species A
C(:,2)=A_0*k(1)/(k(2)-k(1))*(exp(-k(1)*t)-exp(-k(2)*t));  % conc.
                                                          % of B
C(:,3)=A_0-C(:,1)-C(:,2);                % concentrations of C
A=C\Y;                                   % calculation of linear parameters
R=Y-C*A;                                 % residuals
r=R(:);                                  % vectorizing the residual matrix R
```

In Figure 7.9, the results of the data fitting process are illustrated in terms of Beer-Lambert's law in its matrix notation $\mathbf{C} \times \mathbf{A} = \mathbf{Y}$. The individual plots represent the corresponding matrices of the matrix product.

Some care has to be taken in assessing the results if the chemical model consists of several first-order reactions. In such cases there is no unique relationship between observed exponential curves and mechanistic rate constants [25, 26]. In our example, the mechanism $A \xrightarrow{k_1} B \xrightarrow{k_2} C$, an equivalent solution with the same minimal sum of squares, ssq, can be obtained by swapping k_1 and k_2 (i.e., at $k_1 = 1.501 \pm 0.002 \times 10^{-3}$ s^{-1} and $k_2 = 2.996 \pm 0.005 \times 10^{-3}$ s^{-1}). This phenomenon is also known as the "slow-fast" ambiguity. The iterative procedure will converge to one of the two solutions, depending on the initial guesses for the rate constants. This can easily be verified by the reader. Fortunately, results with interchanged rate constants often lead to meaningless (e.g., negative) or unreasonable molar absorptivity spectra for compound B. Simple chemical reasoning and intuition usually allows the resolution of the ambiguity.

7.4.5 NONWHITE NOISE

The actual noise distribution in $\mathbf{Y}$ is often unknown, but generally a normal distribution is assumed. White noise signifies that all experimental standard deviations, $\sigma_{i,j}$, of all individual measurements, $y_{i,j}$, are the same and uncorrelated. The least-squares criterion applied to the residuals delivers the most likely parameters only under the condition of so-called white noise. However, even if this prerequisite is not fulfilled, it is usually still useful to perform the least-squares fit. This makes it the most commonly applied method for data fitting.

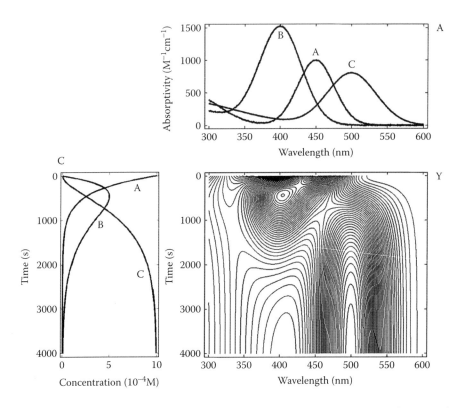

FIGURE 7.9 Reaction $A \to B \to C$. Results of the data-fitting procedure visualized in terms of Beer-Lambert's law in its matrix notation $\mathbf{C} \times \mathbf{A} = \mathbf{Y}$.

If the standard deviations $\sigma_{i,j}$ for all elements of the matrix $\mathbf{Y}$ are known or can be estimated, it makes sense to use this information in the data analysis. Instead of the sum of squares as defined in Equation 7.6, it is the sum over all appropriately weighted and squared residuals that is minimized. This is known as chi-square fitting [15, 16].

$$\chi^2 = \sum_{i=1}^{nt} \sum_{j=1}^{n\lambda} \left(\frac{r_{i,j}}{\sigma_{i,j}} \right)^2 \tag{7.24}$$

If all $\sigma_{i,j}$ are the same (white noise), the calculated parameters of the χ^2 fit will be the same as for least-squares fitting. If the $\sigma_{i,j}$ are not constant across the data set, the least-squares fit will overemphasize those parts of the data with high noise.

In absorption spectrometry, $\sigma_{i,j}$ is usually fairly constant, and χ^2 fitting has no advantages. Typical examples of data with nonconstant and known standard deviations are encountered in emission spectroscopy, particularly if photon counting techniques are employed, which are used for the analysis of very fast luminescence decays [27]. In such cases, measurement errors follow a Poisson distribution instead of a Gaussian or normal distribution, and the standard deviation of the measured emission intensity is a function of the intensity itself [16].

$$\sigma_{i,j} = \sqrt{y_{i,j}} \tag{7.25}$$

The higher the intensity, the higher is the standard deviation. At zero intensity, the standard deviation is zero as well.

Follow we discuss the implementation of the χ^2 analysis in an Excel spreadsheet. It deals with the emission decay of a solution with two emitters of slightly different lifetimes. Measurements are done at one wavelength only. Column C of the Excel spreadsheet shown in Figure 7.10 contains the estimated standard deviation σ_i for each intensity reading y_i; according to Equation 7.25, the standard deviation is simply the square root of the intensity. Column D contains the calculated intensity as the sum of two exponential decays.

$$y_i = amp_1 \cdot e^{-t_i/\tau_1} + amp_2 \cdot e^{-t_i/\tau_2} \tag{7.26}$$

Note that in this context, lifetimes τ are used instead of rate constants k; the relationship between the two is $\tau = 1/k$. Column F contains the squared weighted residuals $(r_i/\sigma_i)^2$, as indicated in Figure 7.10. The sum over all its elements, χ^2, is put into cell B6, and its value is minimized as shown in the Solver window.

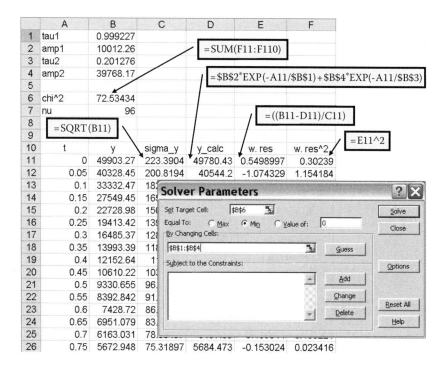

FIGURE 7.10 χ^2 fitting with Excel's Solver.

The parameters to be fitted are in cells B1:B4 and represent the two lifetimes τ_1 and τ_2 as well as the corresponding amplitudes amp_1 and amp_2.

Figure 7.11 shows the results of the χ^2 fit in Figure 7.11a and the normal least-squares fit in Figure 7.11b. The noisy lines represent the residuals from the analysis. The nonweighted residuals in Figure 7.11b clearly show a noise level increasing with signal strength. As can be seen in Table 7.1, the χ^2 analysis results in parameters generally closer to their true values.

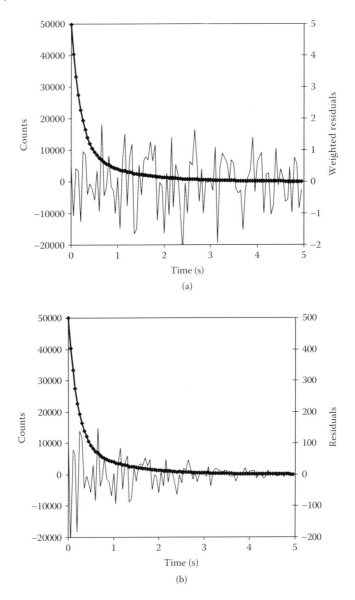

FIGURE 7.11 (a) χ^2 and (b) least-squares fit of emission spectroscopic data.

TABLE 7.1
Results of the δ^2 Analysis of Emission Spectroscopic Data

	True Values	δ^2	Least Squares
τ_1 (s^{-1})	1	0.9992	0.9976
amp_1	10,000	10,012	10,048
τ_2 (s^{-1})	0.2	0.2009	0.2013
amp_2	40,000	39,768	39,749

The implementation in a MATLAB program is straightforward; e.g., MATLAB Box 7.5c needs to be amended in the following way.

```
R=Y-C*A_hat;        % residual matrix R
Chi=R./SigmaY;      % division by sigma_y
r=Chi(:);           % vectorizing the residual matrix R
```

Of course the matrix SigmaY needs to be passed into the functions as an additional parameter.

Another advantage in knowing the standard deviations of the measurements is that we can determine if a fit is sufficiently good. As a rule of thumb, this is achieved if $\chi^2 \cong \nu$, where ν is the degree of freedom, which has earlier been defined in Equation 7.23. With $\chi^2 \cong 72.5$ and $\nu = 96$ (100 − 2 − 2), this condition is clearly satisfied for our example spreadsheet in Figure 7.10. If χ^2 is too big, something is wrong, most likely with the model. If χ^2 is too small, most likely the σ_{ij} have been overestimated.

So far we have shown how multivariate absorbance data can be fitted to Beer-Lambert's law on the basis of an underlying kinetic model. The process of nonlinear parameter fitting is essentially the same for any kinetic model. The crucial step of the analysis is the translation of the chemical model into the kinetic rate law, i.e., the set of ODEs, and their subsequent integration to derive the corresponding concentration profiles.

7.5 CALCULATION OF THE CONCENTRATION PROFILES: CASE II, COMPLEX MECHANISMS

In Section 7.3, we gave the explicit formulae for the calculation of the concentration profiles for a small set of simple reaction mechanisms. Often there is no such explicit solution, or its derivation is rather demanding. In such instances, numerical integration of the set of differential equations needs to be carried out. We start with a simple example:

$$2A \underset{k_-}{\overset{k_+}{\rightleftharpoons}} B \qquad (7.27)$$

The analytical formula for the calculation of the concentration profiles for A and B for the above model is fairly complex, involving the `tan` and `atan` functions

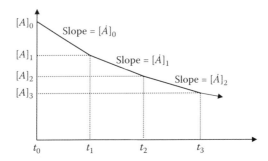

FIGURE 7.12 Euler's method for numerical integration.

(according to MATLAB's symbolic toolbox). However, knowing the rate law and concentrations at any time, one can calculate the derivatives of the concentrations of A and B at this time numerically.

$$[\dot{A}] = -2k_+[A]^2 + 2k_-[B]$$
$$[\dot{B}] = k_+[A]^2 - k_-[B] \tag{7.28}$$

Euler's method [15, 28] represented in Figure 7.12 is the simplest way to perform this task. Because of its simplicity it is ideally suited to demonstrate the general principles of the numerical integration of ordinary differential equations.

Starting at time t_0, the initial concentrations are $[A]_0$ and $[B]_0$; the derivatives $[\dot{A}]_0$ and $[\dot{B}]_0$ are calculated according to Equation 7.28. This allows the computation of new concentrations, $[A]_1$ and $[B]_1$, for the species A and B after a short time interval $\Delta t = t_1 - t_0$.

$$[A]_1 = [A]_0 + \Delta t[\dot{A}]_0$$
$$[B]_1 = [B]_0 + \Delta t[\dot{B}]_0 \tag{7.29}$$

These new concentrations in turn allow the determination of new derivatives and thus another set of concentrations $[A]_2$ and $[B]_2$ after the second time interval $t_2 - t_1$. As shown in Figure 7.12, this procedure is simply repeated until the desired final reaction time is reached.

With Euler's simple method, very small time intervals must be chosen to achieve reasonably accurate profiles. This is the major drawback of this method and there are many better methods available. Among them, algorithms of the Runge-Kutta type [15, 28, 29] are frequently used in chemical kinetics [3]. In the following subsection we explain how a fourth-order Runge-Kutta method can be incorporated into a spreadsheet and used to solve nonstiff ODEs.

7.5.1 FOURTH-ORDER RUNGE-KUTTA METHOD IN EXCEL

The fourth order Runge-Kutta method is the workhorse for the numerical integration of ODEs. Elaborate routines with automatic step-size control are available in MATLAB. We will show their usage in several examples later.

First, without explaining the details [15], we will develop an Excel spreadsheet for the numerical integration of the reaction mechanism $2A \underset{k_-}{\overset{k_+}{\rightleftharpoons}} B$, as seen in Figure 7.13. The fourth-order Runge-Kutta method requires four evaluations of concentrations and derivatives per step. This appears to be a serious disadvantage, but as it turns out, significantly larger step sizes can be taken for the same accuracy, and the overall computation times are much shorter. We will comment on the choice of appropriate step sizes after this description.

We explain the computations for the first time interval Δt (cell E5) between $t_0 = 0$ and $t_1 = 1$, representative of all following intervals. Starting from the initial concentrations $[A]_{t0}$ and $[B]_{t0}$ (cells B5 and C5), the concentrations $[A]_{t1}$ and $[B]_{t1}$ (cells B6 and C6) can be computed in the following way:

1. Calculate the derivatives of the concentrations at t_0:

$$[\dot{A}]_{t_0} = -2k_+[A]_{t_0}^2 + 2k_-[B]_{t_0}$$

$$[\dot{B}]_{t_0} = k_+[A]_{t_0}^2 - k_-[B]_{t_0}$$

   In the Excel language, for A, this translates into =−2*$B$1*B5^2+2*$B$2*C5, as indicated in Figure 7.13. Note, in the figure we only give the cell formulae for the computations of component A; those for B are written in an analogous way.

2. Calculate approximate concentrations at intermediate time point $t = t_0 + \Delta t/2$:

$$[A]_1 = [A]_{t_0} + \frac{\Delta t}{2}[\dot{A}]_0$$

$$[B]_1 = [B]_{t_0} + \frac{\Delta t}{2}[\dot{B}]_0$$

   Again, the Excel formula for component A is given in Figure 7.13.

3. Calculate the derivatives at intermediate time point $t = t_0 + \Delta t/2$:

$$[\dot{A}]_1 = -2k_+[A]_1^2 + 2k_-[B]_1$$

$$[\dot{B}]_1 = k_+[A]_1^2 - k_-[B]_1$$

4. Calculate another set of concentrations at the intermediate time point $t = t_0 + \Delta t/2$, based on the concentrations at t_0 but using the derivatives $[\dot{A}]_1$ and $[\dot{B}]_1$:

$$[A]_2 = [A]_{t_0} + \frac{\Delta t}{2}[\dot{A}]_1$$

$$[B]_2 = [B]_{t_0} + \frac{\Delta t}{2}[\dot{B}]_1$$

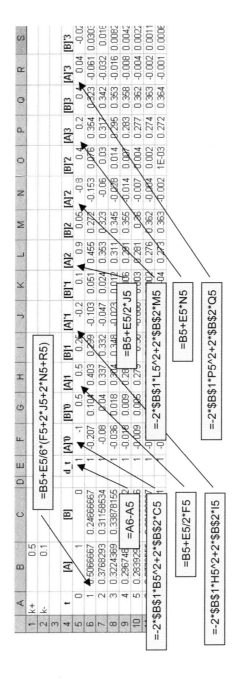

FIGURE 7.13 Excel spreadsheet for the numerical integration of the rate law for the reaction $2A \underset{k_-}{\overset{k_+}{\rightleftharpoons}} B$ using fourth-order Runge-Kutta equations.

Kinetic Modeling of Multivariate Measurements with Nonlinear Regression

5. Compute another set of derivatives at the intermediate time point $t = t_0 + \Delta t/2$:

$$[\dot{A}]_2 = -2k_+[A]_2^2 + 2k_-[B]_2$$

$$[\dot{B}]_2 = k_+[A]_2^2 - k_-[B]_2$$

6. Next, the concentrations at time t_1 after the complete time interval $\Delta t = t_1 - t_0$ are computed based on the concentrations at time t_0 and the derivatives $[\dot{A}]_2, [\dot{B}]_2$, at time $t = t_0 + \Delta t/2$:

$$[A]_3 = [A]_{t_0} + \Delta t[\dot{A}]_2$$

$$[B]_3 = [B]_{t_0} + \Delta t[\dot{B}]_2$$

7. Computation of the derivatives at time t_1:

$$[\dot{A}]_3 = -2k_+[A]_3^2 + 2k_-[B]_3$$

$$[\dot{B}]_3 = k_+[A]_3^2 - k_-[B]_3$$

8. Finally, the new concentrations after the full time interval $\Delta t = t_1 - t_0$ are computed as:

$$[A]_{t_1} = [A]_{t_0} + \frac{\Delta t}{6}([\dot{A}]_{t_0} + 2[\dot{A}]_1 + 2[\dot{A}]_2 + [\dot{A}]_3)$$

$$[B]_{t_1} = [B]_{t_0} + \frac{\Delta t}{6}([\dot{B}]_{t_0} + 2[\dot{B}]_1 + 2[\dot{B}]_2 + [\dot{B}]_3)$$

These concentrations are put as the next elements into cells B6 and C6 and provide the new start concentrations to repeat steps 1 through 8 for the next time interval Δt (cell E6) between $t_1 = 1$ and $t_2 = 2$.

Figure 7.14 displays the resulting concentration profiles for species A and B.

For fast computation, the determination of the best step size (interval) is crucial. Steps that are too small result in correct concentrations at the expense of long computation times. On the other hand, intervals that are too big save computation time but result in poor approximation. The best intervals lead to the fastest computation of concentration profiles within the predefined error limits. The ideal step size is not constant during the reaction and thus needs to be adjusted continuously.

One particular class of ordinary differential equation solvers (ODE-solvers) handles stiff ODEs and these are widely known as stiff solvers. In our context, a system of ODEs sometimes becomes stiff if it comprises very fast and also very slow steps or relatively high and low concentrations. A typical example would be an oscillating reaction. Here, a highly sophisticated step-size control is required to achieve a reasonable compromise between accuracy and computation time. It is well outside the scope of this chapter to expand on the intricacies of modern numerical

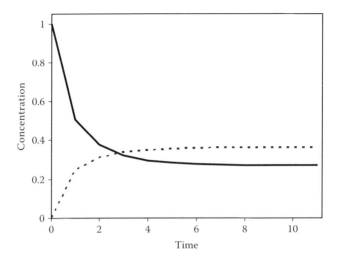

FIGURE 7.14 Concentration profiles for a reaction $2A \underset{k_-}{\overset{k_+}{\rightleftharpoons}} B$ (—— A, ... B) as modeled in Excel using a fourth-order Runge-Kutta for numerical integration.

integration routines. MATLAB provides an excellent selection of routines for this task. For further reading, consult the relevant literature and the MATLAB manuals [15, 28, 29].

7.5.2 INTERESTING KINETIC EXAMPLES

Next, we will look into various kinetic examples of increasing complexity and determine solely concentration profiles (**C**). This can be seen as kinetic simulation, since the calculations are all based on known sets of rate constants. Naturally, in an iterative fitting process of absorbance, data on these parameters would be varied until the sum of the squared residuals between measured absorbances (**Y**) and Beer-Lambert's model (**C** × **A**) is at its minimum.

7.5.2.1 Autocatalysis

Processes are called autocatalytic if the products of a reaction accelerate their own formation. An extreme example would be a chemical explosion. In this case, it is usually not a chemical product that directly accelerates the reaction; rather, it is the heat generated by the reaction. The more heat produced, the faster is the reaction; and the faster the reaction, the more heat that is produced, etc.

A very basic autocatalytic reaction scheme is presented in Equation 7.30.

$$A \xrightarrow{k_1} B$$
$$A + B \xrightarrow{k_2} 2B \quad (7.30)$$

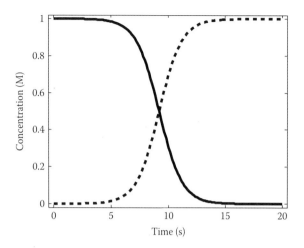

FIGURE 7.15 Concentration profiles for the autocatalytic reaction $A \xrightarrow{k_1} B$; $A + B \xrightarrow{k_2} 2B$.

Starting with component A, there is a relatively slow first reaction to form the product B. The development of component B opens another path for its formation in the second reaction, which is of the order two. Therefore, the higher the concentration of B, the faster is the decomposition of A to form more B.

$$[\dot{A}] = -k_1[A] - k_2[A][B]$$
$$[\dot{B}] = -k_1[A] + k_2[A][B]$$
(7.31)

Figure 7.15 shows the calculated corresponding concentration profiles using the rate constants $k_1 = 10^{-4}$ s^{-1} and $k_2 = 1$ M^{-1}s^{-1} for initial concentrations $[A]_0 = 1\,M$ and $[B]_0 = 0\,M$. We used MATLAB's Runge–Kutta-type ODE-solver ode45. In MATLAB Example 7.6b, the function is given that generates the differential equations. It is repeatedly called by the ODE-solver in MATLAB Example 7.6a.

MATLAB Example 7.6a

```
% autocatalysis
% A --> B
% A + B --> 2 B

c0=[1;0];                        % initial conc of A and B
k=[1e-4;1];                      % rate constants k1 and k2
[t,C]=ode45('ode_autocat',20,c0,[],k); call ode-solver
plot(t,C)                        % plotting C vs t
```

MATLAB Example 7.6b

```
function c_dot=ode_autocat(t,c,flag,k)

% A --> B
% A + B --> 2 B
c_dot(1,1)=-k(1)*c(1)-k(2)*c(1)*c(2);      % A_dot
c_dot(2,1)= k(1)*c(1)+k(2)*c(1)*c(2);      % B_tot
```

7.5.2.2 Zeroth-Order Reaction

Zeroth-order reactions do not really exist; they are always macroscopically observed reactions only where the rate of the reaction is independent of the concentrations of the reactants. Formally, the ODE is:

$$[\dot{A}] = -k[A]^0 = -k \tag{7.32}$$

A simple mechanism that mimics a zeroth-order reaction is the catalytic transformation of A to C. A reacts with the catalyst Cat to form an intermediate activated complex B. B in turn reacts further to form the product C, releasing the catalyst, which in turn continues reacting with A.

$$\begin{aligned} A + Cat &\xrightarrow{k_1} B \\ B &\xrightarrow{k_2} C + Cat \end{aligned} \tag{7.33}$$

The total concentration of catalyst is much smaller than the concentrations of the reactants or products. Note that, in real systems, the reactions are reversible and usually there are more intermediates, but for the present purpose this minimal reaction mechanism is sufficient.

$$\begin{aligned} [\dot{A}] &= -k_1[A][Cat] \\ [\dot{Cat}] &= -k_1[A][Cat] + k_2[B] \\ [\dot{B}] &= k_1[A][Cat] - k_2[B] \\ [\dot{C}] &= k_2[B] \end{aligned} \tag{7.34}$$

The production of C is governed by the amount of intermediate B, which is constant over an extended period of time. As long as there is an excess of A with respect to the catalyst, essentially all of the catalyst exists as complex, and thus this concentration is constant. The crucial differential equation is the last one; it is a zeroth-order reaction as long as $[B]$ is constant.

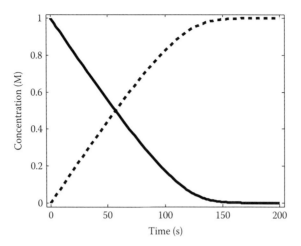

FIGURE 7.16 Concentration profiles for the reaction $A + Cat \xrightarrow{k_1} B$; $B \xrightarrow{k_2} C + Cat$. The reaction is zeroth order for about 100 s.

The kinetic profiles displayed in Figure 7.16 have been integrated numerically with MATLAB's stiff solver ode15s using the rate constants $k_1 = 1000$ M^{-1}s^{-1}, $k_2 = 100$ s^{-1} for the initial concentrations $[A]_0 = 1$ M, $[Cat]_0 = 10^{-4}$ M, and $[B]_0 = [C]_0 = 0$ M. For this model, the standard Runge-Kutta routine is far too slow and thus useless. In MATLAB Example 7.7b, the function is given that generates the differential equations. It is repeatedly called by the ODE-solver in MATLAB Example 7.7a.

MATLAB Example 7.7a

```
% 0th order kinetics
% A + Cat --> B
% B --> C + Cat

c0=[1;1e-4;0;0];            % initial conc of A, Cat, B and C
k=[1000;100];               % rate constants k1 and k2
[t,C] = ode15s('ode_zero_order',200,c0,[],k);  % call ode-solver
plot(t,C)                   % plotting C vs t
```

MATLAB Example 7.7b

```
function c_dot=ode_zero_order(t,c,flag,k)

% 0th order kinetics
% A + Cat --> B
% B --> C + Cat

c_dot(1,1)=-k(1)*c(1)*c(2);                  % A_dot
c_dot(2,1)=-k(1)*c(1)*c(2)+k(2)*c(3);        % Cat_dot
c_dot(3,1)= k(1)*c(1)*c(2)-k(2)*c(3);        % B_dot
c_dot(4,1)= k(2)*c(3);                       % C_dot
```

7.5.2.3 Lotka-Volterra (Sheep and Wolves)

This example is not chemically relevant, but is all the more exciting. It models the dynamics of a population of predators and preys in a closed system. Consider an island with a population of sheep and wolves. In the first "reaction," the sheep are breeding. Note that there is an unlimited supply of grass and that this reaction could go on forever. But there is the second "reaction," where wolves eat sheep and breed themselves. To complete the system, wolves have to die a natural death.

$$\begin{aligned} sheep &\xrightarrow{k_1} 2\ sheep \\ wolf + sheep &\xrightarrow{k_2} 2\ wolves \\ wolf &\xrightarrow{k_3} dead\ wolf \end{aligned} \quad (7.35)$$

The following differential equations have to be solved:

$$\begin{aligned}{} [\dot{sheep}] &= k_1[sheep] - k_2[wolf][sheep] \\ [\dot{wolf}] &= k_2[wolf][sheep] - k_3[wolf] \end{aligned} \quad (7.36)$$

The kinetic population profiles displayed in Figure 7.17 have been obtained by numerical integration using MATLAB's Runge-Kutta solver `ode45` with the rate constants $k_1 = 2$, $k_2 = 5$, $k_3 = 6$ for the initial populations $[sheep]_0 = 2$, $[wolf]_0 = 2$. For simplicity, we ignore the units. In MATLAB Example 7.8b, the function is given that generates the differential equations. It is repeatedly called by the ODE-solver in MATLAB Example 7.8a.

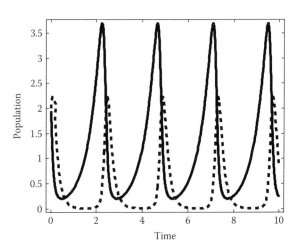

FIGURE 7.17 Lotka-Volterra's predator and prey "kinetics."

Kinetic Modeling of Multivariate Measurements with Nonlinear Regression 251

MATLAB Example 7.8a

```
% lotka volterra
% sheep   --> 2 sheep
% wolf + sheep --> 2 wolves
% wolf   --> dead wolf

c0=[2;2];              % initial 'conc' of sheep and wolves
k=[2;5;6];             % rate constants k1, k2 and k3
[t,C] = ode45('ode_lotka_volterra',10,c0,[],k);
                       %call ode-solver
plot(t,C)              % plotting C vs t
```

MATLAB Example 7.8b

```
function c_dot=ode_lotka_volterra(t,c,flag,k)

% lotka volterra
% sheep   --> 2 sheep
% wolf + sheep --> 2 wolves
% wolf   --> dead wolf

c_dot(1,1)=k(1)*c(1)-k(2)*c(1)*c(2);        % sheep_dot
c_dot(2,1)=k(2)*c(1)*c(2)-k(3)*c(2);        % wolf_dot
```

Surprisingly, the dynamics of such a population is completely cyclic. All properties of the cycle depend on the initial populations and the "rate constants." The "reaction" *sheep* $\xrightarrow{k_1}$ 2 *sheep* contradicts the law of conservation of mass and, thus, cannot directly represent reality. However, as we will see in the next example, oscillating reactions do exist.

7.5.2.4 The Belousov-Zhabotinsky (BZ) Reaction

Chemical mechanisms for real oscillating reactions are very complex and are not understood in every detail. Nevertheless, there are approximate mechanisms that correctly represent several main aspects of real reactions. Often, not all physical laws are strictly obeyed, e.g., the law of conservation of mass.

The Belousov-Zhabotinsky (BZ) reaction involves the oxidation of an organic species such as malonic acid (*MA*) by an acidified aqueous bromate solution in the presence of a metal ion catalyst such as the $Ce^{(III)}/Ce^{(IV)}$ couple. At excess [*MA*], the stoichiometry of the net reaction is

$$2BrO_3^- + 3CH_2(COOH)_2 + 2H^+ \xrightarrow{catalyst} 2BrCH(COOH)_2 + 3CO_2 + 4H_2O \tag{7.37}$$

A short induction period is typically followed by an oscillatory phase visible by an alternating color of the aqueous solution due to the different oxidation states of the metal catalyst. Addition of a colorful redox indicator, such as the $Fe^{II/III}(phen)_3$ couple, results in more dramatic color changes. Typically, several hundred oscillations with a periodicity of approximately a minute gradually die out within a couple of hours, and the system slowly drifts toward its equilibrium state.

In an effort to understand the BZ system, Field, Körös, and Noyes developed the so-called FKN mechanism [30]. From this, Field and Noyes later derived the Oregonator model [31], an especially convenient kinetic model to match individual experimental observations and predict experimental conditions under which oscillations might arise.

$$BrO_3^- + Br^- \xrightarrow{k_1} HBrO_2 + HOBr$$

$$BrO_3^- + HBrO_2 \xrightarrow{k_2} 2HBrO_2 + 2M_{ox}$$

$$HBrO_2 + Br^- \xrightarrow{k_3} 2HOBr \qquad (7.38)$$

$$2HBrO_2 \xrightarrow{k_4} BrO_3^- + HOBr$$

$$MA + M_{ox} \xrightarrow{k_5} \tfrac{1}{2} Br^-$$

M_{ox} represents the metal ion catalyst in its oxidized form. Br^- and BrO_3^- are not protonated at pH $\approx$ 0.

It is important to stress that this model is based on an empirical rate law that clearly does not comprise elementary processes, as is obvious from the unbalanced equations. Nonetheless, the five reactions in the model provide the means to kinetically describe the four essential stages of the BZ reaction [32]:

1. Formation of $HBrO_2$
2. Autocatalytic formation of $HBrO_2$
3. Consumption of $HBrO_2$
4. Oxidation of malonic acid (MA)

For the calculation of the kinetic profiles displayed in Figure 7.18, we used the rate constants $k_1 = 1.28$ M^{-1}s^{-1}, $k_2 = 33.6$ M^{-1}s^{-1}, $k_3 = 2.4 \times 10^6$ M^{-1}s^{-1}, $k_4 = 3 \times 10^3$ M^{-1}s^{-1}, $k_5 = 1$ M^{-1}s^{-1} for $[H]^+ = 0.8$ M at the initial concentrations $[BrO_3^-]_0 = 0.063$ M, $[Ce^{(IV)}]_0 = 0.002$ M ($= [M_{ox}]_0$), and $[MA]_0 = 0.275$ M [3, 32]. We applied again MATLAB's stiff solver `ode15s`. Note that for this example, MATLAB's default relative and absolute error tolerances (`RelTol` and `AbsTol`) for solving ODEs have to be adjusted to increase numerical precision.

For this example, we do not give the MATLAB code for the differential equations. The code can be fairly complex and, thus, its development is prone to error. The problem is even more critical in the spreadsheet application where several cells need to be rewritten for a new mechanism. We will address this problem later when we discuss the possibility of automatic generation of computer code based on traditional chemical equations.

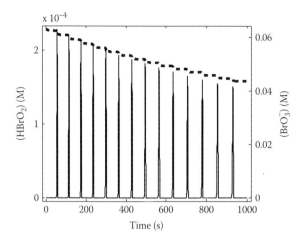

FIGURE 7.18 The BZ reaction as represented by the Oregonator model. Calculated concentration profiles for $HBrO_2$ (—) and BrO_3^- (…) toward the thermodynamic equilibrium. Note the different ordinates for $[HBrO_2]$ and $[BrO_3^-]$.

7.6 CALCULATION OF THE CONCENTRATION PROFILES: CASE III, VERY COMPLEX MECHANISMS

For most academic investigations, reaction conditions are kept under as much control as possible. Solutions are thermostatted and buffered, and investigations are carried out in an excess of an inert salt. This is done to keep temperature, pH, and ionic strength constant. In industrial situations, it is often not possible, nor is it necessarily desirable, to control conditions. Temperature fluctuations within safe limits are not necessarily detrimental and the addition of external buffer or salt is out of the question.

A few developments have been published recently that attempt to incorporate such experimental "inconsistencies" into the numerical analysis of the measurements [33–35]. The central formula, the set of differential equations that needs to be integrated, can be written in a very general way.

$$\dot{\mathbf{C}} = f(\mathbf{C}(\mathbf{k})) \quad (7.39)$$

The differential of the matrix of concentrations with respect to time, $\dot{\mathbf{C}}$, is a function of the matrix of concentrations, $\mathbf{C}$, and both depend on the chemical model with its vector of parameters, in our case the rate constants, $\mathbf{k}$. To accommodate experimental inconsistencies, such as the ones mentioned above, we need to adjust this set of equations appropriately.

Let us start with variable temperature. Rate constants are influenced by temperature, T, and the numerical solutions of the differential equations will be affected. We can write

$$\dot{\mathbf{C}} = f(\mathbf{C}(\mathbf{k}(T))) \quad (7.40)$$

There are two models that quantitatively describe the relationship between temperature and rate constants, the Arrhenius theory and the Eyring theory [2, 3]. Engineers prefer the Arrhenius equation because it is slightly simpler, while kineticists prefer the Eyring equation because its parameters (entropy and enthalpy of activation, $\Delta S^{\ne}$ and $\Delta H^{\ne}$, respectively) can be interpreted more directly. Here, we will use Eyring's equation.

$$\mathbf{k(T)} = \frac{k_B T}{h} e^{\frac{\Delta S^{\ne}}{R} - \frac{\Delta H^{\ne}}{RT}} \tag{7.41}$$

where R is the gas constant, and k_B and h are Boltzmann's and Planck's constants, respectively.

Whenever the ODE-solver calls for the calculation of the differential equations, the actual values for the rate constants have to be inserted into the appropriate equations. Obviously, the temperature has to be recorded during the measurement.

Figure 7.19 compares the concentration profiles for the simple reaction $A \rightarrow B$ at constant and increasing temperature. The concentration profiles for the isothermal reaction are the same as in Figure 7.1 and MATLAB Example 7.1. The nonisothermal reaction is based on the activation parameters $\Delta S^{\ne} = -5$ J mol^{-1} K^{-1} and $\Delta H^{\ne} = 80$ kJ mol^{-1} and a temperature gradient from 5 to 55°C over the same time interval. According to Eyring's equation (Equation 7.41), this leads to rate constants $\mathbf{k}(T)$ between 0.003 s^{-1} (t = 0 s, T = 5°C) and 0.691 s^{-1} (t = 50 s, T = 55°C).

There are clear advantages and also clear disadvantages in this new approach for the analysis of nonisothermal measurements [33]. Now there are two new parameters, $\Delta S^{\ne}$ and $\Delta H^{\ne}$, for each rate constant, i.e., there are twice as many parameters to be fitted. Naturally, this can lead to difficulties if not all of them are well defined. Another problem lies in the fact that molar absorptivity spectra of the species can show significant temperature dependencies. Advantages include the fact that, in

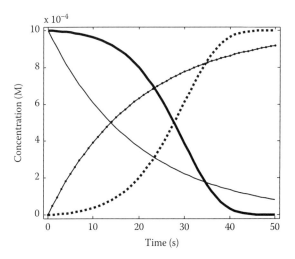

FIGURE 7.19 Concentration profiles for a first-order reaction $A \rightarrow B$ (— A, ... B) at constant (thin lines) and increasing (thick lines) temperature.

data_abc

```
function [t,Y]=data_abc

% absorbance data generation for A -> B -> C

t=[0:25:4000]';                         % reaction times
w=[300:(600-300)/(1024-1):600];%  1024 wavelengths
k=[.003 .0015];                         % rate constants
A_0=1e-3;                               % initial concentration of A

C(:,1)=A_0*exp(-k(1)*t);                % concentrations of species A
C(:,2)=A_0*k(1)/(k(2)-k(1))*(exp(-k(1)*t)-exp (-k(2)*t)); % conc. of B
C(:,3)=A_0-C(:,1)-C(:,2);               % concentrations of C

A(1,:)=1.0e3*exp(-((w-450).^2)/((60^2)/(log(2)*4))) + ...
       0.5e3*exp(-((w-270).^2)/((100^2)/(log(2)*4))); % molar spectrum of A
A(2,:)=1.5e3*exp(-((w-400).^2)/((70^2)/(log(2)*4))) + ...
       0.3e3*exp(-((w-250).^2)/((150^2)/(log(2)*4))); % molar spectrum of B
A(3,:)=0.8e3*exp(-((w-500).^2)/((80^2)/(log(2)*4))) + ...
       0.4e3*exp(-((w-250).^2)/((200^2)/(log(2)*4))); % molar spectrum of C

Y=C*A;               % applying Beer's law to generate Y
randn('seed',0);     % fixed start for random number generator
R=1e-2*randn(size(Y)); % normally distributed noise
Y=Y+R;               % of standard deviation 0.01
```

REFERENCES

1. Benson, S.W., *The Foundations of Chemical Kinetics*, McGraw-Hill, New York, 1960.
2. Wilkins, R.G., *Kinetics and Mechanism of Reactions of Transition Metal Complexes*, VCH, Weinheim, Germany, 1991.
3. Espenson, J.H., *Chemical Kinetics and Reaction Mechanisms*, McGraw-Hill, New York, 1995.
4. Mauser, H. and Gauglitz, G., in *Photokinetics: Theoretical, Fundamentals and Applications*, Compton, R.G. and Hancock, G., Eds., Elsevier Science, New York, 1998, p. 555.
5. *Matlab R11.1*, The Mathworks, Natick, MA, 1999; http://www.mathworks.com.
6. Hood, G., *Poptools*, CSIRO, Canberra, 2003; http://www.cse.csiro.au/poptools.
7. Bunker, D.L., Garrett, B., Kleindienst, T., and Long, G.S., III, Discrete simulation methods in combustion kinetics, *Combust. Flame*, 1974, 23, 373–379.
8. Gillespie, D.T., Exact stochastic simulation of coupled chemical reactions, *J. Phys. Chem.*, 1977, 81, 2340–2361.
9. Turner, J.S., Discrete simulation methods for chemical kinetics, *J. Phys. Chem.*, 1977, 81, 2379–2408.
10. Hinsberg, W. and Houle, F., *Chemical Kinetics Simulator (CKS)*, IBM Almaden Research Center, San Jose, CA, 1996; http://www.almaden.ibm.com/st/msim/ckspage.html.
11. Zheng, Q. and Ross, J., Comparison of deterministic and stochastic kinetics for nonlinear systems, *J. Chem. Phys.*, 1991, 94, 3644–3648.

12. Mathur, R., Young, J.O., Schere, K.L., and Gipson, G.L., A comparison of numerical techniques for solution of atmospheric kinetic equations, *Atmos. Environ.*, 1998, 32, 1535–1553.
13. Rodiguin, N.M. and Rodiguina, E.N., *Consecutive Chemical Reactions; Mathematical Analysis and Development*, D. van Nostrand, Princeton, NJ, 1964.
14. Seber, G.A.F. and Wild, C.J., *Nonlinear Regression*, John Wiley & Sons, New York, 1989.
15. Press, W.H., Vetterling, W.T., Teukolsky, S.A., and Flannery, B.P., *Numerical Recipes in C*, Cambridge University Press, Cambridge, 1995.
16. Bevington, P.R. and Robinson, D.K., *Data Reduction and Error Analysis for the Physical Sciences*, McGraw-Hill, New York, 2002.
17. Lagarias, J.C., Reeds, J.A., Wright, M.H., and Wright, P.E., Convergence properties of the Nelder-Mead simplex method in low dimensions, *SIAM J. Optimization*, 1998, 9, 112–147.
18. Billo, E.J., *Excel for Chemists: A Comprehensive Guide*, John Wiley & Sons, New York, 2001.
19. De Levie, R., *How to Use Excel in Analytical Chemistry and in General Scientific Data Analysis*, Cambridge University Press, Cambridge, 2001.
20. Kirkup, L., *Data Analysis with Excel: An Introduction for Physical Scientists*, Cambridge University Press, Cambridge, 2002.
21. Kirkup, L., *Principles and Applications of Non-Linear Least Squares: An Introduction for Physical Scientists Using Excel's Solver*, 2003; http://www.science.uts.cdu.au/physics/nonlin2003.html.
22. Maeder, M. and Zuberbühler, A.D., Nonlinear least-squares fitting of multivariate absorption data, *Anal. Chem.*, 1990, 62, 2220–2224.
23. Marquardt, D.W., An algorithm for least-squares estimation of nonlinear parameters, *J. Soc. Ind. Appl. Math.*, 1963, 11, 431–441.
24. Levenberg, K.Q., A method for the solution of certain non-linear problems in least squares, *Appl. Math.*, 1949, 2, 164.
25. Vajda, S. and Rabitz, H., Identifiability and distinguishability of first-order reaction systems, *J. Phys. Chem.*, 1988, 92, 701–707.
26. Vajda, S. and Rabitz, H., Identifiability and distinguishability of general reaction systems, *J. Phys. Chem.*, 1994, 98, 5265–5271.
27. O'Connor, D.V. and Phillips, D., *Time-Correlated Single Photon Counting*, Academic Press, London, 1984.
28. Bulirsch, R. and Stoer, J., *Introduction to Numerical Analysis*, Springer, New York, 1993.
29. Shampine, L.F. and Reichelt, M.W., The Matlab Ode Suite, *SIAM J. Sci. Comp.*, 1997, 18, 1–22.
30. Field, R.J., Körös, E., and Noyes, R.M., Oscillations in chemical systems, II: thorough analysis of temporal oscillation in the bromate-cerium-malonic acid system, *J. Am. Chem. Soc.*, 1972, 94, 8649–8664.
31. Field, R.J. and Noyes, R.M., Oscillations in chemical systems, IV: limit cycle behavior in a model of a real chemical reaction, *J. Chem. Phys.*, 1974, 60, 1877–1884.
32. Scott, S.K., *Oscillations, Waves, and Chaos in Chemical Kinetics*, Oxford Chemistry Press, Oxford, 1994.
33. Maeder, M., Molloy, K.J., and Schumacher, M.M., Analysis of non-isothermal kinetic measurements, *Anal. Chim. Acta*, 1997, 337, 73–81.

34. Wojciechowski, K.T., Malecki, A., and Prochowska-Klisch, B., REACTKIN — a program for modeling the chemical reactions in electrolytes solutions, *Comput. Chem.*, 1998, 22, 89–94.
35. Maeder, M., Neuhold, Y.M., Puxty, G., and King, P., Analysis of reactions in aqueous solution at non-constant pH: no more buffers? *Phys. Chem. Chem. Phys.*, 2003, 5, 2836–2841.
36. Bayada, A., Lawrance, G.A., Maeder, M., and Molloy, K.J., ATR-IR spectroscopy for the investigation of solution reaction kinetics — hydrolysis of trimethyl phosphate, *Appl. Spectrosc.*, 1995, 49, 1789–1792.
37. Binstead, R.A., Zuberbühler, A.D., and Jung, B., *Specfit/32*, Spectrum Software Associates, Chapel Hill, NC, 1999.
38. Puxty, G., Maeder, M., Neuhold, Y.-M., and King, P., *Pro-Kineticist II*, Applied Photophysics, Leatherhead, U.K., 2001; http://www.photophysics.com.
39. Dyson, R., Maeder, M., Puxty, G., and Neuhold, Y.-M., Simulation of complex chemical kinetics, *Inorg. React. Mech.*, 2003, 5, 39–46.
40. Missen, R., Mims, W.C.A., and Saville, B.A., *Introduction to Chemical Reaction Engineering and Kinetics*, John Wiley & Sons, New York, 1999.
41. Binstead, R.A., Jung, B., and Zuberbühler, A.D., *Specfit/32*, Spectrum Software Associates, Marlborough, MA, 2003; http://www.bio-logic.fr/rapid-kinetics/specfit/.
42. De Juan, A., Maeder, M., Martinez, M., and Tauler, R., Combining hard- and soft-modelling to solve kinetic problems, *Chemom. Intell. Lab. Syst.*, 2000, 54, 123–141.
43. Diewok, J., De Juan, A., Maeder, M., Tauler, R., and Lendl, B., Application of a combination of hard and soft modeling for equilibrium systems to the quantitative analysis of pH-modulated mixture samples, *Anal. Chem.*, 2003, 75, 641–647.
44. Furusjö, E. and Danielsson, L.-G., Target testing procedure for determining chemical kinetics from spectroscopic data with absorption shifts and baseline drift, *Chemom. Intell. Lab. Syst.*, 2000, 50, 63–73.
45. Jandanklang, P., Maeder, M., and Whitson, A.C., Target transform fitting: a new method for the non-linear fitting of multivariate data with separable parameters, *J. Chemom.*, 2001, 15, 511–522.
46. Windig, W. and Antalek, B., Direct exponential curve resolution algorithm (Decra) — a novel application of the generalized rank annihilation method for a single spectral mixture data set with exponentially decaying contribution profiles, *Chemom. Intell. Lab. Syst.*, 1997, 37, 241–254.

8 Response-Surface Modeling and Experimental Design

Kalin Stoyanov and Anthony D. Walmsley

CONTENTS

8.1 Introduction 264
8.2 Response-Surface Modeling 265
    8.2.1 The General Scheme of RSM..265
    8.2.2 Factor Spaces 268
        8.2.2.1 Process Factor Spaces..268
        8.2.2.2 Mixture Factor Spaces...269
        8.2.2.3 Simplex-Lattice Designs..272
        8.2.2.4 Simplex-Centroid Designs...275
        8.2.2.5 Constrained Mixture Spaces..279
        8.2.2.6 Mixture+Process Factor Spaces 283
    8.2.3 Some Regression-Analysis-Related Notation....................................286
8.3 One-Variable-at-a-Time vs. Optimal Design...288
    8.3.1 Bivariate (Multivariate) Example 288
    8.3.2 Advantages of the One-Variable-at-a-Time Approach 290
    8.3.3 Disadvantages..290
8.4 Symmetric Optimal Designs 290
    8.4.1 Two-Level Full Factorial Designs 290
        8.4.1.1 Advantages of Factorial Designs.......................................290
        8.4.1.2 Disadvantages of Factorial Designs 291
    8.4.2 Three or More Levels in Full Factorial Designs...............................291
    8.4.3 Central Composite Designs 293
8.5 The Taguchi Experimental Design Approach...294
8.6 Nonsymmetric Optimal Designs...298
    8.6.1 Optimality Criteria 298
    8.6.2 Optimal vs. Equally Distanced Designs..299
    8.6.3 Design Optimality and Design Efficiency Criteria 302
        8.6.3.1 Design Measures..303
        8.6.3.2 D-Optimality and D-Efficiency 304
        8.6.3.3 G-Optimality and G-Efficiency 305

	8.6.3.4	A-Optimality 306
	8.6.3.5	E-Optimality 306

8.7 Algorithms for the Search of Realizable Optimal
Experimental Designs 306
    8.7.1 Exact (or N-Point) D-Optimal Designs 307
        8.7.1.1 Fedorov's Algorithm 307
        8.7.1.2 Wynn-Mitchell and van Schalkwyk Algorithms 308
        8.7.1.3 DETMAX Algorithm 308
        8.7.1.4 The MD Galil and Kiefer's Algorithm 309
    8.7.2 Sequential D-Optimal Designs 310
        8.7.2.1 Example 311
    8.7.3 Sequential Composite D-Optimal Designs 313
8.8 Off-the-Shelf Software and Catalogs of Designs of Experiments 316
    8.8.1 Off-the-Shelf Software Packages 316
        8.8.1.1 MATLAB 316
        8.8.1.2 Design Expert 319
        8.8.1.3 Other Packages 319
    8.8.2 Catalogs of Experimental Designs 320
8.9 Example: the Application of DOE in Multivariate Calibration 321
    8.9.1 Construction of a Calibration Sample Set 321
        8.9.1.1 Identifying of the Number of Significant Factors 322
        8.9.1.2 Identifying the Type of the Regression Model 325
        8.9.1.3 Defining the Bounds of the Factor Space 327
        8.9.1.4 Estimating Extinction Coefficients 329
    8.9.2 Improving Quality from Historical Data 330
        8.9.2.1 Improving the Numerical Stability
                of the Data Set 333
        8.9.2.2 Prediction Ability 334
8.10 Conclusion 337
References 337

8.1 INTRODUCTION

The design of experiments (DOE) is part of response-surface modeling (RSM) methodology. The purpose of experimental designs is to deliver as much information as possible with a minimum of experimental or financial effort. This information is then employed in the construction of sensible models of the objects under investigation.

    This chapter is intended to describe the basic methods applied in the construction of experimental designs. An essential part of this chapter is the examination of a formal approach to investigating a research problem according to the "black box" principle and the factor spaces related to it.

    Most of the approaches are illustrated with examples. We also illustrate how experimental designs can serve to develop calibration sample sets — a widely applied method in chemometrics, especially multivariate calibration.

8.2 RESPONSE-SURFACE MODELING

8.2.1 The General Scheme of RSM

One can divide RSM into three major areas: the design of experiments, model fitting, and process or product optimization. One of the major origins of this area of statistical modeling is the classical paper of Box and Wilson [1].

Suppose for example, one wishes to optimize the yield of a batch chemical reaction by adjusting the operating conditions, which include the reaction temperature and the concentration of one of the reagents. The principles of *experimental design* describe how to plan and conduct experiments at different combinations of temperature and reagent concentration to obtain the maximum amount of information (a response surface) in the fewest number of experiments. When properly designed experiments are utilized, the principles of response-surface modeling can then be used to fit a statistical model to the measured response surface. In this example, the response surface is yield as a function of temperature and reagent concentration. Once a statistically adequate model is obtained, it can be used to find the set of optimum operating conditions that produce the greatest yield.

As a general approach in RSM, one uses the "black box" principle (see Figure 8.1a). According to this principle, any technological process can be characterized by its input variables, x_i, $i = 1, \ldots, n$,; the output or response variables, y_i, $i = 1, \ldots, s$; and the noise variables, w_i, $i = 1, \ldots, l$. One then considers two ways of performing an experiment, active or passive. There are several important presumptions for active experiments:

- The set of the noise variables, in comparison with the input variables, is assumed to have insignificant influence on the process.
- During an active experiment, the experimenter is presumed to be able to *control* the values of x_i, with negligibly small error, when compared with the range of the variation of each of the input variables. ("To control" here means to be able to set a desirable value of each of the input variables and to be able maintain this value until the necessary measurement of the process output or response variable(s) has been performed.)
- The experimenter is presumed to be able to measure the output variables, y_i, $i = 1, s$ again with negligibly small error, when compared with the range of their variation.

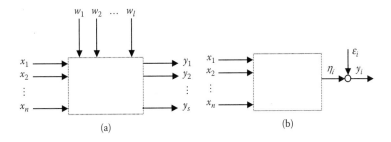

FIGURE 8.1 The "black box" principle.

The active-experiment approach is usually applicable in laboratory conditions. In fact, this is the major area of application of the methods of the experimental design. In the case where the experimenter is not able to control the input variables, one deals with a passive experiment. In this case, the experimenter is presumed able to measure, with negligible error, the values of the input variables (or factors) as well as the values of the output variables, i.e., the responses. In this case, it is also possible for the principles of experimental design to be applied [2]. Passive experiments are common in process analysis where the user has little or no control over the process variables under investigation.

It is assumed that the summation of the noise variables over each of the s responses can be represented by one variable ε_i, $i = 1,..., s$. Without any loss of generality, we assume that values of the error variables are distributed normally having variance σ^2 and mean zero, $\varepsilon_i \sim \mathcal{N}(0, \sigma^2)$. Based on these assumptions, one can represent the "black box" principle in a slightly different way (see Figure 8.1b). Now the measurable response variable, y_i, can be represented as:

$$y_i = \eta_i + \varepsilon_i \tag{8.1}$$

where η_i is the real but unknown value of the response and ε_i is the value of the random error associated with y_i. The response function is:

$$\eta_i = \eta_i(x_1, x_2, ..., x_n), \quad i = 1, s \tag{8.2}$$

It is a general assumption in RSM that, within the operational region (the area of feasible operating conditions), the function η is continuous and differentiable. Let $\mathbf{x}_0 = (x_{10}, x_{20}, ..., x_{n0})^T$ represent the vector of some particular feasible operating condition, e.g., a point in the operational region. It is known that we can expand η around $\mathbf{x}_0$ in a Taylor series:

$$\eta(\mathbf{x}) = \eta(x_0) + \sum_{i=1}^{n} \frac{\partial \eta(\mathbf{x})}{\partial x_i}\bigg|_{x=x_0} (x_i - x_{i0})$$

$$+ \sum_{i=1}^{n-1} \sum_{i=i+1}^{n} \frac{1}{2!} \frac{\partial^2 \eta(\mathbf{x})}{\partial x_i \partial x_j}\bigg|_{x=x_0} (x_i - x_{i0})(x_j - x_{j0}) \tag{8.3}$$

$$+ \sum_{j=i+1}^{n} \frac{1}{2!} \frac{\partial^2 \eta(\mathbf{x})}{\partial x_i^2}\bigg|_{x=x_0} (x_i - x_{i0})^2 + \cdots$$

By making the following substitutions

$$\beta_0 = \eta(\mathbf{x})_0 \quad \beta_i = \frac{\partial \eta(\mathbf{x}_0)}{\partial x_i}\bigg|_{x=x_0} \quad \beta_{ij} = \frac{1}{2!} \frac{\partial^2 \eta(\mathbf{x}_0)}{\partial x_i \partial x_j}\bigg|_{x=x_0} \quad \beta_{ii} = \frac{1}{2!} \frac{\partial^2 \eta(\mathbf{x}_0)}{\partial x_i^2}\bigg|_{x=x_0} \tag{8.4}$$

Equation 8.3 takes on a familiar polynomial form of the type:

$$\eta(\mathbf{x}) = \beta_0 + \sum_{i=1}^{n} \beta_i x_i + \sum_{i=1}^{n-1}\sum_{j=i+1}^{n} \beta_{ij} x_i x_j + \sum_{j=1}^{n} \beta_{ii} x_i^2 + \cdots \quad (8.5)$$

The coefficients β_i in Equation 8.5 describe the behavior of the function η near the point $\mathbf{x}_0$. If one is able to estimate values for β_i, then a model that describes the object can be built. The problem here is that, according to Equation 8.1, we can only have indirect measurements of the real values of η, hence we are unable to calculate the real coefficients $\boldsymbol{\beta}$ of the model described by Equation 8.5. Instead, we can only calculate their estimates, $\hat{\mathbf{b}}$. Also, since the Taylor series is infinite, we must decide how many and which terms in Equation 8.5 should be used.

The typical form of the regression model is

$$\hat{y}_j = \sum_{i=0}^{k-1} b_i f_i(\mathbf{x}), j = 1, s \quad (8.6)$$

Considering Equation 8.3, Equation 8.6 receives its widely used form,

$$\hat{y}_j = b_0 + \sum_{i=1}^{k-1} b_i f_i(\mathbf{x}), j = 1, s \quad (8.7)$$

where $\hat{y}_j$ is the predicted value of the jth response at an arbitrary point $\mathbf{x}$, k is the number of the regression coefficients, b_i is the estimate of the ith regression coefficient, and $f_i(\mathbf{x})$ is the ith regression function.

The method of response-surface modeling provides a framework for addressing the above problems and provides accurate estimates of the real coefficients, $\boldsymbol{\beta}$. The basic steps of RSM methodology are

1. Choose an appropriate response function, η.
2. Choose appropriate factors, $\mathbf{x}$, having a significant effect on the response.
3. Choose the structure of the regression model — a subset of terms from Equation 8.5.
4. Design the experiment.
5. Perform the measurements specified by the experimental design.
6. Build the model and calculate estimates of the regression coefficients, $\hat{\mathbf{b}}$.
7. Perform a statistical analysis of the model to prove that it describes adequately the dependence of the measured response on the controlled factors.
8. Use the model to find optimal operating conditions of the process under investigation. This is done by application of a numerical optimization algorithm using the model as the function to be optimized.
9. Check in practice whether the predicted optimal operating conditions actually deliver the optimal (better in some sense) values of the response.

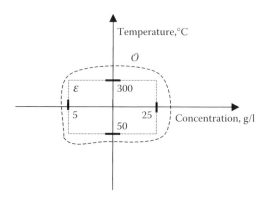

FIGURE 8.2 Two-dimensional factor space showing experimental region, $\mathcal{E}$, and operating region, $\mathcal{O}$.

8.2.2 FACTOR SPACES

8.2.2.1 Process Factor Spaces

We next define the concepts of the operating region, $\mathcal{O}$, and experimental region, $\mathcal{E}$ (see Figure 8.2). The operating region is the set of all theoretically possible operating conditions for the input variables, **x**. For example, in a chemical reactor, the upper and lower bounds of the operating conditions for temperature might be dictated by the reaction mixture's boiling point and freezing point. The reactor simply cannot be operated above the boiling point or below the freezing point. Usually there is only a rough idea about where the boundaries of $\mathcal{O}$ are actually situated. The experimental region $\mathcal{E}$ is the area of experimental conditions where investigations of the process take place. We define $\mathcal{E}$ by assigning some boundary values to each of the input variables. The boundaries of $\mathcal{E}$ can be independent of actual values of the factors or they can be defined by some function of **x**.

All possible operating conditions are represented as combinations of the values of the input variables. Each particular combination $\mathbf{x} = \{x_1, x_2, ..., x_m\}$ is represented as a point in a Descartes coordinate system. It is important that each point included in $\mathcal{E}$ must be feasible. This includes the points positioned in the interior and also on the boundaries of $\mathcal{E}$, which represent extreme operating conditions, those typically positioned at the edges or corners of $\mathcal{E}$.

The input variables shown in Figure 8.1 can be divided into two main groups, process variables and mixture variables. Process variables are mutually independent, thus we can change the value of each of them without any effect on the values of the others. Typical examples for process variables are temperature, speed of stirring, heating time, or amount of reagent. Examples of process factor spaces are shown in Figure 8.2 and Figure 8.3.

It would be convenient if all of the calculations related to the values of the process variables could be performed using their natural values or natural scales; for instance, the temperature might be varied between 100 and 400°C and the speed

Response-Surface Modeling and Experimental Design

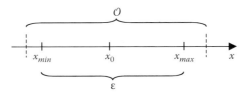

FIGURE 8.3 One-dimensional factor space in natural variables.

between 1 and 3 rpm. Unfortunately, this is not recommended because, practically speaking, most calculations are sensitive to the scale or magnitude of the numbers used. This is why we apply a simple transformation of the process variables. Each variable is coded to have values within the range [−1, 1]. The transformation formula is shown in Equation 8.8

$$x_i = \frac{\tilde{x}_i - \tilde{x}_{0i}}{|\tilde{x}_{i\max} - \tilde{x}_{0i}|}, i = 1, p \tag{8.8}$$

where $\tilde{x}_i$ is the natural value of the ith variable, $\tilde{x}_{0i}$ is its mean, and p is the number of process variables. Factor transformation or factor coding is illustrated graphically in Figure 8.4 for two-dimensional and three-dimensional process factor spaces.

After completing the response-surface modeling process described here, the inverse transformation can be used to obtain the original values of the variables.

$$\tilde{x}_i = (|\tilde{x}_{i\max} - \tilde{x}_{0i}|)x_i + \tilde{x}_{0i}, i = 1, p \tag{8.9}$$

8.2.2.2 Mixture Factor Spaces

Quite frequently, and especially in research problems arising in chemistry and chemistry-related areas, an important type of factor variables is encountered, and these are called "mixture variables." Apart from the usual properties that are common to all factors considered in the "black box" approach (see Figure 8.1), mixture

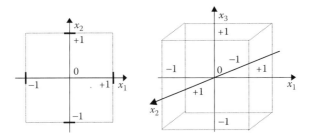

FIGURE 8.4 Two- and three-dimensional factor spaces in coded variables.

variables have some additional features. The most commonly encountered are those involving constraints imposed on the values of two or more variables, as shown in Equation 8.10 and Equation 8.11

$$\sum_{i=1}^{q} x_i = 1 \qquad (8.10)$$

subject to

$$0 \le x_i \le 1, i = 1, \ldots, q \qquad (8.11)$$

The constraints shown in Equation 8.10 and Equation 8.11 are a consequence of the nature of mixture problems. In the example illustrated by these equations, each variable represents the relative proportion of a particular ingredient in a mixture blended from q components. For example, a mixture of three components, where the first component makes up 25% of the total, the second component makes up 15% of the total, and the third component makes up 60% of the total, is said to be a ternary mixture. The respective values of the mixture variables are $x_1 = 0.15$, $x_2 = 0.25$, $x_3 = 0.60$, giving $x_1 + x_2 + x_3 = 1$. Depending on the number of mixture variables, the mixture could be binary, ternary, quaternary, etc.

For a mixture with q variables (i.e., q components), the mixture factor space is a subspace of the respective q-variables in Euclidean space. In Figure 8.5, Figure 8.6, and Figure 8.7, we see the relationship between the mixture coordinate system and the respective Euclidean space. Figure 8.5 illustrates the case of a binary mixture. The constraint described by Equation 8.11 holds for points A, B, and C; however, only point B and all points on the heavy line in Figure 8.5 are points from the mixture space satisfying the conditions described by the constraints in Equation 8.10 as well.

Figure 8.6 illustrates the relationship between Euclidean and mixture-factor space for three variables. Here, we see that the set of the mixture points lying on

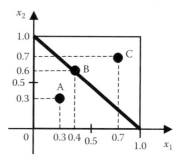

FIGURE 8.5 Relationship between the barycentric and Descartes coordinate systems, two-dimensional example.

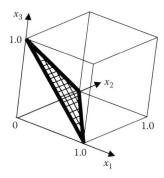

FIGURE 8.6 Relationship between the barycentric and Descartes coordinate systems, three-dimensional example.

the patterned plane inside the cube follows the constraints described by Equation 8.11. Figure 8.7 shows the point with coordinates $x_1 = 0.15$, $x_2 = 0.25$, $x_3 = 0.60$ in a simplex coordinate system (a) and its position in the corresponding Descartes coordinate system (b). The geometric figure in which the points lie in a barycentric coordinate system is called a "simplex." The name originates from the fact that any q-dimensional simplex is the simplest convex q-dimensional figure.

Systematic work on experimental designs in the area of mixture experiments was originated by Henry Scheffé, [3, 4]. Cornell provides an extensive reference on the subject [5].

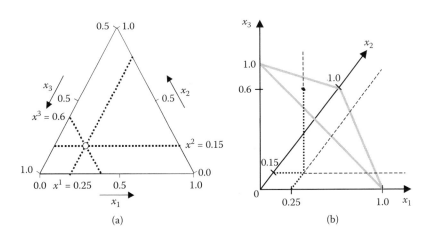

FIGURE 8.7 Coordinates of the point $\mathbf{x} = [x_1 = 0.25, x_2 = 0.15, x_3 = 0.60]$ in a mixture factor space. The position of the point is shown in barycentric (a) and Descartes (b) coordinate systems.

8.2.2.3 Simplex-Lattice Designs

The first designs for mixture experiments were described by Scheffé [3] in the form of a grid or lattice of points uniformly distributed on the simplex. They are called "$\{q, v\}$ simplex-lattice designs." The notation $\{q, v\}$ implies a simplex lattice for q components used to construct a mixture polynomial of degree v. The term "mixture polynomial" is introduced to distinguish it from the polynomials applicable for mutually independent or process variables, which are described later in our discussion of factorial designs (section 8.4). In this way, we distinguish "mixture polynomials" from classical polynomials.

As seen in Equation 8.10, there is a linear dependence between the input variables or controlled factors that create a nonunique solution for the regression coefficients if calculated by the usual polynomials. To avoid this problem, Scheffé [3] introduced the *canonical form* of the polynomials. By simple transformation of the terms of the standard polynomial, one obtains the respective canonical forms. The most commonly used mixture polynomials are as follows:

Linear:

$$\hat{y} = \sum_{i=1}^{q} b_i x_i; \tag{8.12}$$

Quadratic:

$$\hat{y} = \sum_{i=1}^{q} b_i x_i + \sum_{i=1}^{q-1} \sum_{j=i+1}^{q} b_{ij} x_i x_j; \tag{8.13}$$

Full cubic:

$$\hat{y} = \sum_{i=1}^{q} b_i x_i + \sum_{i=1}^{q-1} \sum_{j=i+1}^{q} b_{ij} x_i x_j + \sum_{i=1}^{q-1} \sum_{j=i+1}^{q} c_{ij} x_i x_j (x_i - x_j) + \sum_{i=1}^{q-2} \sum_{j=i+1}^{q-1} \sum_{l=j+1}^{q} b_{ijl} x_i x_j x_l; \tag{8.14}$$

Special cubic:

$$\hat{y} = \sum_{i=1}^{q} b_i x_i + \sum_{i=1}^{q-1} \sum_{j=i+1}^{q} b_{ij} x_i x_j + \sum_{i=1}^{q-2} \sum_{j=i+1}^{q-1} \sum_{l=j+1}^{q} b_{ijl} x_i x_j x_l. \tag{8.15}$$

The simplex-lattice type of experimental design for these models consists of points having coordinates that are combinations of the vth proportions of the variables:

$$x_i = \frac{0}{v}, \frac{1}{v}, \ldots, \frac{v}{v}, \quad i = 1, 2, \ldots, q. \tag{8.16}$$

As an example, the design $\{q = 3, v = 2\}$ for three mixture variables and a quadratic ($v = 2$) model consists of all possible combinations of the values:

$$x_i = \left(\frac{0}{2}, \frac{1}{2}, \frac{2}{2}\right) = \left(0, \frac{1}{2}, 1\right) \quad i = 1, 2, 3. \tag{8.17}$$

The corresponding design consists of the points having the coordinates:

$$\begin{bmatrix} 1,0,0 \\ 0,1,0 \\ 0,0,1 \\ 1/2, 1/2, 0 \\ 1/2, 0, 1/2 \\ 0, 1/2, 1/2 \end{bmatrix}$$

We can quickly verify that all combinations of the values listed in Equation 8.17 are subject to the constraint shown in Equation 8.10. The design constructed in this manner is shown in Figure 8.8b.

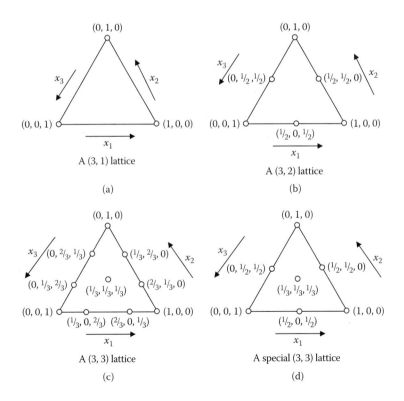

FIGURE 8.8 Examples of simplex lattices for (a) linear, (b) quadratic, (c) full cubic, and (d) special cubic models.

8.2.2.3.1 Advantages of the Simplex-Lattice Designs

Simplex-lattice designs historically were the first designs intended to help in the research of mixtures. They are simple to construct and there are simple formulas to calculate the regression coefficients using a conventional hand calculator. Furthermore, the regression coefficients are easy to interpret. For example, the values of b_i in the models described by Equation 8.12 through Equation 8.15 represent the effect of the individual components (input variables) on the magnitude of the response, where $x_i = 1$, $x_j = 0$, $j = 1, q, i \neq j$. The magnitude of the b_i coefficients thus gives an estimate of the relative importance of the individual components on the outcome (i.e., response) of the experiment. Unfortunately, interpretation of the higher-order coefficients, namely β_{ij}, β_{ijl}, δ_{ij}, is not so straightforward. Each of these coefficients is influenced by several factors.

The simplex-lattice designs are *composite* designs. Usually, at the beginning of a research project, the experimenter does not know the correct order of the model that best describes the relationship between the input factors and the response. If a model is chosen with too high of an order when the true model is of a lower order, then overfitting combined with an unnecessarily large number of experiments is the likely outcome. By using *composite* designs, the experimenter can start with a model of low order, possibly even a linear model, which is the lowest possible order. If the resulting model does not appear to be inadequate, it is possible to simply *add* new observations to the existing ones and fit a higher-order model, giving new regression coefficients. For example, in the case of a three-factor mixture problem, one can start with the first-order {3,1} design shown in Figure 8.8a. After the measurements are performed, the model described by Equation 8.12 can be used to calculate the regression coefficients. If an excessive lack of fit is observed, additional measurements can be added at the points [0.5, 0.5 0.0], [0.5, 0.0, 0.5], and [0.0, 0.5, 0.5], giving a second-order lattice, {3,2}, shown in Figure 8.8b. The augmented experimental data can then be used to fit a model of the type described by Equation 8.13. If the resulting model is still not satisfactory, another measurement at [1/3, 1/3, 1/3] can be added to construct the special cubic model described by Equation 8.15. Unfortunately, the *composite* feature of the simplex-lattice designs does not continue to higher orders beyond the special cubic order. For example, in order to get a *full* cubic design from *special* cubic design, the measurements at the three points of the type {0.5, 0.5, 0} must be discarded and replaced with another six of the type {$^1/_3$, $^2/_3$, 0,}.

Another advantage of simplex-lattice designs having special cubic order or lower order is that they are D-optimal. Namely, they have the maximum value of the determinant of the information matrix in the case of mixtures, $\mathbf{X^TX}$. Another common advantage of simplex-lattice designs is the possibility of generating component contour plots showing the behavior of the model in a three-dimensional space.

8.2.2.3.2 Disadvantages of the Simplex-Lattice Designs

The simplex-lattice designs are applicable only for problems where the condition

$$0 \leq x_i \leq 1, i = 1, \ldots, q \tag{8.18}$$

holds. This means that each of the components has to be varied between 0 and 100%. These designs assume that it is possible to prepare "mixtures" where one or more

Response-Surface Modeling and Experimental Design

of the components are not included (i.e., having 0% concentrations). This may not be practical for some particular investigations. Additionally, phase changes (e.g., solid–liquid) must not occur during the variation of the components. All mixtures must be homogeneous.

8.2.2.3.3 Simplex-Lattice Designs, Example

Suppose we wish to construct an experimental design for a mixture consisting of four components and model it with a cubic polynomial model, as described by Equation 8.14. Our task is to construct a {4,3} simplex-lattice design with $q = 4$ and $v = 3$. The proportions of each of the components are calculated by Equation 8.16, giving the following values:

$$x_i = 0, 1/3, 2/3, 3/3 \approx (0, 0.333, 0.666, 1)$$

The design matrix consists of all permutations of these proportions. The structure of the regression model will be

$$\hat{y} = \beta_1 x_1 + \beta_2 x_2 + \beta_3 x_3 + \beta_4 x_4 + \beta_{12} x_{12} + \beta_{13} x_{13} + \beta_{14} x_{14} + \beta_{23} x_{23} + \beta_{24} x_{24} + \beta_{34} x_{34}$$
$$+ \delta_{12} x_1 x_2 (x_1 - x_2) + \delta_{13} x_1 x_3 (x_1 - x_3)$$
$$+ \delta_{14} x_1 x_4 (x_1 - x_4) + \delta_{23} x_2 x_3 (x_2 - x_3) + \delta_{24} x_2 x_4 (x_2 - x_4)$$
$$+ \delta_{34} x_3 x_4 (x_3 - x_4) + \delta_{123} x_1 x_2 x_3 + \delta_{124} x_1 x_2 x_4 + \delta_{134} x_1 x_3 x_4 + \delta_{234} x_2 x_3 x_4$$

Rows 1 to 4 in Table 8.1 represent all of the possible combinations of the proportions 1 and 0. Rows 5 to 16 include all combinations of (0.666, 0.333, and 0), and rows 17 to 20 include all combinations of (0.333, 0.333, 0.333, and 0). The far right column represents the values where the measurements of the responses should be recorded. The subscripts are added for convenience and denote the numbers of the factors having values different from zero. The values for the responses could be single measurements or mean values of several replicates.

8.2.2.4 Simplex-Centroid Designs

One of the major shortcomings of simplex-lattice designs is that they include blends that consist of only v components, where v is the order of the model. For example, if one intends to explore a five-component system, applying a second-order model would only give mixtures of up to two components in a {5, 2} simplex-lattice design. No mixtures of the type {1/3,1/3,1/3, 0, 0}, {1/4,1/4,1/4,1/4, 0}, or {1/5,1/5,1/5,1/5, 1/5, 0}, appear in this design. Figure 8.8b illustrates the same principle for a {3, 2} simplex-lattice design. In this case, no point of the type {1/3,1/3,1/3} appears. The lack of measurements at blends consisting of a higher number of components decreases chances that the model will describe high-order interactions or sharp changes in the response surface. To improve the distribution of the points within the simplex space, Scheffé [4] introduced *simplex-centroid designs*. These designs are constructed of points where

TABLE 8.1
Four-Component Simplex-Lattice Mixture Design for a Cubic Polynomial Model

No.	Proportions of the Mixture Components				Response
	x_1	x_2	x_3	x_4	y
1	1	0	0	0	y_1
2	0	1	0	0	y_2
3	0	0	1	0	y_3
4	0	0	0	1	y_4
5	0.666	0.333	0	0	y_{112}
6	0.666	0	0.333	0	y_{113}
7	0.666	0	0	0.333	y_{114}
8	0.333	0.666	0	0	y_{122}
9	0.333	0	0.666	0	y_{133}
10	0.333	0	0	0.666	y_{144}
11	0	0.666	0.333	0	y_{223}
12	0	0.666	0	0.333	y_{224}
13	0	0.333	0.666	0	y_{233}
14	0	0.333	0	0.666	y_{244}
15	0	0	0.666	0.333	y_{334}
16	0	0	0.333	0.666	y_{344}
17	0.333	0.333	0.333	0	y_{123}
18	0.333	0.333	0	0.333	y_{124}
19	0.333	0	0.333	0.333	y_{134}
20	0	0.333	0.333	0.333	y_{234}

the nonzero compounds in the blends are of equal proportions. For example, the design for a four-component system consists of:

1. All blends with one nonzero compound, i.e., the vertices of the simplex, $x_i = 1$, $x_j = 0$, $i \neq j$, $j = 1,q$
2. All blends with two nonzero compounds, i.e., $x_i = x_j = 1/2$, $x_l = 0$, $l \neq i$, $l \neq j$, $l = 1, q$
3. All blends with three nonzero compounds, i.e., $x_i = x_j = x_l = 1/3$, $x_l = 0$, $l \neq i$, $l \neq j$, $l \neq k$, $l = 1,q$
4. One blend where all of the four compounds are presented in equal proportions, i.e., $x_1 = x_2 = x_3 = x_4 = 1/4$.

The points of this type of design are positioned at the vertices, the center of the edges of the simplex, the centroids of the planes of the simplex, and the centroid of the simplex. To generalize the notation, we can consider the vertices of the simplex [1, 0, ..., 0] as centroids of a zero-dimensional plane, the points at the edges [1/2,1/2,0,..., 0] as centroids of one-dimensional planes, the points of the type [1/3,1/3,1/3,0,..., 0] as centroids of two-dimensional planes, the points of the type [1/4,1/4,1/4,1/4,0,..., 0] as centroids of three-dimensional planes, and so on. The

Response-Surface Modeling and Experimental Design

simplex-centroid designs, unlike the simplex lattices, are not model dependent. For each number of components there is only one design. Hence, provided with the number of the components, q, one can calculate the number of the points in a simplex-centroid design from the formula in Equation 8.19

$$N = \sum_{r=1}^{q} \frac{q!}{r!(q-r)!} = \sum_{r=1}^{q} \binom{q}{r} = 2^q - 1, \qquad (8.19)$$

where the quantity $\binom{q}{r}$ is the well-known binomial coefficient and represents the number of the centroids on the $r-1$ dimensional planes. Thus $\binom{4}{1} = \frac{4!}{1!(4-1)!} = 4$ is the number of the centroids on $r-1 = 1-1 = 0$-dimensional planes, $\binom{4}{2} = \frac{4!}{2!(4-2)!} = 6$ is the number of the centroids on $r-1 = 2-1 = 1$-dimensional planes, and so on. It is easiest if one thinks of r simply as the number of the nonzero components included in the centroid.

The common structure of the regression model applicable to all simplex-centroid designs is shown in Equation 8.20.

$$\hat{y} = \sum_{i=1}^{q} \beta_i x_i + \sum_{i=1}^{q-1}\sum_{j=i+1}^{q} \beta_{ij} x_i x_j + \sum_{i=1}^{q-2}\sum_{j=i+1}^{q-1}\sum_{l=j+1}^{q} \beta_{ijl} x_i x_j x_j + \ldots + \beta_{12\ldots q} x_1 x_2 \ldots x_q. \qquad (8.20)$$

For example, the model for $q = 4$ is:

$$\begin{aligned}\hat{y} = {}& b_1 x_1 + b_2 x_2 + b_3 x_3 + b_4 x_4 + b_{12} x_1 x_2 + b_{13} x_1 x_3 + b_{14} x_1 x_4 \\ & + b_{23} x_2 x_3 + b_{24} x_2 x_4 + b_{34} x_3 x_4 + b_{123} x_1 x_2 x_3 + b_{124} x_1 x_2 x_4 \\ & + b_{134} x_1 x_3 x_4 + b_{234} x_2 x_3 x_4 + b_{1234} x_1 x_2 x_3 x_4 \end{aligned} \qquad (8.21)$$

8.2.2.4.1 Advantages of Simplex-Centroid Designs

There are two important advantages of the simplex-centroid type of design. Firstly, simplex-centroid designs are D-optimal, which means that they have the maximum value of the determinant of the information matrix. Secondly, simplex-centroid designs can be extended to include new *variables*. For instance, one can perform the experiments specified by the simplex-centroid design for $q = 4$ variables and increase the complexity of the problem at a later time by adding more components, for example $q = 6$. The resulting measurements can be used to augment the *old* design matrix.

8.2.2.4.2 Disadvantages of Simplex-Centroid Designs

The major disadvantage of simplex-centroid designs is that only one type of model can be applied, namely a model having the structure shown in Equation 8.20.

TABLE 8.2
Experimental Design for Lithium Lubricant Study

No.	Blend Proportions				Colloid Stability (%)
	x_1	x_2	x_3	x_4	y
1	1	0	0	0	$y_1 = 11.30$
2	0	1	0	0	$y_2 = 9.970$
3	0	0	1	0	$y_3 = 9.060$
4	0	0	0	1	$y_4 = 7.960$
5	0.5	0.5	0	0	$y_{12} = 8.540$
6	0.5	0	0.5	0	$y_{13} = 5.900$
7	0.5	0	0	0.5	$y_{14} = 8.500$
8	0	0.5	0.5	0	$y_{23} = 12.660$
9	0	0.5	0	0.5	$y_{24} = 6.210$
10	0	0	0.5	0.5	$y_{34} = 11.000$
11	0.333	0.333	0.333	0	$y_{123} = 8.260$
12	0.333	0.333	0	0.333	$y_{124} = 7.950$
13	0.333	0	0.333	0.333	$y_{134} = 7.100$
14	0	0.333	0.333	0.333	$y_{234} = 8.470$
15	0.25	0.24	0.25	0.25	$y_{1234} = 7.665$

8.2.2.4.3 Simplex-Centroid Design, Example

Suppose we wish to investigate the influence of four aliphatic compounds, designated x_1, x_2, x_3, and x_4, on the "colloidal stability" of lithium lubricants. The general aim is to search for blends having a minimum quantity of expensive 12-hydroxystearic acid without decreasing the quality of the lubricant. It is possible to investigate the full range from 0 to 100% for each of the components. Additionally, the research team is interested in investigating blends having two, three, and four components. Based on these criteria, it is decided that a four-component simplex-centroid design should be used. The experimental design and the measured response values are shown in Table 8.2.

The structure of the regression model is shown in Equation 8.21. After calculating the regression coefficients, the resulting regression model is obtained:

$$\hat{y} = 11.3x_1 + 9.97x_2 + 9.06x_3 + 7.96x_4 - 8.38x_1x_2 - 17.12x_1x_3 - 4.52x_1x_4$$
$$+ 12.58x_2x_3 - 11.02x_2x_4 + 9.96x_3x_4 - 11.19b_{123}x_1x_2x_3 + 23.34x_1x_2x_4$$
$$- 28.14x_1x_3x_4 - 48.78x_2x_3x_4 + 66.76x_1x_2x_3x_4$$

The model lack-of-fit was estimated from six additional measurements, shown below at points not included in the original design, where y represents the measured response and $\hat{y}$ represents the model estimated response.

No.	x_1	x_2	x_3	x_4	y	$\hat{y}$	$y - \hat{y}$	$(y - \hat{y})^2$
1	0.666	0.333	0	0	8.70	8.99	−0.29	0.0841
2	0.333	0.666	0	0	8.65	8.54	0.11	0.0121
3	0	0.666	0.333	0	12.75	12.45	0.30	0.09
4	0	0.333	0.666	1	12.01	12.14	−0.13	0.0169
5	0	0	0.666	0.333	11.10	10.89	0.21	0.0441
6	0	0	0.333	0.666	10.40	10.52	−0.12	0.0144

From these six additional experiments, the estimated value of the residual variance or model lack-of-fit is:

$$S_{res}^2 = \frac{\sum_{i=1}^{N}(y_i - \hat{y}_i)^2}{N} = \frac{0.2618}{6} = 0.0436$$

To estimate the variance of error in the measurements, an additional $N_\varepsilon = 4$ measurements were performed at one point in the simplex, giving

$$S_\varepsilon^2 = \frac{\sum_{i=1}^{N_\varepsilon}(y_i - \bar{y})^2}{N_\varepsilon} = 0.0267$$

with degrees of freedom $v_\varepsilon = 4 - 1 = 3$. An F-test can be used to compare the model lack-of-fit with the measurement variance, giving the following F-ratio:

$$F = \frac{S_{res}^2}{S_\varepsilon^2} = \frac{0.436}{0.0267} = 1.633.$$

The critical value of the F-statistic is $F_{(6,3,0.95)} = 8.94$. Comparing the calculated value of F with the critical value $F = 1.633 < F_{(6,3,0.95)} = 8.94$, we conclude that there is no evidence for lack of model adequacy; hence, the model is statistically acceptable. Having produced an acceptable regression model, we can generate a grid of points within the simplex using some satisfactory small step, say $\delta = 0.01$, to calculate the predicted value of the response at each point in the grid. The best mixture is the point that satisfies the initial requirements, i.e., the point having a high response (percent colloid stability) at a low quantity of the expensive ingredient.

8.2.2.5 Constrained Mixture Spaces

It is not uncommon for mixtures having zero amount of one ore more of the components to be of little or no practical use. For example, consider a study aimed

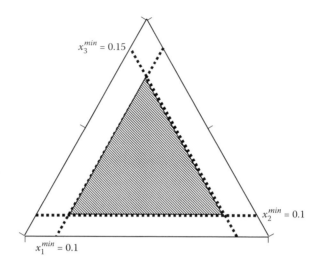

FIGURE 8.9 Constrained region of three mixture variables with lower bounds only.

at determining the proportions of cement, sand, and water giving a concrete mixture of maximal strength. Obviously, it is not practical to study mixtures consisting of 100% water, or 50% water and 50% sand; hence, we must define some subregion in the mixture space where some sensible experiments can be performed and meaningful responses obtained. In cases such as these, we define new constraints in addition to the constraints imposed by Equation 8.10 and Equation 8.11:

$$a_i \leq x_i \leq b_i, \quad i = 1, q. \qquad (8.22)$$
$$0 \leq a_i, b_i \leq 1$$

There are some special but quite widespread cases where only lower or upper bounds are imposed. For the case where only lower bounds define the subregion, Equation 8.22 becomes

$$a_i \leq x_i \leq 1, \quad i = 1, q. \qquad (8.23)$$

which is illustrated graphically in Figure 8.9.

The shaded subregion shown in Figure 8.9 also has the shape of a simplex. To avoid the inconvenience of working with lower bounds, we transform the coordinates of the points in the subregion to achieve lower bounds equal to 0 and upper bounds equal to 1. The original input variables of the mixture design can be transformed into pseudocomponents by using the following formula:

$$z_i = \frac{x_i - a_i}{1 - A}, i = 1, q \qquad (8.24)$$

where

$$A = \sum_{i=1}^{q} a_i < 1 \tag{8.25}$$

is the sum of the lower bounds. By employing this approach, all of the methods applicable for analysis of unconstrained mixture problems can be used. To perform the measurements, we need to have the original values of the input variables (or components). To reconstruct the design in the original coordinates from the pseudocomponents, the inverse transformation described in Equation 8.26 is used:

$$x_i = a_i + (1 - A)z_i \tag{8.26}$$

The case where only upper bounds are applied is a bit more complicated, namely

$$0 \leq x_i \leq b_i, \ i = 1, q. \tag{8.27}$$

Typical examples of the subregion defined only by upper bounds are shown in Figure 8.10. The shape of the subregion is an inverted simplex. The planes and edges of the subregion cross the corresponding planes and edges of the original simplex. In the case shown in Figure 8.10a, the entire subregion lies within the unconstrained simplex. It is possible, however, to have a case such as the one shown in Figure 8.10b, where part of the inverted simplex determined by the upper bounds lies outside the original one. As a result, the feasible region, i.e., the area where the measurements are possible, does not have simplex shape. Only in the cases where the feasible region has the shape of an inverted simplex can we apply the methods applicable to an unconstrained simplex, as described previously. When the entire subrange lies within the unconstrained simplex, we can use a pseudocomponent transformation with slight modifications due to the fact that the sides of the inverted simplex are not parallel to

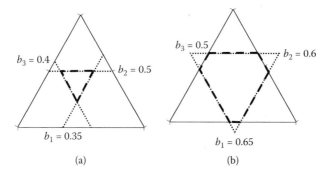

FIGURE 8.10 Constrained region in three-component mixtures. (a) The feasible region (bold lines) shaped as inverted simplex lies entirely within the original simplex. (b) Part of the inverted simplex lies outside the feasible region (bold lines) and has irregular shape.

the sides of the original simplex. The transformation formula applicable for upper-bound constraints is suggested by Crosier [6] and shown in Equation 8.28

$$z_i = \frac{b_i - x_i}{B - 1}, \quad i = 1, \ldots, q \tag{8.28}$$

where

$$B = \sum_{i=1}^{q} b_i > 1. \tag{8.29}$$

The inverse transformation is obvious from Equation 8.28.

There are simple formulas to determine the shape of the subregion in the case of upper bounds. In general, the feasible region will be an inverted simplex that lies entirely within the original one if and only if

$$\sum_{i=1}^{q} b_i - b_{min} \leq 1. \tag{8.30}$$

An example showing a constrained mixture region with lower bounds and upper bounds is illustrated in Figure 8.11. We see that for this particular case, the following constraints apply:

$$0.1 \leq x_1 \leq 0.6$$
$$0.2 \leq x_2 \leq 0.45$$
$$0.15 \leq x_3 \leq 0.65$$

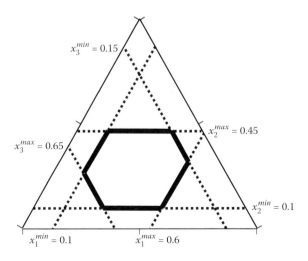

FIGURE 8.11 Constrained mixture region with upper and lower bounds.

In addition to these commonly encountered constraints, there are also multi-component constraints,

$$a_i \leq \alpha_{1i}x_1 + \alpha_{2i}x_2 + \cdots + \alpha_{qi}x_q \leq b_i, i = 1,\ldots \quad (8.31)$$

where $\alpha_{1,i}, \ldots, \alpha_{qi}$ are known coefficients. These types of constraints can be presented alone or in combination with the constraints mentioned above to define even more-complicated subregions.

8.2.2.6 Mixture+Process Factor Spaces

The research problems examined so far have involved process or mixture variables only, hence $n = p$ or $n = q$ (see Figure 8.11). These are special cases of the general problem where both types of variables are present, namely $n = p + q$. The problem where both process and mixture variables are taken into consideration was formulated for the first time by Scheffé [4]. The mixture–process factor space is a cross product of the mixture and process factor spaces. Each vector $\mathbf{x} = [x_1, x_3, \ldots, x_q, x_{q+1}, \ldots, x_{q+p=n}]$ consists of q coordinates, for which the conditions described in Equation 8.10, Equation 8.11, and Equation 8.22 hold. The remaining coordinates represent the values of process variables. The usual practice is to use transformed or coded values of the process variables rather then the natural ones (see Equation 8.8 and Section 8.2.2.1).

Symmetric experimental designs for mixture+process factor spaces are the cross products of symmetric designs for process variables and mixture variables. Figure 8.12 shows an experimental design in mixture+process factor space for a model where both types of variables are of the first order. In both of the examples shown in Figure 8.12, the process variables are described by a two-level full factorial

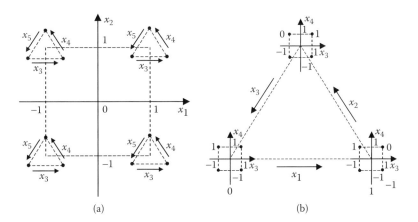

FIGURE 8.12 Two different presentations of a 12-point design for three mixture variables and two process variables. (a) The three-point {3,1} simplex-lattice design is constructed at the position of each of the 22 points of two-level full factorial design. (b) The 22 full factorial design is repeated at the position of each point of the {3,1} simplex-lattice design. The way of representation is related to the order chosen for the variables.

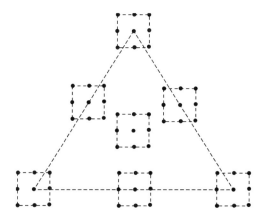

FIGURE 8.13 Mixture+process factor space for incomplete cubic simplex-lattice design combined with a 3^3 full factorial design.

design (FFD). The design for the mixture variables is a simplex lattice of type $\{3,1\}$. The two different representations of the design in Figure 8.12 are identical and represent the same mixture+process factor space.

A much more complicated example for a mixture–process space is shown in Figure 8.13, where an incomplete cubic simplex-lattice design is combined with a 3^3 full factorial experimental design. The matrix of the design for Figure 8.13 is shown in Table 8.3.

One can also combine the process-variable factor space with a constrained mixture space. Figure 8.14 shows an example of the combined space constructed from three mixture variables with lower and upper bounds and one process variable.

The presence of both mutually dependent (mixture) and independent (process) variables calls for a new type of regression model that can accommodate these peculiarities. The models, which serve quite satisfactorily, are combined canonical models. They are derived from the usual polynomials by a transformation on the mixture-related terms. To construct these types of models, one must keep in mind some simple rules: these models do not have an intercept term, and for second-order models, only the terms corresponding to the process variables can be squared. Also, despite the external similarity to the polynomials for process variables only, it is not possible to make any conclusions about the importance of the terms by inspecting the values of the regression coefficients. Because the process variables depend on one another, the coefficients are correlated. Basically, the regression model for mixture and process variables can be divided into three main parts: mixture terms, process terms, and mixture–process interaction terms that describe the interaction between both types of variables. To clearly understand these kinds of models, the order of the mixture and process parts of the model must be specified. Below are listed some widely used structures of combined canonical models. The number of the mixture variables is designated by q, the number of the process variables is designated by p, and the total number of variables is $n = q + p$.

TABLE 8.3
Mixture+Process Design Constructed from a Full Factorial 3^3 Design and Incomplete {3,3} Lattice

No.	x_1	x_2	x_3	x_4	x_5	No.	x_1	x_2	x_3	x_4	x_5
1	0.00	0.00	1.00	−1.00	−1.00	33	0.50	0.00	0.50	0.00	0.00
2	1.00	0.00	0.00	−1.00	−1.00	34	0.50	0.50	0.00	0.00	0.00
3	0.00	1.00	0.00	−1.00	−1.00	35	0.33	0.33	0.33	0.00	0.00
4	0.00	0.50	0.50	−1.00	−1.00	36	0.00	0.00	1.00	1.00	0.00
5	0.50	0.00	0.50	−1.00	−1.00	37	1.00	0.00	0.00	1.00	0.00
6	0.50	0.50	0.00	−1.00	−1.00	38	0.00	1.00	0.00	1.00	0.00
7	0.33	0.33	0.33	−1.00	−1.00	39	0.00	0.50	0.50	1.00	0.00
8	0.00	0.00	1.00	0.00	−1.00	40	0.50	0.00	0.50	1.00	0.00
9	1.00	0.00	0.00	0.00	−1.00	41	0.50	0.50	0.00	1.00	0.00
10	0.00	1.00	0.00	0.00	−1.00	42	0.33	0.33	0.33	1.00	0.00
11	0.00	0.50	0.50	0.00	−1.00	43	0.00	0.00	1.00	−1.00	1.00
12	0.50	0.00	0.50	0.00	−1.00	44	1.00	0.00	0.00	−1.00	1.00
13	0.50	0.50	0.00	0.00	−1.00	45	0.00	1.00	0.00	−1.00	1.00
14	0.33	0.33	0.33	0.00	−1.00	46	0.00	0.50	0.50	−1.00	1.00
15	0.00	0.00	1.00	1.00	−1.00	47	0.50	0.00	0.50	−1.00	1.00
16	1.00	0.00	0.00	1.00	−1.00	48	0.50	0.50	0.00	−1.00	1.00
17	0.00	1.00	0.00	1.00	−1.00	49	0.33	0.33	0.33	−1.00	1.00
18	0.00	0.50	0.50	1.00	−1.00	50	0.00	0.00	1.00	0.00	1.00
19	0.50	0.00	0.50	1.00	−1.00	51	1.00	0.00	0.00	0.00	1.00
20	0.50	0.50	0.00	1.00	−1.00	52	0.00	1.00	0.00	0.00	1.00
21	0.33	0.33	0.33	1.00	−1.00	53	0.00	0.50	0.50	0.00	1.00
22	0.00	0.00	1.00	−1.00	0.00	54	0.50	0.00	0.50	0.00	1.00
23	1.00	0.00	0.00	−1.00	0.00	55	0.50	0.50	0.00	0.00	1.00
24	0.00	1.00	0.00	−1.00	0.00	56	0.33	0.33	0.33	0.00	1.00
25	0.00	0.50	0.50	−1.00	0.00	57	0.00	0.00	1.00	1.00	1.00
26	0.50	0.00	0.50	−1.00	0.00	58	1.00	0.00	0.00	1.00	1.00
27	0.50	0.50	0.00	−1.00	0.00	59	0.00	1.00	0.00	1.00	1.00
28	0.33	0.33	0.33	−1.00	0.00	60	0.00	0.50	0.50	1.00	1.00
29	0.00	0.00	1.00	0.00	0.00	61	0.50	0.00	0.50	1.00	1.00
30	1.00	0.00	0.00	0.00	0.00	62	0.50	0.50	0.00	1.00	1.00
31	0.00	1.00	0.00	0.00	0.00	63	0.33	0.33	0.33	1.00	1.00
32	0.00	0.50	0.50	0.00	0.00	—	—	—	—	—	—

Linear (first order) for both mixture and process variables:

$$\hat{y} = \sum_{i=1}^{q} b_i x_i + \sum_{j=q+1}^{n} b_j x_j. \tag{8.32}$$

Second order for both mixture and process variables:

$$\hat{y} = \sum_{i=1}^{q} b_i x_i + \sum_{i=1}^{q-1}\sum_{j=i+1}^{q} b_{ij} x_i x_j + \sum_{i=1}^{q}\sum_{j=q+1}^{n} b_{ij} x_i x_j + \sum_{i=q+1}^{n-1}\sum_{j=i+1}^{n} b_{ij} x_i x_j + \sum_{i=q+1}^{n} b_{ii} x_i^2. \tag{8.33}$$

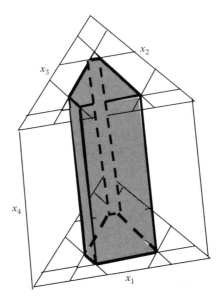

FIGURE 8.14 Combined mixture+process factor space (the shaded area) constructed from three mixture variables x_1, x_2, x_3 and one process variable x_4.

Linear for the mixture and second order for the process variables, which includes mixture–process interactions:

$$\hat{y} = \sum_{i=1}^{q} b_i x_i + \sum_{i=1}^{q-1} \sum_{j=q+1}^{n} b_{ij} x_i x_j + \sum_{i=q+1}^{n-1} \sum_{j=i+1}^{n} b_{ij} x_i x_j + \sum_{j=q+1}^{n} b_{ij} x_i^2. \tag{8.34}$$

Linear for both mixture and mixture–process interactions:

$$\hat{y} = \sum_{i=1}^{q} b_i x_i + \sum_{i=1}^{q} \sum_{j=q+1}^{n} b_{ij} x_i x_j. \tag{8.35}$$

8.2.3 Some Regression-Analysis-Related Notation

At this point, it is helpful to introduce some notation that will be used to further describe experimental designs and response-surface modeling. As was described earlier, all possible operating conditions are represented as combinations of the values of the input variables. Each particular combination is a point in the operating region of a process and is called a treatment. These sets of points can be denoted in matrix form

$$\mathbf{X} = \begin{pmatrix} x_{11} & x_{12} & \cdots & x_{1n} \\ x_{21} & x_{22} & \cdots & x_{2n} \\ \vdots & \vdots & \vdots & \vdots \\ x_{N1} & x_{N1} & \cdots & x_{Nn} \end{pmatrix}$$

where $\mathbf{X}_{N \times n}$ is called the design matrix, N is the number of treatments, and n is the number of factors under consideration. High levels and low levels of a controlled factor are typically coded as values of $x_i = +1$ and $x_i = -1$, respectively. As previously noted, the coded factor levels are related to the experimental measurement scale by a transformation function. An important matrix closely related to the design matrix is the extended design matrix $\mathbf{F}$, which describes the relationship between the coded factor levels and the experimental measurement scale. This matrix plays an important role in the calculations discussed below.

Given a design matrix, $\mathbf{X}$, the next step is to construct the extended design matrix $\mathbf{F}$. The general structure of any polynomial of k coefficients is

$$\hat{y} = \sum_{i=1}^{k} b_i f_i(\mathbf{x}) \tag{8.36}$$

The entries of $\mathbf{F}$ are constructed from the terms of the regression model, hence

$$\mathbf{F} = \begin{bmatrix} f_1(\mathbf{x}_1) & f_2(\mathbf{x}_1) & \cdots & f_k(\mathbf{x}_1) \\ f_1(\mathbf{x}_2) & f_2(\mathbf{x}_2) & \cdots & f_k(\mathbf{x}_2) \\ \vdots & \vdots & \vdots & \vdots \\ f_1(\mathbf{x}_N) & f_2(\mathbf{x}_N) & \cdots & f_k(\mathbf{x}_N) \end{bmatrix}. \tag{8.37}$$

The exact structure of each of the functions $f_1(\mathbf{x}), \ldots, f_k(\mathbf{x})$ depends on the transformation or factor coding used. For example, the $\mathbf{F}$ matrix for a three-level full factorial design for two process variables and a second-order model is shown in Table 8.4.

TABLE 8.4
Structure of the Extended Design Matrix F for a Second-Order Model with Two Process Variables

$f_1 \equiv 1$	$f_2 \equiv x_1$	$f_3 \equiv x_2$	$f_4 \equiv x_1 x_2$	$f_5 \equiv x_1^2$	$f_6 \equiv x_2^2$
1	−1	−1	1	1	1
1	0	−1	0	0	1
1	1	−1	−1	1	1
1	−1	0	0	1	0
1	0	0	0	0	0
1	1	0	0	1	0
1	−1	1	−1	1	1
1	0	1	0	0	1
1	1	1	1	1	1

Note: Here $N = 9$, $m = r = 2$, $k = 6$. The structure of the regression model is $\hat{y} = b_o + b_1 x_1 + b_2 x_2 + b_{12} x_1 x_2 + b_{11} x_1^2 + b_{22} x_2^2$.

The least-squares solution to fitting the regression model is

$$\mathbf{b}_k = \left(\mathbf{F}^T_{N\times k} \mathbf{F}_{N\times k} \right)^{-1} \mathbf{F}^T_{N\times k} \mathbf{y}_N \tag{8.38}$$

Here $\mathbf{b}_k$ is the k-dimensional vector of the regression coefficients, $\mathbf{y}$ is the vector of the measured response values, and $\mathbf{F}$ is the extended design matrix. The matrix product $\mathbf{F}^T\mathbf{F}$ is called the information matrix, and its inverse is called the dispersion matrix. Because the inverse of the information matrix is included in Equation 8.38, its properties play a crucial role in the calculation of the regression coefficients. For instance, if the inverse of $\mathbf{F}^T\mathbf{F}$ does not exist (if it is singular), it will be impossible to calculate the coefficients of the regression model. If $\mathbf{F}^T\mathbf{F}$ is ill-conditioned (the information matrix is *nearly* singular), the inverse matrix can be calculated, but the values of the regression coefficients may be subject to such large errors that they are completely wrong. Because the information matrix depends entirely on the entries of $\mathbf{X}$, it is very important to consider exactly which points of the factor space should be included in the design matrix, $\mathbf{X}$. No matter how good the performance of the measurement technique and the accuracy of the measured response in the vector $\mathbf{y}$, the incorrect choice of the design matrix $\mathbf{X}$, and thus $\mathbf{F}$ and $\mathbf{F}^T\mathbf{F}$, can compromise all of the experimental efforts.

Common sense dictates, therefore, that if it is possible to corrupt the results by choosing the wrong design matrix, then the choice of a better one will improve the quality of the calculations. Following the same logic, if we can improve the calculation of regression coefficients just by manipulating the experimental design matrix, then we can make further improvements by selecting a design with the maximum information (the best information matrix) using the minimum number of the points in $\mathbf{X}$ (i.e., with the fewest number of measurements). This is the topic of the next section of this chapter.

8.3 ONE-VARIABLE-AT-A-TIME VS. OPTIMAL DESIGN

The method of changing one variable at a time to investigate the outcome of an experiment probably dates back to the beginnings of systematic scientific research. The idea is fairly simple. We often need to investigate the influence of several factors. To simplify control and interpretation of the results, we choose to vary only one of the factors by keeping the rest of them at constant values. The method is illustrated in Figure 8.15 and in the following example.

8.3.1 BIVARIATE (MULTIVARIATE) EXAMPLE

Suppose the goal of the experimenter is to apply the one-variable-at-a-time approach to explore the influence of two factors, x_1 and x_2, on the response, y, to find its maximum. The experimenter intends to perform a set of measurements over x_1 by keeping the other factor, x_2, at a constant level until some decrease in the response function is observed (see Figure 8.16). After a decrease is noted, the experimenter applies the same approach to factor x_2 by starting at the best result.

Response-Surface Modeling and Experimental Design

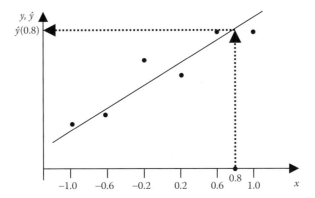

FIGURE 8.15 Illustration of the one-variable-at-a-time approach.

Figure 8.16 illustrates a contour plot showing the shape of the true (but unknown) response surface. After the first step along x_1, the experimenter finds a decrease in the function at point **b**; hence, point **a** is the best one at the moment. By starting from the best point (noted as **a**) and changing the value of x_2, the experimenter finds another decrease in the value of the response at point **c**. The natural conclusion from applying this approach is that the first point **a** = [−1.0, 1.0] is the best one. However, it is clear from the figure that if the experimenter had changed both factors simultaneously, point **d** would have been discovered to have a higher value of the response compared with point **a**. The advantages and disadvantages of the one-variable-at-a-time approach are summarized in the next two subsections.

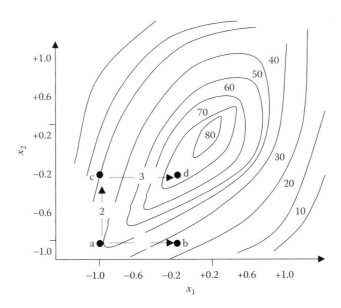

FIGURE 8.16 The one-variable-at-a-time approach applied to a two-factor problem.

8.3.2 Advantages of the One-Variable-at-a-Time Approach

- Easy to perform
- Simple data organization
- Graphical presentation of results
- No need for previous mathematical and statistical knowledge
- Minimal need of complex calculations

8.3.3 Disadvantages

- Poor statistical value of the model fitted to the collected data
- Unnecessarily large number of experiments required
- Significant possibility of missing the extremum when used in optimization studies

8.4 SYMMETRIC OPTIMAL DESIGNS

8.4.1 Two-Level Full Factorial Designs

Returning to the example of Figure 8.16, we see that the four points, namely **a**, **b**, **c**, and **d**, are positioned symmetrically within the experimental region. If we suppose that the region E is constrained in the following manner

$$-1.0 \leq x_1 \leq -0.2$$

$$-1.0 \leq x_1 \leq -0.2$$

then these four points construct what is known as a two-level, two-factor full factorial design. The purpose of these types of experimental designs is to give the experimenter an opportunity to explore the influence of all combinations of the factors. An experimental design organized by combining all possible values of the factors, giving s^m permutations, is called a full factorial design. Here s designates the number of the levels at each factor, and m represents the number of factors.

Figure 8.17 shows an example of a three-factor, two-level full factorial design. The coordinates of the points of the same design are given in Table 8.5. The number of points in a two-level, three-factor full factorial design is $2^3 = 8$. A good tutorial giving basic information about the two-level full factorial design, including an important variation, the fractional factorial design, and information about some basic optimization techniques can be found in the literature [7].

8.4.1.1 Advantages of Factorial Designs

- Simple formulae for calculating regression coefficients
- A classical tool for estimating the mutual significance of multiple factors
- A useful tool for factor screening
- The number of points can be reduced considerably in fractional designs
- Under many conditions, fulfills most of the important optimality criteria, i.e., D-, G-, A-optimality, orthogonality, and rotatability

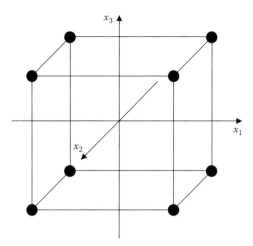

FIGURE 8.17 Two-level, three-factor full factorial design.

8.4.1.2 Disadvantages of Factorial Designs

- Applicable only for linear polynomial models
- Large numbers of treatments or experiments required

8.4.2 THREE OR MORE LEVELS IN FULL FACTORIAL DESIGNS

By extending the approach used in two-level full factorial designs, we can obtain experimental designs that are suitable for polynomial models of second order or higher. A design that is applicable to second-order polynomials is the three-level full factorial design. Figure 8.18 shows an example of a three-level, three-factor full factorial design.

TABLE 8.5
Two-Level, Three-Factor Full Factorial Design

No.	x_1	x_2	x_3
1	−1	−1	−1
2	1	−1	−1
3	−1	1	−1
4	1	1	−1
5	−1	−1	1
6	1	−1	1
7	−1	1	1
8	1	1	1

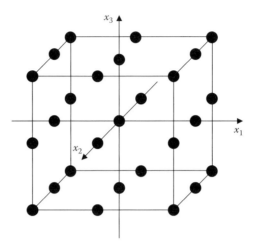

FIGURE 8.18 Three-level, three-factor full factorial design.

The required number of points in a three-level full factorial design is $N = 3^m$. In the case of $n = 3$ factors, we must perform 27 experiments, as seen in Table 8.6. For four factors, 81 experiments are required. Inspection of the three-level design reveals that it also includes a two-level design. This means that the three-level full factorial design is also a composite design. It can be constructed by augmenting a two-level design with additional points, thereby saving the time and expense of replacing the measurements already performed.

In practice, augmentation can be performed after the experimenter has completed a full factorial design and found a linear model to be inadequate. A possible reason is that the true response function may be second order. Instead of starting a completely new set of experiments, we can use the results of the previous design and perform an additional set of measurements at points having one or more zero coordinates. All of the data collected can be used to fit a second-order model.

TABLE 8.6
Three-Level, Three-Factor Full Factorial Design

No.	x_1	x_2	x_3	No.	x_1	x_2	x_3	No.	x_1	x_2	x_3
1	−1	−1	−1	10	−1	−1	0	19	−1	−1	1
2	0	−1	−1	11	0	−1	0	20	0	−1	1
3	1	−1	−1	12	1	−1	0	21	1	−1	1
4	−1	0	−1	13	−1	0	0	22	−1	0	1
5	0	0	−1	14	0	0	0	23	0	0	1
6	1	0	−1	15	1	0	0	24	1	0	1
7	−1	1	−1	16	−1	1	0	25	−1	1	1
8	0	1	−1	17	0	1	0	26	0	1	1
9	1	1	−1	18	1	1	0	27	1	1	1

We can use the same approach to expand the design and obtain a data set applicable for polynomials of third order. The respective full factorial design is constructed by combining all of the factors at five levels, giving a total of $N = 5^n$, or $N = 25$ for $n = 2$, $N = 125$ for $n = 3$, $N = 625$ for $n = 4$, etc. It is apparent that the number of the experiments grows geometrically with the number of factors, and there are not many applications where the performance of 625 experiments to explore four factors is reasonable.

8.4.3 Central Composite Designs

In the previous sections we found that a second-order full factorial design requires an enormous number of measurements. Box and Wilson showed it is possible to have a more economical design while at the same time retaining the useful symmetrical structure of a full factorial design [1].

Figure 8.19 shows an example of two such designs, called central composite designs (CCD), for two variables (Figure 8.19a) and three variables (Figure 8.19b). The idea of central composite designs is to augment a two-level full factorial design by adding so-called axial or star points (see Figure 8.19) and some number of replicate measurements at the center. Each of the star points has coordinates of 0 except those corresponding to the jth factor, $j = 1, n$, where the respective coordinates are equal to $\pm \alpha$. An example of a three-factor central composite design is given in Table 8.7. The number of points for the central composite design is $N = 2^n + 2n + n_c$, where n_c represents the number of the center points.

Central composite designs can be augmented in a sequential manner as well. It is possible to start the investigation by using a full factorial design. After concluding that a linear model is inadequate, one can continue the same investigation by adding additional measurements at the star points and in the center. The choice of the values for α and n_c is very important for the characteristics of the resulting design. Generally, the value of α is in the range from 1.0 to $\sqrt{n}$, depending on the experimental and

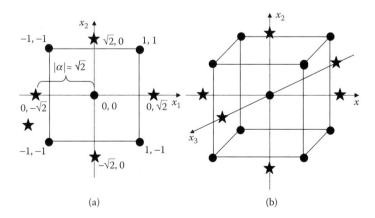

FIGURE 8.19 Central composite designs (CCD) for (a) two factors, where $\alpha = \sqrt{2}$, and (b) three factors.

TABLE 8.7
Three-Factor Central Composite Design with Axial Values α and Four Center Points

No.	x_1	x_2	x_3	No.	x_1	x_2	x_3
1	−1	−1	−1	10	a	0	0
2	1	−1	−1	11	0	-a	0
3	−1	1	−1	12	0	a	0
4	1	1	−1	13	0	0	-a
5	−1	−1	1	14	0	0	a
6	1	−1	1	15	0	0	0
7	−1	1	1	16	0	0	0
8	1	1	1	17	0	0	0
9	-a	0	0	18	0	0	0

operational regions. The initial idea for the choice of α and n_c was to ensure a diagonal structure in the information matrix, $\mathbf{F}^T\mathbf{F}$. Because of the lack of computing facilities in the 1950s, it was necessary to give the investigator the ability to easily calculate the regression coefficients by hand. Later, as inexpensive, powerful computers became available, values for α and n_c were selected to obtain the maximum value of the det $(\mathbf{F}^T\mathbf{F})$ for the region covered by the design. Another consideration in the choice of the values for α and n_c is to ensure the so-called rotatability property of the experimental design. By spacing all the points at an equal distance from the center, a rotatable design is obtained that gives each point equal leverage in the estimation of the regression coefficients.

8.5 THE TAGUCHI EXPERIMENTAL DESIGN APPROACH

During the 1980s, the name of the Japanese engineer Genichi Taguchi became synonymous with "quality." He developed an approach for designing processes that produce high-quality products despite the variations in the process variables. Such processes are called robust because they are insensitive to noise in the processes. The approach was applied with huge success in companies such as AT&T Bell Labs, Ford Motor Co., Xerox, etc. The simplicity of the approach made these methods extremely popular and, in fact, stimulated the development of a new production philosophy.

The idea behind this methodology is to apply a technique of experimental design with the goal of finding levels of the *controlled factors* that make the process robust to the presence of *noise factors*, which are uncontrolled. The controlled factors are process variables that are adjusted during the normal operation of a process. The noise factors are present in combination with controlled factors and have a significant influence on the response or quality of the product. Noise factors are either impossible

Response-Surface Modeling and Experimental Design

TABLE 8.8
Examples of Controlled and Noise Variables

Application	Controlled Variables	Noise Variables
A cake	Amount of sugar, starch, and other ingredients	Oven temperature, baking time, fat content of the milk
Gasoline	Ingredients in the blend, other processing conditions	Type of driver, driving conditions, changes in engine type
Tobacco product	Ingredient and concentrations, other processing conditions	Moisture conditions, storage conditions
Large-scale chemical process	Processing conditions, including the nominal temperature	Deviations from the nominal temperature, deviations from other processing conditions

or not economically feasible to control at some constant level. Table 8.8 shows some examples of controlled variables and noise variables.

In Taguchi methods, we define the quality of the product in terms of the deviation of some response parameter from its target or desirable value. The concept of quality and the idea behind the Taguchi philosophy is illustrated with the example shown in Figure 8.20.

Suppose a hypothetical pharmaceutical company produces tablets of type A, where the important characteristic is the amount of active ingredient present in the tablet. We can imagine an unacceptable process illustrated in Figure 8.20c, where there will be some tablets produced having amounts of the active ingredient below and above the specifications for acceptable tablets. In Figure 8.20a, we see a process that is acceptable according the formal requirements (all of the tablets are within specifications), but there are risks that some small portion of tablets may fall outside the specification in the future. A very high-quality process is shown in Figure 8.20b, where most of the values are concentrated around the target and the risk of producing unacceptable product is very low.

The risks involved in using the process of the type shown in Figure 8.20a are illustrated in the following example. Suppose the products being produced are bolts and nuts. If bolts are produced that are close to the upper acceptance limit (large diameter) and the nuts are produced close to the low limit (small diameter), then the nut simply will not fit on the bolt because its internal diameter will be too small. The opposite situation is possible as well, where the nut has too large an internal diameter compared to the bolt. In this case, the bolt will fit, but it will be too loose to serve as a reliable fastener. The situations described here are a problem for customers, but there are also potential problems for the producer. For the process shown in Figure 8.20a, it is statistically likely that a small number of the products will fall outside the acceptance limits but not be included in the sample used for quality testing and release by the producer. Depending on the sample size, there may also be a significant likelihood that some of these unacceptable products will

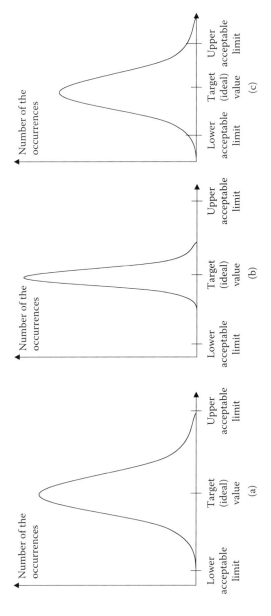

FIGURE 8.20 The concept of quality illustrated with different distributions of some important quality characteristic: (a) a process that produces mostly acceptable product, (b) an "ideal" process, (c) an unacceptable process, with a significant portion of the product outside the acceptable limits.

be in the sample used by the customer for *acceptance* quality testing, causing the entire shipment to be rejected. The tools developed by Taguchi are intended to help the producers keep their processes running so that the quality of their products is as shown in Figure 8.20b despite the presence of the noise factors.

To apply the Taguchi approach, we must define two sets of variables: controlled variables and noise variables. The controlled variables are process variables that can be adjusted or controlled to influence the quality of the products produced by the process. Noise variables represent those factors that are uncontrollable in normal process operating conditions. A study must be conducted to define the range of variability encountered during normal plant conditions for each of the controlled variables and the noise variables. These values of the variables are coded in the range $[-1, 1]$.

To illustrate the idea of a Taguchi design, suppose that there are two controllable variables c_1 and c_2 and three noise factors n_1, n_2, n_3. Here, Taguchi suggests the use of an orthogonal design, so we construct two full factorial designs, one for each of the two groups of factors. The resulting experimental design is illustrated in Table 8.9. The levels of the controlled factors forming the so-called inner array are shown in the first two columns labeled c_1 and c_2. The full factorial design of the noise factors forming the "outer array" is shown in the upper three rows of the table. Each point of the outer array is represented as a column with three entries. The inner array has four rows (two-level, two-factor full factorial design), and the outer array has eight columns (two-level, three-factor full factorial design).

In this way, there are 32 defined experimental conditions. At each set of conditions, the response under investigation is measured. In Table 8.9, the results of the measurements are designated by y_{ij}, $i = 1,4$, $j = 1,8$. Once the measurements are performed, the signal-to-noise ratio (SNR) at each of the points in the inner array (the rows of Table 8.9) is calculated. The combination of levels of the controlled variables that correspond to the highest value of the SNR represents the most robust production conditions within the range of noise factors investigated. For instance, if $SNR_3 = \max\{SNR_1, SNR_2, SNR_3, SNR_4\}$, then the most robust condition for the process corresponds to the levels of the controlled factors equal to $c_1 = -1$ and $c_2 = 1$.

TABLE 8.9
Taguchi Design for Two Controlled Factors and Three Noise Factors

| | | n_3 | -1 | 1 | -1 | 1 | -1 | 1 | -1 | 1 | |
| | | n_2 | -1 | -1 | 1 | 1 | -1 | -1 | 1 | 1 | |
c_1	$c_2 \backslash n_1$		-1	-1	-1	-1	1	1	1	1	
-1	-1		y_{11}	y_{12}	y_{13}	y_{14}	y_{15}	y_{16}	y_{17}	y_{18}	SNR_1
1	-1		y_{21}	y_{22}	y_{23}	y_{24}	y_{25}	y_{26}	y_{27}	y_{28}	SNR_2
-1	1		y_{31}	y_{31}	y_{31}	y_{31}	y_{31}	y_{31}	y_{31}	y_{31}	SNR_3
1	1		y_{41}	y_{41}	y_{41}	y_{41}	y_{41}	y_{41}	y_{41}	y_{41}	SNR_4

There are three formulas for SNR suggested by Taguchi:

1. *Smaller is better*: In situations where the target quality value has to be kept close to zero, the formula is

$$SNR_j^S = -10 \log \sum_{i=1}^{u} \left[\frac{y_{ij}^2}{u} \right], j = 1, g \qquad (8.39)$$

2. *Larger is better*: In the situations where the target quality value has to be as large as possible, the formula is

$$SNR_j^L = -10 \log \left[\frac{1}{u} \sum_{i=1}^{u} \left(\frac{1}{y_{ij}^2} \right) \right], j = 1, g \qquad (8.40)$$

3. *The mean (target) is best*: This formula is applicable for the example illustrated in Figure 8.20

$$SNR_j^T = -10 \log s^2, j = 1, g \qquad (8.41)$$

where

$$s^2 = \sum_{i=1}^{u} \left(\frac{y_{ij} - \bar{y}}{u - 1} \right) \qquad (8.42)$$

The main literature sources for Taguchi methods are his original books [8–10]. A comprehensive study on the use of experimental design as a tool for quality control, including Taguchi methods, can be found in the literature [11]. A good starting point for Taguchi methods and response-surface methodology can also be found in the literature [12].

8.6 NONSYMMETRIC OPTIMAL DESIGNS

8.6.1 OPTIMALITY CRITERIA

Like a sailboat, an experimental design has many characteristics whose relative importance differs in different circumstances. In choosing a sailboat, the relative importance of the characteristics such as size, speed, sea-worthiness, and comfort will depend greatly on whether we plan to sail on the local pond, undertake a trans-Atlantic voyage, or complete America's Cup contest [1].

Box and Draper [13] give a list of desired properties for experimental designs. A good experimental design should:

- Generate a satisfactory distribution of information throughout the region of interest, $\boldsymbol{R}$
- Ensure that the fitted values at x, $\hat{y}(x)$ are as close as possible to the true values at x, $\eta(x)$

Response-Surface Modeling and Experimental Design   299

- Ensure a good ability to detect model lack of fit
- Allow transformations to be estimated
- Allow experiments to be performed in blocks
- Allow designs of increasing order to be built up sequentially
- Provide an internal estimate of error
- Be insensitive to wild observations and violations of the usual assumptions of normal distributions
- Require a minimum number of experimental points
- Provide simple data patterns that allow visual evaluation
- Ensure simplicity of calculations
- Behave well when errors occur in the settings of the predictor variables, the xs
- Not require an impractically large number of predictor variable levels
- Provide a check of the "constancy of variance" assumption

It is obvious that this extensive list of features cannot be adequately satisfied with one design. It is possible, however, to choose a design that fits to the needs of the experimenter and will deliver the necessary comfort or performance.

8.6.2 OPTIMAL VS. EQUALLY DISTANCED DESIGNS

In this section we will look at the *confidence interval* of the predicted value, $\hat{y}$. As we know, the goal of regression analysis is to build a model that minimizes

$$Q = \sum_{i=1}^{N} (y_u - \hat{y})^2, \tag{8.43}$$

thus enabling us to find an estimate of the response value that is as close as possible to the measured one. In regression analysis, the value of $\hat{y}_i$ is actually an estimate of the true (but unknown) value of the response, η_i. To answer the question, "How close is the estimate $\hat{y}_i$ to the response η_i?" we calculate a confidence interval in the region around $\hat{y}$ at the point $\mathbf{x}_0$ by using the formula in Equation 8.44

$$\hat{y}(\mathbf{x}_0) - t_{v,1-\alpha} \left[f(\mathbf{x}_0)^T (\mathbf{F}^{-1}\mathbf{F})^{-1} f(\mathbf{x}_0) \right] S_\varepsilon \leq \eta(\mathbf{x}_0) \leq \hat{y}(\mathbf{x}_0) \\ + t_{v,1-\alpha} \left[f(\mathbf{x}_0)^T (\mathbf{F}^{-1}\mathbf{F})^{-1} f(\mathbf{x}_0) \right] S_\varepsilon \tag{8.44}$$

or equivalently

$$|\hat{y}(\mathbf{x}_0) - \eta(\mathbf{x}_0)| \leq \left| t_{v,1-\alpha} \left[f(\mathbf{x}_0)^T (\mathbf{F}^{-1}\mathbf{F})^{-1} f(\mathbf{x}_0) \right] S_\varepsilon \right| \tag{8.45}$$

where S_ε is the measurement error in y and $t_{v,1-\alpha}$ is Student's t-statistic at v degrees of freedom and probability level α. The product $d(\hat{y}) = f(\mathbf{x}_0)^T (\mathbf{F}^{-1}\mathbf{F})^{-1} f(\mathbf{x}_0)$ is known as a variance of the prediction at point $\mathbf{x}_0$. The width of the confidence interval is a measure of how close the values of $\hat{y}_i$ are to η_i. As can be seen in Equation 8.45, the distance between these two points depends on the extended design matrix, $\mathbf{F}$. If we wish to make the value of $|t_{v,1-\alpha}[f(\mathbf{x}_0)^T(\mathbf{F}^{-1}\mathbf{F})^{-1}f(\mathbf{x}_0)]|$ smaller, which will in turn make the difference $|\hat{y}(\mathbf{x}_0) - \eta(\mathbf{x}_0)|$ smaller, an obvious approach is to manipulate the entries of matrix $\mathbf{F}$ and, therefore, the design points in $\mathbf{X}$. To illustrate the solution to this problem, we begin by exploring the dependence of y on $\mathbf{x}$ and fit the model $\hat{y} = f(\mathbf{x})$, which can be used to predict y at values of $\mathbf{x}$ where no measurements have been made. To conduct the analysis, we choose an experimental region, $\mathcal{E}$, subject to the constraints:

$$\mathcal{E} = \{x: -1 \leq x \leq +1\} \tag{8.46}$$

and some appropriate step size over which to vary $\mathbf{x}$. Choosing a step size $s = 0.4$ gives six measurements, as shown in Figure 8.15. These points represent the measured values of the response, y. The next step is to fit the model illustrated by the line passing through the points and calculate its prediction confidence interval by Equation 8.44 or Equation 8.45, as shown in Figure 8.21.

After examination of Equation 8.45, it is easy to see that the width of the confidence interval depends on three quantities:

Estimate of the measurement error, S_ε
Critical value of the Student statistic, t
Variance of the prediction $d[\hat{y}(x)]$,

The estimate of the measurement error, S_ε, is determined by the measurement process itself and cannot be changed. The critical value of t is also fixed for any selected probability level. Thus, to minimize the width of the confidence interval, it is clear that the only option is to look for some way to minimize of the value of

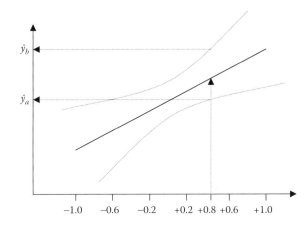

FIGURE 8.21 Influence of the confidence interval on the error of the prediction.

$d[\hat{y}(x)]$. This value depends on the design matrix, so it seems that if we choose better design points for the measurements, we might get a smaller confidence interval and, consequently, a model with better predictive ability.

It is well known that the D-optimal design for one factor and a linear model consists of points positioned at $x^* = -1$ and $x^* = +1$, with an equal number of observations at each of these two design points. In this particular example, the six measurements can be divided into two groups of three measurements, one group at each of the limits. The nonoptimal design matrix is

$$\mathbf{X}^T = [-1.0, -0.6, -0.2, +0.2, +0.6, +1.0]$$

whereas the D-optimal design matrix is:

$$\mathbf{X}^{*T} = [-1.0, -1.0, -1.0, +1.0, +1.0, +1.0]$$

The confidence interval constructed using the D-optimal interval shown in Figure 8.22 is narrower than the confidence interval constructed using the nonoptimal design, giving a reduced range, $[\hat{y}_a^*, \hat{y}_b^*]$, for the prediction of y.

This example shows that it is possible to improve the prediction ability of a model just by rearranging the points where the measurements are made. We replaced the "traditional stepwise" approach to experimental design by concentrating all of the design points at both ends of the experimental region.

We can go further in improving the quality of the model. It was mentioned that the D-optimal design requires equal number of measurements at −1 and +1; however, we have made no mention about the total number of points. In fact, it is possible to reduce the number of measurement to four or even to two without loss of prediction ability. Because it is always useful to have extra degrees of freedom to avoid overfitting and for estimating residual variance, it is apparent that four measurements, two at −1.0 and two at +1, will give the best solution.

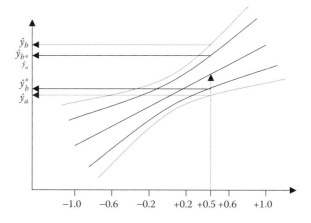

FIGURE 8.22 Comparison of the confidence intervals obtained by use of a nonoptimal design (- - -) and a D-optimal design (---).

8.6.3 DESIGN OPTIMALITY AND DESIGN EFFICIENCY CRITERIA

As described in the previous section, the list of desirable properties for experimental designs is quite long and even controversial. It is impossible to find a design that satisfies all of them. Thus, the experimenter must define precisely his or her goals and resources and choose a satisfactory design. As a more formal quantitative measure of the properties of experimental designs, we might choose optimality criteria. In the search for the most convenient design, we use these criteria to find a proper design that is most suitable for our needs and resources. The list of the optimality criteria is long and continues to increase. There are several criteria that are in wider use, generally because of the considerable amount of theoretical and practical work done with them. In the following subsections, some brief descriptions are given for some of the better known optimality criteria.

A realizable experimental design is a set of N measurements that are to be performed by the experimenter. Each of these measurements is represented by a point in factor space. Hereafter, if it is not mentioned explicitly, the terms *point* and *measurement* will be synonymous. One designates this set by a matrix $\mathbf{X}_{N \times n}$, having N rows and n columns. Each row represents one measurement, which is supposed to be performed at the conditions described by the values of the corresponding row. Each column corresponds to one of the controllable variables that are adjusted during the experiment. The set $S = \{\mathbf{X}\}$ is a subset of $\mathcal{X}$, i.e., the general set of all points, hence

$$S\{\mathbf{X}\} \subset \mathcal{X} \tag{8.47}$$

We define the set $\mathcal{X}_N\{S\}$ as the subset of N points, namely $S_N\{\mathcal{X}_{N \times n}\} \subset \mathcal{X}_N\{S\}$. The one set, $\mathcal{X}_N\{S\}$, that satisfies the stated optimality criteria will be referred as the *optimal design*.

$$S_N^o\{\mathbf{X}_{N \times n}^o\} \subset \mathcal{X}_N(S) \tag{8.48}$$

In other words, having the general set of all points $\mathcal{X}$, we have to find some subset of N points giving the experimental design, $S_N\{\mathbf{X}_{N \times n}\}$. In fact, there may be more than one such N-point subset. All of these N-point subsets form the set, $\mathcal{X}_N\{S\}$, which is the set of all subsets. Hence, the task is to find a member of $\mathcal{X}_N\{S\}$, denoted as $S_N^o\{\mathbf{X}_{N \times m}^o\}$, that satisfies the stated optimality criteria.

Based on one or more optimality criteria, we define an optimality function. The search for the optimal design then becomes a numerical optimization problem. Unfortunately, the search is performed in a space of high dimension using an optimality function that is not usually differentiable. Because of these complications, an iterative process is needed to find an optimal design, and convergence can sometimes be slow. The iterative process can be terminated when the change from one step to the next is sufficiently small; however, we would also like to have some idea how close the selected design is from the best possible design. For some design criteria, the theoretically best values have been derived. A measure of the discrepancy between some optimal design, $\mathbf{X}^*_{N \times m}$, and the theoretically best design, ξ^*, is the difference between the value of its optimality criterion and the corresponding theoretically best value.

The design efficiency is a measure of the difference between the optimality of the locally optimal design and the theoretically proven globally optimal design. The values for the efficiency fall in the range [0, 1], where values closer to 1 represent designs closer to the corresponding theoretical maximum.

8.6.3.1 Design Measures

Suppose we wish to obtain a design $\mathbf{X}_N$ with $N \geq k$. If all of the points, $\mathbf{x}_i$, are different, then the relative weight of each point is $1/N$, denoted by $\psi_i = 1/N$, $i = 1$, N. Applying the concept of relative weights, we obtain the following more general notation for an experimental design, where the design is characterized by its points and their relative weights.

$$\xi(\mathbf{X}, \boldsymbol{\psi}) = \begin{pmatrix} \mathbf{x}_1 & \mathbf{x}_2 & \cdots & \mathbf{x}_N \\ \psi_1 & \psi_2 & \cdots & \psi_N \end{pmatrix}, \forall \psi_i = \frac{1}{N}, \sum_{i=1}^{N} \psi_i = \frac{N}{N} = 1. \quad (8.49)$$

It is possible for some of the design points, $\mathbf{x}_i \subset \{\mathbf{X}_N\}$, $i = 1, \ldots, N$, to coincide, in which case the number of distinct points will be $L < N$. The respective design will be designated as shown in Equation 8.50.

$$\xi(\mathbf{X}, \boldsymbol{\psi}) = \begin{pmatrix} \mathbf{x}_1 & \mathbf{x}_2 & \cdots & \mathbf{x}_L \\ \psi_1 & \psi_2 & \cdots & \psi_L \end{pmatrix}, \exists \psi_i \geq \frac{1}{N}, \sum_{i=1}^{L} \psi_i = \frac{N}{N} = 1 \quad (8.50)$$

Some points will have weights $\psi_i = 2/N, 3/N$, $i = 1, \ldots, L$ and so on, which means that these points will appear twice, three times, etc. in the set of the design points.

Going further, we can extend the definition of weights ψ_i so that each denotes a fractional number of measurements that appear at a particular point. It is obvious from practical considerations that the number of measurements should be an integer. For example, at a particular point we can perform two or three measurements, but not 2.45 measurements. Nevertheless, by assuming that the number of measurements can be a noninteger quantity, Kiefer and Wolfowitz in their celebrated work [14] introduced a continuous design measure. This function, ξ, is assumed to be continuous across the factor space. The locations in the factor space where ξ receives a nonzero value are the points of the experimental design. These points are known as design support points. Hence Equation 8.49 becomes:

$$\xi(\mathbf{X}, \boldsymbol{\psi}) = \begin{pmatrix} \mathbf{x}_1 & \mathbf{x}_2 & \cdots & \mathbf{x}_N \\ \psi_1 & \psi_2 & \cdots & \psi_N \end{pmatrix}, 0 \leq \psi_i \leq 1, \sum_{i=1}^{N} \psi_i = 1. \quad (8.51)$$

Using this approach, we can describe any experimental design by the function $\xi(\psi)$, called the design measure or probability measure. This function is differentiable;

thus we can apply conventional techniques to search for its extrema and find the theoretical maximum value for a given design criterion. The value of ψ_i represents the ideal number of fractional measurements at the ith point, which in turn gives the ideal experimental design. By rounding the fractional numbers to integer values, we can find an approximate design that is not as good but is realizable in practice.

Probably the most important theoretical result from optimal design theory is the general-equivalence theorem [14, 15], which states the following three equivalent assertions:

1. The design ξ^* that maximizes the determinant of the information matrix, det $\mathbf{M}(\xi^*)$
2. minimizes the maximum of the appropriately rescaled variance of the estimated response function $\max_x d(\hat{y}(x))$,
3. which is also equal to k, the number of regression coefficients in the response function, $\max_x d(\hat{y}(x)) = k$.

There are several practical results from the equivalence of these three assertions. By using assertions 1 and 3, we can estimate whether some design is D-optimal or not simply by comparing the value of the maximum variance of prediction to the number of the regression coefficients. The theorem also establishes the equivalence between two design optimality criteria, the maximum determinant (D-optimality), and the minimal maximum variance of the prediction (G-optimality); hence, we can search for D-optimal designs by using the procedure for G-optimal designs, which is much easier. In fact, this approach is the basis of most of the search procedures for producing realizable D-optimal designs, i.e., those having an integer number of measurements. The equivalence between assertions 2 and 3 makes it easy to determine how far some G-optimal design lies from the theoretical maximum.

8.6.3.2 D-Optimality and D-Efficiency

An experimental design with extended design matrix $\mathbf{F}^*$ is referred to as D-optimal if its information matrix fulfills the condition:

$$\det(\mathbf{F}^{*T}\mathbf{F}^*) \max_x \det(\mathbf{F}^T\mathbf{F}). \tag{8.52}$$

The determinant of the information matrix of the D-optimal design has a maximum value among all possible designs. Based on this criterion, the design with information matrix $\mathbf{M}^*$ is better than the design with information matrix $\mathbf{M}$ if the following condition holds:

$$\det(\mathbf{M}^*) > \det(\mathbf{M}). \tag{8.53}$$

The D-optimal design minimizes the volume of the confidence ellipsoid of the regression coefficients. This means that the regression coefficients obtained from D-optimal designs are determined with the highest possible precision.

D-efficiency is defined using Equation 8.54

$$D_{\mathit{eff}} = \left[\frac{\det \mathbf{M}}{\det \mathbf{M}_\xi} \right]^{1/k}, \qquad (8.54)$$

where $\det \mathbf{M}_\xi$ designates the theoretically proven maximum determinant of the respective continuous D-optimal design. It is important to mention that the design with information matrix $\mathbf{M}_\xi$ depends only on the corresponding factor space and the regression model. No other restrictions (i.e., the number of points, required integer numbers of measurements) are imposed. Continuous experimental designs are hypothetical experimental designs where a noninteger number of replicates per experimental point are permitted. These designs have theoretical value only and are listed in specialized reference catalogs [31]. In Equation 8.54, the design having information matrix $\mathbf{M}$ could be any design in the same factor space and regression model. Typical examples are a design having exactly $N = k + 5$ measurements or a design that is supposed to consist of measurements on particular levels of the variables.

8.6.3.3 G-Optimality and G-Efficiency

An experimental design with an information matrix $\mathbf{M}^*$ is G-optimal if the following condition holds,

$$d(\mathbf{M}^*) = \min_{\mathfrak{I}}(\max_{\aleph} d(\mathbf{M})) \qquad (8.55)$$

where $d(\mathbf{M}) = f^T(\mathbf{x})(\mathbf{M})^{-1}\mathbf{f}(\mathbf{x})$ designates the variance of the prediction at point $\mathbf{x}$, $\mathfrak{I}$ is the set of all designs under consideration, and $\aleph$ is the general set of points defining the factor space. Using some experimental design and this expression, we can estimate the variance of prediction at any point in the factor space. This value is a measure of how close a prediction at an arbitrary point $\mathbf{x}$ would be to the true value of the response. This value depends only on the information matrix and the coordinates of the particular point. In fact, the full measure of the prediction ability at $\mathbf{x}$ also depends on the error of the measurement and the distribution of the repeated measurements at the point $\mathbf{x}$.

Considering some experimental design, $\mathbf{X}_1$, we can calculate the maximum variance d_1 over all factor space. Also, we can calculate the maximum variance d_2 over the same factor space by using another design, $\mathbf{X}_2$. The design that gives smaller variance (say $d_2 < d_1$) is said to be the design that minimizes the variance of the prediction where it is maximal. Taking into account that the prediction accuracy at the point with maximum variance is the worst one, the G-optimal design (here design $\mathbf{X}_2$) ensures the maximum prediction accuracy at the worst (in terms of prediction) point.

To calculate the G-efficiency we use the formula in Equation 8.56

$$G_{\mathit{eff}} = \frac{k}{\max_{\aleph}[d(\mathbf{F})]} \qquad (8.56)$$

where $\max_x[d(\mathbf{F})]$ designates the maximum value of the variance of the prediction calculated by using the extended design matrix, $\mathbf{F}$. The formula comes directly from the second and third assertions of the general-equivalence theorem.

8.6.3.4 A-Optimality

A-optimal designs minimize the variance of the regression coefficients. Some design, $\mathbf{M}^*$, is said to be A-optimal if it fulfills the following condition.

$$tr(\mathbf{M}^*)^{-1} = \min_X[tr(\mathbf{M})^{-1}] \qquad (8.57)$$

The term $tr(\mathbf{M})^{-1}$ designates the trace of the dispersion matrix. Because the diagonal elements of $\mathbf{M}^{-1}$ present the variances of the regression coefficients, the trace (e.g., their sum) is a measure of the overall variance of the regression coefficients. The minimization of this measure ensures better precision in the estimation of the regression coefficients.

8.6.3.5 E-Optimality

A criterion that is closely related to D-optimality is E-optimality. The D-optimality criterion minimizes the volume of the confidence ellipsoid of the regression coefficients. Hence, it minimizes the overall uncertainty in the estimation of the regression coefficients. The E-optimality criterion minimizes the length of the longest axis of the same confidence ellipsoid. It minimizes the uncertainty of the regression coefficient that has the worst estimate (highest variance).

An experimental design is referred to as an E-optimal design when the following condition holds,

$$\max_i \delta_i \lfloor (\mathbf{M}^*)^{-1} \rfloor = \min_X \max_i \delta_i[(\mathbf{M})^{-1}], i = 1\ldots R, \qquad (8.58)$$

where by R we designate the rank of the dispersion matrix. A design is E-optimal if it minimizes the maximum eigenvalue of the dispersion matrix. The name of the criterion originates from the first letter of the word "eigenvalue." The eigenvalues of the dispersion matrix are proportional to the main axes of the confidence ellipsoid.

8.7 ALGORITHMS FOR THE SEARCH OF REALIZABLE OPTIMAL EXPERIMENTAL DESIGNS

The theory of D-optimal designs is the most extensively developed, and as a consequence there is quite a long list of works devoted to the construction of practical and realizable D-optimal designs. Good sources describing algorithms for the construction of exact D-optimal designs can be found in the literature [16, 17]. Before describing algorithms for finding D-optimal designs, we first define some nomenclature. Hereafter, by $\mathbf{X}_N$, we denote a matrix of N rows and n columns. Each row designates an experimental point. Each of these points could be in process, mixture, or process+mixture factor spaces. By $\mathbf{F}_N$, we denote the extended design matrix, constructed by using a regression model.

$$\hat{y} = \sum_{i=1}^{k} f_i(\mathbf{x}) \tag{8.59}$$

Also, we use S_L to denote a set of L experimental points in the same factor space, called candidate points. The set of candidate points will be used as a source of points that might possibly be included in the experimental design, $\mathbf{X}_N$. The information matrix of the N-point design, $\mathbf{X}_N$, will be denoted as above by $\mathbf{M}_N = \mathbf{F}^T\mathbf{F}$, where $\mathbf{M}_N$ is the information matrix for some model (Equation 8.59). By the following formula, we denote the variance of the prediction at point $\mathbf{x}_j$.

$$d\left(\mathbf{x}_j, \mathbf{X}_N^{(i)}\right) = \mathbf{f}(\mathbf{x}_j)^T \left(\mathbf{F}_N^{T(i)} \mathbf{F}_N^{(i)}\right)^{-1} \mathbf{f}(\mathbf{x}_j)$$

By the following formula we denote the covariance between $\mathbf{x}_i$ and $\mathbf{x}_j$,

$$d\left(\mathbf{x}_i, \mathbf{x}_j, \mathbf{X}_N^{(i)}\right) = \mathbf{f}(\mathbf{x}_i)^T \left(\mathbf{F}_N^{T(i)} \mathbf{F}_N^{(i)}\right)^{-1} \mathbf{f}(\mathbf{x}_j)$$

where the symbol $(\cdot)^{(i)}$. denotes the result obtained at the ith iteration.

8.7.1 Exact (or N-Point) D-Optimal Designs

8.7.1.1 Fedorov's Algorithm

The algorithm for finding D-optimal designs proposed by Fedorov [18] simultaneously adds and drops a pair of points that result in the maximum increase in the determinant of the information matrix. The algorithm starts with some nonsingular design, $\mathbf{X}_N$. Here, "nonsingular" implies the existence of $\mathbf{M}^{-1}$. During the ith iteration, some point, say $\mathbf{X}_j \in \{\mathbf{X}_N\}$, is excluded from the set of the design points and a different point, $\mathbf{x} \in \{\mathbf{S}\}$, is added to $\mathbf{X}_N$ in such a way that the resulting increase of the det $\mathbf{M}_N$ is maximal. The following ratio of determinants can be used to derive an expression for finding the point that gives the maximum increase,

$$\frac{\det \mathbf{M}_N^{(i+1)}}{\det \mathbf{M}_N^{(i+1)}} = 1 + \Delta_i(\mathbf{x}_j, \mathbf{x}) \tag{8.60}$$

where the so-called "Fedorov's delta function" Δ_i is

$$\Delta_i(\mathbf{x}_j, \mathbf{x}) = \left[-d\left(\mathbf{x}_j, X_N^{(i)}\right)\right] + \left[-d\left(\mathbf{x}, \mathbf{X}_N^{(i)}\right)\right] - d\left(\mathbf{x}, X_N^{(i)}\right) d\left(\mathbf{x}_j, X_N^{(i)}\right) + d^2\left(\mathbf{x}, \mathbf{x}_j, X_N^{(i)}\right),$$

$$\tag{8.61}$$

which can be rewritten as

$$\Delta_i(\mathbf{x}_j, \mathbf{x}) = \left[1 + d\left(\mathbf{x}, \mathbf{X}_N^{(i)}\right)\right]\left[d\left(\mathbf{x}, \mathbf{X}_{N+1}^{(i)}\right) - d\left(\mathbf{x}_j, \mathbf{X}_{N+1}^{(i)}\right)\right] \quad (8.62)$$

To achieve the steepest descent, we choose a pair of points $\mathbf{x}^*$ and $\mathbf{x}_j$ in such a way as to satisfy Equation 8.63.

$$\max_{\mathbf{x}_j \in \{\mathbf{x}_N\}} \max_{\mathbf{x} \in \{S_L\}} \Delta_i(\mathbf{x}_j, \mathbf{x}) = \Delta_i(\mathbf{x}_j, \mathbf{x}^*). \quad (8.63)$$

To find a pair of points $\mathbf{x}^*$ and $\mathbf{x}_j$ fulfilling the condition described in Equation 8.63, we conduct an exhaustive search of all possible combinations of $\mathbf{x}^*$ (additions) and $\mathbf{x}_j$ (deletions). This point-selection procedure proceeds iteratively and terminates when the increase in the determinant between two subsequent iterations becomes sufficiently small.

The method of exchanging points between the design $\mathbf{X}_N$ and the set of candidates S_L is the reason why algorithms based on this idea are called "point exchange algorithms." The basic idea of point-exchange algorithms can be briefly described as follows: given some design, $\mathbf{X}_N$, find one or more points that belong to the set of candidate points, replacing points in $\mathbf{X}_N$. The act of replacement or addition is successful if the optimization criterion is satisfied, e.g., if the determinant rises.

8.7.1.2 Wynn-Mitchell and van Schalkwyk Algorithms

In cases of multifactor ($m > 4,5$) problems, Fedorov's algorithm can become extremely slow. To avoid this shortcoming while retaining some of the useful properties of this approach, two approximations of the original algorithm have been proposed. Both are intended to maximize the delta function by applying fewer calculations.

The first modification to be presented is an algorithm known as the Wynn-Mitchell method. The algorithm was developed by T. Mitchell [19] and was based on the theoretical works of H. Wynn [21, 22]. In this algorithm, at the ith iteration, a point $\mathbf{x} \in \{S_L\}$ maximizing the first bracketed term of Equation 8.62 is added to the design. Then a point $\mathbf{x}_j \in \{\mathbf{X}_N\}$ maximizing the second term of Equation 8.62 is removed from the design. In this way the maximization of the function in Equation 8.62 is divided into two separate steps that reduce the number of calculations needed. The algorithm of van Schalkwyk [23] is similar; however, it adds a point that maximizes the first term of Equation 8.61 and removes a point, $\mathbf{x}_j$, that maximizes the second term. Both algorithms are considerably faster than Fedorov's algorithm and, thus, they can be effectively applied to larger problems. The trade-off, however, is decreased efficiency. Neither algorithm follows the steepest descent of the delta function, instead simply performing a kind of one-variable-at-a-time optimization.

8.7.1.3 DETMAX Algorithm

In 1974 Mitchell published the DETMAX algorithm [19], which, with slight improvement, becomes one of the most effective algorithms described to date.

DETMAX is, in fact, an improved version of the Wynn-Mitchell algorithm. Instead of adding points one at a time, the initial N-point design is augmented with K points. From the resulting $(N + K)$-point design, a group of K points is selected for exclusion, thus returning back to an N-point design. The augmentation/exclusion process is called "excursions of length K." The algorithm starts with excursions having $K = 1$. After reaching optimality, the length of the excursion is increased by 1, and so on. The algorithm stops when the algorithm reaches K_{max}. A value of $K_{max} = 6$ was recommended for discrete factor spaces.

8.7.1.4 The MD Galil and Kiefer's Algorithm

Despite its high efficiency, DETMAX has one major shortcoming. To perform the excursions, we need to calculate the value of the variance of the prediction at each iteration.

$$d\left(\mathbf{x}, \mathbf{X}_N^{(i)}\right) = \mathbf{f}^T(\mathbf{x})\left(\mathbf{F}_N^T \mathbf{F}_N\right)^{-1} \mathbf{f}(\mathbf{x}). \tag{8.64}$$

It can be seen that Equation 8.64 requires calculation of the inverse of the information matrix, an operation that can become the time-limiting factor, even for problems of moderate size. Galil and Kiefer [20] proposed a more effective algorithm by making slight improvements to Mitchell's DETMAX method [19]. They managed to speed it up by replacing the following computationally slow operations with faster updating formulas:

- Construct $\mathbf{F}$
- Calculate $\mathbf{F}^T\mathbf{F}$
- Calculate $(\mathbf{F}^T\mathbf{F})^{-1}$
- Calculate $\det(\mathbf{F}^T\mathbf{F})$ and d

For updating the values of the inverse matrix, the determinant, and the variances of prediction, they used the following formulas:

$$\det\left(\mathbf{F}_{N\pm1}^T \mathbf{F}_{N\pm1}\right) = \det\left(\mathbf{F}_N^T \mathbf{F}_N\right)\left(1 \pm \mathbf{f}^T(\mathbf{x})(\mathbf{F}_N^T \mathbf{F}_N)^{-1}\mathbf{f}(\mathbf{x})\right) \tag{8.65}$$

$$(\mathbf{F}_{N\pm1}^T \mathbf{F}_{N\pm1})^{-1} = (\mathbf{F}_N^T \mathbf{F}_N)^{-1} \mp \frac{\left((\mathbf{F}_N^T \mathbf{F}_N)^{-1}\mathbf{f}(X)\right)\left(\left(\mathbf{F}_N^T \mathbf{F}_N\right)^{-1}\mathbf{f}(X)\right)^T}{\left(1 \pm \mathbf{f}^T(\mathbf{x})\left(\mathbf{F}_N^T \mathbf{F}_N\right)^{-1}\mathbf{f}(\mathbf{x})\right)} \tag{8.66}$$

$$\mathbf{f}^T(\mathbf{x}_i)(\mathbf{F}_{N\pm1}^T \mathbf{F}_{N\pm1})\mathbf{f}^T(\mathbf{x}_i) = \mathbf{f}^T(\mathbf{x}_i)\left(\mathbf{F}_N^T \mathbf{F}_N\right)^{-1}\mathbf{f}(\mathbf{x}_i) \mp \frac{\mathbf{f}^T(\mathbf{x}_i)(\mathbf{F}_N^T \mathbf{F}_N)^{-1}\mathbf{f}(\mathbf{x}_i)}{\left(1 \pm \mathbf{f}^T(\mathbf{x})\left(\mathbf{F}_N^T \mathbf{F}_N\right)^{-1}\mathbf{f}(\mathbf{x})\right)} \tag{8.67}$$

$$d(\mathbf{x}_i, \mathbf{X}_{N+1}) = d(\mathbf{x}_i, \mathbf{X}_N) \mp \frac{d(\mathbf{x}_i, \mathbf{x}, \mathbf{X}_{N+1})^2}{\left(1 \pm d(\mathbf{x}, \mathbf{X}_N)\right)}, i = 1,\ldots, N$$

The resulting increase in speed is sufficient to allow the algorithm to be run several times, starting from different initial designs, and delivering better results.

8.7.2 SEQUENTIAL D-OPTIMAL DESIGNS

All of the experimental designs mentioned here can be classified as exact designs. These designs have different useful features and one common disadvantage: their properties depend on the number of points. For example, if some design is D-optimal for N points, one cannot expect that after the removal of two points the design having $N - 2$ points will be D-optimal as well. This peculiarity has clear practical implications. Suppose the experimenter obtains an N-point D-optimal design. During the course of the experimental work, after $N - 2$ experiments have been performed, the experimenter runs out of some necessary raw materials. By switching batches or suppliers of the raw material, the remaining two measurements will often introduce additional uncontrolled variability. The use of the reduced set of $N - 2$ measurements, even if $N > k$ and the number of the degrees of freedom is high enough, will probably affect the quality of the regression model.

To illustrate the loss of an important feature, recall a full factorial design. It was said that full factorial designs have one very useful feature: we can calculate the regression coefficients just by using a calculator. This is possible because of the diagonal structure of the information matrix, $[\mathbf{F}^T\mathbf{F}]$. However, the removal of just one of the points of the design will disturb its diagonal structure. Similarly, the removal of just one of the measurements from a D-optimal design will make the design no longer D-optimal. Hence, we must be very careful when using exact (or N-point) experimental designs to ensure that all of the necessary resources are available.

The method of sequential quasi-D-optimal designs was developed to avoid these shortcomings of D-optimal designs [21, 24]. The idea behind this methodology is to give the experimenter the freedom to choose the number of the measurements and to be able to stop at any time during the course of the experimental work.

These designs have a structure that is schematically outlined in Figure 8.23. They are constructed of two blocks, noted here as Block A and Block B, where the number

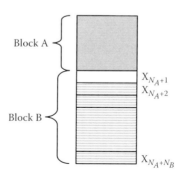

FIGURE 8.23 Schematic representation of a sequential design.

Response-Surface Modeling and Experimental Design

of points in Block A and Block B are denoted as N_A and N_B, respectively. The points in Block A are selected to make up a D-optimal exact experimental design, where $N_A = k$; thus, Block A is an optimal experimental design constructed of the minimum possible number of points. If we complete this design, we will be assured that the information matrix is not singular and we will be able to build a model. In practice, it is not advisable to build a regression model with zero degrees of freedom ($v = N_A - k = 0$).

The first step in the construction of sequential D-optimal designs is to find the points of Block A. The design matrix of this design is denoted as $\mathbf{F}_{N_A}$, and the information matrix is denoted by $\mathbf{M}_N = [\mathbf{F}_{N_A}^T \mathbf{F}_{N_A}]$. The next step is to find a design having an information matrix $[\mathbf{F}_{N_A+1}^T \mathbf{F}_{N_A+1}]$ such that $\mathbf{F}_{N_A+1} = \begin{bmatrix} \mathbf{F}_{N_A} \\ \mathbf{f}_i^T \end{bmatrix}$ and $\det(F_{N_A+1}) = \max_{f_i \subset S} \det[\begin{smallmatrix} \mathbf{F}_{N_A} \\ \mathbf{f}_i^T \end{smallmatrix}]$. The design $\mathbf{F}_{N_A+1}$ maximizes the determinant among all designs having N_A+1 points. This procedure is repeated in an iterative fashion. Starting with the design $\mathbf{F}_{N_A+1}$, we find another design, $\mathbf{F}_{N_A+2}$, using the same procedure. We continue this process to obtain a sequence of designs

$$\mathbf{F}_{N_A} \subset \mathbf{F}_{N_A+1} \subset \mathbf{F}_{N_A+2} \subset \ldots \mathbf{F}_{N_B} \subset \ldots,$$

where $\mathbf{F}_{N_A+j}$ is obtained from $\mathbf{F}_{N_A+j-1}$ by adding a point that maximizes the determinant of its information matrix. The practical value of these designs comes from the fact that the experimenter can choose any one of them because all of them are quasi D-optimal. To go from one to another design means simply to perform one more experiment. If the experimenter for some reason decides not to perform the next experiment, he or she will have already obtained a quasi D-optimal design.

We have already noted that the experimenter can choose the number of experiments in sequential D-optimal designs. The only limitation is that the minimum number of experiments should be larger than or equal to N_A. In practice, the actual number of measurements is determined by the availability of resources, e.g., time, materials, etc. In fact, with this approach, we can choose the number of measurements that provides a predetermined prediction accuracy for our model. This is illustrated in the following example.

8.7.2.1 Example

This example (see Vuchkov [25]) illustrates a method for choosing the number of experiments in a sequential D-optimal design. Table 8.10 shows a sequential (quasi) D-optimal experimental design for two mutually independent process variables. The design is optimal for a second-order polynomial having the following structure:

$$\hat{y} = b_0 + b_1 x_1 + b_2 x_2 + b_{12} x_1 x_2 + b_{11} x_1^2 + b_{22} x_2^2.$$

The points numbered 1 through 10 form the so-called Block A. The rest of the points, numbered 11 through 22, form Block B. In Table 8.10, we can find a total of 13 different designs having 10, 11, …, 22 points. We can start the experiments by using the points in Block A, i.e., the first design with 10 points. After that, we

TABLE 8.10
Sequential D-Optimal Experimental Design for Two Process Variables and a Second-Order Polynomial

No.	x_1	x_2	max d (F) s	y (%)
1	1	1	—	12.2
2	1	−1	—	13.7
3	−1	1	—	7.2
4	−1	−1	—	10.7
5	0	0	—	7.65
6	0	1	—	9.2
7	1	0	1.400	12.3
8	0	−1	1.250	13.8
9	−1	0	0.806	6.65
10	1	1	0.805	9.7
11	1	−1	0.795	14.6
12	−1	1	0.794	9.7
13	−1	−1	0.529	11.35
14	0	0	0.446	10.0
15	1	−1	0.438	16.0
16	−1	−1	0.438	10.2
17	1	1	0.426	11.5
18	−1	1	0.425	—
19	1	0	0.396	—
20	0	−1	0.350	—
21	−1	0	0.342	—
22	0	1	0.285	—

perform sequentially any number of the measurements specified by the entries in points 11, 12, etc.

In the fourth column of this table, we find the estimates of the maximum variance of prediction, calculated for each of the designs. For instance, the value $d = 1.4$ is the maximum variance of prediction for the design consisting of points 1 to 7, whereas the value $d = 0.4446$ corresponds to the design consisting of points 1 to 14. The confidence interval for the predicted value of the response is given by

$$|\hat{y}(\mathbf{x}) - \eta(\mathbf{x})| \leq t_{v, 1-\alpha} s_\varepsilon \sqrt{d(\mathbf{x})} \qquad (8.68)$$

where s_ε is the standard deviation of the observation error estimated at $v + 1$ replicates, $t_{v,(1-\alpha)}$ is the value of Student's t-statistic at v degrees of freedom and confidence level α, and $d(\mathbf{x})$ is the variance of the prediction at point $\mathbf{x}$. If point $\mathbf{x}$ is chosen such that $d(\mathbf{x})$ is at its maximum, then the right-hand side of the inequality in Equation 8.68 becomes the upper bound of the prediction error that can be achieved with the design.

Suppose that we wish to achieve error μ,

$$|\hat{y}(\mathbf{x}) - \eta(\mathbf{x})| \leq \mu \qquad (8.69)$$

which means that we must perform measurements until

$$\mu \geq t_{v,1-\alpha} s_\varepsilon \sqrt{d(\mathbf{x})} \qquad (8.70)$$

The quantity on the right-hand side of Equation 8.70 depends on the measurement error. If the value of s_ε is large, then we will have to perform huge number of measurements to offset this by the value of $d(\mathbf{x})$. On the other hand, if the measurement error is small, we will be assured of obtaining good predictions, even with a small number of measurements.

The data in this example represent an investigation of ammonia production in the presence of a particular catalyst. The measured yield in percent is shown in the far right column of Table 8.10. Suppose we wish to achieve a prediction error less than $\mu = \pm 1.5\%$ in an example, where the standard deviation (measurement error) is $s_\varepsilon^2 = 1.05$ estimated with 11 measurements, i.e., with degrees of freedom $v = 10$. The critical value of Student's t-statistic is found to be $t_{10,0.95} = 2.228$. At $N = 16$ experiments, we check to see if the desired level of accuracy is achieved and obtain $t_{v,1-\alpha}\sqrt{d} = 2.228\sqrt{0.438 \times 1.05} = 1.51 > 1.5$. At $N = 16$ experiments, we obtain $t_{v,1-\alpha}\sqrt{d} = 1.48 < 1.5$; therefore, we can stop at $N = 17$ and be assured that, 95% of the time, we will achieve a prediction error not worse than $\pm 1.5\%$, which is considerably smaller than the range of the variation in the response value.

8.7.3 SEQUENTIAL COMPOSITE D-OPTIMAL DESIGNS

A major shortcoming of the optimal experimental designs discussed so far is that we must assume a particular form for the regression model and then construct an appropriate design for the model. From a practical point of view, this means that the choice of the correct form of the model must be made at a stage in the experimental investigation when we possess little information about the model. In this case, the experimenter has to employ knowledge acquired in previous studies or from surveys of the literature, and then proffer an educated guess at the structure of the model. At the end of the investigation we can decide whether or not this assumption was correct. In cases where the preliminary choice of the model was wrong, the experimenter must select a new experimental design for a different functional form of the regression model, throw out the data collected so far, and conduct a new investigation. The only benefit from the initial and not quite successful investigation is the knowledge that the previously assumed model was found to be incorrect and that a higher-order model is probably necessary.

There is an important class of experimental design that largely avoids these kinds of problems and offers the experimenter the possibility of using the same data in the context of two different models. An example of these types of designs is the central composite design mentioned in Section 8.4.3. They have many useful features, but like all other symmetrical designs, we must perform all of the experiments in the list. If we fail to perform just one of them, the design will lose its desirable properties.

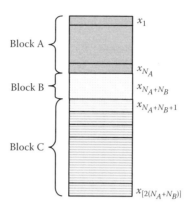

FIGURE 8.24 Schematic representation of an optimal sequential composite design.

To avoid these kinds of problems, some new experimental designs have been proposed [26, 27]. Called optimal composite sequential designs (OCSD), these designs are an extension of optimal sequential designs (OSD) in that they are optimal for more than one type of model. The structure of a typical OCSD is shown in Figure 8.24.

Unlike OSDs, OCSDs are constructed of three blocks, denoted here as Blocks A, B, and C. Suppose that the design shown in Figure 8.24 were constructed to be optimal for some models M_1 and M_2 having coefficients numbers k_1 and k_2, respectively, where $k_2 > k_1$. Block A is constructed as an exact D-optimal design for model M_1, where $N_A = k_1$. The extended design matrix of this design is $\mathbf{F}_{N_A}$. To construct the second part of the OCSD, namely Block B, we assume that the number of points in the design is the minimum required for model M_2, and thus the design for M_2 is embedded in the design for M_1. The extended design matrix for the second model is then $[\begin{smallmatrix}\mathbf{F}_{N_A}\\\mathbf{F}_{N_B}\end{smallmatrix}]$, where $N_B = k_2 - k_1$. As a result, the design for the second model consists of $k_2 = N_A + N_B$ points, the minimum number of points required for model M_2. The procedure for obtaining the points in Block B is the same as for Block A. The only difference is that we have to leave unchanged the points in Block A and manipulate only the points in Block B. The next step is to generate the sequential part of the design, shown in Block C. Here we can apply two approaches:

1. Search for sequential designs that are optimal for model M_2. In this case we apply the same procedure as described for OSD.
2. Search for sequential designs that are optimal for both models M_1 and M_2. In this case we switch alternatively between the two models. Point number $N_A + N_B + 1$ applies to model M_1, and the next point in the sequence with number $N_A + N_B + 2$ applies to M_2.

Whichever approach is adopted for the construction of the sequential designs, at each stage we are able build two types of regression models, based on the structure of M_1 or M_2.

The following example shows a typical OCSD for two factors (see Table 8.11). The design is constructed to be quasi D-optimal for two models. Model M_1 is a

TABLE 8.11
Optimal Sequential Composite Design for Two Process Variables and Two Polynomial Models: M_1, Full Second-Order Model; and M_2, Incomplete Third-Order Model

No.	x_1	x_2	max $d(M_1)$	max $d(M_2)$
1	−1	−1	—	—
2	1	−1	—	—
3	1	0	—	—
4	−1	1	—	—
5	0	1	—	—
6	1	1	2.750	—
7	−0.5	−0.5	1.358	—
8	−0.5	0.5	1.358	7.531
9	0	0.5	1.318	1.668
10	−1	0.5	0.893	1.342
11	0.5	−1	0.847	1.284
12	−0.5	−1	0.795	0.996
13	−1	−0.5	0.795	0.905
14	0.5	0.5	0.766	0.785
15	1	1	0.653	0.754
16	−0.5	1	0.625	0.689

second-order polynomial as shown in Equation 8.71 and model M_2 is an incomplete third-order polynomial as shown in Equation 8.72.

$$M_1 : \hat{y} = b_0 + \sum_{i=1}^{r} b_i x_i + \sum_{i=1}^{r-1} \sum_{i>i}^{r} b_{ij} x_i x_j + \sum_{i=1}^{r} b_{ii} x_i^2 \quad (8.71)$$

$$M_2 : \hat{y} = b_0 + \sum_{i=1}^{r} b_i x_i + \sum_{i=1}^{r-1} \sum_{i>i}^{r} b_{ij} x_i x_j + \sum_{i=1}^{r} b_{ii} x_i^2 + \sum_{i=1}^{r} b_{iii} x_i^3 \quad (8.72)$$

The number of the regression coefficients for the two models is $k_1 = 6$ and $k_2 = 9$, respectively. Thus Block A consists of six points (see rows 1 to 6 in Table 8.11), and Block B has 3 = 9 − 6 points (rows 7 to 9). The remainder of the points (rows 10 to 16) represent Block C, the sequential part of the design.

This design is used in the same manner as the previously described optimal sequential designs, except that now it is possible to use either of the models described in Equation 8.71 and Equation 8.72. We start the investigation by performing experiments 1 to 6. After that, we continue with experiments 7 to 9. Once we have completed more than k_1 experiments, it becomes possible to build a model with structure M_1. For example, if the model built over points 1 to 8 appeared to be inadequate, we could continue the experimental work by adding additional measurements according to the list in Table 8.11. Once we have completed more than k_2 experiments, we can build a model with structure M_2. At the same time, we will have accumulated enough experiments to reapply M_1 and to check whether the new measurements have yielded a better model, M_1. Starting from point 10, we can apply both of the model structures. From point 10 onward, we can apply the rules previously discussed for choosing the number of measurements needed to achieve a desired level of accuracy. If we are constructing a model based on M_1, then it is important to use the values for the maximum variance of prediction given in column 4 of Table 8.11. Alternatively, we use column 5 for model M_2.

8.8 OFF-THE-SHELF SOFTWARE AND CATALOGS OF DESIGNS OF EXPERIMENTS

8.8.1 OFF-THE-SHELF SOFTWARE PACKAGES

There are many software packages that offer varying degrees of support for the construction of optimal experimental designs.

8.8.1.1 MATLAB

MATLAB™ is a product of MathWorks, Inc. Detailed information can be found on the Web site http://www.mathworks.com, and there is also a newsgroup at comp.soft-sys.matlab. Procedures for the construction of experimental designs are included in MATLAB's Statistics Toolbox, which is not included in the base package and must be purchased separately. With this package, it is possible to construct full factorial designs by using the functions `fullfact` and `ff2n`. In fact, the function that generates classical full factorial designs is `ff2n`. By using it, one can construct two-level n-factor designs. The function `fullfact`, despite its name, is in fact a combinatorial function that generates all permutations of n variables, each taken at 1, ..., r levels. For example, constructing a design of two factors, where the first one is varied at four levels and the second one at three levels, is equivalent to generating a permutation of two variables, where the first is varied at four levels and the second at three. The respective MATLAB command is

```
>> D=fullfact([4 3])
```

and the result is:

```
>> D =
       1    1
       2    1
       3    1
       4    1
       1    2
       2    2
       3    2
       4    2
       1    3
       2    3
       3    3
       4    3
```

The next task is to transform the values 1 to 4 and 1 to 3 into coded variables, obtaining the levels [−1, −0.33333, +0.33333, +1] for the first variable and [−1, 0, 1] for the second one.

In MATLAB, there is no straightforward way to generate fractional factorial designs. The *Hadamard transform* function can be used instead, which generates $n \times n$ H_n Hadamard matrices [28]. These matrices have an important feature, in that the columns are pairwise orthogonal, which makes them easy to use as an experimental design for $n - 1$ variables. The Hadamard matrices produced by MATLAB are normalized, which means that all of the entries of the first column are equal to 1, and only the remaining $n - 1$ columns can be treated as variables. Actually, each matrix H_n produced by MATLAB is equivalent to a design

$$\mathbf{H}_n \equiv \begin{bmatrix} 1 & x_{11} & \cdots & x_{(n-1)1} \\ 1 & x_{12} & \cdots & x_{(n-1)2} \\ \vdots & \vdots & \vdots & \vdots \\ 1 & x_{1n} & \cdots & x_{(n-1)n} \end{bmatrix} = \mathbf{F}_n,$$

where $\mathbf{F}_n$ is the extended design matrix for $n - 1$ variables, n experiments, and a linear model. Another peculiarity is that only Hadamard matrices H_n exist where the order is $n = 1$, $n = 2$, or $n = 4t$, where t is a positive integer. Thus one can generate fractional factorial designs for only $r = 1$, 2 and $r = 4t - 1$ variables.

In the MATLAB Statistics Toolbox, there are two functions for generating exact D-optimal designs, `cordexch` and `rowexch`. Both procedures are equivalent from the user's point of view. To use them, one must specify the number of variables, the number of the experiments, and the type of the desired regression model. Four different model choices are provided:

Linear:

$$\hat{y} = b_o + \sum_{l}^{r} b_i x_i; \tag{8.73}$$

Interaction:

$$\hat{y} = b_o + \sum_{i=1}^{r} b_i x_i + \sum_{i=1}^{r-1} \sum_{i<j}^{r} b_i b_j x_{ij}; \tag{8.74}$$

Quadratic:

$$\hat{y} = b_o + \sum_{i=1}^{r} b_i x_i + \sum_{i=1}^{r-1} \sum_{i<j}^{r} b_i b_j x_{ij} + \sum_{i=1}^{r} b_{ii} x_i^2; \tag{8.75}$$

Pure quadratic:

$$\hat{y} = b_o + \sum_{i=1}^{r} b_i x_i + \sum_{i=1}^{r} b_{ii} x_i^2. \qquad (8.76)$$

The output of the function is the design matrix $\mathbf{X}_N$. A typical session for the construction of an exact D-optimal design for $n = 3$ variables, $N = 10$ points, and a quadratic model (Equation 8.75) is shown below:

```
>> Xn=rowexch(3,10,'q')
>> Xn =
        0      -1      -1
        0       0      -1
        1       1       1
        1       1      -1
        0      -1       1
        1      -1       0
       -1       1      -1
       -1      -1       0
       -1       1       1
        0       1       0
```

It is important to note that it is unlikely that the algorithm will generate symmetrical designs (FFD, CCD, etc.), because the algorithms for exact optimal designs do not always converge to the best design. They start with some initial, usually a randomly generated, design and iteratively improve it. Because the search for exact experimental design operates in a highly complicated space, the results from run to run can differ from each other in terms of the obtained optimality (e.g., the value of the determinant for D-optimality). When using these methods, it is advisable to run the procedure several times and use only the best of the generated designs. This can be accomplished in MATLAB with a simple script for calculating the determinant of the design. The script should take the output of cordexch and rowexch (say Xn) and construct the respective extended design matrix, $\mathbf{F}_N$, using the same structure as the model (here quadratic) and calculate the determinant of the information matrix. After calculating this figure of merit for each generated design, it is a simple matter to choose the one that has highest value.

An example of a MATLAB session (three process variables, 10 experiments, quadratic model [see Equation 8.75]) follows:

```
>> Xn=rowexch(3,10,'q');
>> Fn=x2fx(Xn,'q');
>> det(Fn'*Fn)
>> ans = 1048576
```

Our experience with cordexch and rowexch reveals that the algorithm behaves quite well, and it is usually sufficient to perform approximately 10 to 20 runs to achieve the best design. With increasing numbers of the factors, more runs will be required.

There is also a useful function called daugment that is used to construct augmented experimental designs. The idea is similar to the approach used for the construction of

Block B for optimal composite sequential designs, discussed earlier. The only difference is that the initial design (equivalent to Block A) and the augmentation (equivalent to Block B), as discussed in Section 8.7.3 about OCSD, are for one and the same type of model. The rationale here is to use some data (e.g., a previously performed experimental design of N_1 points) and to enrich it by adding some additional, say N_2, experiments. As a result, we will obtain a design of $N_1 + N_2$ experiments just by performing the new N_2 experiments. There is one important precaution to be mentioned here. Before augmenting the original set of experimental data with a new set, it is extremely important for the experimenter to make sure that all of the conditions used for measuring the two sets of experiments are identical. For example, a new or repaired instrument, new materials, or even a new (more/less experienced) technician could cause some shift in the results not related to the process under investigation.

8.8.1.2 Design Expert

Design Expert® version 6 (DX6) offered by Stat-Ease, Inc., (http://www.statease.com) is a typical off-the-shelf software package. The software is designed to guide the experimenter through all of the steps in response-surface modeling up to the numerical (also graphically supported) optimization of the response function. Apart from the rich choice of experimental designs available, DX6 calculates the regression model along with a comprehensive table of ANOVA results. The package also has good graphical tools, especially for contour plots for both Descartes and barycentric (including constrained) coordinate systems.

The DX6 program generates most of the symmetric designs, Taguchi orthogonal designs, and exact D-optimal designs for process, mixture, and process+mixture combined spaces. DX6 can also handle constrained mixtures. It can also produce the respective two- and three-dimensional-contour Descartes and mixture plots. There is considerable flexibility provided in model construction and their modification.

The package also includes a multiresponse optimization function, based on a desirability function [12] that reflects the desirable ranges or target values for each response. The desirable ranges are from 0 to 1 (least to most desirable, respectively). One can also define the importance of the different responses and the program can produce and graphically represent the optimum of the desirability function.

The main shortcoming of the program is that the experimenter is unable to import measured response values and their corresponding (probably nonoptimal, but existing and already performed) experimental designs. This problem can be circumvented by using the clipboard (copy/paste functions in Microsoft® Windows). Practically speaking, the user is expected to use only the experimental designs provided by the package. Another drawback is that the user cannot access the graphical and optimization facilities by entering regression coefficients calculated with another program (e.g., Microsoft Excel or MATLAB).

8.8.1.3 Other Packages

Other packages offer tools for constructing experimental designs, including the DOE add-in of Statistica, http://www.statsoft.com/; SPSS, http://www.spss.com; Minitab, http://www.minitab.com; MultiSimplex, http://www.multisimplex.com; MODDE, from Umetrics, Inc., http://www.umetrics.com/; and SAS/QC and JMP, which are products of the SAS Institute Inc., http://www.sas.com.

8.8.2 CATALOGS OF EXPERIMENTAL DESIGNS

The best software packages can be quite expensive, but they provide the user with flexibility and convenience. Still, the user is expected to have some computer and programming skills, depending on the particular package, and the user must have some knowledge in the construction of optimal designs. However, in some particular applications of experimental design (see Section 8.9 below), it is sometimes impossible to use design of experiment (DOE) software.

As an alternative to DOE software, it is possible to use published tables or catalogs of experimental designs. The formalization of a technological process with inputs, outputs, and noise variables, as presented at the beginning of this chapter, provides a framework for generalization that makes it possible to apply the principles of experimental design to many different kinds of problems. Catalogs of optimal experimental designs (COED) can be found in the literature [29], which include tables of previously generated experimental designs. A typical design taken from a catalog is shown in Table 8.12 [29]. This is a D-optimal composite sequential design (see Section 8.7.3) for three mixture variables and one process variable. The design is an optimal composite for the following two models:

$$\hat{y} = \sum_{i=1}^{q} b_i x_i + \sum_{i=1}^{q+1-1} \sum_{j<i}^{q+r} b_{ij} x_i x_j + \sum_{i=q+1}^{q+r} b_{ii} x_i^2 \tag{8.77}$$

$$\hat{y} = \sum_{i=1}^{q} b_i x_i + \sum_{i=1}^{q+r} \sum_{j<i}^{q+r} b_{ij} x_i x_j + \sum_{i=q+1}^{q+r} b_{ii} x_i^2 + \sum_{i=1}^{q-1} \sum_{i<j}^{q} c_{ij} x_i x_j (x_i - x_j)$$

$$+ \sum_{i=1}^{q-2} \sum_{j<i}^{q-1} \sum_{l<j}^{q} b_{ijl} x_i x_j x_l \tag{8.78}$$

In the note to Table 8.12, we find the basic information for the design, including the number of variables and the number of regression coefficients for the two models. The total number of points in the design (here 28) is determined by the value of k_2, the number of coefficients in the larger model. As mentioned previously, the experimenter is supposed to perform at least k_1 experiments to be able to estimate the coefficients of a model of the type shown in Equation 8.77 and at least k_2 for a model of the type shown in Equation 8.78. After the 10th and the 14th point, we see two thinner lines, which emphasize these boundaries.

This design was generated by using a grid constructed in the mixture+process factor space. Each point of this grid was a combination of [0, 0.5, 1, 0.212, 0.788, −1, 1], subject to the restriction $\sum_{i=1}^{q} x_i = 1$ for the mixture coordinates. The two columns at the right contain the values of the maximum prediction variance. These values can be used to determine the number of experiments needed to provide the desired level of prediction accuracy. This approach was described earlier in this chapter. In addition to catalogs with OCSD, there are also catalogs of OSD [30] and exact optimal designs (EOD) [31].

TABLE 8.12
Optimal Composite Sequential Design for Three Mixture Variables and One Process Variable, Optimal for Models Having Structures in Equations 8.77 and 8.78

No 4.1. (3.1), q = 3, r = 1, K_1 = 10, K_2 = 14, T = 0.5, U = 0.212, V = 0.788, W = 0.333

No.	x_1	x_2	x_3	x_4	Max. d'	Max. d'
1	0	1	0	1	—	—
2	0	1	0	−1	—	—
3	0	0	1	−1	—	—
4	T	0	T	0	—	—
5	1	0	0	1	—	—
6	1	0	0	−1	—	—
7	0	T	T	0	—	—
8	0	0	1	1	—	—
9	W	W	W	−1	—	—
10	T	T	0	0	1.834	—
11	V	0	U	1	1.722	—
12	0	V	U	1	1.515	—
13	0	U	V	1	1.497	—
14	V	U	0	0	1.422	8.590
15	U	0	V	−1	1.418	3.482
16	U	V	0	0	1.375	2.708
17	T	T	0	−1	0.996	1.868
18	V	0	U	−1	0.971	1.563
19	W	W	W	0	0.970	1.508
20	U	0	V	1	0.962	1.476
21	0	V	U	−1	0.937	1.315
22	0	U	V	−1	0.934	1.125
23	U	V	0	1	0.753	1.125
24	V	U	0	1	0.728	0.886
25	0	0	1	0	0.667	0.759
26	0	1	0	−1	0.667	0.759
27	1	0	0	−1	0.531	0.684
28	W	W	W	1	0.530	0.665

8.9 EXAMPLE: THE APPLICATION OF DOE IN MULTIVARIATE CALIBRATION

8.9.1 CONSTRUCTION OF A CALIBRATION SAMPLE SET

One of the key areas in which experimental design is used in analytical chemistry is in the development of spectroscopic calibration models. Clearly, when the number of factors (chemical components) is small, the task is trivial, but it is also common to encounter applications that have many different components. These can include active components, inactive additives, dyes, etc., and the concentration levels of these components can vary widely. In such cases, the method of experimental design can be used to determine the optimum set of standards required to prepare calibration models. One of the difficulties in developing a new calibration model is that the mixtures must be within an appropriate range for the instrumental measurement method. Using mixtures that are too concentrated gives rise to a nonlinear response (e.g., in UV spectroscopy, for peaks that no longer obey Beer's Law) and, as such, the samples must be diluted. Conversely, using mixtures that are too dilute will cause a loss in the signal-to-noise ratio, and hence introduce unnecessary noise into the model. Given these constraints, it is possible to use the method of experimental design with some reference spectra to build up a simulated calibration set with proposed standards that are in the correct range for the chosen analytical measurement method, without the need to perform any preliminary experiments. These simulated calibration data can be used to build calibration models using PLS, etc. to test that there is sufficient variability in the calibration set to be useful for modeling purposes. The net result of this approach is that the user can very quickly develop a calibration model and test it prior to performing any actual experiments, which thus maximizes productivity and reduces waste. A further advantage of the approach described here is that one can perform some screening to discover inactive or nonabsorbing components in the mixtures. The combination of screening, DOE, and simulation of calibration mixture spectra proposed here can significantly reduce the resources required for performing the actual measurements.

To illustrate the use of experimental designs in an analytical chemistry application, we will examine a problem taken from the agrochemical industry. The problem under investigation was to develop a robust calibration model for several commercial products based on UV spectral measurements. By the term "robust model," we assume that the model will be able to give acceptable predictions even if there is some moderate variation in the controlled and uncontrolled variables.

The successful construction of any calibration model depends to a great extent on the set of calibration points. Considering the components of the products as independent factors, we can construct the respective factor space in the terms discussed earlier in this chapter (see Section 8.2.2). The calibration set will consist of points distributed within this factor space, and the best distribution of these points will be achieved by employing the experimental-design approach. Provided that the number of significant factors and that the type of the required regression model are known, we should be able to construct a successful experimental design. To implement the calibration design and perform the necessary measurements, we also need to know the boundaries of the factor space. The following discussion is directed toward these points.

8.9.1.1 Identifying of the Number of Significant Factors

In this example, 12 products, P1 to P12, were to be considered, each consisting of one to nine components, coded here as C1 to C9. The list of products, their ingredients (components), and the amount of each are listed in Table 8.13. For reasons of commercial confidentiality, we omitted the actual names of the products and

TABLE 8.13
List of Products under Consideration and the Respective Quantities of the Components Included in Each Product

P1		P2		P3		P4	
C6	100	C6	200	C6	200	C6	200
C9	50	C7	0.5	C7	0.5	C7	0.5
C7	2	C1	35	C1	35	C1	35
C1	24	C8	116.7	C8	117	C8	116.7
C8	80	C2	0.25	C2	0.25	C2	0.25
C2	0.4	C3	2.5	C3	2.5	C3	5
C3	2	C5	5	C5	1	C5	10
C5	10	—	—	—	—	—	—

P5		P6		P7		P8	
C6	120	C6	200	C6	120	C6	100
C9	80	6	6	C9	80	C7	0.5
C7	0.5	6	6	C7	0.5	C1	10
C1	35	6	6	C1	10	C2	0.5
C8	117	6	6	C2	0.5	C3	1
C2	0.25	6	6	C3	2.5	C5	10
C3	2.5	6	6	C4	40	C4	40
C5	10	—	—	—	—	—	—

P9		P10		P11		P12	
C6	8.77	C9	200	C9	200	C6	7.93
C9	8.77	C2	0.6	C2	0.6	C9	7.93
C7	0.07	C4	12.5	C4	150	C7	0.128
C4	4.6	—	—	—	—	C4	4.2

ingredients and slightly changed the amount of the ingredients in each product. These changes will not affect the generality of the approach described here.

Figure 8.25 shows the pure-component UV spectra of all nine components. The respective concentrations of the solutions used to measure the pure-component spectra are shown in Table 8.14. It is reasonable to assume that components having large absorption values in the wavelength range of interest will have considerable influence on the calibration model. Conversely, inactive components that have very weak absorption values in the wavelength range of interest will have a weak influence on the model while introducing some additional noise.

The level of UV absorption was used as a screening tool to divide the components into two sets, active (Figure 8.26) and inactive (Figure 8.27), with the active components having strong UV absorption signals in the range from 250 to 360 nm and inactive components having weak or insignificant UV absorption signals in this range.

Removing the UV-inactive components reduced the number of the components to be considered from nine to five. The products and their corresponding UV-active

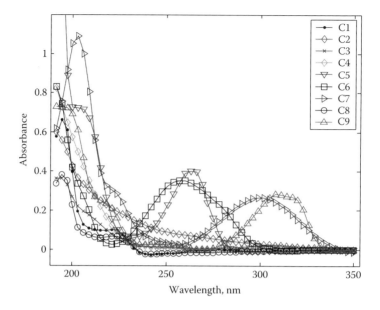

FIGURE 8.25 UV spectra of all nine pure components.

components are listed in Table 8.15. The number of components present in each formulation is also shown.

Examining Table 8.15, we observe that products P6, P10, and P11 have only one UV-active component; P9 and P12 have three UV-active components; P3, P4, P5, P7, and P8 have four UV-active components; and P1 and P2 have five UV-active components. Thus four types of experimental designs are needed for 1, 3, 4, and 5 independent (process) variables.

TABLE 8.14
Concentration of Components C1 to C9 Used to Measure Pure-Component Spectra

Concentration No.	Concentration, ppm
C1	10
C2	10
C3	10
C4	10
C5	100
C6	100
C7	10
C8	5
C9	1000

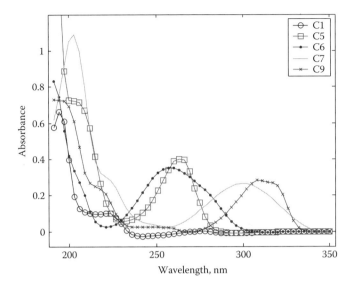

FIGURE 8.26 Spectra of the UV-active components.

8.9.1.2 Identifying the Type of the Regression Model

As the goal is to build a calibration model based on spectral data, we assume that Beer-Lambert's law is valid,

$$A_w = \sum_{i=1}^{p} \varepsilon_i c_i l, \; w = 1, \ldots \tag{8.79}$$

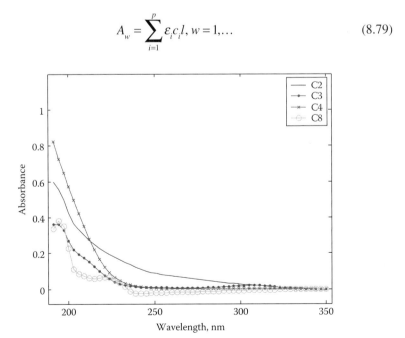

FIGURE 8.27 Spectra of the UV-inactive components.

TABLE 8.15
The "Reduced" Product Formulations after the Removal of the UV-Inactive Components Shown in Figure 8.27

P1		P2		P3		P4	
1	C1	1	C1	1	C1	1	C1
2	C5	2	C2	2	C5	2	C5
3	C6	3	C5	3	C6	3	C6
4	C7	4	C6	4	C7	4	C7
5	C9	5	C7				
P5		P6		P7		P8	
1	C1	1	C9	1	C1	1	C1
2	C5			2	C6	2	C5
3	C6			3	C7	3	C6
4	C7			4	C9	4	C7
P9		P10		P11		P12	
1	C6	1	C9	1	C9	1	C6
2	C7					2	C7
3	C9	—	—	—	—	3	C9

where l is the sample cell path length, ε_i is the extinction coefficient, and c_i is the concentration of the ith component at the wth wavelength. Noting the theoretical linear nature of the response described in Equation 8.79, we assume that a linear polynomial model for p independent variables will be adequate; thus, the model selected for our experimental designs has the following structure

$$\hat{y}_j = b_o + \sum_{i=1}^{p_j} b_i x_i, \ j = 1, 12 \qquad (8.80)$$

where p_j, $j = 1, 12$ represents the number of the jth product, e.g., $p_1 = 5$, $p = 4$, etc. As previously noted, the products fall into one of four categories depending on the number of UV-active components. Thus, four different types of models are needed, one each for one, three, four, and five process variables. Respectively, the number of regression coefficients will be $k_1 = 2$, $k_2 = 4$, $k_3 = 5$, and $k_4 = 6$. The minimum number of points in the designs for each of these models will be determined by the corresponding number of regression coefficients.

Having the number and the type of the variables (components) and the type of the regression model required, we can begin the task of constructing the appropriate experimental designs. In this project it was decided to use exact D-optimal designs having $N_i = k_i + 5$ points. The number of the points selected provides sufficient degrees of freedom to calculate the regression coefficients. The resulting D-optimal designs are shown in Table 8.16.

TABLE 8.16
Catalog of Four Exact D-Optimal Experimental Designs for the Spectroscopic Calibration Problem

$\xi_1(1,8)$	$\xi_2(3,9)$	$\xi_3(4,10)$	$\xi_4(5,11)$
x_1^c	$x_1^c,\ x_2^c,\ x_3^c$	$x_1^c,\ x_2^c,\ x_3^c,\ x_4^c$	$x_1^c,\ x_2^c,\ x_3^c,\ x_4^c,\ x_5^c$
−1	1, −1, −1	−1, −1, −1, −1	1, 1, −1, −1, −1
−1	−1, −1, 1	−1, 1, −1, 1	1, −1, 1, −1, 1
−1	−1, −1, −1	−1, −1, −1, −1	1, 1, 1, 1, −1
−1	−1, 1, −1	−1, 1, 1, 1	−1, 1, 1, 1, 1
1	−1, 1, 1	−1, −1, 1, 1	1, −1, −1, −1, 1
1	1, 1, −1	1, 1, 1, −1	−1, −1, −1, 1, −1
1	1, 1, 1	1, −1, −1, 1	−1, 1, 1, −1, 1
1	1, −1, 1	1, 1, 1, −1	1, −1, 1, 1, −1
—	1, −1, 1	1, 1, −1, 1	1, 1, −1, 1, 1
—	—	1, −1, 1, 1	−1, 1, −1, −1, −1
—	—	—	−1, −1, 1, −1, −1

Note: The numbers in parentheses at the top of the table represent, respectively, the number of variables and the number of measurements.

8.9.1.3 Defining the Bounds of the Factor Space

The coded values for the two levels of the controlled factors in these designs are +1 and −1, which represent the upper and lower boundaries for each variable. To implement the designs, we transform these two levels into the real values. By finding the lower and upper boundaries of the variables for each product, the four generic designs (in coded values) will be transformed to 12 calibration sets (in real values), one for each product.

To define the boundaries, we assume that the models should be valid over a working range of up to ±10% of each component's target value in the formulated products. Considering each of the product formulations individually, we calculate the bounds using Equation 8.81,

$$x_i^{\min} = 0.90 p_i^*$$
$$x_i^{\max} = 1.10 p_i^*$$
(8.81)

where p_i^* designates the target value of the ith component, and, $x_i^{\min}$ and $x_i^{\max}$ are the lower and upper bounds, respectively. For example, if the target value of the ith factor is $x_i^c = 200$, the respective boundaries will be $x_i^{\min} = 0.9 \times 200 = 180$ and $x_i^{\max} = 1.1 \times 200 = 220$. The general formula for the transformation from coded to natural (real) variables and vice versa is

$$x_i^c = \frac{x_i - x_i^c}{x_i^{\max} - x_i^c}.$$

TABLE 8.17
List of Components Included in Product P1, with UV-Inactive Components Shaded

Component	Quantity
C6	100
C9	50
C7	2
C1	24
C8	80
C2	0.4
C3	2
C5	10

For the example given here, the formula becomes:

$$-1 = \frac{x_i - 200}{220 - 200} = \frac{180 - 200}{220 - 200} = \frac{-20}{20}$$

The reverse transformation is obvious.

The process of translating the coded values to real values is illustrated in detail for the construction of the calibration set for product P1. The target values for each component in product P1 are shown in Table 8.17, with UV-inactive components shown as shaded rows.

By taking the entries of Table 8.17 as the target values, we calculate the respective upper and lower bounds for each component. The results are shown in Table 8.18.

Now, using the correspondence between the real and coded upper and lower bounds shown in Table 8.18, we can choose the appropriate design from Table 8.16 and replace the coded entries with the real ones. The set of the calibration points in coded and real values, constructed using design $\xi_4(5,11)$ (5 variables, 11 measurements), is shown in Table 8.19.

TABLE 8.18
Translation of Coded Factor Levels to Real Experimental Levels for Product P1

Component	Lower Bound		Upper Bound	
	Real	Coded	Real	Coded
C6	90	−1	110	1
C9	45	−1	55	1
C7	1.8	−1	2.2	1
C1	21.6	−1	26.4	1
C5	9	−1	11	1

TABLE 8.19
Translated Experimental Design for Product P1

					C6	C9	C7	C1	C5
1,	1,	−1,	−1,	−1	110	55	1.8	21.6	9
1,	−1,	1,	−1,	1	110	45	2.2	21.6	11
1,	1,	1,	1,	−1	110	55	2.2	26.4	9
−1,	1,	1,	1,	1	90	55	2.2	26.4	11
1,	−1,	−1,	−1,	1	110	45	1.8	21.6	11
−1,	−1,	−1,	1,	−1	90	45	1.8	26.4	9
−1,	1,	1,	−1,	1	90	55	2.2	21.6	11
1,	−1,	1,	1,	−1	110	45	2.2	26.4	9
1,	1,	−1,	1,	1	110	55	1.8	26.4	11
−1,	1,	−1,	−1,	−1	90	55	1.8	21.6	9
−1,	−1,	1,	−1,	−1	90	45	2.2	21.6	9

8.9.1.4 Estimating Extinction Coefficients

Using the formulations of the calibration set listed in Table 8.19 and the pure-component spectra measured earlier, we can generate a set of simulated calibration spectra without performing any experimental work and investigate some important properties of the calibration set. The first step is to estimate the matrix of extinction coefficients, **E**, using the pure-component spectra. Assuming the path length, $l = 1$, the ith pure-component spectrum can be represented by

$$A_i = \varepsilon_i c_i^*, \quad i = 1, m_t \tag{8.82}$$

where $\mathbf{A}_i$ is the vector of measured absorbances for the ith component at concentration c_i^*, and ε_i is the respective vector of extinction coefficients. Solving for the vector of extinction coefficients, ε_l, gives Equation 8.83.

$$\varepsilon_i = \frac{A_i}{c_i^*}, \quad i = 1, m_t \tag{8.83}$$

The matrix of extinction coefficients for the components of product P1 can be assembled by arranging the vectors of extinction coefficients into the rows of $E_{P1} = [\varepsilon_1, \varepsilon_5, \varepsilon_6, \varepsilon_7, \varepsilon_9]$. The matrix of concentrations, **C**, for the subset of active species in product P1 is given in the right-hand side of Table 8.19, or in matrix form, $\mathbf{C} = [\mathbf{C}_1, \mathbf{C}_5, \mathbf{C}_6, \mathbf{C}_7, \mathbf{C}_9]$. According to the Beer-Lambert law in Equation 8.79, the product of these two matrices gives the matrix of simulated mixture spectra, **A**, for the calibration set, where the path length, l, is assumed equal to 1.

$$\mathbf{A} = \mathbf{CE} \tag{8.84}$$

Figure 8.28 shows the predicted calibration spectra listed in Table 8.19 for product P1.

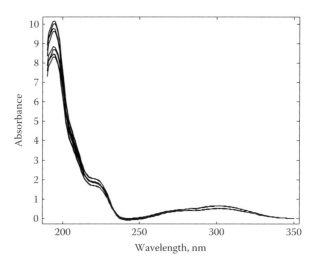

FIGURE 8.28 Predicted calibration spectra for product P1.

For calibration work in the UV range, spectra should have a maximum absorbance less than 1 for Beer's law to be obeyed and to obtain good linear response. Clearly, since this condition does not hold for the simulated calibration spectra shown in Figure 8.28, a simple dilution of the calibration samples should be performed before measuring their UV spectra. This is applicable to any sample type, as the dilution only affects the analysis method and not the final result.

8.9.2 Improving Quality from Historical Data

As was mentioned previously, in process analytical applications we are usually limited in how we do experiments and collect data. Sometimes we are not able to adjust the controlled factors of a process according to the principles of experimental design because it would cause production of product that fails to meet quality standards. In such cases, the only option is to measure the process and deal with the data as received. Experiments performed in this manner are called passive experiments. The values of the measured variables change according to normal variation in the production process. This can cause correlation in the measurements, which in turn can affect the numerical stability of fitting regression models. In cases where it is desirable to achieve on-line or at-line control with a regression model derived from measurements of the process, a procedure is needed to avoid making unnecessary measurements and improve the accuracy of the resulting models.

As a practical example, we consider data provided by BP Amoco from their naphtha processing plant in Saltend, Hull, U.K. Briefly, naphtha is a mixture of hydrocarbons and aromatics. The most important components in the feedstock are naphthalenes and aromatics. Periodically, samples are collected. The near-infrared (NIR) spectra of these samples are measured, and the amount of naphthalenes and aromatics is measured by gas chromatography (GC). A calibration model is constructed using PCR or PLS (see Chapter 6), and the predicted values of naphthalene and aromatic content are used to control the process. Here the goal is to replace costly, time-consuming off-line GC measurements with rapid, on-line NIR measurements. It is possible to collect hundreds or even thousands of NIR spectra at relatively low cost, while it would be cost prohibitive to perform GC analysis on each one. By analysis of the design matrix, **X**, which can be cheaply and quickly measured, we can select a small subset of samples for GC analysis that will give an optimal design, thus minimizing the time and expense of performing GC analysis while maximizing the information we gather as well as the performance of the regression model that we will build from these measurements.

Once the initial model is developed and placed on-line, a large historical database of measurements and predictions can be accumulated. If the process or the measurement instrument drifts over time, the usual practice is to recalibrate the NIR model periodically by collecting new plant samples and performing NIR and GC measurements. To avoid performing costly GC analysis on a large set of samples during normal process operation, some method of using the inexpensive NIR data is needed to select the most informative samples for off-line GC analysis. The resulting historical data can be used in this way to augment the original experimental design with maximum information and minimum effort. As a side effect, better performance of the calibration model could be expected.

Following commonly accepted terminology, **X** represents an $N \times m$ data matrix of NIR spectra with N rows (samples) measured at m variables. The predicted value $\hat{y}_i, i = 1...N$ of the response (naphthalene content or aromatic content) y_i, $i = 1, ..., N$ can be estimated using some appropriate form of a regression model,

$$\hat{y}_i = \sum_{j=1}^{k} b_j f_j(\mathbf{x}), i = 1, ..., N \tag{8.85}$$

By applying regression analysis, a $k \times 1$ vector of the regression coefficients, **b**, is calculated using the formula in Equation 8.86.

$$\mathbf{b} = (\mathbf{F^T F})^{-1} \mathbf{F}^T \mathbf{y} \tag{8.86}$$

Using the notation of experimental design, **F** represents the extended design matrix, where the elements of its $k \times 1$ row-vectors, **f**, are known functions of **x**. The matrix $(\mathbf{F^T F})$ is the Fisher information matrix and its inverse, $(\mathbf{F}^T \mathbf{F})^{-1}$, is the dispersion matrix of the regression coefficients.

As previously noted, in a typical process analytical application, the measured data set might consist of spectral data recorded at a number of wavelengths much higher than the number of samples. The rank, R, of the measured matrix of spectra will be equal to or smaller than the number of the samples N. This causes rank deficiency in **X**, and the direct calculation of a regression or calibration model by use of the matrix inverse using Equation 8.85 and Equation 8.86 is problematic.

This problem can be solved using the multivariate calibration approach of principal component regression (PCR) or partial least squares (PLS), described in Chapter 6. In PCR, the matrix of spectra is decomposed into a matrix of principal component *scores*, **S**, consisting of the vectors $[\mathbf{s}_1, \mathbf{s}_2, \ldots, \mathbf{s}_R]$, and *loadings*, **P**, consisting of the eigenvectors $[\mathbf{p}_1, \mathbf{p}_2, \ldots, \mathbf{p}_R]$ of **X** [32]. During the process of principal component analysis, we retain an appropriate number of principal components (latent variables), i.e., those that describe statistically significant variation of the data. By deleting eigenvectors and scores associated with undesirable noise, a new matrix, **X**′, is calculated

$$\mathbf{X}' = \mathbf{s}_1\mathbf{p}_1^T + \mathbf{s}_2\mathbf{p}_2^T + \cdots + \mathbf{s}_{pc}\mathbf{p}_{pc}^T, \quad pc \leq R \tag{8.87}$$

so that the rank deficiency problem is resolved.

At the core of this approach is the improvement of the *condition number* of the data matrix, **X**. The condition number of a matrix, $cond(\mathbf{X})$, is the ratio of the largest and smallest eigenvalue of **X**. It takes on values from 1 to +infinity, and can be used as a measure of the numerical stability with which the inverse of **X** can be computed. Values in the range from 1 to 1000 usually indicate that the matrix inverse calculation will be very stable. In the limit, as the smallest eigenvalue of **X** goes to zero, $cond(\mathbf{X})$ tends toward infinity, indicating that matrix **X** is singular, i.e., it has a determinant equal to zero, in which case the corresponding regression problem is rank deficient and the inverse of **X** does not exist. When the condition number of **X** is extremely large, the matrix **X** is close to being singular, which means computation of its inverse will be numerically unstable. In PCA and PCR, the rank-deficiency problem is solved by transforming the original variable space into PCA space and deleting principal components corresponding to the smallest, closest to zero, eigenvalues. The result is that the condition number of the new matrix **X**′ is better (lower) than the condition number of the original matrix, **X**.

$$cond(\mathbf{X}) = \frac{\lambda_1}{\lambda_R} > cond(\mathbf{X}') = \frac{\lambda_1}{\lambda_{pc}}; \quad \lambda_R < \lambda_{pc} \tag{8.88}$$

Finally, we turn to the problem of selecting the best experimental design, i.e., a subset of samples for passive experiments, as was outlined in the naphtha example. To construct an optimal design that is robust against ill conditioning of the design matrix, **X**, we use the E-optimality criterion. A design is E-optimal if it minimizes the maximum eigenvalue of the dispersion matrix, $\mathbf{M}^{-1} = (\mathbf{F}^T\mathbf{F})^{-1}$. The name of the criterion originates from the first letter of the word "eigenvalue."

$$\max_i \delta_i\left[(\mathbf{M}^*)^{-1}\right] = \min_\mathbf{x} \max_i \delta_i\left[(\mathbf{M})^{-1}\right], i = 1, \ldots, R, \tag{8.89}$$

where $\delta_i[\mathbf{X}]$ represents the eigenvalues of **X**, and R designates the rank of the dispersion matrix.

By minimizing the maximum eigenvalue of the information matrix, **M**, we will be assured that **M** will be invertible. This would be impossible if the design matrix **X** and, subsequently, the information matrix **M** were ill conditioned. A variant of the previously mentioned E-optimality criterion is shown in Equation 8.90,

$$\max_i \delta_i\left[(\mathbf{F}^T\mathbf{F}^*)^{-1}\right] = \min_{\mathbf{x}} \max_i \delta_i\left[(\mathbf{F}^T\mathbf{F})^{-1}\right], i = 1, \ldots, R, \quad (8.90)$$

which can be simplified to become

$$\min_i \sigma_i[\mathbf{F}] = \max_{\mathbf{x}} \min_i \sigma_i[\mathbf{F}], i = 1, \ldots, R. \quad (8.91)$$

Instead of minimizing the maximum eigenvalue of $\mathbf{M}^{-1} = (\mathbf{F}^T\mathbf{F})^{-1}$, an algorithm can be constructed to maximize the minimum singular value of **F**, i.e., maximize the condition number of **F**. To search for the E-optimal subset, it is possible to develop an algorithm based on the method of DETMAX [19] or the update method of Galil and Kiefer's MD algorithm [20]. Instead of using the D-optimality criterion (maximizing the determinant of the information matrix), we use the E-optimality criterion (maximizing the minimum singular value of **F**).

To conduct a search for the E-optimal design directly to the NIR data, we need a methodology that is robust with respect to correlation between the variables (wavelengths). As previously noted, by using the principal component scores, it is possible to use the E-optimal approach to reduce the number of samples and minimize the number of time-consuming GC measurements while also improving the quality of the calibration model.

8.9.2.1 Improving the Numerical Stability of the Data Set

To compare the selection of E-optimal subsets with a complete data set, an E-optimal subset of ten points was generated. The respective condition numbers, based on 1 to 10 latent variables, were calculated and are shown in Figure 8.29. We see that the condition numbers of the E-optimal subsets are lower and are more stable than the condition number for the whole set. We also observe that at six or more latent variables, the condition number of the complete set increases much more rapidly than the condition numbers of the E-optimal subsets. Calculation of regression coefficients at six or more latent variables may be considerably more stable by using the E-optimal subset of 10 points compared with the whole set of 102 points.

Figure 8.30 shows the behavior of the condition number of an E-optimal subset consisting of 30 points compared with the complete data set. We see that the condition number of the E-optimal subset tends to increase, following the pattern of the complete set, but because of reduced collinearity, its level is lower.

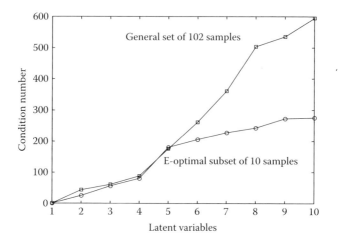

FIGURE 8.29 Effect of using of a ten-point E-optimal subset during the latent-variable extraction.

8.9.2.2 Prediction Ability

To minimize the number of costly GC measurements and obtain better prediction of the naphthalene and aromatic content of the process feedstock, NIR calibration models were used for prediction. The usual practice was followed by building an initial model using a data set acquired over a limited period of time. To keep the model up to date, a campaign for collecting new data was periodically organized, and a new model was built to replace the old one. A typical example of a data set is shown in Figure 8.31. The same autoscaled data is shown in Figure 8.32, where filled circles indicate an E-optimal subset of ten points. The complete set consists

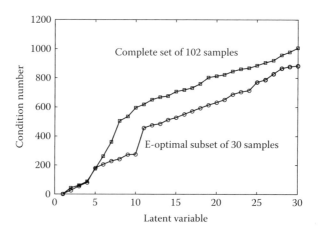

FIGURE 8.30 Effect of using of a 30-point E-optimal subset during the latent-variable extraction.

Response-Surface Modeling and Experimental Design 335

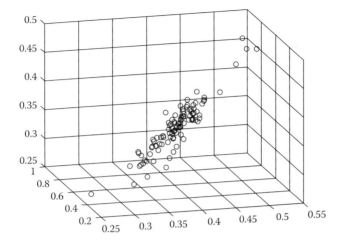

FIGURE 8.31 Illustration of an experimental region covered by a set of 102 NIR spectra projected into a three-dimensional space formed by the first three principal components.

of 102 samples, measured over a 1-year period. Each point represents an NIR spectrum measured at 1299 wavelengths. The points in the figure represent the spectra projected into a three-dimensional space constructed by the first three principal components. Reference values were measured by gas chromatography to determine the quantities of naphthalenes and aromatics present in each. More information about the data set can be found in the literature [2].

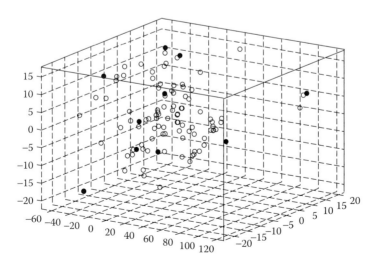

FIGURE 8.32 The autoscaled data shown in Figure 8.31 with the E-optimal subset shown (filled circles).

The predictive ability of the model built using the E-optimal calibration subset was compared with randomly selected subsets. For this purpose, the C_p statistic was used to estimate the predictive ability of a model after its expansion by adding new parameters

$$C_p = \sqrt{\frac{\sum_{i=1}^{N}\left[\hat{y}_i^{(p)} - y_i\right]^2}{\left\{\sum_{i=1}^{N}\left[\hat{y}_i^{(p)} - y_i\right]^2\right\}/(N-2)}} - (N-2p), \quad p = 1 \cdots lv.$$

where $\hat{y}_i^{(p)}$ represents the predicted value of the response at the ith sample using a PCR or PLS model with p latent variables, y_i represents the measured value, N is the number of objects in the calibration set, and p is the number of the parameters in the model including β_0.

Direct comparison using a figure of merit for calibration error between a model built using an E-optimal subset and a model built using the complete set is impossible because of the different number of points in both sets. To ensure an objective comparison between the E-optimal model, we chose ten random sets of $N = 10$ points and computed the average C_p. Comparing these values with the C_p values for the E-optimal set in Table 8.20, it can be seen that E-optimal calibrations are in most cases close to or better (i.e., lower) than the average C_p values of the random sets.

In addition, ten calibration sets of 92 points each were selected, and the respective C_p values were averaged. The results are shown in the last two columns of Table 8.20.

TABLE 8.20
Comparison of C_p Values for Ten-Point E-Optimal Calibrations with Calibrations of Ten Randomly Selected Points

No. Latent Variables	C_p for E-Optimal 10-Point Calibration Set		Average C_p for Random 10-Point Calibration Sets		Average C_p for 92-Point Calibration Sets	
	y_1	y_2	y_1	y_2	y_1	y_2
1	391.06	46.89	153.83	48.59	4680.73	5703.43
2	248.42	33.35	65.58	25.03	2231.24	3807.26
3	43.24	13.56	28.41	15.90	447.84	813.96
4	29.04	13.17	21.27	15.45	138.79	460.05
5	19.23	10.31	11.84	10.72	41.76	136.69
6	16.14	9.65	13.00	10.37	31.74	44.42
7	15.63	12.59	20.09	12.02	24.84	31.79
8	15.35	12.12	21.89	11.91	21.15	30.43
9	16.00	16.00	16.00	16.00	16.00	16.00

Note: Average C_p results for two components, y_1 and y_2, are shown.

Here we see that after selecting the proper number of the latent variables (approximately five to seven), the value of C_p remains higher despite the fact that the calibration set is nine times larger than the calibration set generated using the E-optimal approach. The poorer predictive ability of this calibration set might be due to the existence of clusters of samples. The existence of points that are close to each other is equivalent to giving a higher relative weight to this group of points compared with the others. This increased weight might shift the calibration curve without a good statistical reason.

8.10 CONCLUSION

In this chapter, a brief outline of methods for constructing experimental designs was presented. Throughout this chapter we have attempted to highlight the important connection between response-surface methodology (RSM) and the design of experiments (DOE). Basic information was presented about the most commonly used types of experimental design, as well as some information about a few novel types of experimental design, such as sequential composite designs. We also reviewed some of the basic software packages equipped to deal with experimental designs.

In several cases, we highlighted theoretical discussions with supporting examples taken from the field of chemistry and process analytical chemistry. In particular, a simple calibration example using UV spectroscopy was selected to provide the reader with a familiar point of reference. In this way, the topics of experimental design and response-surface methodology were presented in a fashion that should help the nonexpert see the benefits of this approach prior to implementation. For the user who is unfamiliar with DOE methods, we hope our approach has provided a useful introduction.

REFERENCES

1. Box, G.E.P. and Wilson, K.B., On the experimental attainment of optimum conditions, *J. R. Stat. Soc. Ser. B,* 13, 1–45, 1951.
2. Stoyanov, K.S. and Walsmley, A.D., Development and Maintenance of a Calibration Model for Process Analysis, BPCPACT1, Center for Analytical Technology/technical report, Glasow, 2000.
3. Scheffé, H., Experiment with mixtures, *J. R. Stat. Soc. Ser. B,* 20, 344–366, 1958.
4. Scheffé, H., The simplex-centroid design for experiment with mixtures, *J. R. Stat. Soc. Ser. B,* 25, 235–263, 1965.
5. Cornell, J., *Experiments with Mixtures: Designs, Models, and the Analysis of Mixture Data,* 2nd ed., John Wiley & Sons, New York, 1990.
6. Crosier, R.B., Mixture experiments: geometry and pseudo-components, *Technometrics,* 26, 209–216, 1984.
7. Lundstedt, T. et al., Experimental design and optimization, *Chemom. Intell. Lab. Syst.,* 42, 3–40, 1998.
8. Taguchi, G., *System of Experimental Design: Engineering Methods to Optimize Quality and to Minimize Cost,* UNIPUB/Kraus International, White Plains, NY, 1987.
9. Taguchi, G. and Wu, Y., *Introduction to Off-Line Quality Control,* Central Japan Quality Control Association, Nagoya, 1980.

10. Taguchi, G., *Introduction to Quality Engineering: Designing Quality into Products and Processes,* UNIPUB/Quality Resources, White Plains, NY, 1986.
11. Logothetis, N. and Wynn, H., *Quality through Design,* Oxford Science Publications, Clarendon Press, Oxford, 1989.
12. Myers, R.H. and Montgomery, D.C., *Response Surface Methodology: Process and Product Optimization Using Designed Experiments,* John Wiley & Sons, New York, 1995.
13. Box, G.E.P. and Draper, N., *Empirical Model-Building and Response Surfaces,* John Wiley & Sons, New York, 1997.
14. Kiefer, J. and Wolfowitz, J., The equivalence of two extremum problems, *Can. J. Math.,* 12, 363–366, 1960.
15. Kiefer, J., General equivalence theory for optimum designs (approximate theory), *Ann. Stat.,* 5, 849–879, 1974.
16. Cook, R.D. and Nachtsheim, C.J., A comparison of algorithms for constructing exact D-optimal designs, *Technometrics,* 22, 315–324, 1980.
17. Atkinson, A.C. and Donev, A.N., *Optimum Experimental Designs,* Oxford University Press, Oxford, 1992.
18. Fedorov, V.V., *Theory of Optimal Experiments,* Academic Press, New York, 1972.
19. Mitchell, T.J., An algorithm for the construction of D-optimal experimental designs, *Technometrics,* 16, 203–211, 1974 (reprint Technometrics, 42, 48–56, 2000).
20. Galil, Z. and Kiefer, J., Time- and space-saving computer methods, related to Mitchell's DETMAX, for finding D-optimum designs, *Technometrics,* 22, 301–313, 1980.
21. Wynn, H.P., The sequential generation of D-optimum experimental designs, *Ann. Math. Stat.,* 41, 1655–1664, 1970.
22. Wynn, H.P., Results in the theory and construction of D-optimum experimental designs, *J. R. Stat. Soc. Ser. B,* 34, 133–147, 1972.
23. Van Schalkwyk, D.J., *On the Design of Mixture Experiments,* Ph.D. thesis, University of London, London, 1971.
24. Atwood, C.L., Sequences converging to D-optimal designs of experiments, *Ann. Stat.,* 1, 342–352, 1973.
25. Vuchkov, I.N., *Optimal Planning of the Experimental Investigations,* Technika, Sofia, Bulgaria, 1978 (in Bulgarian).
26. Yonchev, H.A., New computer procedures for constructing D-optimal designs: optimal designs and analysis of experiments, in *Proceedings of the International Conference Workshop,* Neuchatel, Switzerland, Dodge, Y., Fedorov, V.V., and Wynn, H.P., Eds., Elsevier Science Publishers B.V., North-Holland, 1988, pp. 71–80.
27. Yonchev, H.A. and Stoyanov, K., Optimal composite designs for experiments with multicomponent systems, in *International Conference on Systems Science: Systems Science* X, Wroclaw, Poland, 17, 2, 1991, pp. 29–35.
28. Hedayat, A.W. and Wallis, D., Hadamard matrices and their application, *Ann. Stat.,* 6, 1184–1238, 1978.
29. Iontchev, H.A. and Stoyanov, K.A., *Catalogue of Response Surface Design,* University of Chemical Technology and Metallurgy, Sofia, Bulgaria, 1998.
30. Vuchkov, I.N., et al., *Catalogue of Sequentially Generated Designs,* Higher Institute of Chemical Technology, Department of Automation, Sofia, Bulgaria, 1978 (in Bulgarian).
31. Golikova, T.I., et al., *Catalogue of Second Order Designs,* Publishing House of Moscow University, Moscow, 1974 (in Russian).
32. Vandeginste, B.G.M. et al., *Handbook of Chemometrics and Qualimetrics: Part B,* Elsevier, New York, 1998.

9 Classification and Pattern Recognition

Barry K. Lavine and Charles E. Davidson

CONTENTS

9.1 Introduction 339
9.2 Data Preprocessing 341
9.3 Mapping and Display 342
9.4 Clustering 347
9.5 Classification 351
    9.5.1 K-Nearest Neighbor 352
    9.5.2 Partial Least Squares 352
    9.5.3 SIMCA 353
9.6 Practical Considerations 354
9.7 Applications of Pattern-Recognition Techniques 355
    9.7.1 Archaeological Artifacts 356
    9.7.2 Fuel Spill Identification 358
    9.7.3 Sorting Plastics for Recycling 365
    9.7.4 Taxonomy Based on Chemical Constitution 371
References 374

9.1 INTRODUCTION

For most of the 20th century, chemistry has been a data-poor discipline, relying on well-thought-out hypotheses and carefully planned experiments to develop solutions to real-world problems. With the advent of sophisticated instrumentation commonly under computer control, chemistry is slowly evolving into a data-rich field, thereby opening up the possibility of data-driven research. According to Lavine and Workman [1, 2], this, in turn, has led to a new approach for solving scientific problems: (1) measure a phenomenon or chemical process using instrumentation that generates data inexpensively, (2) analyze the multivariate data, (3) iterate if necessary, (4) create and test the model, and (5) develop fundamental multivariate understanding of the process or phenomenon. The new approach does not involve a thought ritual; rather, it is a method involving many inexpensive measurements, possibly a few simulations, and multivariate analysis. It constitutes a true paradigm shift, since multiple experiments and data analysis are used as a vehicle to examine the world from a multivariate perspective. Mathematics is not used for modeling *per se*, but more for discovery, and it is thus a data microscope to sort, probe, and look for hidden relationships in data.

This new approach, which more thoroughly explores the implications of data so that hypotheses are developed with a greater awareness of reality, often produces large quantities of data. To analyze these larger data sets, analytical chemists have turned to pattern-recognition methods because of the advantageous attributes of these procedures [3, 4]. First, methods are available that seek relationships that provide definitions of similarity or dissimilarity between diverse groups of data, thereby revealing common properties among the objects in a data set. Second, a large number of features can be studied simultaneously. Third, techniques are available for selecting important features from a large set of measurements, thereby allowing studies to be performed on systems where the exact relationships are not fully understood. Examples of pattern-recognition methods that have been used by analytical chemists for multivariate data analysis include neural networks, discriminant analysis, clustering, and principal component analysis.

Pattern-recognition methods were originally designed to solve the class membership problem. In a typical pattern-recognition study, samples are classified according to a specific property by using measurements that are indirectly related to the property of interest. An empirical relationship or classification rule is developed from a set of samples for which the property of interest and the measurements are known. The classification rule is then used to predict the property of samples that are not part of the original training set. Developing a classification rule from spectroscopic or chromatographic data may be desirable for several reasons, including the identification of the source of pollutants [5, 6], detection of odorants [7, 8], presence or absence of disease in a patient from which a sample has been taken [9, 10], and food quality testing [11, 12], to name a few.

The set of samples for which the property of interest and measurements are known is called the training set, whereas the set of measurements that describe each sample in the data set is called a pattern. The determination of the property of interest by assigning a sample to its respective class is called recognition, hence the term "pattern recognition."

For pattern-recognition analysis, each sample (e.g., individual test object or aliquot of material) is represented by a data vector $x = (x_1, x_2, x_3, \ldots, x_j, \ldots, x_n)$, where x_j is the value of the jth descriptor or measurement variable (e.g., absorbance of sample at a specific wavelength). Such a vector can be considered as a point in a high-dimensional measurement space. The Euclidean distance between a pair of points in the measurement space is inversely related to the degree of similarity between the objects. Points representing objects from one class tend to cluster in a limited region of the measurement space separate from the others. Pattern recognition is a set of numerical methods for assessing the structure of the data space. The data structure is defined as the overall relation of each object to every other object in the data set.

Pattern recognition has its origins in the field of image and signal processing. The first study to appear in the chemical literature on pattern recognition was published in 1969 and involved the interpretation of low-resolution mass spectral data using the linear learning machine [13]. Modern computers now enable these techniques to be applied routinely to a wide variety of chemical problems such as chemical fingerprinting [14–16], spectral data interpretation [17–19], molecular structure-biological activity correlations [20–22], and cancer classification from microarray data [23–25]. Over the past 2 decades, numerous books and review articles on this subject have been published [26–35].

In this chapter, the three major subdivisions of pattern-recognition methodology are discussed: (1) mapping and display, (2) clustering, and (3) classification. The procedures that must be implemented to apply pattern-recognition methods are also enumerated. Specific emphasis is placed on the application of these techniques to problems in biological and environmental analyses.

9.2 DATA PREPROCESSING

The first step in any pattern-recognition study is to convert the raw data into computer-compatible form. Normally, the raw data are arranged in the form of a table, a data matrix:

$$\begin{bmatrix} x_{11} & x_{12} & x_{13} & \cdots & x_{1N} \\ x_{21} & x_{22} & x_{23} & \cdots & x_{2N} \\ \cdot & \cdot & \cdot & & \cdot \\ \cdot & \cdot & \cdot & & \cdot \\ x_{M1} & x_{M2} & x_{M3} & & x_{MN} \end{bmatrix} \tag{9.1}$$

The rows of the matrix represent the observations, and the columns are the values of the descriptors. In other words, each row is a data or pattern vector, and the components of the data vector are physically measurable quantities called descriptors. It is essential that descriptors encode the same information for all samples in the data set. If variable 5 is the area of a gas chromatographic (GC) peak for phenol in sample 1, it must also be the area of the GC peak for phenol in samples 2, 3, etc. Hence, peak matching is crucial when chromatograms or spectra are translated into data vectors.

The next step involves scaling. The objective is to enhance the signal-to-noise ratio of the data. In the applications discussed herein, two scaling techniques have been used: normalization and autoscaling. The procedures that should be used for a given data set, however, are highly dependent upon the nature of the problem.

Normalization involves setting the sum of the components of each pattern vector equal to some arbitrary constant. For chromatographic data, this constant usually equals 100, so each peak is usually expressed as a fraction of the total integrated peak area. In mass spectrometry, the peak with the largest intensity is assigned a value of 100, and the intensities of the other peaks are expressed as percentages of this fragment peak. In near-infrared and vibrational spectroscopy, the data vectors are normalized to unit length. This is accomplished by dividing each vector by the square root of the sum of the squares of the components composing the vector. Normalization compensates for variation in the data due to differences in the sample size or optical path length. However, normalization can also introduce dependence between variables that could have an effect on the results of the investigation. Thus, one must take into account both of these factors when deciding whether or not to normalize data [36].

Autoscaling involves standardizing the measurement variables so that each descriptor or measurement has a mean of zero and a standard deviation of unity, that is,

$$x_{i,\text{new}} = \frac{(x_{i,\text{orig}} - \bar{x}_{i,\text{orig}})}{s_{i,\text{orig}}} \tag{9.2}$$

where $\bar{x}_{i,\text{orig}}$ is the mean and $s_{i,\text{orig}}$ is the standard deviation of the original measurement variable. If autoscaling is not applied, the larger-valued descriptors tend to dominate the analysis. Autoscaling removes inadvertent weighting of the variables that would otherwise occur. Thus, each variable has an equal weight in the analysis. Autoscaling affects the spread of the data, placing the data points inside a hypercube. However, it does not affect the relative distribution of the data points in the high-dimensional measurement space.

9.3 MAPPING AND DISPLAY

Physical scientists often use graphical methods to study data. If there are only two or three measurements per sample, the data can be displayed as a graph or plot. By examining the plot, a scientist can search for similarities and dissimilarities among samples, find natural clusters, and even gain information about the overall structure of the data set. If there are n measurements per sample ($n > 3$), a two- or three-dimensional representation of the measurement space is needed to visualize the relative position of the data points in n-space. This representation must accurately reflect the structure of the data. One such approach is using a mapping and display technique called principal component analysis (PCA) [37, 38]. A detailed treatment of PCA is provided in Chapter 5; however, some aspects of PCA related to the topic of pattern recognition are summarized here.

Principal component analysis is the most widely used multivariate analysis technique in science and engineering. It is a method for transforming the original measurement variables into new, uncorrelated variables called principal components. Each principal component is a linear combination of the original measurement variables. Using this procedure, a set of orthogonal axes that represents the direction of greatest variance in the data is found. (Variance is defined as the degree to which the data are spread in the n-dimensional measurement space.) Usually, only two or

Classification and Pattern Recognition

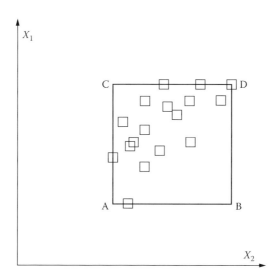

FIGURE 9.1 Fifteen samples projected onto a two-dimensional data space. Because x_1 and x_2 are correlated, the data points are restricted to a small region of the measurement space defined by the vertices A–D of the rectangle. (Adapted from Mandel, J., *NBS J. Res.*, 190 (6), 465–476, 1985. With permission.)

three principal components are necessary to explain a significant fraction of the information present in multivariate data. Hence, principal component analysis can be applied to multivariate data for dimensionality reduction, to identify outliers, to display data structure, and to classify samples.

Dimensionality reduction or data compression is possible with principal component analysis because of correlations between measurement variables. Consider Figure 9.1, which shows a plot of 15 samples in a two-dimensional space. The coordinate axes of this measurement space are defined by the variables x_1 and x_2. Both x_1 and x_2 are correlated, since fixing the value of x_1 limits the range of values possible for x_2. If x_1 and x_2 were uncorrelated, the enclosed rectangle shown in Figure 9.1 would be completely filled by the data points. Because of this correlative relationship, the data points occupy only a fraction of the measurement space.

Information can be defined as the scatter of points in a measurement space. Correlations between measurement variables decrease the scatter and subsequently the information content of the space [39] because the data points are restricted to a small region of the measurement space because of correlations among the measurement variables. If the measurement variables are highly correlated, the data points could even reside in a subspace. This is shown in Figure 9.2. Each row of the data matrix is an object, and each column is a measurement variable. Here x_3 is perfectly correlated with x_1 and x_2, since x_3 (third column) equals x_1 (first column) plus x_2 (second column). Hence, the seven data points lie in a plane (or two-dimensional subspace), even though each point has three measurements associated with it. Because x_3 is a redundant variable, it does not contribute any additional information, which is why the data points lie in two dimensions, not three dimensions.

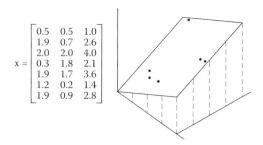

$$x = \begin{bmatrix} 0.5 & 0.5 & 1.0 \\ 1.9 & 0.7 & 2.6 \\ 2.0 & 2.0 & 4.0 \\ 0.3 & 1.8 & 2.1 \\ 1.9 & 1.7 & 3.6 \\ 1.2 & 0.2 & 1.4 \\ 1.9 & 0.9 & 2.8 \end{bmatrix}$$

FIGURE 9.2 In the case of strongly correlated measurement variables, the data points may even reside in a subspace of the original measurement space. (Adapted from Brereton, R.G., Ed., *Multivariate Pattern Recognition in Chemometrics*, Elsevier Science Publishers, Amsterdam, 1992. With permission.)

Variables that contain redundant information are said to be collinear. High collinearity between variables is a strong indication that a new coordinate system can be found that is better at conveying the information present in the data than one defined by the original measurement variables. The new coordinate system for displaying the data is based on variance. (The scatter of the data points in the measurement space is a direct measure of the data's variance.) The principal components of the data define the variance-based axes of this new coordinate system. The first principal component is formed by determining the direction of largest variation in the original measurement space of the data and modeling it with a line fitted by linear least squares (see Figure 9.3) that passes through the center of the data. The second largest principal component lies in the direction of next largest

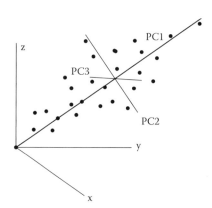

FIGURE 9.3 Principal component axes defining a new set of basis vectors for the measurement space defined by the variables X, Y, and Z. The third principal component describes only noise in the data. (From Brown, S., *Appl. Spectrosc.*, 49 (12), 14A–30A, 1995. With permission.)

variation. It passes through the center of the data and is orthogonal to the first principal component. The third largest principal component lies in the direction of next largest variation. It also passes through the center of the data; it is orthogonal to the first and second principal component, and so forth. Each principal component describes a different source of information because each defines a different direction of scatter or variance in the data. The orthogonality constraint imposed by the mathematics of principal component analysis also ensures that each variance-based axis is independent.

A measure of the amount of information conveyed by each principal component is the variance of the data explained by it, which is expressed in terms of its eigenvalue. For this reason, principal components are arranged in order of decreasing eigenvalues. The most informative or largest principal component is the first, and the least informative or smallest is the last. The amount of information contained in a principal component relative to the original measurement variables, i.e., the fraction of the total cumulative variance explained by the principal component, is equal to the eigenvalue of the principal component divided by the sum of all the eigenvalues. The maximum number of principal components that can be extracted from the data is the smaller of either the number of samples or the number of variables in the data set, as this number defines the largest possible number of independent axes in the data.

If the data are collected with due care, one would expect that only the larger principal components would convey information about clustering, since most of the information in the data should be about the effect of interest. However, the situation is not always so straightforward as implied. Each principal component describes some amount of signal and some amount of noise in the data because of accidental correlation between signal and noise. The larger principal components contain information primarily about signal, whereas the smaller principal components describe primarily noise. By discarding the smaller principal components, noise is discarded, but so is a small amount of signal. However, the gain in signal to noise usually more than compensates for the biased representation of the data that occurs when plotting only its largest principal components. The approach of describing a data set in terms of important and unimportant variation is known as soft modeling in latent variables [40].

Principal component analysis takes advantage of the fact that a large amount of data generated in scientific studies has a great deal of redundancy and therefore a great deal of collinearity. Because the measurement variables are correlated, 800-point spectra do not require 800 independent orthogonal axes to define the position of a sample point in the measurement space. Using principal component analysis, the original measurement variables that constitute a correlated-axes system can be converted into an orthogonal-axes system, which dramatically reduces the dimensionality of the data, since only a few independent axes are needed to describe the data. Spectra of a set of samples often lie in a subspace of the original measurement space, and a plot of the two or three largest principal components of the data can help one to visualize the relative position of the spectra in this subspace.

Another approach for solving the problem of representing data points in an n-dimensional measurement space involves using an iterative technique known as the Kohonen neural network [41, 42] or self-organizing map (SOM). A Kohonen neural network consists of a layer of neurons arranged in a two-dimensional grid or plane. Each neuron is represented by a weight, which is a vector of the same dimension as the input signal. The input is passed to each neuron in the network, and the neuron whose weight vector is closest to the input signal is declared the "winner." For the winning neuron, the weight vector is adjusted to more closely resemble the input signal. The neurons surrounding the winner are also adjusted, but to a lesser degree. This process, when completed, causes similar input signals to respond to neurons that are near each other. In other words, the neural network is able to learn and display the topology of the data.

The procedure for implementing the Kohonen neural network is as follows. First, all of the weight vectors in the network are initialized, that is, their components are assigned random numbers. Next, the random weight vectors and the sample vectors in the data set are normalized to unit length. Training is performed by presenting the data one pattern at a time to the network. The Euclidean distance is computed between the pattern vector and each weight vector in the network. Because competitive learning is used, the sample is assigned to the neuron whose weight vector is closest. The weight vector of the winning neuron and its neighbors are then adjusted to more closely resemble the sample using the Kohonen learning rule:

$$w_i(t+1) = w_i(t) + \eta(t) * \alpha(d_{ic})(x_i - w_{i,\text{old}}) \tag{9.3}$$

where $w_i(t+1)$ is the ith weight vector for the next iteration, $w_i(t)$ is the ith weight vector for the current iteration, $\eta(t)$ is the learning rate function, $\alpha(d_{ic})$ is the neighborhood function, and x_i is the sample vector currently passed to the network. The comparison of all sample data vectors to all weight vectors in the network and the modification of those weights occurs during a single iteration. After the learning rule has been applied, the weights are renormalized to unit length.

The learning rate is a positive real number less than 1, which is chosen by the user. It is a function of time and usually decreases during the training of the network. The neighborhood function determines the magnitude of the weight adjustment based on d_{ic}, which is the link distance in the grid between the central neuron and the neuron currently being adjusted. The magnitude of the adjustment is inversely proportional to the distance between the neuron in question and the central neuron. The neighborhood function can also decrease linearly with the number of iterations, reducing the number of neurons around the winner being adjusted.

In the initial iterations, when η is relatively large and α is wide in scope, the neurons order themselves globally in relation to each other in the data. As these parameters decrease in value, the passing of each sample to the network results in a relatively small change of only a few neurons. Over multiple iterations, the neurons converge to an approximation of the probability distribution of the data. A large number of iterations is required for the weight vectors to converge toward a good approximation of the more numerous sample vectors.

Using this algorithm, the network is able to map the data so that similar data vectors excite neurons that are very near each other, thereby preserving the topology of the original sample vectors. Visual inspection of the map allows the user to identify outliers and recognize areas where groups of similar samples have clustered.

An advantage of self-organizing maps is that outliers, observations that have a very large or very small value for some or all of the original measurement variables, affect only one map unit and its neighborhood, while the rest of the display is available for investigating the remaining data. In all likelihood, a single outlier has little effect on the ending weights, since the neurons are more likely to represent areas of largest sample density. By comparison, outliers can have a drastic and disproportionate effect on principal component plots because of the least-squares property of principal components. A sample far from the other data points can pull the principal components toward it—away from the direction of "true" maximum variance—thus compressing the remaining data points into a very small region of the map. Removing the outlier does not always solve the problem, for as soon as the worst outliers are deleted, other data points may appear in this role.

One disadvantage of self-organizing maps is that distance is not preserved in the mapping. Outliers do not distort the map, but identifying them is more difficult, since an investigation of the neuron weights is necessary. Another disadvantage is the massive number of computations often needed to reach a stable network configuration, which limits the size of the network. It can also lead to a dilemma. Using too few neurons results in clusters overlapping in the map, with information about class structure lost. If too many neurons are used, then training becomes prohibitive. In such a network, most of the neurons would probably be unoccupied. Our own experience, as well as that of other workers, has shown that a network composed of 20×20 units, which can be trained in a reasonable period of time, often possesses the necessary resolution to delineate class structure in multivariate data.

Principal component analysis and Kohonen self-organizing maps allow multivariate data to be displayed as a graph for direct viewing, thereby extending the ability of human pattern recognition to uncover obscure relationships in complex data sets. This enables the scientist or engineer to play an even more interactive role in the data analysis. Clearly, these two techniques can be very useful when an investigator believes that distinct class differences exist in a collection of samples but is not sure about the nature of the classes.

9.4 CLUSTERING

Exploratory data analysis techniques are often quite helpful in elucidating the complex nature of multivariate relationships. In the preceding section, the importance of using mapping and display techniques for understanding the structure of complex multivariate data sets was emphasized. In this section, some additional techniques are discussed that give insight into the structure of a data set. These methods attempt to find sample groupings or clusters within data using criteria developed from the data itself: hence the term "cluster analysis."

A major problem in cluster analysis is defining a cluster (see Figure 9.4). There is no measure of cluster validity that can serve as a reliable indicator of the quality of a proposed partitioning of the data. Clusters are defined intuitively, depending on the context

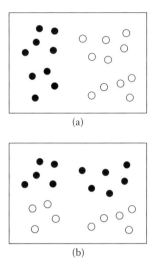

FIGURE 9.4 Are there two or four clusters in the data? (Adapted from Brereton, R.G., Ed., *Multivariate Pattern Recognition in Chemometrics*, Elsevier Science Publishers, Amsterdam, 1992. With permission.)

of the problem, and not mathematically, which limits their utility. Therefore, prior knowledge about the problem is essential when using these methods. Because the threshold value for similarity is developed directly from the data, criteria for similarity are often subjective and depend to a large degree on the nature of the problem investigated, the goals of the study, the number of clusters in the data sought, and previous experience.

Cluster analysis is based on the principle that distances between pairs of points (i.e., samples) in the measurement space are inversely related to their degree of similarity. Although several different types of clustering algorithms exist, e.g., K-means [43], FCV [44], and Patrick Jarvis [45], by far the most popular is hierarchical clustering [46–48], which is the focus here. The starting point for a hierarchical clustering experiment is the similarity matrix, which is formed by first computing the distances between all pairs of points in the data set. Each distance is then converted into a similarity value

$$s_{ik} = 1 - \frac{d_{ik}}{d_{max}} \tag{9.4}$$

where s_{ik} is the measure of similarity between samples i and k, d_{ik} is the Euclidean distance between samples i and k, and d_{max} is the distance between the two most dissimilar samples, which is also the largest distance in the data set.

These similarity values, which vary from 0 to 1, are organized in the form of a table or square symmetric matrix. The similarity matrix is scanned for the largest value, which corresponds to the most similar point pair, and the two samples (comprising the point pair) are combined to form a new point located midway between the two original data points. After the rows and columns corresponding to the original

Classification and Pattern Recognition

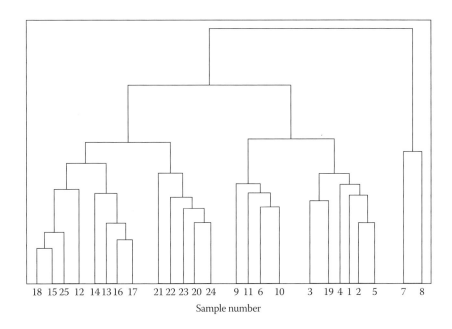

FIGURE 9.5 Dendogram using the single-linkage method developed from the gas chromatograms of the scent marks of Marmoset monkeys. There are four different types of samples in this data set: redhead females (sample nos. 1–5), redhead males (sample nos. 6–11), blackhead females (sample nos. 12–19), and blackhead males (sample nos. 20–24). (From Haswell, S.J., Ed., *Practical Guide to Chemometrics*, 1st ed., Marcel Dekker, New York, 1992. With permission.)

two data points are removed, the symmetry matrix is then updated to include information about the similarity between the new point and every other point in the data set. The matrix is again scanned, the new nearest-point pair is identified and combined to form a single point, the rows and columns of the two data points that were combined are removed, and the matrix is recomputed. This process is repeated until all points have been linked. The result of this procedure is a diagram called a dendogram, which is a visual representation of the relationships between samples in the data set (see Figure 9.5). Interpretation of the results is intuitive, which is the major reason for the popularity of these methods. For example, the dendogram shown in Figure 9.5 suggests that gas chromatograms from the Marmoset monkey data set can be divided into two groups (cluster 1 = samples 18, 15, 25, 12, 14, 13, 16, 17, 21, 22, 23, 20, 24; cluster 2 = samples 9, 11, 6, 10, 3, 19, 4, 1, 2, 5) or four groups (cluster 1 = samples 18, 15, 25, 12, 14, 13, 16, 17; cluster 2 = samples 21, 22, 23, 20, 24; cluster 3 = samples 9, 11, 6, 10; cluster 4 = samples 3, 19, 4, 1, 2, 5). Samples 7 and 8 are judged to be outliers by the dendogram.

The Euclidean distance is the best choice for a distance metric in hierarchical clustering because interpoint distances between the samples can be computed directly (see Figure 9.6). However, there is a problem with using the Euclidean distance, which arises from inadvertent weighting of the variables in the analysis that occurs

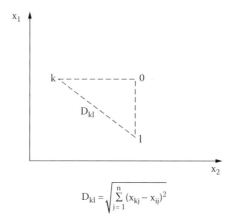

FIGURE 9.6 Euclidean distance between two data points in a two-dimensional measurement space defined by the measurement variables x_1 and x_2. (Adapted from Massart, D.L. and Kaufman, L., *The Interpretation of Analytical Chemical Data by the Use of Cluster Analysis*, John Wiley & Sons, New York, 1983. With permission.)

because of differences in magnitude among the measurement variables. Inadvertent weighting of the variables arising from differences in scale can be eliminated by autoscaling the data. For cluster analysis, it is best to autoscale the data. Autoscaling ensures that all features contribute equally to the distance calculation.

There are a variety of ways to compute distances between data points and clusters in hierarchical clustering. A few are shown in Figure 9.7. The single-linkage method

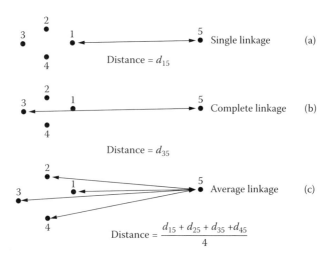

FIGURE 9.7 Computation for the distance between a data cluster and a point using (a) single linkage, (b) complete linkage, and (c) average linkage. (From Haswell, S.J., Ed., *Practical Guide to Chemometrics*, 1st ed., Marcel Dekker, New York, 1992. With permission.)

assesses the similarity between a data point and a cluster by measuring the distance to the nearest point in the cluster. The complete-linkage method assesses similarity by computing the distance to the farthest point in the cluster. Average linkage assesses similarity by computing the distance between all pairs of points where a member of each pair belongs to the cluster, with the average of these distances being a measure of similarity between a cluster and a data point.

All hierarchical clustering procedures yield the same results for data sets with well-separated clusters. However, the results differ when the clusters overlap because of space distorting effects. Single linkage favors the formation of large linear clusters instead of the usual elliptical or spherical clusters. As a result, poorly separated clusters are often chained together. Complete linkage, on the other hand, favors the formation of small spherical clusters. That is why it is a good idea to use at least two different clustering algorithms when studying a data set. If the dendograms are in agreement, then a strong case can be made for partitioning the data into distinct groups. If the cluster memberships differ, the data should be further investigated using principal component analysis or other mapping and display techniques. As a general rule, it is recommended that hierarchical methods be used in tandem with principal component or self-organizing maps to detect clusters in multivariate data. Hierarchical methods are exploratory tools: the absolute validity of a dendogram is less important than insights and suggestions gained by the user about the data structure.

9.5 CLASSIFICATION

So far, only exploratory data analysis techniques, e.g., cluster and principal component analysis, have been discussed. These techniques attempt to analyze multivariate data without directly using the information about the class assignment of the samples to develop projections of the data. Although mapping and display techniques and cluster analysis are powerful methods for uncovering relationships in large multivariate data sets, they are often not sufficient for developing a classification rule. However, the overall goal of a pattern-recognition study is the development of a classification rule that can accurately predict the class membership of an unknown sample.

This section discusses classification methods. The focus is on k-nearest neighbor (K-NN), partial least squares (PLS), and soft independent modeling by class analogy (SIMCA). Multilayered feed-forward neural networks [49, 50] and support vector machines [51, 52] are not discussed in this section. Each method attempts to divide a data space into different regions. In the simplest case, that of a binary classifier, the data space is divided into two regions. Samples that share a common property are found on one side of the decision surface, whereas those samples composing the other category are found on the other side. The decision surface is developed using a training procedure in which the internal structure of the network or machine is adjusted empirically to obtain a best match between the output and the desired result for a set of input data that serves as the training set. Although neural networks and support vector machines can discern subtle patterns in noisy and nonlinear data, overtraining is a problem. Spurious or chance classification is another issue that is of concern. To successfully exploit the inherent advantages of neural networks and support vector machines, it is necessary to use a training set with a large sample point density and the appropriate point distribution in the data space. Because of the difficulty in generating such data sets, similarity-based classifiers, e.g., K-NN and SIMCA, may be preferred in many applications.

9.5.1 K-Nearest Neighbor

K-NN [53, 54] is a powerful classification technique. A sample is classified according to the majority vote of its k-nearest neighbors, where k is an odd number, e.g., 1, 3, or 5. For a given sample, the Euclidean distance is computed from the sample to every other point in the data set. These distances are arranged from smallest to largest to define the sample's k-nearest neighbors. Based on the class label of the majority of the sample's k-nearest neighbors, the sample is assigned to a class in the data set. If the assigned class and the actual class label of the sample match, the test is considered to be a success. The overall classification success rate, calculated over the entire set of points, is a measure of the degree of sample clustering on the basis of class in the data set. Because the data is usually autoscaled for K-NN, a majority vote of the k-nearest neighbors can only occur if the majority of the measurement variables concur.

K-NN cannot furnish a statement about the reliability of a particular classification. However, for training sets that have a large number of samples in each class, the 1-nearest neighbor rule has an error rate that is twice as large as the Bayes classifier, which is the optimum classifier for any set of data [55]. To implement a Bayes classifier, one must have knowledge of the underlying probability distribution functions of each class in the data set. Usually, this knowledge is not available. Clearly, any other classification method, no matter how sophisticated, can at best only improve on the performance of K-NN by a factor of 2. For this reason, K-NN is often used as a benchmark against which to measure other methods.

9.5.2 Partial Least Squares

Soft modeling in latent variables is central to many of the more popular data analysis methods in pattern recognition. For example, a modeling method called partial least squares (PLS) is routinely used for classification because of the quality of the models produced and the ease of their implementation due to the availability of PLS software. Examples of PLS applications include detection of food oil adulteration [56], analyte discrimination by sensor arrays [57], fingerprinting soil microbial communities [58], classifying wastewater samples [59], and typing coffee beans [60], to name a few. Only a summary of the PLS method is provided here. For a detailed description of PLS, the reader is referred to Chapter 6 and the literature [61–65].

PLS is a regression method originally developed by Herman Wold [66] as an alternative to classical least squares for mining collinear data. Motivation for the development of PLS was simple enough: approximate the design space of the original measurement variables with one of lower dimension. The latent variables in PLS are determined iteratively using both the response (target) variable and the measurement variables. Each PLS component is a linear combination of the original measurement variables. These latent variables are rotated to ensure a better correlation with the target variable. The goal of PLS is to seek a weight vector that maps each sample to a desired target value. Because of the rotation, which attempts to find an appropriate compromise between explaining the measurement variables and predicting the response variable, confounding of the desired signal by interference is usually less of a problem than in other soft-modeling methods.

For classification of multivariate data, PLS is implemented in the usual way except that a variable, which records the class membership of each sample, is used as the target variable. The results of the PLS regression are usually presented as pairwise plots of the PLS components, allowing visual assessment of class separation (akin to principal component analysis). If class separation in the PLS plot is pronounced, an unknown sample can be categorized by projecting it onto the plot and assigning the sample to the class whose center is closest. PLS enhances differences between the sample classes and is used when information about class differences does not appear in the largest principal components of the data. A formal explanation as to why PLS is so successful in locating and emphasizing group structure has recently been proposed. According to Rayens [67], PLS is related to linear discriminant analysis when PLS is used for classification, which is why PLS can be expected to perform reasonably well in this role. The performance of PLS tracks very closely that of linear discriminant analysis [68].

9.5.3 SIMCA

Principal component analysis is central to many of the more popular multivariate data analysis methods in chemistry. For example, a classification method based on principal component analysis called SIMCA [69, 70] is by the far the most popular method for describing the class structure of a data set. In SIMCA (soft independent modeling by class analogy), a separate principal component analysis is performed on each class in the data set, and a sufficient number of principal components are retained to account for most of the variation within each class. The number of principal components retained for each class is usually determined directly from the data by a method called cross validation [71] and is often different for each class model.

The variation in the data not explained by the principal component model is called the residual variance. Classification in SIMCA is made by comparing the residual variance of a sample with the average residual variance of those samples that make up the class. This comparison provides a direct measure of the similarity of a sample to a particular class and can be considered as a measure of the goodness of fit of a sample for a particular class model. To provide a quantitative basis for this comparison, an F-statistic is used to compare the residual variance of the sample with the mean residual variance of the class [72]. The F-statistic can also be used to compute an upper limit for the residual variance of those samples that belong to the class, with the final result being a set of probabilities of class membership for each sample.

There are several advantages in using SIMCA to classify data. First, an unknown sample is only assigned to the class for which it has a high probability. If the sample's residual variance exceeds the upper limit for every class in the training set, the sample would not be assigned to any of these classes because it is either an outlier or is from a class not represented in the training set. Secondly, some of the classes in the data set may not be well separated. Hence, a future sample might be assigned to two or more groups by SIMCA, which would make sense given the degree of overlap between theses classes. By comparison, K-NN and many other classification methods would forcibly assign the sample to a single class, which would be a mistake in this particular situation. Thirdly, SIMCA is sensitive to the quality of the data used to generate the principal component models for each class in the training set. There are diagnostics to assess the quality of the data as a result of using principal component analysis, including (a) modeling power [73], which describes how well a variable helps each principal component to model variation in the data, and (b) discriminatory power [74], which describes how well a variable helps each principal component model to classify samples. Variables with low modeling power and low discriminatory power usually can be deleted from the analysis, since they contribute only noise to the principal component models.

Friedman and Frank [75] have shown that SIMCA is similar in form to quadratic discriminant analysis. The maximum-likelihood estimate of the inverse of the covariance matrix, which conveys information about the size, shape, and orientation of the data cloud for each class, is replaced by a principal component estimate. Because of the success of SIMCA, statisticians have recently investigated methods other than maximum likelihood to estimate the inverse of the covariance matrix, e.g., regularized discriminant analysis [76]. For this reason, SIMCA is often viewed as the first successful attempt by scientists to develop robust procedures for carrying out statistical discriminant analysis on data sets where maximum-likelihood estimates fail because there are more features than samples in the data set.

9.6 PRACTICAL CONSIDERATIONS

The choice of the training set is important in any pattern-recognition study. Each class must be well represented in the training set. Experimental variables must be controlled or otherwise accounted for by the selection of suitable samples that take into account all sources of variability in the data, for example, lot-to-lot variability. Experimental artifacts such as instrumental drift or sloping baseline must be minimized. Features containing information about differences in the source profile of each class must be present in the data. Otherwise, the classifier is likely to discover rules that do not work well on test samples, i.e., samples that are not part of the original data.

The choice of the classifier is also important. There are several methods available, and the choice depends strongly on the kind of data and its intended use. Should the underlying relationship discovered in the data be readily understandable, or is accuracy the most important consideration? Interpretation can be assisted by using mapping and display methods. Testing the classifier with a validation set can assess its accuracy.

Finally, feature selection is crucial to ensure a successful pattern-recognition study, since irrelevant features can introduce so much noise that a good classification of the data cannot be obtained. When these irrelevant features are removed, a clear and well-separated class structure is often found. The deletion of irrelevant variables is therefore an important goal of feature selection. For averaging techniques such as K-NN, partial least squares, or SIMCA, feature selection is vital, since signal is averaged with noise over a large number of variables, with a loss of discernible signal amplitude when noisy features are not removed from the data. With neural networks, the presence of irrelevant measurement variables may cause the network to focus its attention on the idiosyncrasies of individual samples due to the net's ability to approximate a variety of complex functions in higher-dimensional space, thereby causing it to lose sight of the broader picture, which is essential for generalizing any relationship beyond the training set.

Feature selection is also necessary because of the sheer enormity of many classification problems, for example, DNA microarray data, which consists of thousands of descriptors per observation but only 50 or 100 observations distributed equally between two classes. Feature selection can improve the reliability of a classifier because noisy variables increase the chances of false classification and decrease classification success rates on new data. It is important to identify and delete features from the data set that contain information about experimental artifacts or other systematic variations in the data not related to legitimate chemical differences between the classes represented in the study. For many studies, it is inevitable that relationships among sets of conditions are used to generate the data and the patterns that result. One must realize this in advance when approaching the task of analyzing such data. Therefore, the problem confronting the data analyst is, "How can the information in the data characteristic of the class profile be used without being swamped by the large amount of qualitative and quantitative information contained in the chromatograms or spectra due to experimental conditions?" If the basis of classification for samples in the training set is other than legitimate group differences, unfavorable classification results will be obtained for the prediction set, despite a linearly separable training set. (If the samples can be categorized by a linear classifier, the training set is said to be linearly separable.) The existence of these complicating relationships is an inherent part of multivariate data.

Pattern recognition is about reasoning, using the available information about the problem to uncover information contained within the data. Autoscaling, feature selection, and classification are an integral part of this reasoning process. Each plays a role in uncovering information contained within the data.

9.7 APPLICATIONS OF PATTERN-RECOGNITION TECHNIQUES

Pattern-recognition analyses are usually implemented in four distinct steps: data preprocessing, feature selection, mapping and display or clustering, and classification. However, the process is iterative, with the results of classification or mapping and display often determining further preprocessing steps and reanalysis of the data. Usually, principal component analysis, self-organizing maps, or PLS is used to explore class structure in a data set. Classification rules are then developed, with the membership of each class reflecting the composition of the clusters detected in the data.

Although the procedures selected for a given problem are highly dependent upon the nature of the problem, it is still possible to develop a general set of guidelines for applying pattern-recognition techniques to real data sets. In this final section, a framework for solving the class-membership problem is presented based on a review of data from four previously published studies.

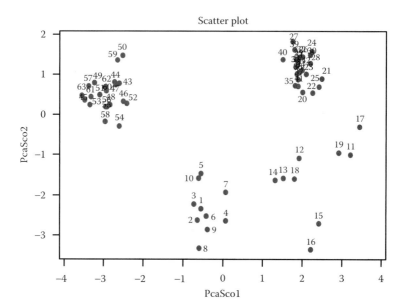

FIGURE 9.8 Plot of the two largest principal components of the 10 metals for the 63 obsidian samples. The two largest principal components account for 78% of the total cumulative variance in the data. Each volcanic glass sample is represented by its sample number in the plot. (Samples 1–10 are from Quarry 1; samples 11–19 are from Quarry 2; samples 20–42 are from Quarry 3; samples 43–63 are from Quarry 4.)

9.7.1 ARCHAEOLOGICAL ARTIFACTS

This study involves obsidian, which is a volcanic glass used by ancient people to construct weapons, tools, and jewelry. Because the composition of this glass is quite homogenous, it is reasonable to assume that one could trace Indian artifacts recovered from archaeological sites to the quarry from where the volcanic glass was originally obtained.

To assess this hypothesis, 63 samples of volcanic glass were collected from four quarries that correspond to the natural geological sources of obsidian in the San Francisco Bay area. (Samples 1 to 10 are from Quarry 1; samples 11 to 19 are from Quarry 2; samples 20 to 42 are from Quarry 3; samples 43 to 63 are from Quarry 4.) The investigators analyzed the 63 glass samples for ten elements: Fe, Ti, Ba, Ca, K, Mn, Rb, Sr, Y, and Zn. Next, a principal component analysis was performed on the data. Figure 9.8 shows a plot of the two largest principal components of the data. From the principal component map, it is evident that our volcanic glass samples can be divided into four groups, which correspond to the quarry sites from which the volcanic glass was obtained. To confirm the presence of these four clusters in the data, the investigators also used single linkage to analyze the data. The dendogram shown in Figure 9.9 also indicates the presence of four clusters in the data (at similarity 0.40, samples 1 to 10 form a cluster; samples 11 to 19 form another

Classification and Pattern Recognition

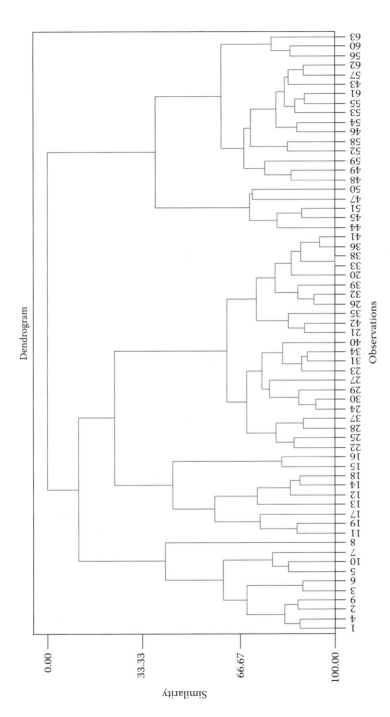

FIGURE 9.9 Single-linkage dendogram of the obsidian data: samples 1–10 are from Quarry 1, samples 11–19 are from Quarry 2, samples 20–42 are from Quarry 3, and samples 43–63 are from Quarry 4.

TABLE 9.1
Training Set for Jet Fuel Example

Number of Samples	Fuel Type
54	JP-4 (fuel used by USAF fighters)
70	Jet-A (fuel used by civilian airliners)
32	JP-7 (fuel used by SR-71 reconnaissance plane)
29	JPTS (fuel used by TR-1 and U-2 aircraft)
43	JP-5 (fuel used by navy jets)

cluster; samples 20 to 42 form a third cluster; and samples 43 to 63 form the fourth cluster). This result was significant because it provided archaeologists with an important tool to acquire information about the migration patterns and trading routes of the Indians in this region using x-ray fluorescence data obtained from the artifacts. Further details about the obsidian data can be found in the literature [77].

9.7.2 FUEL SPILL IDENTIFICATION

In this study, gas chromatography and pattern-recognition techniques were used to develop a potential method for classifying different types of weathered and unweathered jet fuels [78]. The data for the training set consisted of 228 fuel samples representing five different types of jet fuels: JP-4, Jet-A, JP-7, JPTS, and JP-5. The 228 neat jet fuel samples obtained from Wright Patterson and Mulkilteo Energy Management Laboratory were splits from regular quality control standards that were purchased over a 3-year period to verify the authenticity of manufacturer's claims about the properties of these fuels.

Prior to GC analysis, the fuel samples were stored in sealed containers at $-20°C$. The gas chromatograms of the neat jet fuel samples were used as the training set (see Table 9.1). The prediction set consisted of 25 gas chromatograms of weathered jet fuels (see Table 9.2). Eleven of the 25 weathered jet fuel samples were collected from sampling wells as a neat oily phase floating on the top of well water; 7 of the 25 fuel samples were recovered fuels extracted from the soil near various fuel spills. The other seven fuel samples had been subjected to weathering in the laboratory.

Each fuel sample was diluted with methylene chloride and then injected onto a capillary column using a split-injection technique. High-speed GC profiles were obtained using a high-efficiency fused-silica capillary column 10 m in length with an internal diameter of 0.10 mm and coated with 0.34 mm of a bonded and cross-linked 5% phenyl-substituted polymethylsiloxane stationary phase. The column was temperature programmed from 60 to 270°C at 18°/min using an HP-5890 gas chromatograph equipped with a flame-ionization detector, a split-splitless injection

TABLE 9.2
Prediction Set for Jet Fuel Example

Sample Number	Type	Source
PF007	JP-4	Sampling well at Tyndall [a]
PF008	JP-4	Sampling well at Tyndall [a]
PF009	JP-4	Sampling well at Tyndall [a]
PF010	JP-4	Sampling well at Tyndall [a]
PF011	JP-4	Sampling well at Tyndall [a]
PF012	JP-4	Sampling well at Tyndall [a]
PF013	JP-4	Sampling well at Tyndall [a]
KSE1M2	JP-4	Soil extract near a sampling well [b]
KSE2M2	JP-4	Soil extract near a sampling well [b]
KSE3M2	JP-4	Soil extract near a sampling well [b]
KSE4M2	JP-4	Soil extract near a sampling well [b]
KSE5M2	JP-4	Soil extract near a sampling well [b]
KSE6M2	JP-4	Soil extract near a sampling well [b]
KSE7M2	JP-4	Soil extract near a sampling well [b]
MIX1	JP-4	Weathered fuel added to sand
MIX2	JP-4	Weathered fuel added to sand
MIX3	JP-4	Weathered fuel added to sand
MIX4	JP-4	Weathered fuel added to sand
STALE-1	JP-4	Weathered in laboratory [c]
STALE-2	JP-4	Weathered in laboratory [c]
STALE-3	JP-4	Weathered in laboratory [c]
PIT1UNK	JP-5	Sampling pit at Key West air station [d]
PIT1UNK	JP-5	Sampling pit at Key West air station [d]
PIT2UNK	JP-5	Sampling pit at Key West air station [d]
PIT2UNK	JP-5	Sampling pit at Key West air station [d]

[a] Sampling well was near a previously functioning storage depot. Each well sample was collected on a different day.
[b] Dug with a hand auger at various depths. Distance between sampling well and soil extract was approximately 80 yards.
[c] Old JP-4 fuel samples that had undergone weathering in a laboratory refrigerator.
[d] Two pits were dug near a seawall to investigate a suspected JP-5 fuel leak.

port, and an HP-7673A autosampler. Gas chromatograms representative of the five fuel types are shown in Figure 9.10.

The first step in the pattern-recognition study was peak matching the gas chromatograms using a computer program [79] that correctly assigns peaks by first computing Kovat's retention indices [80] for compounds eluting off the column. Because the n-alkane peaks are the most prominent features present in these gas chromatograms, it was a simple matter to compute these indices. Using this procedure was equivalent to placing the GC peaks on a retention-index scale based on the predominant peaks in the chromatograms.

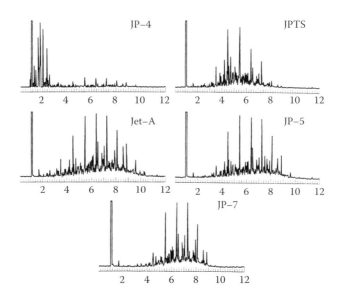

FIGURE 9.10 High-speed gas chromatographic profiles of JP-4, Jet-A, JP-7, JPTS, and JP-5 fuels. (From Lavine, B.K. et al., *Anal. Chem.*, 67, 3846–3852, 1995. With permission.)

The peak-matching program analyzed the GC data by first developing a template of peaks. Integration reports were examined and peaks were added to the template if they did not match the retention indices of previously reported peaks. A preliminary data vector was then produced for each gas chromatogram by matching the retention indices of peaks in each gas chromatogram with the retention indices of the features present in the template. A feature was assigned a value corresponding to the normalized area of the GC peak in the chromatogram. The number of times that a particular feature was found to have a nonzero value was calculated, and peaks below a user-specified number of nonzero occurrences were deleted from the data set. The peak-matching software yielded a final cumulative reference file containing 85 peaks, although not all peaks were present in all gas chromatograms. Using the peak-matching software, each gas chromatogram was transformed into an 85-dimensional data vector, $x = (x_1, x_2, x_3, \ldots, x_j, \ldots, x_{85})$ for pattern-recognition analysis, where x_j is the area of the jth peak normalized to constant sum using the total integrated peak area.

In GC data, outliers — data points lying far from the main body of the data — may exist. Because outliers have the potential to adversely influence the performance of many pattern-recognition techniques, outlier analysis was performed on each fuel class in the training set using the generalized distance test [81]. Three Jet-A and four JP-7 fuel samples were found to be outliers and were removed from the database. The remaining set of data — 221 gas chromatograms of 85 peaks each — was normalized to 100 and autoscaled to ensure that each peak had equal weight in the analysis.

Principal component analysis was used to examine the trends present in the training-set data. Figure 9.11 shows a plot of the two largest principal components of the 85 GC peaks obtained from the 221 gas chromatograms. Each chromatogram is represented as a point in the principal component map. The JP-4, JP-7, and JPTS

Classification and Pattern Recognition

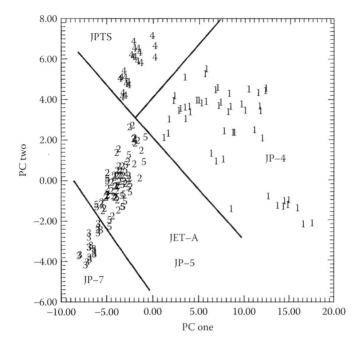

FIGURE 9.11 Plot of the two largest principal components of the 85 GC peaks for the 221 neat jet fuel samples. The map explains 72.3% of the total cumulative variance of the data: 1 = JP-4, 2 = Jet-A, 3 = JP-7, 4 = JPTS, and 5 = JP-5. (From Lavine, B.K. et al., *Anal. Chem.*, 67, 3846–3852, 1995. With permission.)

fuels are well separated from each other and from Jet-A and JP-5 fuels in the two-dimensional map of the data. Evidently, there is information characteristic of fuel type in the gas chromatograms of neat jet fuels.

The overlap of Jet-A and JP-5 fuel samples in the principal component map is not surprising because of the similarity in the physical and chemical properties of these two fuels [82]. An earlier study reported that gas chromatograms of Jet-A and JP-5 fuels were more difficult to classify than the gas chromatograms of other jet fuels because of the similarity in the overall hydrocarbon composition of these two fuel materials. Nevertheless, the investigators concluded that fingerprint patterns existed within the gas chromatograms of Jet-A and JP-5 fuels characteristic of fuel type, which was consistent with the plot obtained for the second- and third-largest principal components of the training-set data (see Figure 9.12). The plot in Figure 9.12 indicates that differences do indeed exist between the hydrocarbon profiles of Jet-A and JP-5 fuels.

To better understand the problems involved with classifying gas chromatograms of Jet-A and JP-5 fuels, it was necessary to focus attention on these two fuels. Figure 9.13 shows a plot of the two largest principal components of the 85 GC peaks obtained from the 110 Jet-A and JP-5 gas chromatograms. An examination of the principal component plot revealed that Jet-A and JP-5 fuel samples lie in different regions of the principal component map. However, the data points

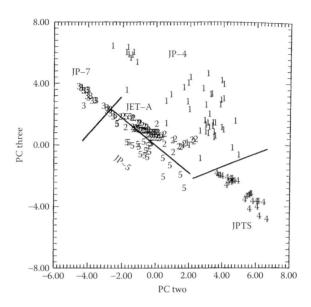

FIGURE 9.12 Plot of the second- and third-largest principal components developed from the 85 GC peaks for the 221 neat jet fuel samples. The map explains 23.1% of the total cumulative variance of the data: 1 = JP-4, 2 = Jet-A, 3 = JP-7, 4 = JPTS, and 5 = JP-5. (From Lavine, B.K. et al., *Anal. Chem.*, 67, 3846–3852, 1995. With permission.)

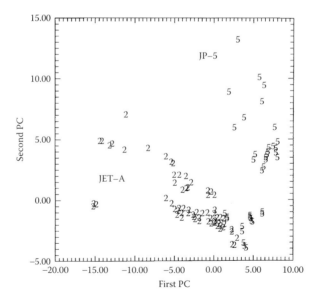

FIGURE 9.13 Principal component map of the 110 neat Jet-A and JP-5 fuel samples developed from the 85 GC peaks. The map explains 80% of the total cumulative variance. The JP-5 fuels are divided into two distinct subgroups: fuel samples that lie close to the Jet-A fuels and fuel samples located in a region of the map distant from Jet-A fuels: 2 = Jet-A and 5 = JP-5. (From Lavine, B.K. et al., *Anal. Chem.*, 67, 3846–3852, 1995. With permission.)

representing the JP-5 fuels form two distinct subgroups in the map. This clustering can pose a problem, since an important requirement for a successful pattern-recognition study is that each class in the data set be represented by samples that are similar in some way. Subclustering of the JP-5 fuel samples suggests a lack of similarity among the samples representing the JP-5 fuels. This lack of similarity might be due to the presence of experimental artifacts in the data. Therefore, it is important that GC peaks responsible for the subclustering of JP-5 fuels be identified and deleted.

The following procedure was used to identify GC peaks strongly correlated with the subclustering. First, the JP-5 fuel samples were divided into two categories on the basis of the observed subclustering. Next, peaks strongly correlated with this subclustering were identified using the variance weight [83, 84], which is a measure of the ability of a feature to discriminate between two different sample groups. Variance weights were also computed for the following category pairs: JP-4 vs. Jet-A, Jet-A vs. JP-5, JP-7 vs. JP-5, and JPTS vs. JP-5. A GC peak was retained for further analysis only if its variance weight for the subclustering dichotomy was lower than for any of the other category pairs. The 27 GC peaks that produced the best individual classification results when the gas chromatograms were classified as JP-4, Jet-A, JP-7, JPTS, or JP-8 were retained for further study. The 27 peaks selected using this procedure spanned the entire gas chromatogram.

Figure 9.14 shows a plot of the two largest principal components of the 27 GC peaks obtained from the 221 neat jet fuel gas chromatograms. Table 9.3 lists the results of K-NN, which was also used to analyze the data. On the basis of K-NN

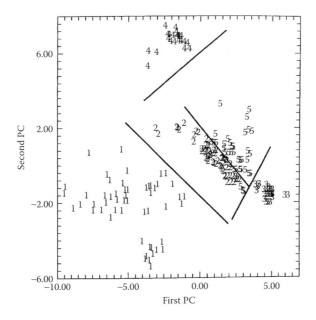

FIGURE 9.14 Principal component map of the 221 neat jet fuels developed from the 27 GC peaks and representative of 75% of the total cumulative variance: 1 = JP-4, 2 = Jet-A, 3 = JP-7, 4 = JPTS, and 5 = JP-5. (From Lavine, B.K. et al., *Anal. Chem.*, 67, 3846–3852, 1995. With permission.)

TABLE 9.3
K-NN Classification Results for Jet Fuel Example

Class	Number in Class	1-NN	3-NN	5-NN	7-NN
JP-4	54	54	54	54	54
Jet-A	67	67	67	67	67
JP-7	28	28	28	28	28
JPTS	29	29	29	29	29
JP-5 [a]	43	43	41	36	37
Total	221	221	219	214	215

[a] Misclassified JP-5 fuel samples were categorized as JET-A. This result is not surprising because of the similarities in the hydrocarbon composition of these two fuel materials.

and the principal component map, it is evident that in the 27-dimensional data space, the five fuel classes are well separated. Furthermore, the principal component map does not indicate the existence of subclustering within any fuel class, which suggests that each fuel class is represented by a homogenous set of objects.

A five-way classification study involving JP-4, Jet-A, JP-7, JPTS, and JP-5 in the 27-dimensional data space was also undertaken using SIMCA. Principal component models were developed from the 27 GC peaks for each of the five fuel classes in the training set. The number of principal components retained for each fuel class was determined by cross validation. Each gas chromatogram in the training set was then classified on the basis of its residual or goodness of fit. The probability of a gas chromatogram belonging to a particular class was determined from its residual variance for the corresponding principal component model by way of an F-test, which involves comparing the sample's residual variance of fit to the average residual variance for those samples that make up the class using a variance ratio. Each sample was assigned to the class for which it had the smallest variance ratio. If the variance ratio exceeded the critical F-value for that class, it was not assigned to it. Results from the five-way classification study are shown in Table 9.4. The recognition rate for the principal component models developed from the 27 GC peaks was very high.

A prediction set of 25 gas chromatograms was employed to test the predictive ability of the 27 GC peaks and the class models associated with them. Because the gas chromatograms in the prediction set were run a few months before the neat jet fuel gas chromatograms, they constituted a true prediction set. Table 9.5 summarizes the results. All of the weathered jet fuel gas chromatograms were correctly classified.

The high classification success rate obtained for the weathered fuel samples suggests that information about fuel type is present in the gas chromatograms of weathered jet fuels. This is a significant finding, since the changes in composition that occur after a jet fuel is released into the environment can be a serious problem in fuel spill identification. These changes arise from microbial

TABLE 9.4
SIMCA Training-Set Results for Jet Fuel Example

Class	Number of Principal Components	F-Criterion [a] Number in Class	Number Correct	Percent
JP-4	1	54	54	100%
Jet-A [b]	1	67	61	91.1%
JP-7 [b]	1	28	27	96.4%
JPTS	1	29	29	100%
JP-5	1	43	43	100%
Total	—	221	214	96.8%

[a] Classifications were made on the basis of the variance ratio: $F = [s_p/s_0]^2[N_q - N_c - 1]$, where s_p^2 is the residual of sample p for class i, s_0^2 is the variance of class i, N_q is the number of samples in the class, and N_c is the number of principal components used to describe the class. A sample is assigned to the class for which it has the lowest variance ratio. If the sample's variance ratio exceeds the critical F-value for that class, then it is not assigned to the class. The critical F-value for each sample in the training set is $F \leq [(M - N_c, (M - N_c)(N_q - N_c - 1)]$, where M is the number of measurement variables or GC peaks used to develop the principal-component model.

[b] Misclassified Jet-A and JP-7 fuel samples were categorized as JP-5 by SIMCA.

degradation, loss of water-soluble compounds due to dissolution, and evaporation of lower-molecular-weight alkanes. However, the weathered jet fuel samples used in this study were recovered from a subsurface environment. In such an environment, evaporation is severely retarded [85], and the loss of water-soluble compounds is not expected to be a serious problem [86]. In all likelihood, the predominant weathering factor in subsurface fuel spills is biodegradation, which does not appear to have a pronounced effect on the overall GC profile of the fuels based on these results. Evidently, the weathering process for aviation turbine fuels in subsurface environments is greatly retarded compared with surface spills, with the subsurface spills preserving the fuel's identity for a longer period of time.

9.7.3 Sorting Plastics for Recycling

In this study, Raman spectroscopy and pattern-recognition techniques were used to develop a potential method to differentiate common household plastics by type [87–89], which is crucial to ensure the economic viability of recycling. The test data consisted of 188 Raman spectra of six common household plastics: high-density polyethylene (HDPE), low-density polyethylene (LDPE), polyethylene terephthalate (PET), polypropylene (PP), polystyrene (PS), and polyvinylchloride

TABLE 9.5
Prediction-Set Results (F-Values) for Jet Fuel Example

Samples	JP-4	Jet-A	JP-7	JPTS	JP-5
PF007	3.82	10.04	58.9	12.4	7.43
PF008	3.69	9.62	57.6	12.5	7.14
PF009	3.71	9.84	57.6	12.6	7.32
PF010	3.30	16.7	73.7	11.8	10.8
PF011	3.57	9.64	58.9	12.8	7.39
PF012	4.11	7.74	78.2	1.3.5	12.04
PF013	4.33	8.19	79.8	12.6	12.3
KSE1M2	2.83	24.4	63.9	30.4	11.21
KSE2M2	2.25	16.2	70.8	21.6	11.09
KSE3M2	2.51	9.41	71.0	17.3	10.2
KSE4M2	2.40	10.11	71.3	17.83	10.4
KSE5M2	2.33	7.76	56.4	17.9	7.61
KSE6M2	1.87	13.4	69.3	20.8	10.4
KSE7M2	2.21	9.85	67.3	18.3	9.78
MIX1	1.33	34.9	71.3	38.2	13.3
MIX2	1.33	11.93	53.3	20.9	7.37
MIX3	1.44	12.3	55.2	20.6	7.71
MIX4	1.59	9.51	48.6	19.9	6.27
STALE-1	1.72	73.7	151.9	54.7	31.5
STALE-2	0.58	28.7	123.8	30.9	22.6
STALE-3	0.541	28.7	127.3	29.9	22.6
PIT1UNK	6.62	1.19	6.11	33.02	0.504
PIT1UNK	6.57	1.15	6.03	32.9	0.496
PIT2UNK	6.51	1.14	6.14	32.8	0.479
PIT2UNK	6.51	1.14	6.27	32.7	0.471

Note: A sample is assigned to the fuel class yielding the smallest variance ratio. If the variance ratio exceeds the critical F-value for that particular fuel class, the sample is not assigned to that class. Critical F-values for prediction-set samples are obtained using one degree of freedom for the numerator and $N_q - N_c - 1$ degrees of freedom for the denominator (see [19]). For JP-4, the critical F-value at $\alpha_{0.975}$ is $F(1,52) = 5.35$, and for JP-5 it is $F(1,41) = 5.47$.

(PVC). The plastic containers used in this study were collected from residential homes and from BFI Recycling in Pocatello, ID. Each plastic sample was cut from collected containers. Spectra of the cut plastics were obtained with a Spex 500M (1/2) meter Raman spectrometer, which incorporated Spex Model 1489 collection optics, an Omnichrome Model 160 T/B air-cooled Ar$^+$ laser, and a liquid-nitrogen-cooled charge-coupled detector. Figure 9.15 shows Raman spectra representative of the six types of plastic. Each Raman spectrum was an average of sixteen 1-s scans collected over the wave-number range 850 to 1800 cm^{-1} for 1093 points.

Classification and Pattern Recognition

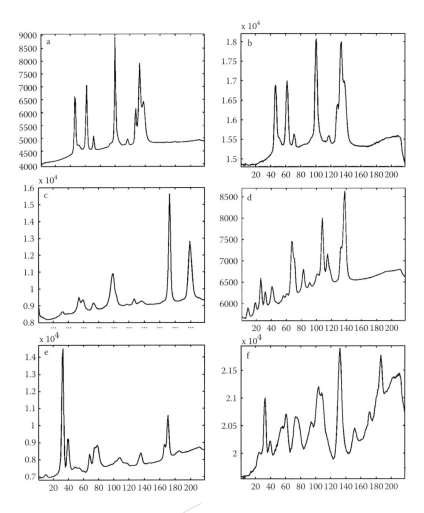

FIGURE 9.15 Raman spectra of the plastics: (a) high-density polyethylene (HDPE), (b) low-density polyethylene (LDPE), (c) polyethylene terephthalate (PET), (d) polypropylene (PP), (e) polystyrene (PS), and (f) polyvinylchloride (PVC). (From *Pattern Recognition, Chemometrics, and Imaging for Optical Environmental Monitoring*, Siddiqui, K. and Eastwood, D., Eds., Proceedings of SPIE, 1999, pp. 103–112. With permission.)

The spectra were boxcar averaged every 10 points to yield 218-point spectra. The averaged spectra were baseline corrected for offsets using a linear polynomial and then normalized to unit length to adjust for variations in the optical path length. The spectra were divided into a training set of 169 spectra (see Table 9.6) and a prediction set of 19 spectra (see Table 9.7). Spectra in the prediction set were chosen by random lot. For pattern-recognition analysis, each plastic sample was represented by a data vector, $x = (x_1, x_2, x_3, \ldots, x_j, \ldots, x_{218})$, where x_j is the

TABLE 9.6
Training Set for Plastics Example

Number of Spectra	Plastic Type
33	High-density polyethylene (HDPE)
26	Low-density polyethylene (LDPE)
35	Polyethylene terephthalate (PET)
26	Polypropylene (PP)
32	Polystyrene (PS)
17	Polyvinylchloride (PVC)
169	Total

Raman intensity of the jth point from the baseline-corrected normalized Raman spectrum.

The first step in the study was to apply principal component analysis to the spectra in the training set. Figure 9.16 shows a plot of the two largest principal components of the 218-point spectra that compose the training set. Prior to principal component analysis, spectra were normalized to unit length and then mean centered. Each spectrum is represented as a point in the principal component map. Clustering of spectra by sample type is evident. When the prediction-set samples were projected onto the principal component map, 17 of the 19 samples were found to lie in a region of the map with plastic samples that bore the same class label (see Figure 9.17). One HDPE sample, however, was misclassified as LDPE, and one LDPE sample was misclassified as HDPE. This result is not surprising in view of the overlap between these two classes (HDPE and LDPE) in the map due to the similarity of their Raman spectra.

The next step was to use a Kohonen two-dimensional self-organizing map to represent the spectral data in the 218-dimensional measurement space. Self-organizing maps are, for the most part, used to visualize high-dimensional data. However, classification and prediction of multivariate data can also be performed with these

TABLE 9.7
Prediction Set for Plastics Example

Number of Spectra	Plastic Type
5	High-density polyethylene (HDPE)
2	Low-density polyethylene (LDPE)
5	Polyethylene terephthalate (PET)
2	Polypropylene (PP)
5	Polystyrene (PS)
0	Polyvinylchloride (PVC)
19	Total

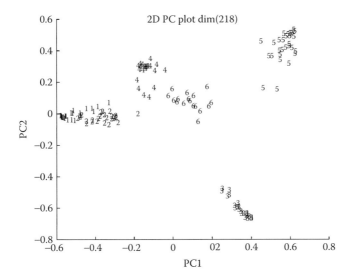

FIGURE 9.16 Plot of the two largest principal components of the 218-point, normalized, and mean-centered spectra that compose the training set. Each spectrum is represented as a point in the principal component map: 1 = HDPE, 2 = LDPE, 3 = PET, 4 = PP, 5 = PS, and 6 = PVC.

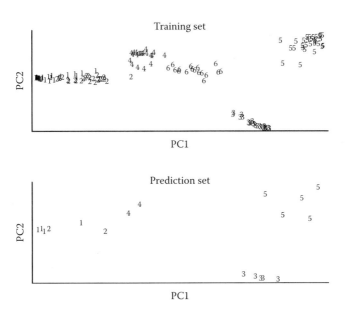

FIGURE 9.17 Prediction-set samples projected onto the principal component map developed from the 218-point spectra composing the training set. Each spectrum is represented as a point in the map: 1 = HDPE, 2 = LDPE, 3 = PET, 4 = PP, 5 = PS, and 6 = PVC.

maps. For the plastic data set, the learning rate of the Kohonen self-organizing map was varied from 0.05 to 0.001 during training. Each spectrum in the training set was trained 50 times. The number of neurons used to study the training-set data was set to 64 (i.e., an 8×8 grid).

Selection of the map's grid size was crucial to ensure an accurate representation of the data. From our previous experience, the number of neurons used should be between one-third and one-half of the number of samples in the training set. Because of the similarity among spectra in the data set, it is logical for the number of neurons used to be less than the number of samples in the data set. Using too few neurons, however, produces a map of the data that does not faithfully reflect the distribution of the points in the high-dimensional measurement space, whereas using too many neurons can generate a map of the data that is not aesthetically pleasing, since there are many neurons that do not respond to any of the spectra, making it more difficult to discern clustering in the data. These conclusions admittedly are subjective in nature, since they depend on the data itself, with different workers having different ideas on what are the best parameters.

After the weights for the neurons were trained, the distances between each sample and the neurons were calculated. Samples were assigned to the nearest neuron, with each neuron given a class label based on the class assignment of the samples responding to it. In all cases, only samples with the same class label responded to a given neuron. Figure 9.18 shows a self-organizing map generated from the training-set data, with the prediction-set samples projected onto the map. Each data point on the map was entered as a 1 (HDPE), 2 (LDPE), 3 (PET), 4 (PP), 5 (PS), or 6 (PVC). The self-organizing map was developed from the training-set data, which consisted of 169 Raman spectra and 218 spectral variables. The large centered numbers in Figure 9.18

Self-organizing map

1^1	2	2^2					
1^1	2	2	2		5^5		
1^1	2^2	2	6		5	5	
1^1	1^1		6		5^5		
			6	5^5	5	5	
3^3	3^3	3					
3^3	3^3	3		4	4^4	4	
			4	4^4	4	4	

FIGURE 9.18 Self-organizing map generated from the training-set data with the prediction-set samples projected onto the map. Each point from the training and prediction sets was entered on the map: 1 = HDPE, 2 = LDPE, 3 = PET, 4 = PP, 5 = PS, and 6 = PVC.

represent the class assignments of these samples. On the basis of these assignments, clustering of the spectra by plastic type is again evident.

Prediction-set samples were then passed to the network and assigned to the nearest neuron. They are represented as superscripts in Figure 9.18. All 19 prediction-set samples were assigned to neurons that had spectra with the same class label. The validation of the self-organizing maps using these 19 prediction-set samples implies that information is contained in the Raman spectra of the plastics characteristic of sample type. Classification of the plastics by the self-organizing map was as reliable as that obtained by other methods, and more importantly, the classification was obtained without any preassumptions about the data.

Self-organizing maps in conjunction with principal component analysis constitute a powerful approach for display and classification of multivariate data. However, this does not mean that feature selection should not be used to strengthen the classification of the data. Deletion of irrelevant features can improve the reliability of the classification because noisy variables increase the chances of false classification and decrease classification success rates on new data. Furthermore, feature selection can lead to an understanding of the essential features that play an important role in governing the behavior of the system or process under investigation. It can identify those measurements that are informative and those measurements that are uninformative. However, any approach used for feature selection should take into account the existence of redundancies in the data and be multivariate in nature to ensure identification of all relevant features.

9.7.4 TAXONOMY BASED ON CHEMICAL CONSTITUTION

This study involved the development of a potential method to differentiate Africanized honeybees from European honeybees. The test data consisted of 164 gas chromatograms of cuticular hydrocarbon extracts obtained from the whole bodies of Africanized and European honeybees. The Africanized and European honeybee samples used in this study were from two social castes: foragers and nest bees. Africanized honeybees from Brazil, Venezuela, Argentina, Panama, and Costa Rica were collected as returning foragers or from within colonized (bait) hives. These bees were identified on site as Africanized based on a field test for colony defense behavior. European honeybees were collected from managed colonies in Florida and Venezuela and represented a variety of commercially available stocks found in the U.S.

Cuticular hydrocarbons were obtained by rinsing the dried, pinned, or freshly frozen bee specimens in hexane. The hydrocarbon fraction analyzed by gas chromatography was isolated from the concentrated washings by means of a silicic acid column with hexane used as the eluant. The extracted hydrocarbons (equivalent to 1/25th of a bee) were coinjected with authentic n-paraffin standards onto a glass column (1.8 m × 2 mm) packed with 3% OV-17 on Chromosorb WAW DMCS packing (120–140 mesh). Kovat retention indices (KI) were assigned to compounds eluting off the GC column. These KI values were used for peak identification.

TABLE 9.8
Training Set for Honeybee Example

Number of Samples	Specimen Type
13	Africanized foragers from Central America
30	Africanized foragers from Venezuela
6	Africanized nest bees from Central America
30	European foragers from Venezuela
30	European foragers from Florida
109	Total

Packed GC columns afforded moderately well-resolved, reproducible profiles of the hydrocarbon fraction. Each gas chromatogram contained 40 peaks corresponding to a set of standardized retention time windows. Further details about the collection of this data can be found in the literature [90].

The chromatograms were translated into data vectors by measuring the areas of the 40 peaks. Each gas chromatogram was initially represented by a data vector $x = (x_1, x_2, x_3, x_j, ..., x_{40})$ where x_j is the area of the jth peak normalized to constant sum using the total integrated peak area. However, only eight of the peaks were considered for pattern-recognition analysis. These peaks were identified in a previous study as being strongly correlated to subspecies. Furthermore, the compounds composing these peaks were present in the wax produced by nest bees, and the concentration pattern of the wax constituents is believed to convey genetic information about the honeybee colony.

Principal component analysis was used to investigate the data. The training set (see Table 9.8) consisted of 109 gas chromatograms of cuticular hydrocarbon extracts obtained from Africanized and European honeybees (60 European and 49 Africanized). A principal component plot developed from the 8 GC peaks did not show clustering of the bees on the basis of subspecies. We attributed this to the fact that a coordinate system based on variance did not convey information about subspecies for this data set. Therefore, PLS was used to generate a map of the data. PLS uses variance-based axes that are rotated to ensure a better correlation with the target variable, which is subspecies. The rotation provides an appropriate compromise between explaining the measurement variables and predicting the class of each sample.

Figure 9.19 shows a plot of the two largest PLS components developed from the eight GC peaks and 109 training-set samples. The European honeybees are designated as "1," and the Africanized honeybees are "2." Separation of the honeybees by class is evident in the PLS plot of the data. The fact that class discrimination is only associated with the largest PLS component of the data is encouraging.

To assess the predictive ability of these eight GC peaks, a prediction set (see Table 9.9) of 55 gas chromatograms was employed. We chose to map the 55 gas

Classification and Pattern Recognition

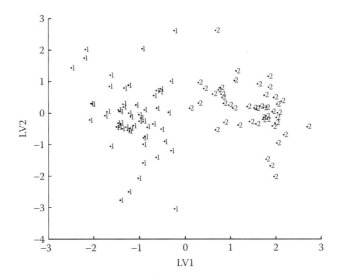

FIGURE 9.19 Plot of the two largest PLS components developed from the 8 GC peaks and 109 chromatograms in the training set. Each chromatogram is represented as a point in the map: 1 = European honeybees, and 2 = Africanized honeybees.

chromatograms directly onto the PLS score plot defined by the 109 chromatograms and eight GC peaks. Figure 9.20 shows the prediction-set samples mapped onto the PLS plot developed from the training-set data. Of the 55 gas chromatograms, 53 lie in a region of the map with bee samples that carry the same class label. Evidently, there is a direct relationship between the concentration pattern of these compounds and the identity of the bees' subspecies.

The results of this study demonstrate that gas chromatography and pattern-recognition techniques can be applied successfully to the problem of identifying Africanized honeybees. Classifiers capable of achieving very high classification success rates within the present data were found, and these classifiers were able to classify bees, a task that was not part of the original training set. Because of these

TABLE 9.9
Prediction Set for Honeybee Example

Number of Samples	Specimen Type
10	Africanized foragers from Central America
5	Africanized nest bees from Central America
10	European workers from Florida
30	European nest bees from Florida
55	Total

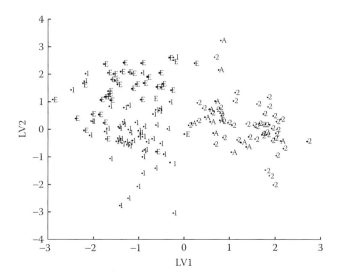

FIGURE 9.20 PLS map generated from the training-set data with the 55 prediction-set samples projected onto the map. Each chromatogram is represented as a point in the map: 1 = European honeybees, 2 = Africanized honeybees, E = European honeybee prediction-set sample, and A = Africanized honeybee prediction-set sample.

results, a larger study [91] was undertaken, where capillary-column GC and pattern-recognition techniques were used to develop a method to differentiate Africanized honeybees from European honeybees based on differences in chemical constitution.

REFERENCES

1. Lavine, B.K. and Workman, J., Jr., Fundamental review of Chemometrics, *Anal. Chem.,* 74 (12), 2763–2770, 2002.
2. Workman, J., Jr., The state of multivariate thinking for scientists in industry, *Chemolab,* 60 (1–2), 161–172, 2002.
3. Jurs, P.C. and Isenhour, T.L., *Chemical Applications of Pattern Recognition,* John Wiley & Sons, New York, 1975.
4. Stuper, A.J., Brugger, W.E., and Jurs, P.C., *Computer Assisted Studies of Chemical Structure and Biological Function,* John Wiley & Sons, New York, 1979.
5. Ludvigsen, L., Albrechtsen, H.J., Holst, H., and Christiansen, T.H., Correlating phospholipid fatty acids (PLFA) in a landfill leachate polluted aquifer with biogeochemical factors by multivariate statistical methods, *FEMS Microbiol. Rev.,* 20 (3–4), 447–460, 1997.
6. Ravichandran, S., Ramanibai, R., and Pundarikanthan, N.V., Ecoregions for describing water quality patterns in Tamiraparani Basin, South India, *J. Hydrol.,* 178 (1–4), 257–276, 1996.
7. Vernat-Rossi, V., Garcia, C., Talon, R., Denoyer, C., and Berdague, J.L., Rapid discrimination of meat products and bacterial strains using semiconductor gas sensors *Sens. Actuators B,* B37 (1–2), 43–48, 1996.

8. Hanaki, S., Nakamoto, T., and Morizumi, T., *Sens. Actuators A,* A57 (1), 65–71, 1996.
9. Hahn, P., Smith, I.C.P., Leboldus, L., Littman, C., Somorjai, R.L., and Bezabeh, T., Artificial odor-recognition system using neural network for estimating sensory quantities of blended fragrance. The classification of benign and malignant human prostate tissue by multivariate analysis of ^{1}H magnetic resonance spectra, *Canc. Res.,* 57 (16), 3398–3401, 1997.
10. Benninghoff, L., von Czarnowski, D., Denkhaus, E., and Lemke, K., Analysis of human tissues by total reflection x-ray fluorescence. Application of chemometrics for diagnostic cancer recognition, *Spectrochim. Acta, Part B,* 52B (17), 1039–1046, 1997.
11. Goodacre, R., *Appl. Spectrosc.,* 51 (8), 1144–1153, 1997.
12. Briandet, R., Kemsley, E.K., and Wilson, R.H., *J. Agric. Food Chem.,* 44 (1), 170–174, 1996.
13. Jurs, P.C., Kowalski, B.R., Isenhour, T.L., and Reiley, C.N., *Anal. Chem.,* 41, 1945–1948, 1969.
14. Lavine B.K. and Carlson, D.A., *Microchem. J.,* 42, 121–125, 1990.
15. Smith, A.B., Belcher, A.M., Epple, G., Jurs, P.C., and Lavine, B.K., *Science,* 228, 175–177, 1985.
16. Pino, J.A., McMurry, J.E., Jurs, P.C., Lavine, B.K., and Harper, A.M., *Anal. Chem.,* 57 (1), 295–302, 1985.
17. Gemperline, P.J. and Boyer, N.R., *Anal. Chem.,* 67 (1), 160–166, 1995.
18. Shah N.K. and Gemperline, P.J., *Anal. Chem.,* 62 (5), 465–470, 1990.
19. Gemperline, P.J., Webber, L.D., and Cox, F.O., *Anal. Chem.,* 61 (2), 138–144, 1989.
20. Van De Waterbeemd, H., Ed., *Chemometric Methods in Molecular Design (Methods and Principles in Medicinal Chemistry,* Vol. 2, John Wiley & Sons, New York, 1995.
21. McElroy, N.R., Jurs, P.C., and Morisseau, C., *J. Med. Chem.,* 46 (6), 1066–1080, 2003.
22. Bakken, G.A. and Jurs, P.C., *J. Med. Chem.,* 43 (23), 4534–4541, 2000.
23. Cho, J.H., Lee, D., Park, J.H., Kunwoo, K., and Lee, I.B., *Biotech. Prog.,* 18 (4), 847–854, 2002.
24. Golub, T.R., Slonim, D.K., Tamayo, P., Huard, C., Gaasenbeek, M., Mesirov, J.P., Coller, H., Loh, M.L., Downing, J.R., Caliguiri, M.A., Bloomfield, C.D., and Lander, E.S., *Science,* 286, 531–537, 1999.
25. Alon, U., Barkai, N., Notterman, D.A., Gish, K., Ybarra, S., Mack, D., and Levine, A.J., *Proc. Nat. Acad. Sci.,* 96, 6745–6750, 1999.
26. Brereton, R.G., Ed., *Multivariate Pattern Recognition in Chemometrics,* Elsevier Science Publishers, Amsterdam, 1992.
27. Fukunaga, K., *Statistical Pattern Recognition,* 2nd ed., Academic Press, New York, 1990.
28. McLachlan, G.J., *Discriminant Analysis and Statistical Pattern Recognition,* John Wiley & Sons, New York, 1992.
29. Duda, R., Hart, P.E., and Stork, D.G., *Pattern Classification and Scene Analysis,* 2nd ed., Wiley-Interscience, New York, 2000.
30. Webb, A.R., *Statistical Pattern Recognition,* 2nd ed., John Wiley & Sons, New York, 2002.
31. Siddiqui, K.J. and Eastwood, D., Eds., *Pattern Recognition, Chemometrics, and Imaging for Optical Environmental Monitoring,* SPIES Proceedings, Vol. 3854, September 1999.
32. Beebe, K.R., Pell, R.J., and Seasholtz, M.B., *Chemometrics: A Practical Guide,* John Wiley & Sons, New York, 1998.
33. Geladi, P. and Grahn, H., *Multivariate Image Analysis,* John Wiley & Sons, New York, 1996.

34. Jurs, P.C., *Science,* 232, 1219–1224, 1986.
35. Ripley, B.D., *Pattern Recognition and Neural Networks,* Cambridge University Press, U.K., 1996.
36. Johansson, E., Wold, S., and Sjodin, K., *Anal. Chem.,* 56, 1685–1688, 1984.
37. Jackson, J.E., *A User's Guide to Principal Component Analysis,* John Wiley & Sons, New York, 1991.
38. Jolliffe, I.T., *Principal Component Analysis, Springer-Verlag,* New York, 1986.
39. Mandel, J., *NBS J. Res.,* 190 (6), 465–476, 1985.
40. Brown, S., *Appl. Spectrosc.,* 49 (12), 14A–30A, 1995.
41. Kohonen, T., *Self-Organizing Maps,* 3rd ed., Springer-Verlag, Berlin, 2000.
42. Zupan, J. and Gasteiger, J., *Neural Networks for Chemists,* VCH Publishers, New York, 1993.
43. Johnson, R.A. and Wichern, D.W., *Applied Multivariate Statistical Analysis,* 2nd ed., Prentice-Hall, London, 1982.
44. Bezdek, J.C., Coray, C., Gunderson, R., and Watson, J., *SIAM J. Appl. Math.,* 40 (2), 358–372, 1981.
45. Jarvis, R.A. and Patrick, E.A., *IEEE Trans. Comput.,* C22, 1025–1032, 1973.
46. Massart, D.L. and Kaufman, L., *The Interpretation of Analytical Chemical Data by the Use of Cluster Analysis,* John Wiley & Sons, New York, 1983.
47. Everitt, B.S., *Cluster Analysis,* Heinemen Educational Books, London, 1980.
48. Hartigan, J.S., *Clustering Algorithms,* John Wiley & Sons, New York, 1975.
49. Wassserman, P.D., *Neural Computing: Theory and Practice,* Van Nostrand Reinhold, New York, 1989.
50. Curry, B. and Rumelhart, D.E., *Tetrahed. Comput. Method.,* 3, 213–237, 1990.
51. Cristianini, N. and Shawe-Taylor, J., *An Introduction to Support Vector Machines and Other Kernel-Based Learning Methods,* Cambridge University Press, UK, 2000.
52. Kecman, V., *Learning and Soft Computing: Support Vector Machines, Neural Networks, and Fuzzy Logic Models (Complex Adaptive Systems),* MIT Press, Cambridge, MA, 2001.
53. Tou, J.T. and Gonzalez, R.C., *Pattern Recognition Principles,* Addison Wesley, Reading, MA, 1974.
54. James, M., *Classification,* John Wiley & Sons, New York, 1985.
55. Cover, T.M. and Hart, P.E., *IEEE Trans. Inf. Theory,* IT-13, 21–35, 1967.
56. Ozen, B.F. and Mauer, L.J., *J. Agric. Food Chem.,* 50, 3898–3901, 2002.
57. Jurs, P.C., Bakken G.A., and McClelland, H.E., *Chem. Rev.,* 100 (7), 2649–2678, 2000.
58. Song, X., Hopke, P.K., Burns, M.A., Graham, K., and Scow, K., *Environ. Sci. Technol.,* 33 (20), 3524–3530, 1999.
59. Saaksjarvi, E., Khalighi, M., and Minkkinen, P., *Chemolab,* 7, 171–180, 1989.
60. Briandet, R., Kemsley, E., and Wilson, R., *J. Agric. Food Chem.,* 44, 170–174, 1996.
61. Geladi, P. and Kowalski, B.R., *Anal Chim. Acta,* 185, 1–17, 1986.
62. Geladi, P. and Kowalski, B.R., *Anal Chim. Acta,* 185, 19–32, 1986.
63. Frank, I.E. and Friedman, J.H., *Technometrics,* 35 (2), 109–135, 1993.
64. Thomas, E., *Anal. Chem.,* 66 (15), 795A–804A, 1994.
65. Martens, H. and Naes, T., *Multivariate Calibration,* John Wiley & Sons, Chichester, U.K., 1989.
66. Wold, H., Soft modeling by latent variables: the nonlinear iterative partial least squares approach, in *Perspectives in Probability and Statistics,* Papers in Honor of M.S. Bartlett, Gani, J., Ed., Academic Press, London, 1975.
67. Barker, M. and Rayens, W., *J. Chem.,* 17 (3), 166–173, 2003.

68. Vaid, T.P., Burl, M.C., and Lewis, N.S., *Anal. Chem.*, 73, 321–331, 2001.
69. Wold, S., *Patt. Recog.*, 8, 127–139, 1976.
70. Blomquist, G., Johansson, E., Soderstrom, B., and Wold, S., *J. Chromat.*, 173, 19–32, 1979.
71. Stone, M., *J. R. Stat. Soc.*, 36, 111–121, 1974.
72. Wold, S., *Technometrics*, 20, 397–405, 1978.
73. Wold, S., Albano, C., Dunn, W.J., III, Edlund, U., Esbensen, K., Hellberg, S., Johansson, E., Lindberg, W., and Sjostrom, M., Multivariate data analysis in chemistry, in *Chemometrics: Mathematics and Statistics in Chemistry*, Kowalski, B.R., Ed., D. Reidel Publishing, Dordrecht, The Netherlands, 1984.
74. Wold, S. and Sjostrom, M., SIMCA, a method for analyzing chemical data in terms of similarity and analogy, in *Chemometrics, Theory and Application*, Kowalski, B.R., Ed., American Chemical Society, Washington, D.C., 1977.
75. Frank, I.E. and Friedman, J.H., *J. Chem.*, 3, 463–475, 1989.
76. Friedman, J., *JASA*, 84 (405), 165–175, 1989.
77. Kowalski, B.R., Schatzki, T.F., and Stross, F.H., *Anal. Chem.*, 44, 2176–2182, 1972.
78. Lavine, B.K., Mayfield, H.T., Kromann, P.R., and Faruque, A., *Anal. Chem.*, 67, 3846–3852, 1995.
79. Mayfield, H.T. and Bertsch, W., *Comput. Appl. Lab.*, 1, 130–137, 1983.
80. Kovats, E., in *Advances in Chromatography*, Vol. 1, Giddings, J.C. and Keller, R.A., Eds., Marcel Dekker, New York, p. 230, 1965.
81. Schwager, S.J. and Margolin, B.H., *Ann. Stat.*, 10, 943–953, 1982.
82. Coordinating Research Council, *Handbook of Aviation Fuel Properties*, Atlanta, 1983.
83. Scharaf, M., Illman, D., and Kowalski, B.R., *Chemometrics*, John Wiley & Sons, New York, 1986, p. 195.
84. Harper, A.M., Duewer, D.L., Kowalski, B.R., and Fasching, J.L., ARTHUR and experimental data analysis: the heuristic use of polyalgorithms, in *Chemometrics: Theory & Applications*, Kowalski, B.R., Ed., ACS Symposium Series 52, American Chemical Society, Washington, D.C., 1977.
85. Spain, J.C., Sommerville, C.C., Butler, L.C., Lee, T.J., and Bourquin, A.W., Degradation of Jet Fuel Hydrocarbons in Aquatic Communities, USAF report ESL-TR-83-26, AFESC, Tyndall AFB, FL, 1983.
86. Coleman, W.E., Munch, J.W., Streicher, R.P., Ringhand, H.P., and Kopfler, W.F., *Arch. Environ. Contam. Toxicol.*, 13, 171–180, 1984.
87. Allen, V., Kalivas, J.H., and Rodriguez, R.G., *Appl. Spec.*, 53 (6), 672–681, 1999.
88. Lavine, B.K., Davidson, C.E., and Moores, A.J., *Vibrat. Spectrosc.*, 842, 1–13, 2002.
89. Lavine, B.K., Davidson, C.E., and Westover, D., Spectral pattern recognition using self organizing maps, *J. Chem. Inf. Comp. Sci.*, 44(3), 1056–1064, 2004.
90. Lavine, B.K. and Carlson, D.A., *Anal. Chem.*, 59 (6), 468A–470A, 1987.
91. Lavine, B.K., Ward, A.J.I., Smith, R.K., and Taylor, O.R., *Microchem. J.*, 1989, 39, 308–317.

10 Signal Processing and Digital Filtering

Steven D. Brown

CONTENTS

10.1 Introduction ... 379
10.2 Noise Removal and the Problem of Prior Information ... 380
    10.2.1 Signal Estimation and Signal Detection ... 381
10.3 Reexpressing Data in Alternate Bases to Analyze Structure ... 382
    10.3.1 Projection-Based Signal Analysis as Signal Processing ... 383
10.4 Frequency-Domain Signal Processing ... 385
    10.4.1 The Fourier Transform ... 386
    10.4.2 The Sampling Theorem and Aliasing ... 386
    10.4.3 The Bandwidth-Limited, Discrete Fourier Transform ... 388
    10.4.4 Properties of the Fourier Transform ... 389
10.5 Frequency Domain Smoothing ... 394
    10.5.1 Smoothing ... 394
    10.5.2 Smoothing with Designer Transfer Functions ... 395
10.6 Time-Domain Filtering and Smoothing ... 398
    10.6.1 Smoothing ... 398
    10.6.2 Filtering ... 400
    10.6.3 Polynomial Moving-Average (Savitsky-Golay) Filters ... 403
10.7 Wavelet-Based Signal Processing ... 406
    10.7.1 The Wavelet Function ... 406
    10.7.2 Time and Frequency Localizations of Wavelet Functions ... 408
    10.7.3 The Discrete Wavelet Transform ... 409
    10.7.4 Smoothing and Denoising with Wavelets ... 412
References ... 414
Further Reading ... 416

10.1 INTRODUCTION

In any attempt at collecting data, even with the most advanced instrumentation, the measured signal is corrupted with noise. The noise amplitude may be small, and it may not change the observed signal shape or amplitude significantly, but often the noise contribution is large enough to obscure the true shape and amplitude of the signal.

This chapter is concerned with mathematical methods that are intended to enhance data by decreasing the contribution of the noise relative to the desired signal and by recovering the "true" signal response from one altered by instrumental or other effects that distort the shape of the true response. Noise, it should be understood, encompasses all aspects of the observed signal not related to the desired response. This definition includes the usual high-frequency, random responses commonly regarded as noise in introductory treatments of signal processing, but it also includes responses of low and mid frequency.

Data enhancement can be done in concert with data acquisition, a process known as "filtering." Many real-time data processing schemes begin with filtering to remove some of the noise, and follow with some sort of decision based on the results of the filtering. This sort of filtering for data enhancement and "on-line" decisions is useful for process monitoring or instrument control, two areas where chemometric methods are growing in popularity. Filtering data that are acquired at high rates demands special care, since the speed of the data reduction step is very important. The computational burden imposed by a potential method must be considered, because the time spent in filtering may decrease the data throughput unacceptably.

More often the data-enhancement step is done later, to simplify the data processing and to lower the computational burden placed on the instrumental computer. When data are enhanced either "at-line" or "off-line," after data acquisition is complete, the process is called *smoothing*. Smoothing methods are much more varied than ones used exclusively for filtering because the time and computational constraints are not as demanding. It should be noted, however, that most filtering methods can also be used for smoothing.

Efficient noise removal, by either filtering or smoothing, is a key part of any preprocessing of the data to be subjected to most chemometric methods. There is more to be gained from noise removal than just cosmetic improvement in the data. Even though many chemometric methods themselves produce a degree of noise reduction, the presence of significant amounts of noise can frustrate the application of those mathematical methods that make use of variations in peak amplitude and shape, such as principal component analysis and pattern classification methods. Large amounts of noise can also degrade the results obtained from calibration methods. Thus, noise removal by smoothing is often done prior to classification or calibration. In these cases, the noise reduction step is part of the preprocessing of the data.

10.2 NOISE REMOVAL AND THE PROBLEM OF PRIOR INFORMATION

Consider the noisy data given in Figure 10.1. Our task is to enhance the signal-to-noise ratio, if we can. To process these spectra, we must consider what is already known about the data. When a chemical measurement $r(t)$ is obtained, we presume that this measurement consists of a true signal $s(t)$ corrupted by noise $n(t)$. For simplicity, the linear additivity of signal and noise is usually assumed, as depicted by the measurement in Equation 10.1

$$r(t) = s(t) + n(t) \tag{10.1}$$

Signal Processing and Digital Filtering

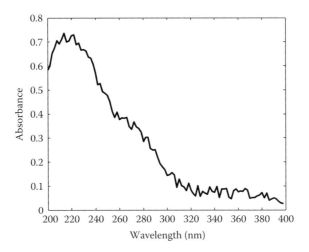

FIGURE 10.1 A noisy spectrum to be enhanced.

In this equation, the parameters r, s, and n are all given as functions of time, the independent variable, but the relation is equally true when r, s, and n are functions of other independent variables that define the channels of measurement, such as potential or wavelength, as in the example here. For the discussion here, equations will be given in terms of time, but it is understood that other independent variables could be substituted.

The goal of data enhancement is the extraction of the true signal $s(t)$, given the measured sequence $r(t)$. A simple measure of the success of this effort is the increase in the signal-to-noise ratio, S/N, where

$$\text{S/N} = (\text{maximum peak height})/(\text{root-mean-square of noise}) \qquad (10.2)$$

Generally this ratio is maximized by attenuating the noise term $n(t)$ and retaining the true signal $s(t)$.

10.2.1 SIGNAL ESTIMATION AND SIGNAL DETECTION

If it happens that $s(t)$, the true form of the signal, is known prior to data enhancement, the process of noise reduction is simplified greatly. In this case, a *known* signal is sought from within noise. This is a problem in signal *estimation*. The observed quantity r can be taken as a multiple of the known signal plus noise, so that

$$r(t) = as(t) + n(t) \qquad (10.3)$$

If the noise term is random, with zero mean, the multiplicative factor a is easily found by regression of $r(t)$ onto the signal model given by Equation 10.3. Simple application of least-squares regression gives

$$a = \sum_{t=1}^{N}(r(t)-\bar{r})(s(t)-\bar{s}) \Big/ \sum_{t=1}^{N}(r(t)-\bar{r})^2 \qquad (10.4)$$

The random noise is attenuated in the regression because of the least-squares averaging process. The signal-to-noise ratio can be increased substantially without danger of signal distortion in this case. In fact, once the multiplicative factor a is found, the noise-free version of the observed signal $r(t)$ is easily generated from Equation 10.3, taking $n(t)$ as zero.

If, on the other hand, the signal form $s(t)$ is not exactly known, it must be identified in the noise. This is an example of signal *detection*. In this case, a suitable model for the desired quantity is not available, and the separation of signal and noise is not as straightforward as with estimation. In attenuating noise, there is a danger of simultaneously altering the characteristics of the signal. Maximizing the S/N may produce an observed signal with a shape that is significantly distorted from that of the true signal $s(t)$, while an effort aimed at minimizing distortion in the observed signal may improve the observed S/N only a little. A decision must be made as to the relative importance of noise reduction vs. inaccuracy in the signal shape or amplitude resulting from signal distortion. This decision is an example of the variance-bias trade-off that is a key factor throughout the fields of data analysis. Here, the variance reduction associated with noise attenuation must be balanced with the bias in our results for the estimated signal that results from distortion or corruption of the true signal in the data processing. Thus, the previous definition of the S/N ratio as a measure of the success of data enhancement must be qualified by the degree to which the true signal $s(t)$ is distorted in the enhancement process, which might be reflected in a sum of squared error of signal estimation, for example. The root mean squared error of estimation (RMSEE) can be used to describe the contributions of noise variance and signal bias to the result,

$$\text{RMSEE} = \sqrt{\left(\sum_{t=1}^{N}\left(r(t)-\hat{s}(t)\right)^2\right)\Big/N-1} \qquad (10.5)$$

where there are N channels in the data and $\hat{s}(t)$ describes the estimated signal.

The reader will note that the RMSEE is closely related to the RMSEP term used in calibration. Martens and Naes have discussed the significance of error estimates in depth in their book [1].

10.3 REEXPRESSING DATA IN ALTERNATE BASES TO ANALYZE STRUCTURE

To examine the distortion of signal and the reduction of the noise, it is necessary to consider the "structure" — that is, the amount, location, and nature — of signal and noise in the observed responses. Each channel used to measure the data's response can be regarded mathematically as an axis, and because of the channel-to channel similarity (or correlation) in the data's response, these axes are correlated. Thus, an ultraviolet spectrum measured at 100 wavelength channels can be regarded as a 100-point vector or, equivalently, as a point in a 100-dimensional space. Because our goal is to understand the signal and noise content of the spectral response, the fact that many of the measurement axes are correlated means that we cannot look at just a few to get at most of the signal and noise information contained in the spectral response. It is far more useful to express the response data in another way, one in which each of the axes expresses *independent* information about signal and noise. To meet our goal that each axis express independent information, the axes we use to reexpress our data must be independent themselves, and so we need an orthogonal axis system to describe our response data.

The new axis system we choose for reexpression of our response data depends on what we want to examine in the data set. For example, we could choose to focus our attention on the way in which the data varied by decomposing the set of time-domain data according to its component sources of variance using principal component analysis. Or, we could examine the various frequency components contained in the data by decomposing the data according to its component frequencies using the Fourier transform. It is also possible to examine the data set according to localized frequencies by using a wavelet transform. Each of these methods for reexpression of a data set converts the original correlated set of time- (or wavelength-) axis data into an equivalent set of data expressed in a new *basis*, an axis system with orthogonal axes, allowing us to express independent sources of information in each dimension of the reexpressed data. The reexpressed data contain the same signal and noise as in the original, time-domain representation, but present that information in a different way because of the changed axis system. Thus, part of signal processing involves finding a way to express the data so that the maximum separation of the signal and noise components can be achieved, and then using mathematical tools to perform the removal of noise while retaining signal.

10.3.1 PROJECTION-BASED SIGNAL ANALYSIS AS SIGNAL PROCESSING

The most common way to reexpress a data set in chemometrics makes use of the principal components of the data. Here, the data are expressed in terms of the components of the variance-covariance matrix of the data. To get a variance-covariance matrix, we need not one spectrum, as shown in Figure 10.1, but a set of similar spectra for which the same underlying effects are operative. That is, we must have the same true signal and the same noise effects. Neither the exact contributions of signal nor the noise need be identical from spectrum to spectrum, but the same basic effects should be present over the set of data if we are to use variance-covariance matrices to discern how to retain signal and attenuate noise.

Even though principal component analysis (PCA) of data might not normally be considered a form of signal processing, it is, although it operates on a set of data rather than on an individual spectrum. In PCA, a decision is made to regard the signal component as the major source of variation in the responses, and the noise component as a minor source. The truncation of the minor principal components of the reexpressed data amounts to a *smoothing operation* carried out by projection of the full data set into a subspace with lessened contributions from information judged to be noise. We can evaluate the success of the smoothing operation by any of a number of measures, as discussed by Malinowski [2]. The smoothing in PCA is also a data-compression step aimed at retaining information and reducing the dimension of the data, recognizing that there will usually be far fewer scores than measurement channels.

Simple truncation of some unimportant principal component axes describing the data set is not the only way to alter the nature of signal and noise components in the responses, however. Two other methods use the idea of projection to alter signal and noise contributions in a signal.

Let us start with the simpler method. If the signal vector **s** is known, we can use this knowledge to project out the signal component of the response while removing all else (noise). It is instructive to see how that is done. Suppose that we have a ($n \times m$) matrix of responses **R** such that each row (an $n \times 1$ spectrum) of **R** is described by

$$\mathbf{r}_i = \mathbf{Mc}_i + \mathbf{n} \tag{10.6}$$

where **M** is a ($n \times j$) matrix of j spectral signatures, **c** is a ($1 \times j$) concentration vector describing the amount of each component of **M**, and **n** is a ($n \times 1$) vector of zero-mean, white noise. Then we can rewrite Equation 10.6 as

$$\mathbf{r}_i = \mathbf{sc}_s + \mathbf{Uc}_u + \mathbf{n} \tag{10.7}$$

where **s** is the signal vector, **U** is a ($n \times j-1$) matrix of spectral signatures excluding **s**, $\mathbf{c}_u$ is the concentration vector corresponding to the spectral signatures in **U**, and $\mathbf{c}_s$ is the amount of component **s**. If we define $\mathbf{U}^{\#}$ as the pseudo-inverse of the nonsquare matrix **U**, namely $(\mathbf{U}^T\mathbf{U})^{-1}\mathbf{U}^T$, and **I** as the ($n \times n$) identity matrix, we can use the projection operator $\mathbf{P}_\perp = (\mathbf{I} - \mathbf{U}\mathbf{U}^{\#})$ to remove **U** from the response **r**, leaving

$$\mathbf{P}_\perp \mathbf{r}_i = \mathbf{P}_\perp \mathbf{sc}_s + \mathbf{P}_\perp \mathbf{n} \tag{10.8}$$

an equation defining the detection of a known signal **s** in the presence of white noise **n**. Equation 10.8, like Equation 10.3, is easily solved by least-squares regression using the known signal **s** and the projected response $\mathbf{P}_\perp \mathbf{r}_i$ to obtain $\mathbf{c}_s$. Note that this signal-processing method, called orthogonal subspace projection (OSP) [3], attenuates the white-noise component and isolates the desired signal from any unspecified, but measured, noise component in **U** through projection of the measured response **r** to a basis where the spectral components of **U** have no contribution. A similar idea has been used to extract the net analyte signal in calibration in both conventional [4] and inverse regression modeling [5].

If our goal is to do signal processing in support of a subsequent calibration (Chapter 6), we might take the idea of signal processing through projection a step farther, and project our response data to a space where any of the systematic variation in the response that is unrelated to the target property is annihilated. That is the basic idea behind the methodology known as orthogonal signal correction (OSC) [6–9].

FEARN'S ALGORITHM FOR OSC SMOOTHING [6]:

1. Obtain a matrix **M** containing a majority of the variation in **R** *not* associated with the property **Y**.
   $$M = 1 - R^T Y (Y^T R R^T Y)^{-1} Y^T R$$
2. Premultiply **M** by **R** to yield the matrix **Z**, such that ZZ^T is symmetric.
   $$Z = RM$$
3. Calculate the first principal component of ZZ^T and compute the normalized weight vector **w** from the PCA loading vector **p** (the first eigenvector of ZZ^T with eigenvalue λ).
   $$w = (\lambda)^{-1/2} M R^T p$$
4. Calculate the score vector from **R** and **w**.
   $$t = Rw$$
5. Orthogonalize **t** to **Y** and compute a new loading vector **p***. This defines a basis for the part of the response **R** that is orthogonal to the property **Y**.
   $$t^* = t - Y(Y^T Y)^{-1} Y^T t$$
   $$p^* = R^T t^* / (t^{*T} t^*)$$
6. Compute the residual matrix **E** by subtracting the orthogonal component defined by **t*** and **p*** from **R**.
   $$E = R - t^* p^{*T}$$

To remove another OSC component, use **E** as **R** and repeat steps 1 to 6. New test samples can be corrected with the regression matrix **W** and loading matrix **P** of the OSC-corrected data.
$$R^*_{test} = R_{test} - R_{test} W (P^T W)^{-1} P^T$$

The performance of algorithms that implement OSC varies a good deal [7, 8]. The reader will note that the Fearn algorithm uses a PCA rather than a partial least squares (PLS) calculation to set the direction of the orthogonal components, a difference that makes the variation removed by this version of OSC not completely orthogonal to the property. Once again, the variance-bias trade-off is at work; this version of OSC slightly emphasizes variance removal over bias reduction. The assumption underlying the signal processing here is that most of the variation in the response is produced by desired signal and is therefore relevant to the property. If so, the first PCA loading will closely resemble the first PLS loading for these data.

10.4 FREQUENCY-DOMAIN SIGNAL PROCESSING

On occasion, the set of signals needed for modeling on the basis of data variance and covariance is not available or, perhaps, the interest is focused on a single signal, for which covariance information is lacking. One especially convenient way of describing the structure of a signal makes use of Fourier series. According to the theory of Fourier series, any signal can be described in terms of a weighted sum of sine and cosine functions of different frequencies [10]. Finding what frequencies are present in the Fourier series description of an observation can provide information on the signal and noise components of the noisy observations.

10.4.1 THE FOURIER TRANSFORM

Any continuous sequence of data $h(t)$ in the time domain can also be described as a continuous sequence in the frequency domain, where the sequence is specified by giving its amplitude as a function of frequency, $H(f)$. For a real sequence $h(t)$ (the case for any physical process), $H(f)$ is series of complex numbers. It is useful to regard $h(t)$ and $H(f)$ as two representations of the *same sequence*, with $h(t)$ representing the sequence in the time domain and $H(f)$ representing the sequence in the frequency domain. These two representations are called transform pairs. The frequency and time domains are related through the Fourier transform equations

$$H(f) = \int_{-\infty}^{+\infty} h(t) e^{2\pi i f t} dt \qquad (10.9a)$$

$$h(t) = \int_{-\infty}^{+\infty} H(f) e^{-2\pi i f t} df \qquad (10.9b)$$

The notation employed here for the Fourier transform follows the "system 1" convention of Bracewell [10]. Equations given in this chapter will follow this convention, since it requires fewer 2π factors in some equations. In some texts, Equation 10.9 is given in terms of the angular frequency ω, where $\omega = 2\pi f$ (the "system 2" convention of Bracewell), so that any comparisons between equations given here and those given elsewhere should take differences in the notational conventions into account. The units of the transform pair also deserve comment. If $h(t)$ is given in seconds, $H(f)$ has units of seconds^{-1} (the usual units of frequency), but, as noted above, $h(t)$ may be in any other units: if $h(t)$ is given in units of volts, $H(f)$ will be in units of volts^{-1}.

10.4.2 THE SAMPLING THEOREM AND ALIASING

The continuous, infinite Fourier transform defined in Equation 10.9, unfortunately, is not convenient for signal detection and estimation. Most physically significant data are recorded only at a fixed set of evenly spaced intervals in time and not between these times. In such situations, the continuous sequence $h(t)$ is approximated by the discrete sequence hn

$$hn = h(n\delta), \text{ with } n = \ldots, -2, -1, 0, 1, 2, \ldots \qquad (10.10)$$

where δ is the sampling interval. Associated with the sampling interval δ is a frequency called the *Nyquist frequency*, fc, which is defined by the relation

$$fc = 1/2\delta \qquad (10.11)$$

The Nyquist frequency is the maximum frequency that can be present in the continuous sequence $h(t)$, if $h(t)$ is to be perfectly represented by the sampled

Signal Processing and Digital Filtering

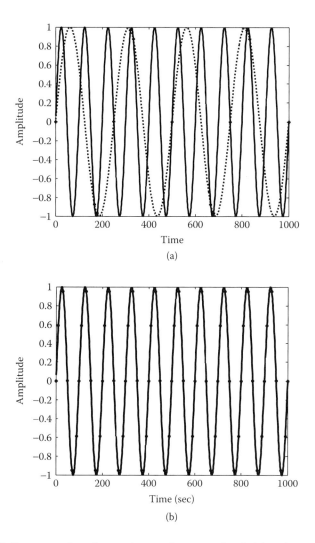

FIGURE 10.2 Reconstruction of a continuous sine-wave signal: (a) undersampled (4/10×) continuous curve (solid line) with the aliased curve reconstructed from the samples (dotted line); (b) adequately sampled (4×) continuous curve.

sequence hn. In essence, at least two samples must be taken per cycle for a sine wave to reproduce that sine wave. A typical reconstruction is shown in Figure 10.2.

As demonstrated for a simple sine wave, if the sampling (the dotted sequence in Figure 10.2a and Figure 10.2b) is done with at least twice the frequency of the highest frequency present in the sampled wave, perfect reconstruction of the continuous sequence from the discrete sequence of samples (the starred points) is possible. This remarkable fact is known as the *sampling theorem*.

If it happens that the continuous sequence $h(t)$ contains only a finite number of frequencies, that sequence is said to be bandwidth limited. This term is often shortened to the equivalent term "bandlimited." The sampling theorem states that a signal must be bandwidth limited to frequencies less than *fc* for all signal information to be preserved through the sampling process. If signal information is lost, distortion of the signal will result.

When a sequence that contains frequencies above *fc* is sampled at interval δ, a phenomenon called *aliasing* occurs. In aliasing, the sampling process falsely assigns all of the power from frequencies outside the frequency range $-fc < f < fc$ into that range. To avoid aliasing effects, it is necessary to know the natural (true) bandwidth of a signal, prior to sampling, or to limit the signal bandwidth to a preset value by analog filtering prior to the sampling step. The sampling must then be carried out at a rate so that the highest frequency component of the bandlimited signal is sampled at least twice per cycle. Thus, aliasing limits the number of frequencies that can be examined when the time-domain signal is transformed into the Fourier domain. The effect of aliasing is seen in the top of Figure 10.2, where undersampling of the sin $10\pi t$ wave produces an aliased response with apparent frequency of $4\pi t$. The power from the $10\pi t$ frequency has been falsely assigned to a lower frequency because sampling was too slow; only frequencies less than $8\pi t$ could be properly represented. When the sampling rate is increased to $20\pi t$, full sampling is possible on the $10\pi t$ frequency, and the sine wave is correctly reproduced, as shown in Figure 10.2b.

10.4.3 THE BANDWIDTH-LIMITED, DISCRETE FOURIER TRANSFORM

The matter of sampling and limited representation of frequencies requires a second look at the representation of data in the time and frequency domains, as well as the transformation between those domains. Specifically, we need to consider the Fourier transformation of bandwidth-limited, finite sequences of data so that the S/N enhancement and signal distortion of physically significant data can be explored. We begin with an evaluation of the effect of sampling, and the sampling theorem, on the range of frequencies at our disposal for some set of time-domain data.

Suppose a sequence of samples is taken on some continuous sequence $h(t)$, so that

$$h_k = h(t_k) \text{ and } t_k = k\delta \quad (10.12)$$

where k is some integer from 0 to $N-1$ and δ is the sampling interval. As we saw above, the Nyquist criterion limits the range of frequencies available from the Fourier transform of h_k. With N time samples input, only N frequencies will be available, namely

$$fn = n/(N\delta) \text{ with } n = -N/2, \ldots, 0, \ldots, N/2 \quad (10.13)$$

By convention, this range of frequencies, where n varies from $-N/2$ to $N/2$, is mapped to an equivalent set running from 0 to N. Because of this mapping, zero frequency now corresponds to $n = 0$, positive frequencies from $0 < f < fc$ correspond to values $1 \leq n \leq N/2 - 1$, and negative frequencies from $-fc < f < 0$ correspond to values $N/2 + 1 \leq n \leq N - 1$. The alert reader will note that in the Nyquist mapping of frequencies, the value $n = N/2$ corresponds to both $-fc$ and fc; in fact, sampled sequences are *cyclic*, being joined at the extreme ends. This is why, no matter what mapping is used, only N frequencies are available from N samples, and $N/2 - 1$ of these frequencies will be negative frequencies.

The integral transforms given in Equation 10.9 can now be approximated by discrete sums, so that the Fourier transform pairs now are described by the equations

$$Hn = \sum_{k=0}^{N-1} h_k e^{2\pi i k n/N} \qquad (10.14a)$$

$$hk = 1/N \sum_{n=0}^{N-1} H_n e^{-2\pi i k n/N} \qquad (10.14b)$$

Calculation of these discrete Fourier transforms (DFT) is not at all rapid if done directly. If, on the other hand, the data are sampled so that N is a multiple of 2, a fast Fourier transform (FFT) algorithm can be used to perform the above transform operations, permitting a significant savings in time. For this reason, the FFT algorithm is most often used to accomplish the DFT process. A number of FFT algorithms now exist; two popular ones are the Cooley-Tukey transform and the related Sande-Tukey transform. The mechanistic details of these transforms are covered in many textbooks. Good discussions of FFT algorithms are provided by Bracewell [10].

Figure 10.3 shows the plot of Hn for the spectral data given in Figure 10.1. Note that the units of the abscissa are in wavelength^{-1}, the units of "frequency" for this member of the transform pair. Note also the negative frequencies, which give the "mirror image" effect to the transform.

Many FFT algorithms store data so that the frequencies run from 0 to the largest positive value, then from the largest negative value to the smallest negative value, as this improves calculation speed. In these, a "mirror image" effect is noticeable when the frequency domain data are plotted, but the plot differs from those shown in Figure 10.3: now the left half of the plots shown in Figure 10.3 is appended to the right half so that the plot starts and ends at frequency 0. Recalling the cyclic nature of sampled sequences will help the reader in realizing that both ways of describing the frequency sequences are equivalent. It is also common that the mirror image of the transformed data (the part with negative frequencies) is removed for plotting or for additional data analysis. To keep the proper intensity information, the positive frequencies (but never zero frequency) are multiplied by 2 when negative frequencies are deleted.

10.4.4 PROPERTIES OF THE FOURIER TRANSFORM

The interconversion of transform pairs by the Fourier transformation has several unique properties that will be important in analyzing the structure of signals and noise. For the transform pairs $h(t)$ and $H(f)$, important relations are summarized in Table 10.1. The reader should take note that, even though continuous functions are used to illustrate the properties of the transformation, all of the properties listed in Table 10.1 are identical for the continuous and discrete Fourier transforms.

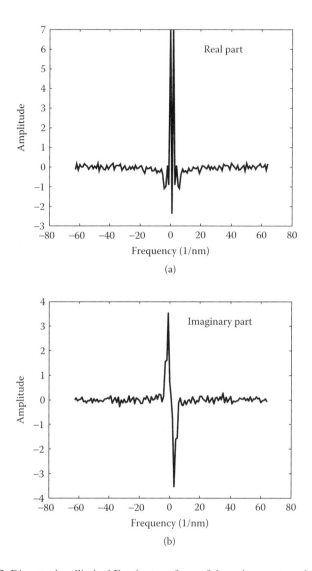

FIGURE 10.3 Discrete, bandlimited Fourier transform of the noisy spectrum in Figure 10.1.

The first property listed in Table 10.1, called the similarity property, concerns the scaling of the transform pairs. Because the transform pairs describe the same function, the scaling described by the similarity property can refer to "time scaling," as given in Table 10.1, or it can describe "frequency scaling," namely

$$\{1/|b|\}\ h(t/b) <> H(bf) \tag{10.15}$$

Similarly, the shift property can describe "time shifting," as given in Table 10.1, or it can describe "frequency shifting," where

$$h(t)\ e - 2\pi i f_0 t <> H(f - f_0) \tag{10.16}$$

TABLE 10.1
Some Properties of the Fourier Transform

Property	Time Domain {h(t)}	Frequency Domain {H(f)}
Similarity	$x(at)$	$1/|a| \, X(f/a)$
Translation (shift)	$x(t-a)$	$e^{-2\pi i a f} X(f)$
Derivative	$x'(t)$	$2\pi i f X(f)$
Symmetry	$x(-t)$	$X(-f) = X(N-f)$
Complex conjugate	$x^*(t)$	$X^*(-f)$
Power (Parseval's theorem)	$\sum_{t=0}^{N-1} x(t) * x^*(t)$	$\sum_{t=0}^{N-1} X(f) * X^*(f)$
Linearity	$a[x(t)] + b[y(t)]$	$a[X(f)] + b[Y(f)]$
Convolution	$x(t) * y(t)$	$X(f)Y(f)$
Autocorrelation	$x(t) * x^*(t-a)$	$X^*(f)X(f) = |X(f)|^2$
Correlation	$x^*(t) * y(t-a)$	$X^*(f)Y(f)$

Note: In this table, a and b are arbitrary constants. Also, the symbol "*" used between variables denotes a convolution operation, such that the notation $a*b = \sum_{k=1}^{N} a(k)b(r-k)$. When used as a superscript, the symbol "*" denotes complex conjugation, such that $(a + ib)^* = (a - ib)$.

Parseval's theorem demonstrates that the total power is the same whether it is computed in the time domain or the frequency domain. It is often of interest to determine the power in some small interval $(f + df)$, as a function of the frequency. This is called the *spectrum* of the signal $h(t)$.

For the pairs of functions $\{x(t), y(t)\}$ and $\{X(f), Y(f)\}$, which are related by the Fourier transform, other unique properties exist. These are listed in the lower half of Table 10.1 and are briefly summarized below.

The addition property demonstrates that the sum of functions is preserved in the transformation. This, together with the similarity property, indicates that the Fourier transformation is a linear operation. This means that the statistical moments of the signal (e.g., mean, variance, skewness, etc.) are preserved across the transform.

One of the most useful relations in signal processing is the convolution property. The convolution of two discrete (sampled) functions $x(t)$ and $y(t)$ is defined as

$$x(t) * y(t) = \sum_{k=0}^{N-1} x(k) y(t-k) \tag{10.17}$$

where $x(t) * y(t)$ is a function in the time domain, and where $x*y = y*x$. According to the convolution property, the convolution of two functions has a transform pair that is just the product of the individual Fourier transforms. A property related to convolution is cross correlation. The cross correlation of two functions $x(t)$ and $y(t)$, sampled over N points, is defined as $\gamma_{xy}(\tau)$, where

$$\gamma xy(\tau) = \text{corr}(x,y) = \sum_{k=0}^{N-1} x^*(ky(k+\tau)) \tag{10.18}$$

and where τ is called the lag. This function is related to its transform pair by the correlation property. The cross correlation of x and y has a transform pair that is just the product of the complex conjugate of the individual Fourier transform of one function multiplied by the Fourier transform of the second function. A special case of the correlation property arises when a function is correlated with itself, an operation called autocorrelation. In this case corr (x, x) has the transform pair $|X(f)|^2$.

As already noted, the properties of convolution and correlation are the same, whether or not a continuous or discrete transformation is used, but because of the cyclic nature of sampled sequences discussed previously, the mechanics of calculating correlation and convolution of functions are somewhat different. The discrete convolution property is applied to a periodic signal s and a finite, but periodic, sequence r. The period of s is N, so that s is completely determined by the N samples $s_0, s_1, \ldots, s_N$. The duration of the finite sequence r is assumed to be the same as the period of the data: N samples. Then, the convolution of s and r is

$$s*r = \sum_{k=-N/2+1}^{N/2} s_{j-k} r_k = SnRn \tag{10.19}$$

where Sn is the discrete Fourier transform of the sequence s, and Rn is the discrete Fourier transform of the sequence r. The sequence r_k maps the input s to the convolved output channel k. For example, r_1 tells the amount of signal in sample j that is placed in output channel $j + 1$, while r_3 maps input signals in sample j to output channels $j + 3$, and so forth. Figure 10.4 demonstrates the convolution process.

Note that in Figure 10.4, the response function for negative times (r_{-1} and r_{-2} here) is wrapped around and is placed at the extreme right end of the convolving sequence. To avoid an effect called "convolution leakage," it is important to keep in mind the requirements for convolution, namely that sequences s and r have the same length and that both are treated as periodic. The latter requirement means that the convolution operation will be cyclic, and we must take care to protect the ends of sequence s. To do so, we must ensure that the first and last k elements of sequence s are zero, where k is the *larger* of the number of nonzero, negative elements of sequence r, or the number of nonzero, positive elements of sequence r. This "padding" of the sequence to be convolved avoids convolution leakage (which occurs in the padded areas, but these are of no interest to us and can be discarded). Further discussion of the details of discrete convolution is given in the texts by Bracewell [10], by Press et al. [11], and by Bose [12].

Signal Processing and Digital Filtering

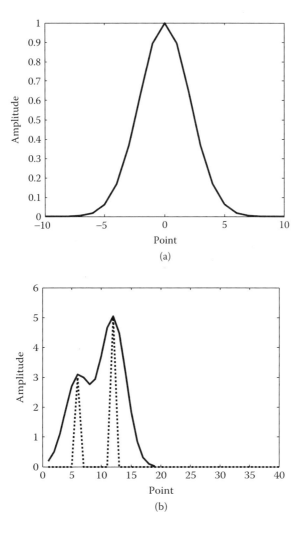

FIGURE 10.4 Discrete convolution of two functions: (a) Gaussian broadening function; (b) true signal (dotted line) and broadened result (solid line) of convolution with the Gaussian function.

With a background in the mechanics and the properties of the Fourier transform, we are now ready to investigate ways of enhancing signal and removing noise while minimizing signal distortion. Two approaches are possible. We could deal with the signal directly, in the time domain; or we could take advantage of the simplicity of some operations in the frequency domain in processing the signal there, after a suitable Fourier transformation. Both methods are commonly used in processing of chemical data. Each is discussed separately in Sections 10.5 and 10.6.

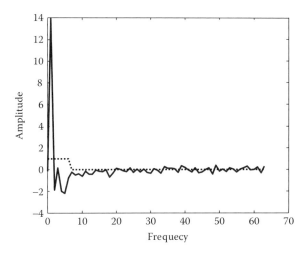

FIGURE 10.5 Window function (dotted line) and the real part of the Fourier transform. The window is applied to both real and imaginary parts of the transformed data after removal of negative frequencies and scaling by 2. The location of the window cutoff is indicated at point 7.

10.5 FREQUENCY DOMAIN SMOOTHING

10.5.1 SMOOTHING

Consider again the noisy time-domain signal given in Figure 10.1. For the examples discussed below, the true signal $s(t)$ is unknown. We would like to improve the S/N of this signal. In the time domain, noise and signal seem to coexist, but in the frequency domain, the signal seems to exist as a set of low frequencies, and the noise as a set of high frequencies, as shown in Figure 10.3. To enhance the signal, we need only to decrease the amplitude of the noise-containing frequencies while leaving the signal-containing frequencies unchanged.

Figure 10.5 shows one way to reduce noise without altering the signal significantly. If the frequency-domain representation of the noisy signal is multiplied by a "window" function, the resulting frequency-domain representation has zero amplitude at frequencies where the window function is 0, and unchanged amplitude at frequencies where the window function is 1. Picking the window transition point (where the function changes from 1 to 0) to be in the region between signal frequencies and noise frequencies and then multiplying the noisy frequency-domain representation by this window function will result in an enhanced frequency-domain representation, which can then be converted back to a time-domain representation, as shown in Figure 10.6. The enhanced signal has considerably less noise than was contained in the original signal. This signal has been *smoothed* by the application of the window function in the frequency domain. The window function is called the *transfer function* of the smoother.

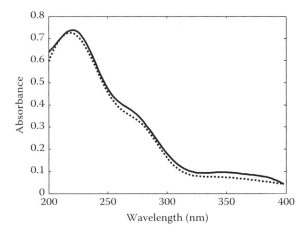

FIGURE 10.6 Smoothed data after inverse Fourier transformation. The rectangular smoother shown in Figure 10.5 was used.

A very important consideration in this method is the location of the *cutoff frequency* for the transfer function. This point is the frequency where the transfer function changes from a value of nearly 1 to a value of nearly 0; frequencies above the cutoff frequency are mainly attenuated, while frequencies below this point are mainly passed through the filtering operation. In the example presented here, it was easy to see where the cutoff frequency should be placed, because signal and noise were well separated in the frequency domain. This convenient separation of signal and noise in the frequency domain is not always true, however. Noise and signal often overlap in the frequency domain as well as in the time domain.

10.5.2 SMOOTHING WITH DESIGNER TRANSFER FUNCTIONS

When significant overlap of noise and signal occurs in the frequency domain, the successful enhancement of the data depends on the selection of a suitable smoother transfer function. It is generally not satisfactory to zero part of the frequency domain representation while leaving other parts intact, as was done in the previous example. Usually, this smoothing operation will distort the signal. One reason that this smoothing method fails to produce a smoothed signal close to the true signal can be seen from the properties of the impulse response function used in the time-domain convolution. As shown in Figure 10.7, the transform pair for a step-shaped transfer function produces an impulse function of the form $\sin(x)/x$. When convolved with a noisy signal, this function produces spurious oscillations in the smoothed data. The oscillations, or "feet" on a peak, arise from the abrupt change made in a step-shaped transfer function [13–16].

A "designer" transfer function can be crafted to obtain a specific type of smoothing, one that is, for example, maximally flat over a range of frequencies or with a steep transition from the passband to the stop band. Some common "designer" transfer

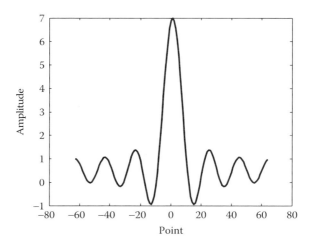

FIGURE 10.7 Impulse response function of the seven-point rectangular smoother window function used in Figure 10.6. Note that the Fourier transform of a step function has the form $\sin(x)/x$.

functions are based on Chebyshev polynomials, elliptic functions, and the Butterworth filter equation [12, 16]. These particular transfer functions were originally developed for use in hard-wired circuits but are now conveniently implemented digitally. A designer transfer function can be made empirically. For example, a trapezoidal function might be devised to smooth a noisy spectrum, as shown in Figure 10.8. A smoother with this transfer function smoothes data in a way very

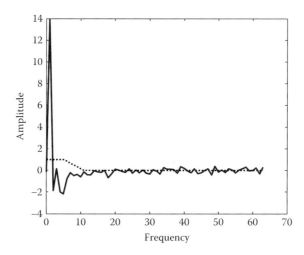

FIGURE 10.8 Smoothing in the frequency domain with a trapezoidal smoothing function. Here a trapezoid was used with vertices at frequency points 0, 6, and 12.

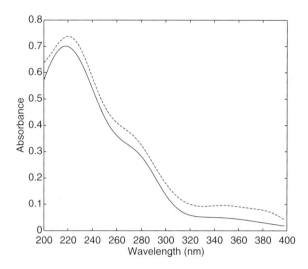

FIGURE 10.9 Smoothed data from application of the trapezoidal smoother shown in Figure 10.8. The true, noise-free signal is shown as a dotted line.

similar to that done by the step-shaped, rectangular transfer function, but it reduces the slight, spurious oscillations on the peaks in the smoothed data, as demonstrated in Figure 10.9. Although the oscillations are quite small in this example, and the use of a trapezoidal window is not essential, often these oscillations can be much larger, particularly when the cutoff point is selected in a region where the frequency-domain representation has nonzero values. Altering the transfer function in this way is sometimes called *apodization*, because it removes spurious "feet" from the smoothed data [14, 15].

If the true signal shape is known before smoothing, a special designer transfer function can be created. In this case, the transfer function can be the complex conjugate of the frequency response of the signal itself. The result is a *matched smoother*, because the transfer function is matched to the characteristics of the signal. When the complex conjugate of the Fourier transform of the desired signal shape is used as the transfer function, the smoother operation is equivalent to a time-domain cross correlation of the noisy signal with the true, noise-free signal, which, at lag 0, reduces the noise without altering signal shape [17]. This smoother permits recovery of the desired (known) signal from large amounts of white noise, just as in the regression examples given previously. In fact, the correlation operation is equivalent to the regression method outlined in Equation 10.3 and again in Equation 10.5. In fact, matched smoothing is a least-squares optimal result, because that operation amounts to regressing the unknown signal onto the known signal, as discussed previously. Figure 10.10 demonstrates the noise removal that is obtained with matched smoothers. Note, however, that a *convolution* of the noise-free and noisy signals would not be useful, since the convolution operation alters the shape of the smoothed signal.

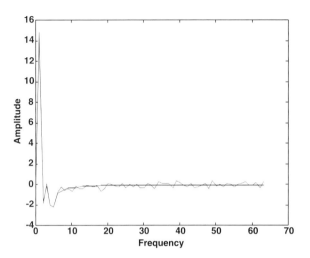

FIGURE 10.10 The frequency response of the noisy signal and that of its corresponding, matched smoother. The real part of the transfer function of the matched smoother for the noisy spectrum (solid line) is shown dotted.

10.6 TIME-DOMAIN FILTERING AND SMOOTHING

10.6.1 SMOOTHING

It is possible to smooth data in the time domain. Generally, when the signal shape is known, signal smoothing is done by a regression step, where the known signal shape is regressed on the noisy signal. This is matched smoothing, and it is generally done in the time domain for convenience. In this case, the whole noisy signal is processed

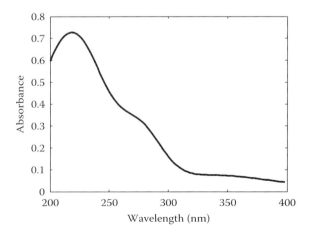

FIGURE 10.11 Results from matched smoothing of the noisy signal in Figure 10.1. The true signal is shown dotted. The fit is good because the noisy signal is zero-mean with constant noise variance.

in the single regression step, and any zero-mean noise effects are significantly attenuated by the least-squares averaging. Nonzero mean effects, such as baseline and interfering, unknown components, are not removed by simple matched smoothing. In these cases, an alternative signal-processing method must be employed.

When the signal shape is not known in advance, or when there are interfering components that are not of known shape, a convenient way to remove some of the noise is to convolve a smoother function with the noisy data. Multiplication of two functions in the Fourier domain is, in terms of theory, exactly equivalent to convolution of their transforms in the time domain. Generally, it is far simpler to do the smoothing step in the frequency domain, but occasionally, there is a need to have immediate access to the smoothed, time-domain data for additional processing. Thus, rather than using a simple multiplication of two frequency-domain representations, we will need to convolve two time-domain representations. One time-domain representation is simple: it is the noisy, time-domain signal, for example, the signal given in Figure 10.1. The other, according to the properties of the Fourier transform, should be the time-domain representation of the smoother, which is known as the smoother *impulse response function*. Here, for purpose of illustration, we might try time-domain smoothing with the rectangular transfer function shown in Figure 10.5. The impulse function of this smoother is shown in Figure 10.7. The discrete convolution of the noisy signal with the time-domain impulse response function, defined by Equation 10.20

$$s(t) = h(t)*x(t) \qquad (10.20)$$

is shown in Figure 10.12.

Comparison of Figure 10.6 and Figure 10.12 indicates that there is a significant difference between the two theoretically equivalent approaches. They differ for two

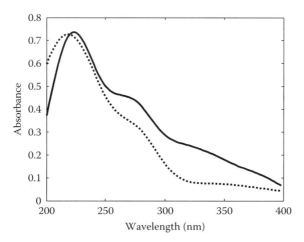

FIGURE 10.12 Time-domain smoothing of the noisy data in Figure 10.1 with the impulse response function of Figure 10.7, processed from left to right in this spectrum. The true signal is shown as a dotted line. Note the significant filter lag in this example.

reasons. First, the theory that connects convolution and multiplication presumes continuous, infinite signals with perfect Fourier transforms, and does not consider practical aspects of using the DFT for the transform, where round-off and other effects can lead to changes in the transformed result as compared with the convolved signal. Usually, however, the differences in results arising from the use of the DFT to perform transforms are small. Second, and more important, is the fact that, in the frequency domain, all of the signal is processed at once, but in the time domain, the convolution operation ensures that the noisy signal is processed *serially*: the convolution operation involved with the smoothing occurs from left to right or from right to left. The fact that the signal processing obtained by convolution of the impulse function and the noisy signal is not simultaneous in the time domain leads to *filter lag*, which often manifests itself as a smearing of signal. An example of filter lag is seen in Figure 10.12. Filter lag is a common effect in filtering of data by a time-domain impulse function. One way to reduce the filter lag is to convolute the data with the smoother first in one direction, then repeat the convolution of the smoother and the output signal, but this time in the opposite direction. This process takes time and is generally done off-line, but it produces data with relatively little distortion from lag effects as compared with a single filtering pass.

Because multiplication is simpler to perform than the convolution operation and because frequency-domain smoothing avoids filter lag effects, it is easier to perform this smoothing operation in the frequency domain. Having a fast transform between the two domains is an important consideration, because smoothing a time-domain signal will require two transformations: one to convert the time-domain signal to the frequency domain, and one to convert the results of the smoothing back to the time domain. Use of the FFT or another fast transform algorithm generally makes this consideration unimportant, except in real-time applications, where very high data throughput is essential. The small round-off error generated by the transforms is usually not an important impediment to the analysis.

10.6.2 FILTERING

When very rapid enhancement of data is desired, it might be necessary to enhance the data as they are collected. For this reason, the data-enhancement step is carried out in the time domain, both to avoid spending the time needed for the two transformation processes and to avoid any delays due to the need for complete observations in any data-enhancement method based on smoothing. As discussed previously, the removal of noise will be done by a convolution of an impulse response function with the noisy data, a serial process that is suited to the real-time improvement of noisy data. Real-time enhancement of incomplete data is called *filtering*, and it is always done in the time domain if the data are collected in the time domain. If the data were analyzed in the frequency domain, complete observations would be needed to prevent problems in representing all signal frequencies: the number of frequencies available cannot exceed the number of time-domain data. In time-domain filtering or smoothing, the basic operation is the cyclic convolution of the filter (or smoother) impulse-response function with the noisy data to generate a signal with attenuated noise characteristics. As discussed previously, the cyclic convolution operation is

Signal Processing and Digital Filtering

more commonly expressed as a discrete sum over the number of measurements available, N, so that the kth filtered datum is determined by the relation

$$s(k) = \sum_{n=1}^{N} h(n)x(n-k) \qquad (10.21)$$

where $h(n)$ is the filter impulse-response function defined over the set of N points, and where the input data are given by the sequence $x(n)$. This difference equation is the defining relation for a *finite impulse response* (FIR) *filter*, so named because the filter impulse-response function has finite (meaning noninfinite) values everywhere. With FIR filters, the present response $s(k)$ is not dependent on previous values of s [12, 18].

To discover the way in which FIR filters enhance data, consider the simple FIR filter given by the impulse-response equation

$$h(n) = 1/N \qquad (10.22)$$

This impulse response defines the simple moving-average filter, sometimes known as the "boxcar averager." As discussed previously, the convolution operation defined by Equation 10.21 can be regarded as a moving of the impulse-response function $h(n)$, which operates over a predetermined "window," through the data $s(n)$. Here the impulse response is a simple average of the values of the noisy signal $x(n)$ inside an N-point "window," as shown in Figure 10.13.

Averaging is a least-squares process that reduces the effects of noise, if the noise is zero-mean and fairly random [10], and the moving-average filter removes high-frequency noise well. It is less successful at removing low-frequency noise, since these nonzero-mean variations are less likely to be affected by the averaging. It also

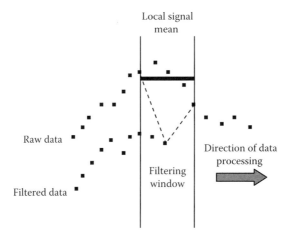

FIGURE 10.13 FIR signal processing with a simple moving-average filter.

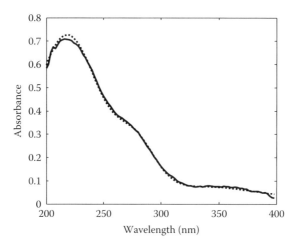

FIGURE 10.14 Time-domain processing of noisy data with a ten-point, moving-average filter. The moving-average-filtered data are indicated as a solid line. The true signal is shown as a dotted line.

is not able to filter the extreme "ends" of the data set, since a fixed-sized window of data is averaged to estimate a new value for the *middle point of the window*. Starting a five-point window in a spectrum means that the first two points cannot be improved; similarly, the last two points are never reached by the moving window. Typically, for a window of $2N + 1$ points, $2N$ points will be left unfiltered. These unfiltered points can be seen at the ends in Figure 10.14, where the operation of the simple ten-point, moving-average filter is demonstrated on the noisy data of Figure 10.1. This fast operation, which is easily done in real time, has effectively removed the high-frequency noise, but some of the low-frequency noise remains, and there may be a small amount of filter lag.

The size of the filter window determines the number of points averaged by the filter. As one might expect, the number of data averaged has a profound effect on the noise reduction process. It also strongly affects the amount of distortion of the enhanced data due to filter lag effects. The effect of window size on noise reduction and on distortion is best seen through examination of the filter transfer function. Three of these functions, for different window sizes, are shown in Figure 10.15. Clearly, choice of window size drastically affects the range of frequency components that pass the filter. Larger windows pass a considerably smaller range of frequencies. The collapse of the filter transfer function as the filter window is increased has some disadvantages, too. A large window size may give a filter transfer function that does not pass frequencies associated with signal. When signal frequencies are lost, distortion results. It is possible to remove the signal entirely, leaving only background, with a large enough filter window. How large of a window can be used without significant distortion? The window size should not be larger than the *smallest* peak half-width present in the signal. This rule ensures that the narrowest peaks (with the highest component frequencies) will not be clipped off by the moving-average filter.

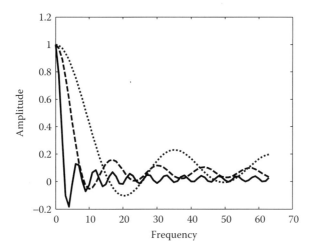

FIGURE 10.15 Transfer functions for FIR filters shown as a function of window size. Transfer functions for simple, moving-average filters with windows of 5 (…), 10 (- - -), and 25 (—) are shown.

The moving-average filter can be compared with the other smoothers discussed previously by transformation of the filter impulse function to the frequency domain, which produces the filter transfer functions shown in Figure 10.15. The general shape of the moving-average filter transfer function is similar to those "designer" transfer functions discussed previously, but with some important differences. One difference is the presence of negative regions in the moving-average filter transfer function. These regions will produce a filtered, frequency-domain representation with *negative* intensities at some frequencies after multiplication of the filter transfer function by the frequency-domain representation of the noisy signal. These negative intensities in the frequency domain lead to phase errors [13] in the filtered output, so that a decrease in the number of negative regions in the transfer function is desirable, if possible. There are also a number of humps in the transfer function, where poorly attenuated frequencies can "leak" through to the filtered output. Again, a decrease in the number of these positive areas is desirable, if possible.

10.6.3 POLYNOMIAL MOVING-AVERAGE (SAVITSKY-GOLAY) FILTERS

From Figure 10.13 and Figure 10.14, it is clear that a moving-average filter should provide significant noise reduction. However, signal distortion also occurs, as can be seen in Figure 10.14, where the smoothed signal obtained from application of a moving-average filter using a ten-point window and the true signal $s(t)$ are compared. The distortion produced by the filtering process is apparent. To reduce this distortion while retaining noise attenuation, simple polynomials can be fitted to data in a manner analogous to the moving-average filter discussed previously.

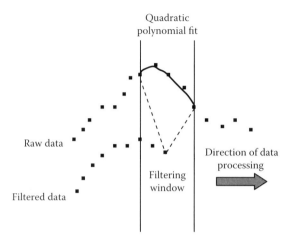

FIGURE 10.16 Polynomial least-squares filtering. A quadratic, five-point polynomial filter is shown.

Polynomial moving-average filtering involves fitting a polynomial to a set of noisy data within a window of points, as shown in Figure 10.16. Using higher-order polynomials should allow fitting of data that changes rapidly within the filter window. So long as the noise continues to change more rapidly than the data, good noise rejection will be possible. Any set of polynomials could be used to define a polynomial filter [19]. Selecting the set is analogous to choosing a frequency-domain designer transfer function. For example, Chebyshev polynomials can be used to filter data in the time domain. One especially convenient set of filter polynomials is based on power functions. Savitsky and Golay first introduced chemists to the use of moving-average filtering with orthogonal polynomials based on these power functions, and many texts in chemistry now refer to the method as "Savitsky-Golay" filtering. [20, 21–24]. As with other FIR filters, the polynomial filter function is convolved with the noisy time-domain data to produce an enhanced representation of the true signal. Generation of the filter polynomial from the power series is simple [24], and many authors provide listings of the polynomials [21–23]. It is worth noting that the matrix approach published in chemistry by Bialkowski [24] permits filtering of the ends of the signal as well, and its implementation is considerably easier than the direct convolution code originally published.

By selecting the order of the convolving polynomial, the user is able to alter the filtering process. With low-order polynomials, more noise is filtered, but at the cost of increased signal distortion. In fact, the moving-average filter is just a very low-order (order 1) polynomial moving-average filter. Increasing the window used in the filtering also decreases noise, as with the moving-average filter, but at the cost of significant signal distortion [13]. Filtering with low-order polynomial functions and large filter windows is called strong filtering, and it is used when obtaining high S/N is more important than preserving signal shape. When filtering is done with small filter windows and higher-order polynomials, less noise is removed because

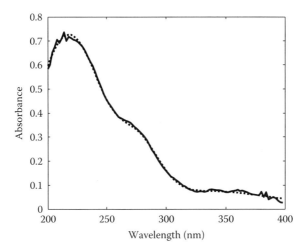

FIGURE 10.17 Strong filtering of noisy data using quadratic polynomials. Filtering was done with a five-point smoothing window. The true signal is shown as a dotted trace.

the higher-order polynomials will describe rapidly varying signal (and, of course, noise) better than low-order polynomials, but less signal distortion will be caused in filtering. Filtering with high-order polynomials is known as weak filtering, which is mainly of use when preservation of signal shape, not S/N ratio, is most important [25, 26]. Figure 10.17 and Figure 10.18 demonstrate the results of strong and weak filtering on the noisy signal shown in Figure 10.1.

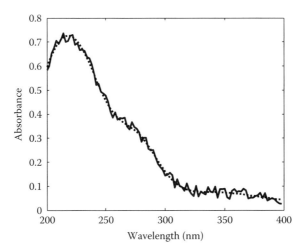

FIGURE 10.18 Weak filtering of noisy data using quadratic polynomials. Weak filtering was done with a 20-point smoothing window. The true signal is shown as a dotted trace.

10.7 WAVELET-BASED SIGNAL PROCESSING

The time-frequency relationships discussed previously for the removal of noise from responses have great power for the attenuation of those components containing specific frequencies, but they also have a serious drawback: in transforming time-domain data to the frequency domain, the time information on the signal is lost. A Fourier analysis cannot provide information on *when* a particular frequency event occurred. There is not much need to locate events in time when the signal shape is not changing with time (or with another independent variable) — that is, for stationary signals. However, when the observed response or the desired signal contains nonstationary components such as drift, abrupt changes, peaks, the beginning and ends of events, then a Fourier analysis is not well suited. In recognition of this limitation, Gabor [27] developed what he called the *short-term Fourier transform* (STFT), where only a portion of the response contained in a preset transform window at some center point b is subjected to Fourier analysis. This approach provides both time and frequency information by repeating the transform for different values of the window center, but the precision of that information is limited by the size of the transform window; and because the window is fixed, there is no ability to vary the window size to improve the precision of the analysis of either time or frequency.

Wavelet analysis takes Gabor's idea one step further: it defines a windowing transform technique with variably sized window regions. The continuous wavelet transform of the sequence h(t) is defined by Equation 10.23

$$H(f, b, s) = \int_{-\infty}^{+\infty} h(t) \sqrt{\frac{1}{s}} \psi^* \left(\frac{t-b}{s} \right) dt \qquad (10.23)$$

where b is the location of the window in time, s is the wavelet *support* (the finite time-spread over which the wavelet function is nonzero and therefore has action on the signal), and ψ is a function of location a and support s called a *wavelet*. Note that this transform is related to the Fourier transform defined in Equation 10.9.

10.7.1 THE WAVELET FUNCTION

A wavelet function is a waveform of limited duration that has an average value of zero, as seen in Figure 10.19. Only a very few wavelet families have shapes that can be expressed in closed-form equations. Wavelets may seem to be of arbitrary and somewhat unsatisfying shapes, but in fact their shape derives from the properties desired from the associated digital filter bank that implements the discrete wavelet transform. One cannot just "create" a wavelet of arbitrary shape and expect it to have many of the desirable properties of the conventional wavelet families, but the details go beyond the scope of this chapter. The text by Strang and Ngyuen discusses the somewhat complicated issue of wavelet creation and filter banks in great detail [28], as do books by Mallat [29] and Kaiser [30]. There are also many useful Internet tutorials encompassing a wide range of levels of sophistication [31]. As Figure 10.19 shows, the shape of a wavelet can be close to symmetric, as in the symlet or coiflet families, or very asymmetric, like the family of Daubechies wavelets. Some wavelets

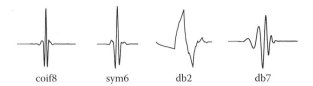

FIGURE 10.19 Some example mother wavelet functions. From left to right: a coiflet (coif), a symlet (sym), and two Daubechies (db) wavelets. The numbers relate to the number of vanishing moments of the wavelet.

(e.g., the db-2) may even appear "noisy," but wavelet functions are not of uncertain shape, nor do they contain noise.

Wavelet analysis consists of breaking any observed signal response into its projections onto a set of shifted and scaled versions of the *mother wavelet* function. The shifted and scaled versions of the mother wavelet function are known as child wavelets, and the collection is called a *wavelet family*, designated by the name and the number indicating the number of vanishing moments associated with the mother wavelet function. Thus a db4 wavelet defines a basis set described by the Daubechies-type wavelet with four *vanishing moments*. The basic db4 wavelet (the mother) and shifted and stretched versions of the basic db4 wavelet shape (her children) are shown in Figure 10.20. The two aspects of wavelets that are important for a signal-processing step are vanishing moments, the symmetry, and the orthogonality of analysis. The vanishing moments associated with a wavelet family determine the degree of polynomial functions that are suppressed by the wavelet; thus, a db4 wavelet analysis would suppress quartic and lower-order polynomials in any data subjected to a db4 wavelet analysis because these functions will have db4 wavelet coefficients equal to zero. A wavelet family usually forms an orthogonal basis in which a signal can be reexpressed, just as the set of sines

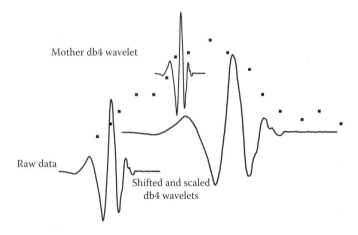

FIGURE 10.20 A mother Daubechies wavelet and two of its shifted and scaled children fitting a noisy signal. Note that the interval of the fit depends on the scale of the wavelet.

and cosines do in Fourier analysis, or the set of eigenvectors do in a principal component analysis. The projections generated by fitting the family of wavelets to the signal provides coefficients that define the signal in the wavelet domain.

Wavelet analysis is related to Fourier analysis, and many of the properties of the Fourier transform transfer to the wavelet transform. However, there are significant differences between the two approaches to generating bases for reexpression of response data. The Fourier decomposition is based on using harmonics and 90-degree phase shifts of the sine function to yield an orthogonal basis set. In many wavelet families, scaling and shifting of the mother wavelet function also generates a new orthogonal basis set for reexpression of the response data. Here, scaling means compressing (using a small scale) or stretching (with a large scale factor) the mother wavelet. Shifting is done by simply delaying the start of the wavelet basis function by a set amount. In the Fourier transform, where a response is reexpressed in terms of its projections onto sine and cosine waves of various frequencies, the basis functions of the reexpression are active over the entire signal response because the sine function is global rather than local in influence. In a wavelet analysis, however, the reexpressions of the response are local, because the region of influence of the mother and children wavelets defining the basis are also local. Thus, there is a localization of the projection of the signal onto a wavelet in both scale (the fit to a particular child wavelet) and location, so the wavelet decomposition expresses a signal in terms of the projections onto a set scale as a function of time. Figure 10.21 demonstrates the similarity of the Fourier and wavelet transforms and the relationship of frequency and scale.

10.7.2 TIME AND FREQUENCY LOCALIZATIONS OF WAVELET FUNCTIONS

A critical difference between the Fourier transform defined in Equation 10.9 and the wavelet transform defined in Equation 10.22 is the fact that the latter permits localization in both frequency and time; that is, we can use the equation to determine what frequencies are active at a specific time interval in a sequence. However, we cannot get *exact* frequency information and exact time information simultaneously because of the Heisenberg uncertainty principle, a theorem that says that for a given signal, the variance of the signal in the time domain σ^2_t and the variance of the signal in the frequency (e.g., Fourier) domain σ^2_F are related

$$\sigma^2_t \, \sigma^2_F \geq 1/2 \tag{10.24}$$

In the limiting case of the sinusoid (the Fourier transform), σ^2_F is zero and σ^2_t is infinite. Using the Heisenberg principle as a guide, we can see that because wavelets operate on a sequence over a finite range of time (the time-spread function, which depends on the wavelet scale, as shown in Figure 10.20), a particular wavelet function is not representative of a single frequency nor even of a finite range of frequencies. As we scale the wavelet function by a factor of 2, the time spread goes down by a factor of 2 and the frequency range goes up by a factor of 2. Thus, the frequency resolution of a wavelet is dependent on the scaling used. Because the wavelet basis set is based on the scaling of the mother wavelet, the wavelet transform is a time-scale transform.

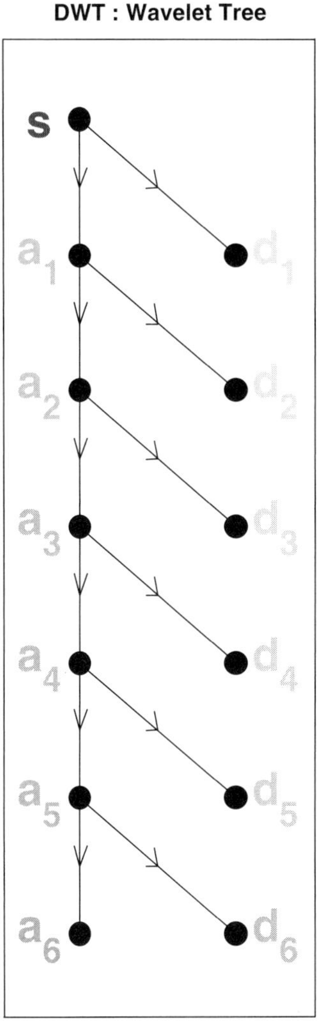

DWT : Wavelet Tree

FIGURE 10.21 Wavelet and sine basis functions for orthogonal transforms.

10.7.3 THE DISCRETE WAVELET TRANSFORM

Just as the discrete Fourier transform generates discrete frequencies from sampled data, the discrete wavelet transform (often abbreviated as DWT) uses a discrete sequence of scales a^j for $j < 0$ with $a = 2^{1/v}$, where v is an integer, called the *number of voices* in the octave. The wavelet support — where the wavelet function is nonzero — is assumed to be $[-K/2, K/2]$. For a signal of size N and $1 \le a^j \le N/K$, a *discrete wavelet* ψ is defined by sampling the scale at a^j and time (for scale 1) at its integer values, that is

$$\psi_j[n] = \frac{1}{\sqrt{a^j}} \psi\left(\frac{n}{a^j}\right) \qquad (10.25)$$

Both the signal and wavelet are N-periodized. Then, the discrete wavelet transform of t, $\mathrm{Wt}[n, a^j]$, is defined by the relation

$$\mathrm{Wt}[n, a^j] = \sum_{m=0}^{N-1} t[m]\psi^*[m-n] \qquad (10.26)$$

which is a circular convolution between the signal t and $\psi_1[n] = \psi[-n]$. These circular convolutions are computed with an FFT operation. The Mallat pyramid algorithm [32] is generally used to perform the discrete wavelet transform.

Decomposition of a response by the discrete wavelet transform is generally represented by a filter bank combining a low-pass filter and a high-pass filter that, together, may be considered to implement a discrete wavelet transform for a specific family of wavelets. The low-pass filter generates a representation with low frequencies, or high scales, called the *approximation*. The high-pass filter generates a representation with high frequencies, or low scales, called the *detail*. We can iteratively apply the filter bank to decompose any approximation representation into a new approximation and detail. This set of nested approximation and detail representations is the usual output of the DWT. The iterative decomposition process is often represented in a tree structure, as shown in Figure 10.22.

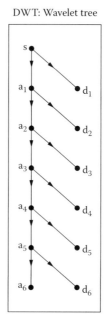

FIGURE 10.22 The tree of wavelet components that results from an input signal vector **s** using a filter-bank implementation of the discrete wavelet transform calculated for six scales. Here **a** is an approximation component vector (obtained from low-pass filtering), and **d** is a detail component vector (from high-pass filtering) from the filter bank. Reconstruction of the signal from the wavelet coefficients occurs by retracing the tree from level 6 up to **s**.

Signal Processing and Digital Filtering

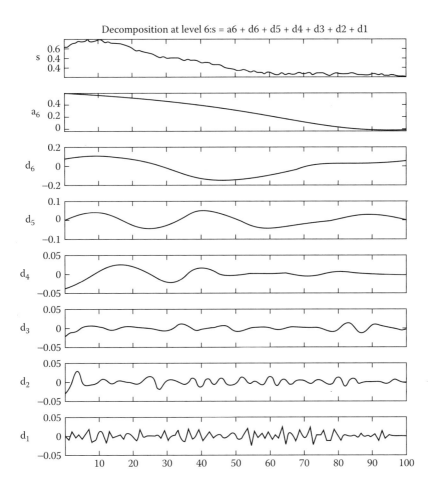

FIGURE 10.23 The cascade of wavelet coefficient vectors output from the wavelet tree filter banks defining the discrete wavelet transform in Figure 10.22. A db-7 mother wavelet was used for the decomposition of the noisy signal in Figure 10.1.

The representation of data in scales permits us to examine the time occurrence of rapidly changing aspects of the response (at low scales) and slowly changing aspects (at high scales) by simply examining different scale plots. Figure 10.23 shows a wavelet decomposition of the noisy signal in Figure 10.1 done according to the wavelet decomposition tree of Figure 10.22.

Figure 10.23 demonstrates one aspect of discrete wavelet transforms that shows similarity to discrete Fourier transforms. Typically, for an N-point observed signal, the points available to decomposition to approximation and detail representations decrease by (about) a factor of 2 for each increase in scale. As the scale increases, the number of points in the wavelet approximation component decreases until, at very high scales, there is a single point. Also, like a Fourier transform, it is possible to reconstruct the observed signal by performing an inverse wavelet transform, known as reconstruction, on the set of wavelet representations of the data. Unlike the inverse Fourier transform, however, not all wavelet families have a reconstruction operation defined for the set of filter banks that implement their DWT. A signal transformed using one of these wavelet families cannot be inverse transformed from its set of wavelet representations. The wavelet functions used most often, including Daubechies, coiflet, and symlet wavelet families, all have inverse transforms. Performing an inverse wavelet transform for these wavelets amounts to traversing up the wavelet decomposition tree.

10.7.4 SMOOTHING AND DENOISING WITH WAVELETS

The wavelet representation shown in Figure 10.23 suggests that we might be able to improve the signal-to-noise ratio of the noisy signal by simply removing one or more detail components used to represent the signal, especially the detail component at high scales. Truncation of this sort is exactly analogous to that done in PCA: we delete one or more basis vectors from the reexpression of the response because we believe that they mainly describe a noise component of our data. In doing so, we reduce the variance of the response data at a cost of some increase in bias, because we distort the signal through the truncation step. A strong truncation, where many wavelet components are truncated, can also be used to compress a signal, in this case because we may be able to convey most of the information from a response with many variables in far fewer wavelet components, which can be considered as a projection of the original signal into a subspace of much smaller dimension. Truncated wavelet reexpressions can be used as inputs

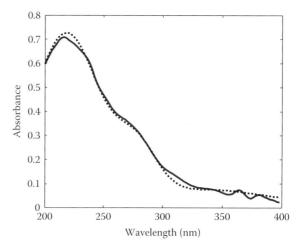

FIGURE 10.24 Smoothing of the noisy signal in Figure 10.1 by using the db-7 wavelet transform at six levels of decomposition with reconstruction. The resulting signal reconstructed from wavelet coefficients from levels 6 to 2 is shown, along with the true signal (dotted line). Note that the smoothing is obtained by not including wavelet coefficients from detail levels d1 and d2 in the reconstruction.

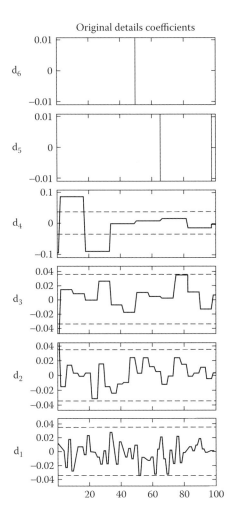

FIGURE 10.25 Denoising of the wavelet components of the noisy signal in Figure 10.1 using an entropy threshold and db-7 wavelets. Thresholds for the decomposition are shown as dashed lines. Wavelet coefficients contained inside the thresholds are set to zero in this form of denoising.

to other chemometric methods in a way similar to truncated PCA reexpressions of data.

The truncation process can also be regarded as smoothing with wavelets because we have selected a scale corresponding to a range of frequency components as describing noise, and we have attenuated (zeroed) it. The effect of an aggressive smoothing of the noisy data in Figure 10.1 with wavelet analysis is shown in Figure 10.24. As in Fourier analysis, the effectiveness of smoothing depends on the degree of separation of signal and noise, but here what matters is separation of signal and noise in time and scale.

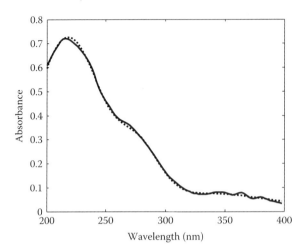

FIGURE 10.26 The denoised signal. The true signal is shown as a dotted line. Compare with Figure 10.24 to see the benefit of denoising for this signal.

With wavelet analysis, there is another way to reduce noise in a signal without truncation of the set of wavelet coefficients. Suppose that we put a threshold on each of the wavelet representations shown in Figure 10.23. Then, any coefficients that remain inside of the threshold values could be reduced as just noise on a real value of zero, and all the scales could be retained in the reconstruction step to reduce bias. In this way, only some few wavelet coefficient amplitudes are attenuated, and these are not attenuated to zero but to some limiting value. This process, called *denoising*, reduces noise but retains signal shape. Its effectiveness is determined in the ability of the analyst to set those threshold values, a process that, not surprisingly, has received a lot of attention in the wavelet literature. Often, the entropy of the signal is used as a way to establish thresholds for denoising, but the reader is referred to the literature for more on this subject [33–39]. Figure 10.26 shows the result of one denoising approach on the noisy signal (Figure 10.1) that we have examined throughout this chapter.

REFERENCES

1. Martens, H. and Naes, T., *Multivariate Calibration,* John Wiley & Sons, Chichester, U.K., 1989.
2. Malinowski, E.R., *Factor Analysis in Chemistry,* 3rd ed., Wiley-Interscience, New York, 2002.
3. Harsanyi, J.C. and Chang, C.-I., Hyperspectral image classification and dimensionality reduction: an orthogonal subspace projection approach, *IEEE Trans. Geosci. Rem. Sens.,* 32, 779, 1994.
4. Booksh, K. and Kowalski, B.R., Theory of analytical chemistry, *Analyt. Chem.,* 66, 782A, 1994.

5. Ferré, J., Brown, S.D., and Rius, X., Improved calculation of the net analyte signal in inverse multivariate calibration, *J. Chemom.*, 15, 537, 2001.
6. Fearn, T., On orthogonal signal correction, *Chemom. Intell. Lab. Syst.*, 50, 47, 2000.
7. Svensson, O., Kourti, T., and MacGregor, J.F., An investigation of orthogonal signal correction algorithms and their characteristics, *J. Chemom.*, 16, 176, 2002.
8. Westerhuis, J.A., de Jong, S., and Smilde, A.K., Direct orthogonal signal correction, *Chemom. Intell. Lab. Syst.*, 56, 13, 2001.
9. Trygg, J. and Wold, S., O2PLS, a two-block (x-y) latent variable regression (LVR) method with an integral OSC filter, *J. Chemom.*, 17, 53, 2003.
10. Bracewell, R.N., *The Fourier Transform and Its Applications*, 2nd ed., John Wiley & Sons, New York, 1972.
11. Press, W.H., Flannery, B.P., Teukolsky, S.A., and Vettering, W.T., *Numerical Recipes: The Art of Scientific Computing*, Cambridge University Press, Cambridge, MA, 1986.
12. Bose, N.K., *Digital Filters: Theory and Applications*, Elsevier Science, New York, 1985.
13. Willson, P.D. and Edwards, T.H., Sampling and smoothing of spectra, *Appl. Spectrosc. Rev.*, 12, 1, 1976.
14. Horlick, G., Digital data handling of spectra utilizing Fourier transforms, *Anal. Chem.*, 44, 943, 1972.
15. Betty, K.R. and Horlick, G., A simple and versatile Fourier domain digital filter, *Appl. Spectrosc.*, 30, 23, 1976.
16. Kaiser, J.F. and Reed, W.A., Data smoothing using low-pass digital filters, *Rev. Sci. Instrum.*, 48, 1447, 1977.
17. Dyer, S.A. and Hardin, D.S., Enhancement of Raman spectra obtained at low signal-to-noise ratios: matched filtering and adaptive peak detection, *Appl. Spectrosc.*, 39, 655, 1985.
18. Bialkowski, S.E., Real-time digital filters: FIR filters, *Anal. Chem.*, 60, 355A, 1988.
19. Willson, P.D. and Polo, S.R., Polynomial filters of any degree, *J. Opt. Soc. Am.*, 71, 599, 1981.
20. Enke, C.G. and Nieman, T.A., Signal-to-noise enhancement by least-squares polynomial smoothing, *Anal. Chem.*, 48, 705A, 1976.
21. Savitsky, A. and Golay, M.J.E., Smoothing and differentiation of data by simplified least-squares procedures, *Anal. Chem.*, 36, 1627, 1964.
22. Steiner, J., Termonia, Y., and Deltour, J., Comments on smoothing and differentiation of data by simplified least square procedure, *Anal. Chem.*, 44, 1906, 1972.
23. Madden, H.H., Comments on the Savitsky-Golay convolution method for least-squares fit smoothing and differentiation of digital data, *Anal. Chem.*, 50, 1383, 1978.
24. Bialkowski, S.E., Generalized digital smoothing filters made easy by matrix calculations, *Anal. Chem.*, 61, 1308, 1989.
25. Bromba, M.U.A., Application hints for Savitsky-Golay digital smoothing filters, *Anal. Chem.*, 53, 1583, 1981.
26. Bromba, M.U.A. and Ziegler, H., Variable filter for digital smoothing and resolution enhancement of noisy spectra, *Anal. Chem.*, 56, 2052, 1984.
27. Gabor, D., Theory of communication, *J. IEE*, 93, 429, 1946.
28. Strang, G. and Nguyen, T., *Wavelets and Filter Banks*, Wellesley-Cambridge Press, Wellesley, MA, 1996.
29. Mallat, S., *A Wavelet Tour of Signal Processing*, 2nd ed., Academic Press, New York, 1999.

30. Kaiser, G., *A Friendly Guide to Wavelets,* Birkhauser, Boston, 1994.
31. Amara's wavelet page, http://www.amara.com/current/wavelet.html.
32. Mallat, S.G., A theory for multiresolution signal decomposition: the wavelet representation, *IEEE Trans. Patt. Anal. Mach. Intell.* , 2 (7), 674–693, 1989.
33. Donoho, D.L., De-noising by soft thresholding, *IEEE Trans. Inf. Theory,* 41, 613, 1995.
34. Mittermayr, C.R., Nikolov, S.G., Hutter, H., and Grasserbauer, M., Wavelet denoising of Gaussian peaks: a comparative study, *Chemom. Intell. Lab. Syst.,* 34, 187, 1996.
35. Mittermayr, C.R., Frischenschlager, H., Rosenberg, E., and Grasserbauer, M., Filtering and integration of chromatographic data: a tool to improve calibration?, *Fresenius' J. Anal. Chem.,* 358, 456, 1997.
36. Barclay, V.J., Bonner, R.F., and Hamilton, I.P., Application of wavelet transform to experimental spectra: smoothing, denoising, and data set compression, *Anal. Chem.,* 69, 78, 1997.
37. Nikolov, S., Hutter, H., and Grasserbauer, M., De-noising of SIMS images via wavelet shrinkage, *Chemom. Intell. Lab. Syst.,* 34, 263, 1996.
38. Wolkenstein, M., Hutter, H., Nikolov, S.G., and Grasserbauer, M., Improvement of SIMS image classification by means of wavelet de-noising, *Fresenius' J. Anal. Chem.,* 357, 783, 1997.
39. Alsberg, B.K., Woodward, A.M., Winson, M.K., Rowland, J., and Kell, D.B., Wavelet denoising of infrared spectra, *Analyst,* 645, 1997.

FURTHER READING

Childers, D.G., Ed., *Modern Spectrum Analysis* , IEEE Press, New York, 1978.
Haykin, S., *Adaptive Filter Theory*, 2nd ed., Prentice Hall, Upper Saddle River, NJ, 1991.
Haykin, S., Ed., *Advances in Spectrum Analysis and Array Processing* , Vol. 2, Prentice Hall, Upper Saddle River, NJ, 1991.
MATLAB, *Signal Processing Toolbox for MATLAB — User's Guide,* Mathworks, Natick, MA, 2000.
Misiti, M., Misiti, Y., Oppenheim, G., and Poggi, J.-M., *Wavelet Toolbox for MATLAB Manual*, Mathworks, Natick, MA, 1996.
Rabiner, L.R. and Gold, B., *Theory and Applications of Digital Signal Processing* , Prentice Hall, Upper Saddle River, NJ, 1975.

11 Multivariate Curve Resolution

Romà Tauler and Anna de Juan

CONTENTS

11.1 Introduction: General Concept, Ambiguities, Resolution Theorems 418
11.2 Historical Background 422
11.3 Local Rank and Resolution: Evolving Factor Analysis and Related Techniques 423
11.4 Noniterative Resolution Methods 426
    11.4.1 Window Factor Analysis (WFA) 427
    11.4.2 Other Techniques: Subwindow Factor Analysis (SFA) and Heuristic Evolving Latent Projections (HELP) 429
11.5 Iterative Methods 431
    11.5.1 Generation of Initial Estimates 432
    11.5.2 Constraints, Definition, Classification: Equality and Inequality Constraints Based on Chemical or Mathematical Properties 433
        11.5.2.1 Nonnegativity 434
        11.5.2.2 Unimodality 434
        11.5.2.3 Closure 434
        11.5.2.4 Known Profiles 435
        11.5.2.5 Hard-Modeling Constraints: Physicochemical Models 435
        11.5.2.6 Local-Rank Constraints, Selectivity, and Zero-Concentration Windows 435
    11.5.3 Iterative Target Transformation Factor Analysis (ITTFA) 437
    11.5.4 Multivariate Curve Resolution-Alternating Least Squares (MCR-ALS) 439
11.6 Extension of Self-Modeling Curve Resolution to Multiway Data: MCR-ALS Simultaneous Analysis of Multiple Correlated Data Matrices 440
11.7 Uncertainty in Resolution Results, Range of Feasible Solutions, and Error in Resolution 446
11.8 Applications 448
    11.8.1 Biochemical Processes 449

	11.8.1.1	Study of Changes in the Protein Secondary Structure..451
	11.8.1.2	Study of Changes in the Tertiary Structure 453
	11.8.1.3	Global Description of the Protein Folding Process 453
11.8.2	Environmental Data..454
11.8.3	Spectroscopic Images...461
11.9 Software 465
References..467

11.1 INTRODUCTION: GENERAL CONCEPT, AMBIGUITIES, RESOLUTION THEOREMS

The resolution of a multicomponent system involves the description of the variation of measurements as an additive model of the contributions of their pure constituents [1–10]. To do so, relevant and sufficiently informative experimental data are needed. These data can be obtained by analyzing a sample with a hyphenated technique (e.g., HPLC-DAD [diode array detection], high-performance liquid chromatography–DAD) or by monitoring a process in a multivariate fashion. In these and similar examples, all of the measurements performed can be organized in a table or data matrix where one direction (the elution or the process direction) is related to the compositional variation of the system, and the other direction refers to the variation in the response collected. The existence of these two directions of variation helps to differentiate among components (Figure 11.1).

In this context, it is important to broadly define the concept of a component as any entity giving a distinct and real, pure response. This includes examples as diverse as a chemical compound, a conformational state [11–12], or a pollution source [13–16] whose response could be a profile that includes the relative apportionment of its different pollutants.

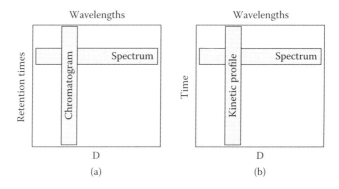

FIGURE 11.1 Examples of data sets coming from multicomponent systems: (a) HPLC-DAD chromatographic run and (b) spectroscopic monitoring of a kinetic process [10].

In the resolution of any multicomponent system, the main goal is to transform the raw experimental measurements into useful information. By doing so, we aim to obtain a clear description of the contribution of each of the components present in the mixture or the process from the overall measured variation in our chemical data. Despite the diverse nature of multicomponent systems, the variation in their related experimental measurements can, in many cases, be expressed as a simple composition-weighted linear additive model of pure responses, with a single term per component contribution. Although such a model is often known to be followed because of the nature of the instrumental responses measured (e.g., in the case of spectroscopic measurements), the information related to the individual contributions involved cannot be derived in a straightforward way from the raw measurements. The common purpose of all multivariate resolution methods is to fill in this gap and provide a linear model of individual component contributions using solely the raw experimental measurements. Resolution methods are powerful approaches that do not require a lot of prior information because neither the number nor the nature of the pure components in a system need to be known beforehand. Any information available about the system may be used, but it is not required. Actually, the only mandatory prerequisite is the inner linear structure of the data set. The mild requirements needed have promoted the use of resolution methods to tackle many chemical problems that could not be solved otherwise.

All resolution methods mathematically decompose a global instrumental response of mixtures into the contributions linked to each of the pure components in the system [1–10]. This global response is organized into a matrix **D** containing raw measurements about all of the components present in the data set. Resolution methods allow for the decomposition of the initial mixture data matrix **D** into the product of two data matrices **C** and **S**^{T}, each of them containing the pure response profiles of the n mixture or process components associated with the row and the column directions of the initial data matrix, respectively (see Figure 11.2). In matrix notation, the expression for all resolution methods is:

$$\mathbf{D} = \mathbf{CS}^T + \mathbf{E} \qquad (11.1)$$

where **D** ($r \times c$) is the original data matrix, **C** ($r \times n$) and **S**^{T} ($n \times c$) are the matrices containing the pure-component profiles related to the data variation in the row direction and in the column direction, respectively, and **E** ($r \times c$) is the error matrix, i.e., the residual variation of the data set that is not related to any chemical contribution. The variables r and c represent the number of rows and the number of columns of the original data matrix, respectively, and n is the number of chemical components in the mixture or process. **C** and **S**^{T} often refer to concentration profiles and spectra (hence their abbreviations and the denomination we will adopt often in this chapter), although resolution methods are proven to work in many other diverse problems [13–20].

From the early days in resolution research, the mathematical decomposition of a single data matrix, no matter the method used, has been known to be subject to ambiguities [1, 2]. This means that many pairs of **C**- and **S**^{T}-type matrices can be found that reproduce the original data set with the same fit quality. In plain words,

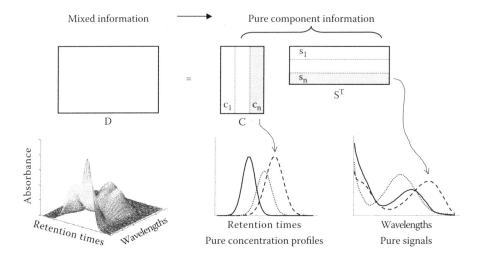

FIGURE 11.2 Resolution of a multicomponent chromatographic HPLC-DAD run (**D** matrix) into their pure concentration profiles (**C** matrix, chromatograms) and responses (**S**^{T} matrix, spectra) [10].

the correct reproduction of the original data matrix can be achieved by using component profiles differing in shape (rotational ambiguity) or in magnitude (intensity ambiguity) from the sought (true) ones [21].

These two kinds of ambiguities can be easily explained. The basic equation associated with resolution methods, **D** = **CS**^{T}, can be transformed as follows:

$$\mathbf{D} = \mathbf{C} \, (\mathbf{T} \, \mathbf{T}^{-1}) \, \mathbf{S^T} \tag{11.2}$$

$$\mathbf{D} = (\mathbf{CT}) \, (\mathbf{T^{-1}} \, \mathbf{S^T}) \tag{11.3}$$

$$\mathbf{D} = \mathbf{C'} \, \mathbf{S'^T} \tag{11.4}$$

where **C**⊕ = **CT** and **S**⊕^{T} = (**T**$^{-1}$ **S**^{T}) describe the **D** matrix as correctly as the true **C** and **S**^{T} matrices do, though **C**⊕ and **S**⊕^{T} are not the sought solutions. As a result of the rotational ambiguity problem, a resolution method can potentially provide as many solutions as **T** matrices can exist. Often this may represent an infinite set of solutions, unless **C** and **S** are forced to obey certain conditions. In a hypothetical case with no rotational ambiguity, that is, in the case where the shapes of the profiles in **C** and **S** are correctly recovered, the basic resolution model could still be subject to intensity ambiguity, as shown in Equation 11.5

$$\mathbf{D} = \sum_{i=1}^{n} \left(\frac{1}{k_i} \mathbf{c}_i \right) \left(k_i \, \mathbf{s}_i^T \right) \tag{11.5}$$

where k_i are scalars and n refers to the number of components. Each concentration profile of the new $\mathbf{C}\oplus$ matrix (Equation 11.4) would have the same shape as the real one, but it would be k_i times smaller, whereas the related spectra of the new $\mathbf{S}\oplus^T$ matrix (Equation 11.4) would be equal in shape to the real spectra, though k_i times more intense.

The correct performance of any curve-resolution (CR) method depends strongly on the complexity of the multicomponent system. In particular, the ability to correctly recover dyads of pure profiles and spectra for each of the components in the system depends on the degree of overlap among the pure profiles of the different components and the specific way in which the regions of existence of these profiles (the so-called concentration or spectral windows) are distributed along the row and column directions of the data set. Manne stated the necessary conditions for correct resolution of the concentration profile and spectrum of a component in the 2 following theorems [22]:

1. The true concentration profile of a compound can be recovered when all of the compounds inside its concentration window are also present outside.
2. The true spectrum of a compound can be recovered if its concentration window is not completely embedded inside the concentration window of a different compound.

According to Figure 11.3, the pure concentration profile of component B can be recovered because A is inside and outside B's concentration window; however, B's pure spectrum cannot be recovered because its concentration profile is totally embedded under the major compound, A. Analogously, the pure spectrum of A can be obtained, but not the pure concentration profile because B is present inside its concentration window, but not outside.

The same formulation of these two theorems holds when, instead of looking at the concentration windows in rows, the "spectral" windows in columns are considered. In this context, the theorems show that the goodness of the resolution results depends more strongly on the features of the data set than on the mathematical background of the CR method selected. Therefore, a good knowledge of the properties of the data sets before carrying out a resolution calculation provides a clear idea about the quality of the results that can be expected.

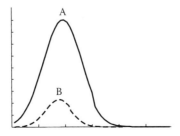

FIGURE 11.3 Concentration profiles for a two-component system (see comments in text related to resolution).

11.2 HISTORICAL BACKGROUND

The field of curve resolution was born in response to the need for a tool to analyze multivariate experimental data from multicomponent dynamic systems. The common goal of all curve-resolution methods is to mathematically decompose the global instrumental response into the pure-component profiles of each of the components in the system. The use of these methods has become a valuable aid for resolving complex systems, especially when obtaining selective signals for individual species is not experimentally possible, too complex, or too time consuming.

Two pioneering papers on curve resolution were published by Lawton and Sylvestre early in the 1970s [1, 2]. In particular, a mixture analysis resolution problem was described in mathematical terms for the case of a simple two-component spectral mixture. Interestingly, several concepts introduced in these early papers were the precursors of the ideas underlying most of the curve-resolution methods developed afterward. For instance, the concept of pure-component solutions as a linear combination of the measured spectra and vice versa was presented; the concept of a subspace spanned by "true" solutions in relation to the subspace spanned by PCA (principal component analysis) solutions was presented; and the concept of a range or band of feasible solutions, and how to reduce the width of this band by means of constraints, such as nonnegativity and closure (mass balance) equations, was presented. Later on, these ideas were reformulated more precisely using the concepts of rotational and intensity ambiguities [23], which are found ubiquitously in all factor-analysis matrix bilinear decomposition methods.

The extension of Lawton and Sylvestre's curve resolution from two- to three-component systems was presented by Borgen et al., [3, 4] focusing on the optimization of ranges of feasible solutions. At the same time, the first edition of Malinowski's book [24] *Factor Analysis in Chemistry* appeared [25], which presented a review of updated concepts and applications. In a way, Malinowski's book could be considered for many researchers in this field as the consolidation of the incipient subject of chemometrics, at a time when this term was still not widely accepted.

The main goal of factor analysis, i.e., the recovery of the underlying "true" factors causing the observed variance in the data, is identical to the main goal of curve-resolution methods. In factor analysis, "abstract" factors are clearly distinguished from "true" factors, and the key operation is to find a transformation from abstract factors to the true factors using rotation methods. Two types of rotations are usually used, orthogonal rotations and oblique rotations. Principal component analysis, PCA, (or principal factor analysis, PFA) produces an orthogonal bilinear matrix decomposition, where components or factors are obtained in a sequential way to explain maximum variance (see Chapter 4, Section 4.3, for more details). Using these constraints plus normalization during the bilinear matrix decomposition, PCA produces unique solutions. These "abstract" unique and orthogonal (independent) solutions are very helpful in deducing the number of different sources of variation present in the data. However, these solutions are "abstract" solutions in the sense that they are not the "true" underlying factors causing the data variation, but orthogonal linear combinations of them. On the other hand, in curve-resolution methods, the goal is to unravel the "true" underlying sources of data variation. It is not only a question of how many different sources are present and how they can be interpreted, but to find out how they are in reality. The price to pay is that unique solutions are not usually obtained by means of curve-resolution methods unless external information is provided during the matrix decomposition.

Different approaches have been proposed during recent years to improve the solutions obtained by curve-resolution methods, and some of them are summarized in the next sections. The field is already mature and, as it has been recently pointed out [26], multivariate curve resolution can be considered as a "sleeping giant of chemometrics," with a slow but persistent growth.

Whenever the goals of curve resolution are achieved, the understanding of a chemical system is dramatically increased and facilitated, avoiding the use of enhanced and much more costly experimental techniques. Through multivariate-resolution methods, the ubiquitous mixture analysis problem in chemistry (and other scientific fields) is solved directly by mathematical and software tools instead of using costly analytical chemistry and instrumental tools, for example, as in sophisticated "hyphenated" mass spectrometry-chromatographic methods.

11.3 LOCAL RANK AND RESOLUTION: EVOLVING FACTOR ANALYSIS AND RELATED TECHNIQUES

Manne's resolution theorems clearly stated how the distribution of the concentration and spectral windows of the different components in a data set could affect the quality of the pure profiles recovered after data analysis [22]. The correct knowledge of these windows is the cornerstone of some resolution methods, and in others where it is not essential, information derived from this knowledge can be introduced to generally improve the results obtained.

Setting the boundaries of windows of the different components can only be done if we are able to know how the number and nature of the components change in the data set. Obtaining this information is the main goal of local-rank analysis methods, which are used to locate and describe the evolution of each component in a system. This is accomplished by combining the information obtained from multiple rank analyses performed locally on limited zones (row or column windows) of the data set.

Some of the local-rank analysis methods, such as evolving-factor analysis (EFA) [27–29], are more process oriented and rely on the sequential evolution of the components as a function of time or any other variable in the data set, while others, such as fixed-size moving-window–evolving-factor analysis (FSMW-EFA) [30, 31], can be applied to processes and mixtures. EFA and FSMW-EFA are the two pioneering local-rank analysis methods and can still be considered the most representative and widely used.

Evolving-factor analysis was born as the chemometric way to monitor chemical-evolving processes, such as HPLC diode-array data, batch reactions, or titration data [27–28]. The evolution of a chemical system is gradually measured by recording a new response vector at each stage of the process under study. Mimicking the experimental protocol, EFA performs principal component analyses on submatrices of gradually increasing size in the process direction, enlarged by adding a row (response), one at a time. This procedure is performed from top to bottom of the data set (forward EFA) and from bottom to top (backward EFA) to investigate the emergence and the decay of the process contributions, respectively. Figure 11.4b displays the information provided by EFA for an HPLC-DAD example and how to interpret the results.

Each time a new row is added to the expanding submatrix (Figure 11.4b), a PCA model is computed and the corresponding singular values or eigenvalues are saved. The forward EFA curves (thin solid lines) are produced by plotting the saved singular values or log (eigenvalues) obtained from PCA analyses of the submatrix expanding in the forward direction. The backward EFA curves (thin dashed lines) are produced by plotting the singular values or log (eigenvalues) obtained from the PCA analysis of the submatrix expanding in the backward direction. The lines connecting corresponding singular values (s.v.), i.e., all of the first s.v., the second s.v., the ith s.v., indicate the evolution of the singular values along the process and, as a consequence, the variation of the process components. Emergence of a new singular value above the noise level delineated by the pool of nonsignificant singular values indicates the emergence of a new component (forward EFA) or the disappearance of a component (backward EFA) in the process.

Figure 11.4b also shows how to build initial estimates of concentration profiles from the overlapped forward and backward EFA curves as long as the process evolves in a sequential way (see the thick lines in Figure 11.4b). For a system with n significant components, the profile of the first component is obtained combining the curve representing the first s.v. of the forward EFA plot and the curve representing the nth s.v. of the backward EFA plot. Note that the nth s.v. in the backward EFA plot is related to the disappearance of the first component in the forward EFA plot. The profile of the second component is obtained by splicing the curve representing the second s.v. in the forward EFA plot to the curve representing $(n − 1)$th s.v. from the backward EFA plot, and so forth. Combining the two profiles into one profile is easily accomplished in a computer program by selecting the minimum value from the two s.v. lines to be combined. It can be seen that the resulting four elution profiles obtained by EFA are good approximations of the real profiles shown in Figure 11.4a.

The information provided by the EFA plots can be used for the detection and location of the emergence and decay of the compounds in an evolving process. As a consequence, the concentration window and the zero-concentration region for each component in the system are easily determined for any process that evolves such that the emergence and decay of each component occurs sequentially. For example, the concentration window of the first component to elute is shown as a shadowed zone in Figure 11.4b. Uses of this type of information have given rise to most of the noniterative resolution methods, explained in Section 11.4 [32–39]. Iterative resolution methods, explained in Section 11.5, use the EFA-derived estimates of the concentration profiles as a starting point in an iterative optimization [40, 41]. The location of selective zones and zones with a number of compounds smaller than the total rank can also be introduced as additional information to minimize the ambiguity in the resolved profiles [21, 41, 42].

As mentioned earlier, FSMW-EFA is not restricted in its applicability to evolving processes, although the interpretation of the final results is richer for this kind of problem.

Multivariate Curve Resolution

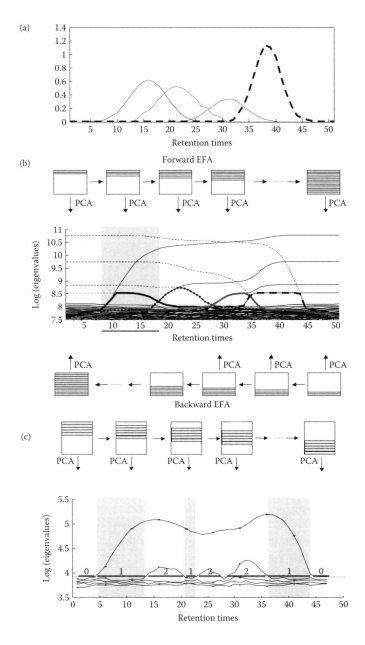

FIGURE 11.4 (a) Concentration profiles of an HPLC-DAD data set. (b) Information derived from the data set in Figure 11.4a by EFA: scheme of PCA runs performed. Combined plot of forward EFA (solid black lines) and backward EFA (dashed black lines). The thick lines with different line styles are the derived concentration profiles. The shaded zone marks the concentration window for the first eluting compound. The rest of the elution range is the zero-concentration window. (c) Information derived from the data set in Figure 11.4a by FSMW-EFA: scheme of the PCA runs performed. The straight lines and associated numbers mark the different windows along the data set as a function of their local rank (number). The shaded zones mark the selective concentration windows (rank 1).

FSMW-EFA does not focus on a description of the evolution of the different components in a system as EFA does; rather, it focuses on the local rank of windows in the concentration domain (rows) or the local rank of windows in the spectral response domain (columns).

FSMW-EFA is carried out by conducting a series of PCA analyses on submatrices obtained by moving a window of a fixed size through the data set, starting at the top of the matrix and moving downward, one row at a time. The singular values or eigenvalues from the repeated analyses are saved, and a plot is constructed by connecting the corresponding singular values as done in EFA. Visual examination of these plots gives a local-rank map of the data set, i.e., a representation of how many components are simultaneously present in the different zones of the data set (Figure 11.4c). For each window analyzed, the number of singular values exceeding the noise level threshold is used to determine the local rank. The local-rank map helps to identify selective zones in the data set (e.g., zones where the local rank is 1) and to know the degree of compound overlap in the data set. The unambiguous determination of the number of compounds present and their identities is only possible in processes where components evolve sequentially or when more external information is available. The window size is a parameter that has an effect on the information obtained (e.g., local-rank maps). Wider windows increase the sensitivity for detecting minor components, including components completely embedded under major compounds. Narrower windows can provide more accurate resolution of boundaries between zones of different rank.

New algorithms based on EFA and FSMW-EFA have refined the performance of the parent methods [43, 44] and have widened their applicability to the study of systems with concurrent processes [45] or complex spatial structure, such as spectroscopic images [46].

11.4 NONITERATIVE RESOLUTION METHODS

Resolution methods are often divided in iterative and noniterative methods. Most noniterative methods are one-step calculation algorithms that focus on the one-at-a-time recovery of either the concentration or the response profile of each component. Once all of the concentration (C) or response (S) profiles are recovered, the other member of the matrix pair, C and S, is obtained by least-squares according to the general CR model, $D = CS^T$ [32–38].

Noniterative methods use information from local-rank maps or concentration windows in a characteristic way. In mathematical terms, these windows define subspaces where the different compounds are present or absent. The subspaces can be combined in clever ways through projections or by extraction of common vectors (profiles) to obtain the profiles sought.

As mentioned in Section 11.3, the cornerstone of these procedures is the correct location of concentration windows of the compounds of interest. Limitations of these methods are linked to this point. Thus, data sets where the compositional evolution of the compounds does not follow any clear pattern, such as in a series of mixtures or image pixels, cannot be resolved by these methods because it is practically impossible to determine the concentration windows of components. Evolving processes are

Multivariate Curve Resolution

the most suitable systems to be analyzed but, again, attention should be paid to situations where the pattern by which components emerge and decay is not sequentially ordered. Some examples that violate this requirement are nonunimodal concentration profiles or small embedded peaks under major peaks. In cases such as these, specialized EFA derivations [39] should be used to avoid incorrect assignment of component windows. Other problems associated with locating window boundaries are due to the presence of noise that can blur the extremes of the concentration windows. Errors from this source can also affect the quality of the final results.

Noniterative methods are fast, but they have clear limitations in their applicability because of the difficulties associated with correct definition of concentration windows and local rank. Their use is practically restricted to processes with sequentially evolving components like chromatography, the components of which fulfill the conditions required by Manne's theorems, to ensure a correct component resolution [22].

11.4.1 Window Factor Analysis (WFA)

Window factor analysis (WFA) was described by Malinowski and is likely the most representative and widely used noniterative resolution method [34, 35]. WFA recovers the concentration profiles of all components in the data set one at a time. To do so, WFA uses the information in the complete original data set and in the subspace where the component to be resolved is absent, i.e., all rows outside of the concentration window. The original data set is projected into the subspace spanned by where the component of interest is absent, thus producing a vector that represents the spectral variation of the component of interest that is uncorrelated to all other components. This specific spectral information, combined appropriately with the original data set, yields the concentration profile of the related component. To ensure the specificity of this spectral information, all other components in the data set should be present outside of the concentration window of the component to be resolved. This means, in practice, that component peaks with embedded peak profiles under them cannot be adequately resolved.

Figure 11.5 illustrates the scheme followed in the WFA resolution. The steps of the WFA method are listed below, followed by a description clarifying their meaning.

1. A PCA model of the original data matrix, **D**, is computed.
2. The concentration windows of each component in the data set are determined.

For each component:

3. A PCA model of a submatrix, **D°**, is computed where the rows related to the concentration window of the nth component to be resolved have been removed.
4. The vector, p_n^{oT}, is computed, which is the part of the spectrum of the nth component orthogonal to the spectra of all other components in the original matrix.
5. The true concentration profile of the nth component is recovered using p_n^{oT} and **D**.

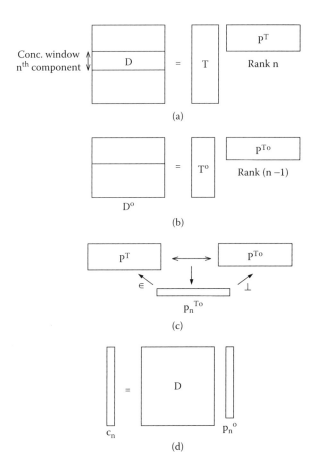

FIGURE 11.5 Recovery of the concentration profile of the nth compound by window factor analysis. (a) PCA of the raw data matrix and determination of the concentration window, **D** (steps 1 and 2); (b) PCA of the matrix formed by suppression of the concentration window of the nth component, **D**° (step 3); (c) recovery of the part of the spectrum of the nth component orthogonal to all the spectra in **D**°, $\mathbf{p}_n^{T_o}$ (step 4); and (d) recovery of the concentration profile of the nth component (step 5).

As a general last step after obtaining the concentration profiles of all components:

6. The pure-spectrum data matrix $\mathbf{S}^T$ is estimated by least squares using **D** and **C**.

WFA starts with the PCA decomposition of the **D** matrix, giving the product of scores and loadings, $\mathbf{TP}^T$. In general, the **D** matrix will have n components, i.e., rank n. The determination of the location of concentration windows for each component is carried out using EFA (see Figure 11.4b) or other methods. Steps 3 to 5 are the core of the WFA method and should be performed as many times as compounds are present in matrix **D** to recover the concentration profiles of the **C** matrix, one at a time.

For each component, a $\mathbf{D}^o$ submatrix is constructed by removing the rows related to its concentration window. Then, a PCA model is computed and the product $\mathbf{T}^o\mathbf{P}^{oT}$ is obtained. Note that $\mathbf{D}^o$ has rank $n - 1$ because the variation due to the component of interest disappears when its corresponding window (rows) in the data matrix $\mathbf{D}$ is deleted. The loading matrices, $\mathbf{P}^T$ and $\mathbf{P}^{oT}$, describe the space of the n pure spectra in $\mathbf{D}$ and the $(n - 1)$ pure spectra in $\mathbf{D}^o$, respectively. The rows in these loading matrices are actually "abstract spectra," and the real spectra can be expressed as a linear combination of them. Using these two loading matrices, $\mathbf{P}^T$ and $\mathbf{P}^{oT}$, it is possible to calculate a vector $\mathbf{p}_n^{oT}$ that is orthogonal to the $(n - 1)$ $\mathbf{p}_i^{oT}$ vectors and that belongs to the space defined by $\mathbf{P}^T$. This vector completes the set of vectors in $\mathbf{P}^{oT}$ and contains the part of the spectra of the removed component uncorrelated to the spectra of the other $(n - 1)$ components in the data matrix. Using this vector with information exclusively related to the removed component, the true concentration profile of this compound can be calculated as follows:

$$\mathbf{D}\mathbf{p}_n^o = \mathbf{c}_n \tag{11.6}$$

The complete $\mathbf{C}$ matrix is then formed by appending row-wise the column concentration profiles found for each component in the $\mathbf{D}$ matrix. The matrix of spectra, $\mathbf{S}^T$, is obtained by least squares using the $\mathbf{D}$ and $\mathbf{C}$ matrices and the basic equation of CR methods, $\mathbf{D} = \mathbf{C}\mathbf{S}^T$:

$$\mathbf{S}^T = (\mathbf{C}^T\mathbf{C})^{-1}\mathbf{C}^T\mathbf{D} \tag{11.7}$$

Recent modifications of the WFA method attempt to solve some of the problems caused by poorly defined boundaries for concentration windows [35].

11.4.2 OTHER TECHNIQUES: SUBWINDOW FACTOR ANALYSIS (SFA) AND HEURISTIC EVOLVING LATENT PROJECTIONS (HELP)

Following the idea of using concentration windows and the subspaces that can be derived, other noniterative methods are focused on the recovery of the response profiles (spectra). This is the case of subwindow factor analysis (SFA), proposed by Manne [38], and other derivations of this method, like parallel vector analysis (PVA) [39]. Unlike WFA, SFA recovers the pure response profile of each component. The individual row response profiles are appended in a columnwise fashion, until the complete $\mathbf{S}^T$ matrix is built. The $\mathbf{C}$ matrix is easily derived by least-squares according to the CR model, $\mathbf{D} = \mathbf{C}\mathbf{S}^T$, as follows:

$$\mathbf{C} = \mathbf{D}\mathbf{S}(\mathbf{S}^T\mathbf{S})^{-1} \tag{11.8}$$

In SFA, the knowledge of the concentration windows is used in such a way that each pure spectrum is calculated as the intersection of two subspaces that have only the compound to be resolved in common. Figure 11.6 illustrates the

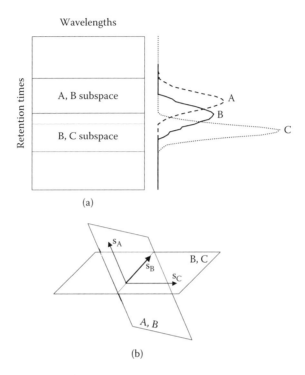

FIGURE 11.6 Application of subwindow factor analysis (SFA) for resolution. (a) Concentration profiles of A, B, and C and subwindows used for the resolution of component B (first containing A and B compounds and second containing B and C compounds). (b) The A,B plane is defined by the pure spectra of A and B (s_A, s_B) and the plane B,C by the pure spectra of B and C (s_B, s_C). The intersection of both planes must be necessarily the pure spectrum of B.

idea behind SFA for a three-component HPLC-DAD system (A, B, and C). Once the concentration windows of the three components are known, one subwindow can be constructed with rows including only A and B and another one with rows where only B and C are present. The intersection of the two planes derived from these subspaces must necessarily give the pure spectrum of B as an answer. The same strategy would be applied to recover the spectra of the rest of the compounds in a general example.

To conduct SFA in practice, the singular-value decomposition (SVD, see Chapter 4) of the two subwindows yields a basis of orthogonal vectors spanning the (A,B) subspace, called $\{e_i\}$, and another basis for the (B,C) subspace, called $\{f_i\}$. The spectrum of B, s_B, can be obtained from these two sets of basis vectors as shown in Equation 11.9,

$$s_B = \sum_i a_i e_i = \sum_i b_i f_i \tag{11.9}$$

The SFA algorithm computes the a_i and b_i values that minimize Equation 11.10,

$$\min_{a_i,b_i}\left\|\sum_i a_i e_i - \sum_i b_i f_i\right\|^2 \qquad (11.10)$$

after which s_B can be obtained by using any of the resulting linear combinations, $\sum_i a_i e_i$ or $\sum_i a_i e_i$.

The spectra of C and A can be obtained in a straightforward fashion, since these components have selective zones in their elution profiles.

The HELP method is another pioneering noniterative method using local-rank information [36, 37] and based on the local-rank analysis of the data set and focuses on finding selective concentration or response windows. When these selective zones exist, the resolution of the system is clear. Thus, for an HPLC-DAD data set, a row related to a selective elution time directly provides the shape of the spectrum of the only component present at that stage of the chromatographic elution. In a similar manner, a column related to a selective wavelength directly provides the chromatographic peak of the only absorbing compound at that wavelength.

HELP works by exploring both the concentration and spectral response spaces with a powerful graphical tool (the so-called datascope) to visually detect potential selective zones in the scores and then loading plots of the data matrix, which are seen as points (representing rows or columns of the original data set) lying on straight lines centered near the origin. A statistical method to confirm the presence of selectivity in the concentration or spectral windows is based on the use of an *F*-test to compare the magnitude of eigenvalues related to potential selective zones of the data set with eigenvalues related to noise zones of the data matrix, i.e., those regions where no chemical components are supposed to be present. The confirmation of a selective zone in the data set, which is actually a rank-one window in the data matrix, will then be obtained when no significant differences are found between the first eigenvalue of a noise-related zone of the data matrix and the second eigenvalue of the potential selective zone. Components with selective concentration and response zones are straightforwardly resolved. Subtraction of the $c_i s_i^T$ contribution of the resolved components from the raw data set can facilitate the resolution from components originally lacking selectivity.

11.5 ITERATIVE METHODS

Iterative resolution methods obtain the resolved concentration and response matrices through the one-at-a-time refinement or simultaneous refinement of the profiles in **C**, in **S**^{T}, or in both matrices at each cycle of the optimization process. The profiles in **C** or **S**^{T} are "tailored" according to the chemical properties and the mathematical features of each particular data set. The iterative process stops when a convergence criterion (e.g., a preset number of iterative cycles is exceeded or the lack of fit goes below a certain value) is fulfilled [21, 42, 47–50].

Iterative resolution methods are in general more versatile than noniterative methods. They can be applied to more diverse problems, e.g., data sets with partial or incomplete selectivity in the concentration or spectral domains, and to data sets with concentration profiles that evolve sequentially or nonsequentially. Prior knowledge about the data set (chemical or related to mathematical features) can be used in the optimization process, but it is not strictly necessary. The main complaint about iterative resolution methods has often been the longer calculation times required to obtain optimal results; however, improved fast algorithms and more powerful PCs have overcome this historical limitation.

The next subsection deals first with aspects common to all resolution methods. These include (1) issues related to the initial estimates, i.e., how to obtain the profiles used as the starting point in the iterative optimization, and (2) issues related to the use of mathematical and chemical information available about the data set in the form of so-called constraints. The last part of this section describes two of the most widely used iterative methods: iterative target transformation factor analysis (ITTFA) and multivariate curve resolution–alternating least squares (MCR-ALS).

11.5.1 Generation of Initial Estimates

Starting the iterative optimization of the profiles in $\mathbf{C}$ or $\mathbf{S}^T$ requires a matrix or a set of profiles sized as $\mathbf{C}$ or as $\mathbf{S}^T$ with rough approximations of the concentration profiles or spectra that will be obtained as the final results. This matrix contains the *initial estimates* of the resolution process. In general, the use of nonrandom estimates helps shorten the iterative optimization process and helps to avoid convergence to local optima different from the desired solution. It is sensible to use chemically meaningful estimates if we have a way of obtaining them or if the necessary information is available. Whether the initial estimates are either a $\mathbf{C}$-type or an $\mathbf{S}^T$-type matrix can depend on which type of profiles are less overlapped, on which direction of the matrix (rows or columns) has more information, or simply on the will of the chemist.

There are many chemometric methods to build initial estimates: some are particularly suitable when the data consists of the evolutionary profiles of a process, such as evolving factor analysis (see Figure 11.4b in Section 11.3) [27, 28, 51], whereas some others mathematically select the purest rows or the purest columns of the data matrix as initial profiles. Of the latter approach, key-set factor analysis (KSFA) [52] works in the FA abstract domain, and other procedures, such as the simple-to-use interactive self-modeling analysis (SIMPLISMA) [53] and the orthogonal projection approach (OPA) [54], work with the real variables in the data set to select rows of "purest" variables or columns of "purest" spectra, that are most dissimilar to each other. In these latter two methods, the profiles are selected sequentially so that any new profile included in the estimate is the most uncorrelated to all of the previously selected ones.

Apart from using chemometric methods, a matrix of initial estimates can always be formed with the rows or columns of the data set that the researcher considers most representative because of chemical reasons, and it can also include external information, such as some reference spectra or concentration profiles, when available.

11.5.2 Constraints, Definition, Classification: Equality and Inequality Constraints Based on Chemical or Mathematical Properties

Although resolution does not require previous information about the chemical system under study, additional knowledge, when it exists, can be used to tailor the sought pure profiles according to certain known features and, as a consequence, to minimize the ambiguity in the data decomposition and in the results obtained.

The introduction of this information is carried out through the implementation of constraints. A constraint can be defined as any mathematical or chemical property systematically fulfilled by the whole system or by some of its pure contributions [55]. Constraints are translated into mathematical language and force the iterative optimization to model the profiles while respecting the conditions desired.

The application of constraints should always be prudent and soundly grounded, and constraints should only be set when there is an absolute certainty about the validity of the constraint. Even a potentially useful constraint can play a negative role in the resolution process when factors like experimental noise or instrumental problems distort the related profile or when the profile is modified so roughly that the convergence of the optimization process is seriously damaged. When well implemented and fulfilled by the data set, constraints can be seen as the driving forces of the iterative process to the right solution and, often, they are found not to be active in the last part of the optimization process.

The efficient and reliable use of constraints has improved significantly with the development of methods and software that allow them to be easily used in flexible ways. This increase in flexibility allows complete freedom in the way combinations of constraints can be used for profiles linked in the different concentration and spectral domains. This increase in flexibility also makes it possible to apply a certain constraint with variable degrees of tolerance to cope with noisy real data. For example, the implementation of constraints often allows for small deviations from ideal behavior before correcting a profile [7, 21, 55]. Methods for correcting the profile to be constrained have evolved into smoother methodologies, which modify the poorly behaving profile so that the global shape is retained as much as possible and the convergence of the iterative optimization is minimally upset [56–61].

There are several ways to classify constraints: the main ones relate either to the nature of the constraints or to the way they are implemented. In terms of their nature, constraints can be based on either chemical or mathematical features of the data set. In terms of implementation, we can distinguish between equality constraints or inequality constraints [56]. An equality constraint sets the elements in a profile to be equal to a certain value, whereas an inequality constraint forces the elements in a profile to be unequal (higher or lower) than a certain value. The most widely used types of constraints will be described using the classification scheme based on the constraint nature. In some of the descriptions that follow, comments on the implementation (as equality or inequality constraints) will be added to illustrate this concept. Figure 11.7 shows the effects of some of these constraints on the correction of a profile.

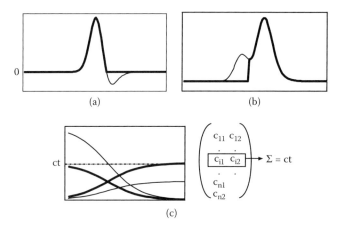

FIGURE 11.7 Effects of some constraints on the shape of resolved profiles. The thin and the thick lines represent the profiles before and after being constrained, respectively. Constraints shown are (a) nonnegativity, (b) unimodality, and (c) closure.

11.5.2.1 Nonnegativity

The nonnegativity constraint is applied when it can be assumed that the measured values in an experiment will always be positive. For example, it can be applied to all concentration profiles and to many experimental responses, such as UV (ultraviolet) absorbances and fluorescence intensities [42, 47, 48, 56, 59]. This constraint forces the values in a profile to be equal to or greater than zero. It is an example of an inequality constraint (see Figure 11.7).

11.5.2.2 Unimodality

The unimodality constraint allows the presence of only one maximum per profile (see Figure 11.7) [42, 55, 60]. This condition is fulfilled by many peak-shaped concentration profiles, like chromatograms or some types of reaction profiles, and by some instrumental signals, like certain voltammetric responses. It is important to note that this constraint does not only apply to peaks, but to profiles that have a constant maximum (plateau) or a decreasing tendency. This is the case for many monotonic reaction profiles that show only the decay or the emergence of a compound [47, 48, 51, 61], such as the most protonated and deprotonated species in an acid-base titration, respectively.

11.5.2.3 Closure

The closure constraint is applied to closed reaction systems, where the principle of mass balance is fulfilled. With this constraint, the sum of the concentrations of all of the species involved in the reaction (the suitable elements in each row of the $\mathbf{C}$ matrix) is forced to be equal to a constant value (the total concentration) at each stage in the reaction [27, 41, 42]. The closure constraint is an example of an equality constraint.

11.5.2.4 Known Profiles

Partial chemical information in the form of known pure response profiles, such as pure-component reference spectra or pure-component concentration profiles for one or more species, can also be introduced in the optimization problem as additional equality constraints [5, 42, 62, 63, 64]. The known profiles can be set to be invariant along the iterative process. The known profile does not need to be complete to be used. When only selected regions of profiles are known, they can also be set to be invariant, whereas the unknown parts can be left loose. This opens up the possibility of using resolution methods for quantitative purposes, for instance. Thus, data sets analogous to those used in multivariate calibration problems, formed by signals recorded from a series of calibration and unknown samples, can be analyzed. Quantitative information is obtained by resolving the system by fixing the known concentration values of the analyte(s) in the calibration samples in the related concentration profile(s) [65].

11.5.2.5 Hard-Modeling Constraints: Physicochemical Models

The most recent progress in chemical constraints refers to the implementation of a physicochemical model into the resolution process [64, 66–73]. In this manner, the concentration profiles of compounds involved in a kinetic or a thermodynamic process are shaped according to the suitable chemical law (see Figure 11.8). A detailed description of methods for fitting kinetic models to multivariate data is provided in Chapter 7.

Such a strategy has been used to reconcile the separate worlds of hard- and soft-modeling and has enabled the mathematical resolution of chemical systems that could not be successfully tackled by either of these two pure methodologies alone. The strictness of the hard-model constraints dramatically decreases the ambiguity of the constrained profiles and provides fitted parameters of physicochemical and analytical interest [64, 66–73], such as equilibrium constants, rate constants, and total analyte concentrations. The soft part of the algorithm allows for modeling of complex systems, where the central reaction system evolves in the presence of absorbing interferents [69, 73].

11.5.2.6 Local-Rank Constraints, Selectivity, and Zero-Concentration Windows

Local-rank constraints are related to mathematical properties of a data set and can be applied to all data sets, regardless of their chemical nature. These types of constraints are associated with the concept of local rank, which describes how the number and distribution of components varies locally along the data set. The key constraint within this family is selectivity. Selectivity constraints can be used in concentration and spectral windows where only one component is present to completely suppress the ambiguity linked to the complementary profiles in the system. Selective concentration windows provide unique spectra of the associated components, and vice versa. The powerful effect of these type of constraints and their direct link with the corresponding

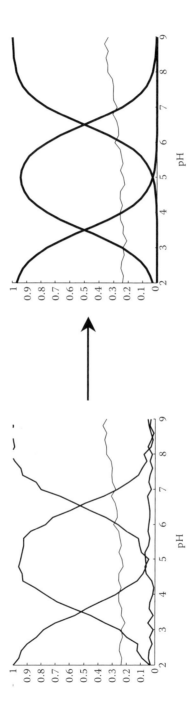

FIGURE 11.8 Effect of the hard-modeling constraint on a set of concentration profiles representing a protonation process in the presence of an interference (left profiles, unconstrained; right profiles, constrained). Only the compounds involved in the protonation are constrained according to the physicochemical law.

concept of chemical selectivity explain their early and wide application in resolution problems [1, 2, 21, 42, 74]. Not so common, but equally recommended is the use of other local-rank constraints in iterative resolution methods [21, 42, 74]. These types of constraints can be used to describe which components are absent in data set windows by setting the number of components inside the windows lower than the total rank. This approach always improves the resolution of profiles and minimizes ambiguity in the final results. Of general applicability, local-rank constraints can be particularly helpful in multicomponent systems like spectroscopic images or mixtures, where more process-related constraints (unimodality, closure, etc.) cannot be used. These rank-related constraints can be set as equality or inequality constraints. Thus, in selective or small rank windows, the profile elements of absent compounds can be set equal to zero (equality constraint). In practice, given the noise of real data, these constraints are more effective if the values of absent species are forced to be lower than a very small threshold value that represents the noise level. This latter implementation corresponds to an inequality constraint.

11.5.3 Iterative Target Transformation Factor Analysis (ITTFA)

As the name suggests, ITTFA is based on target factor analysis (TFA). In TFA, some target vectors with chemical meaning and perfectly characterized, e.g., reference spectra, can be tested to see whether they lie in the space spanned by the data set. If they are found to lie within the space, they can be identified as real sources of variation. In a resolution problem, the appropriate targets should be either prospective concentration profiles or instrumental responses (spectra). However, in practice, the user seldom knows these real profiles or, if he does, the knowledge applies only to some of the components in the system. Therefore, the straightforward application of TFA to resolve a real problem completely is not possible. ITTFA borrows two main ideas from TFA: (1) the space spanned by either the concentration profiles or the spectra can be perfectly known, and (2) the use of an initial target can be used to obtain a true profile of the $\mathbf{C}$ or the $\mathbf{S}^T$ matrix. Essentially, what ITTFA does is to modify a target vector by applying appropriate constraints until it lies within the space spanned by the pure concentration profiles or spectra. This process of testing individual target vectors is repeated with as many target vectors as components in the data set [47, 48].

ITTFA works by optimizing, one at a time, the profiles in either the $\mathbf{C}$ matrix or the $\mathbf{S}^T$ matrix. The direction of optimization can depend on the information available or on the potential ease of resolution. The following explanation holds for a data set where ITTFA is applied to obtain the profiles in the $\mathbf{C}$ matrix. Transposing the original data matrix, the same process would be appropriate for obtaining the profiles in $\mathbf{S}^T$. ITTFA calculates each concentration profile according to the following steps:

1. The score matrix of the data set is calculated by PCA.
2. Estimated concentration profiles are used as initial target vectors.
3. The initial target vectors are projected into the space spanned by the scores.

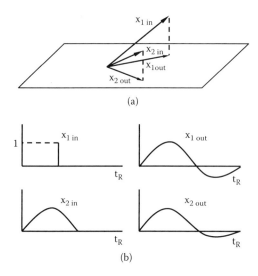

FIGURE 11.9 (a) Geometrical representation of the optimization of a chromatographic profile by ITTFA from an initial needle target. The example represents a two-compound data set. Thick lines represent targets out of the score plane; single lines are targets on the score plane. (b) Evolution in the shape of the chromatographic profile through the ITTFA process.

4. Constraints are applied to the projected target vector.
5. The constrained target vector is projected into the space spanned by the scores.
6. Steps 4 and 5 are repeated until convergence is achieved.

ITTFA starts calculating a PCA model of the original data matrix, $\mathbf{D}$. There is a formal analogy between the PCA decomposition, i.e., $\mathbf{D} = \mathbf{TP}^T$, and the CR decomposition, i.e., $\mathbf{D} = \mathbf{CS}^T$, of a data matrix. The scores matrix, $\mathbf{T}$, and the loadings matrix, $\mathbf{P}^T$, span the same data space as the $\mathbf{C}$ and the $\mathbf{S}^T$ matrices; thus, their profiles can be described as abstract concentration profiles and abstract spectra, respectively. This means that any real concentration profile of $\mathbf{C}$ belongs to the score space and can be described as a linear combination of the abstract concentration profiles in the $\mathbf{T}$ matrix.

The next step is to choose approximate concentration profiles as targets. There are many ways to select these initial vectors, and any method used to provide initial CR estimates can be useful for this purpose. Historically, the vectors obtained after performing VARIMAX rotation onto the scores were used [47] and also the needle targets (i.e., vectors with only one non-null element equal to 1), which are the simplest representation of a peak-shaped profile [48, 75].

The next step is the projection of the initial target ($\mathbf{x}_{1in}$) onto the space spanned by the scores. The projected target ($\mathbf{x}_{1out}$) belongs to the space of the real concentration profiles and, from a purely mathematical point of view, gives an acceptable description of the data set. However, when this profile is plotted, the chemist may not like some of the features that are present (e.g., negative parts, secondary

Multivariate Curve Resolution

maxima, etc.). When this happens, the projected target x_{1out} is modified by using the appropriate constraints. The application of constraints satisfies the chemical features required of the profile but, as a consequence of the modification, pushes the target out of the score plane. The constrained target, x_{2in}, is projected again into the space spanned by the scores, and the newly projected target, x_{2out}, is modified by use of appropriate constraints if necessary. The process goes on until the projected target makes sense from both mathematical and chemical points of view, i.e., until the constrained profile belongs to the score space or until it is very close to it. Figure 11.9 shows the ITTFA optimization of a chromatographic profile for a system with two compounds.

The above process of selecting a target vector and modifying it to match constraints is repeated for each component in the system. Once all of the concentration profiles obtained are appended to form the C matrix, the S^T matrix can be calculated by least squares from D and C.

11.5.4 MULTIVARIATE CURVE RESOLUTION-ALTERNATING LEAST SQUARES (MCR-ALS)

Multivariate curve resolution–alternating least squares (MCR-ALS) uses an alternative approach to iteratively find the matrices of concentration profiles and instrumental responses. In this method, neither the C nor the S^T matrix have priority over each other, and both are optimized at each iterative cycle [7, 21, 42]. The general operating procedure of MCR-ALS includes the following steps:

1. Determine the number of compounds in D.
2. Calculate initial estimates (e.g., C-type matrix).
3. Using the estimate of C, calculate the S^T matrix under appropriately chosen constraints.
4. Using the estimate of S^T, calculate the C matrix under appropriately chosen constraints.
5. From the product of C and S^T found in the above steps of an iterative cycle, calculate an estimate of the original data matrix, D.
6. Repeat steps 3, 4, and 5 until convergence is achieved.

The number of compounds in D can be determined using PCA or can be known beforehand. In any case, the number obtained must not be considered a fixed parameter, and resolution of the system considering different numbers of components is a usual and recommended practice. In contrast to ITTFA, MCR-ALS uses complete C- or S^T-type matrices during the ALS optimization instead of optimizing the profiles one at a time. The core of the method consists of solving the following two least-squares problems under appropriately chosen constraints:

$$\min_{\hat{C}} \left\| \hat{D}_{PCA} - \hat{C}\hat{S}^T \right\| \qquad (11.11)$$

$$\min_{\hat{S}^T} \left\| \hat{D}_{PCA} - \hat{C}\hat{S}^T \right\| \qquad (11.12)$$

In these two equations, the norm of the residuals between the PCA-reproduced data, $\hat{D}_{PCA}$, using the selected number of components, and the ALS-reproduced data using the least-squares estimates of $\mathbf{C}$ and $\mathbf{S^T}$ matrices, $\hat{C}$ and $\hat{S}^T$, is alternatively minimized by keeping constant $\hat{C}$ (Equation 11.11) or $\hat{S}^T$ (Equation 11.12). The least-squares solution of Equation 11.11 is:

$$\hat{S}^T = \hat{C}^+ D_{PCA} \quad (11.13)$$

where $\hat{C}^+$ is the pseudoinverse of the concentration matrix, which for a full-rank matrix gives:

$$\hat{S}^T = (\hat{C}^T\hat{C})^{-1}\hat{C}^T D_{PCA} \quad (11.14)$$

and the least-squares solution of Equation 11.12 is:

$$\hat{C} = D_{PCA}(\hat{S}^T)^+ \quad (11.15)$$

where $(\hat{S}^T)^+$ is the pseudoinverse of the spectra matrix, which for a full-rank matrix gives:

$$\hat{C} = D_{PCA}\hat{S}(\hat{S}^T\hat{S})^{-1} \quad (11.16)$$

Equation 11.11 and Equation 11.12 are solved sequentially, i.e., in each iterative cycle, the concentration matrix $\mathbf{C}$ is calculated and used to get the spectral matrix $\mathbf{S^T}$. Both $\mathbf{C}$ and $\mathbf{S^T}$ are solved under constraints, which can be implemented within the least-squares step [57, 58, 60] or external to the least-squares step. In current software implementations of the MCR-ALS method, different constraints can be applied to the $\mathbf{C}$ or the $\mathbf{S^T}$ matrix and, within each of these matrices, all or some of the profiles can be constrained.

The convergence criterion in the alternating least-squares optimization is based on the comparison of the fit obtained in two consecutive iterations. When the relative difference in fit is below a threshold value, the optimization is finished. Sometimes a maximum number of iterative cycles is used as the stop criterion. This method is very flexible and can be adapted to very diverse real examples, as shown in Section 11.7.

11.6 EXTENSION OF SELF-MODELING CURVE RESOLUTION TO MULTIWAY DATA: MCR-ALS SIMULTANEOUS ANALYSIS OF MULTIPLE CORRELATED DATA MATRICES

The methods presented in previous sections are suitable for use with a data matrix from a single experiment and give results related to the two different directions of the data matrix, i.e., profiles related to the variation along the rows and profiles

Multivariate Curve Resolution

related to the variation along the columns of the data matrix. This is also the reason why a data matrix is called a two-way data set. A data matrix is not the most complex data set that can be found in chemistry. Let us consider a chemical process monitored fluorimetrically. At each reaction time, a series of emission spectra recorded at different excitation wavelengths are obtained. This means that we collect a data matrix at each stage of the reaction. Because the goal is to obtain a picture of the global process, the matrices should be considered altogether. The information about the whole chemical process could be organized into a cube of data (tensor) with three informative directions, i.e., in a three-way data set. Another usual example is coupling data matrices from different HPLC-DAD runs that share all or some of their compounds. In this case, the third direction of the data set accounts for the quantitative differences among runs. Specialized methods have been developed by chemometricians to treat these kinds of problems, and these are covered in greater detail in Chapter 12. An overview of some of the methods useful for multivariate curve resolution is provided here.

Though there is a clear gain in the quality and quantity of information when going from two- to three-way data sets, the mathematical complexity associated with the treatment of three-way data sets can seem, at first sight, a drawback. To avoid this problem, most of the three-way data analysis methods transform the original cube of data into a stack of matrices, where simpler mathematical methods can be applied. This process is often known as *unfolding* (see Figure 11.10).

A cube of data sized $(m \times n \times p)$ can be unfolded in three different directions: along the row space, along the column space, and along the third direction of the cube, also called the tube space. The three unfolding procedures give a row-wise

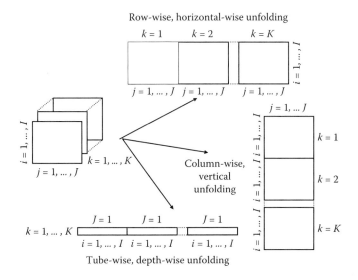

FIGURE 11.10 Three-way data array (cube) unfolding or matricization. Two-way data matrix augmentation.

augmented matrix $\mathbf{D_r}$ ($m \times np$), a columnwise augmented matrix $\mathbf{D_c}$ ($n \times mp$), or a tubewise augmented matrix $\mathbf{D_t}$ ($p \times mn$), respectively (see Figure 11.10). When rank analysis of the three augmented matrices is carried out, the number of components obtained for the three different directions (modes) of the data set may be the same or not. When $\mathbf{D_r}$, $\mathbf{D_c}$, and $\mathbf{D_t}$ have the same rank, the three-way data set is said to be *trilinear*, and when their ranks are different from each other, the data set is *nontrilinear*. (Please note that this definition holds for by far most of the chemical data sets, except those for which phenomena of rank deficiency or rank overlap are present [76].) The resolution of a three-way data set into the matrices $\mathbf{X}$, $\mathbf{Y}$, and $\mathbf{Z}$, which contain the pure profiles related to each of the directions of the three-way data set, changes for trilinear and nontrilinear systems. For trilinear systems, $\mathbf{X}$, $\mathbf{Y}$, and $\mathbf{Z}$ have the same number of profiles (nc), and the three-way core, $\mathbf{C}$, is an identity cube ($nc \times nc \times nc$) whose unity elements are found on the superdiagonal. In this case, the three-way core is often omitted because it does not modify numerically the reproduction of the original tensor. Each element in the original three-way data set can be reproduced as follows:

$$d_{ijk} = \sum_{f=1}^{nc} x_{if} y_{jf} z_{kf} \qquad (11.17)$$

Equation 11.17 is the fundamental expression of the PARAFAC (parallel factor analysis) model [77], which is used to describe the decomposition of trilinear data sets. For nontrilinear systems, the core $\mathbf{C}$ is no longer a regular cube ($ncr \times ncc \times nct$), and the non-null elements are spread out in different manners, depending on each particular data set. The variables ncr, ncc, and nct represent the rank in the row-wise, columnwise, and tubewise augmented data matrices, respectively. Each element in the original data set can now be obtained as shown in Equation 11.18:

$$d_{ijk} = \sum_{f=1}^{ncr} \sum_{g=1}^{ncc} \sum_{h=1}^{nct} x_{if} y_{jg} z_{kh} c_{fgh} \qquad (11.18)$$

Equation 11.18 defines the decomposition of nontrilinear data sets and is the underlying expression of the Tucker3 model [78]. Detailed descriptions of the PARAFAC and Tucker3 models are given in Chapter 12, Section 12.4.

Decompositions of three-way arrays into these two different models require different data analysis methods; therefore, finding out if the internal structure of a three-way data set is trilinear or nontrilinear is essential to ensure the selection of a suitable chemometric method. In the previous paragraphs, the concept of trilinearity was tackled as an exclusively mathematical problem. However, the chemical information is often enough to determine whether a three-way data set presents this feature. How to link chemical knowledge with the mathematical structure of a three-way data set can be easily illustrated with a real example.

Let us consider a three-way data set formed by several HPLC-DAD runs. If the data set is trilinear, $\mathbf{X}$, $\mathbf{Y}$, and $\mathbf{Z}$ will have as many profiles as chemical compounds in the original data set, and this number will be equal to the rank of the data set.

For each chemical compound, there will be only one profile in **X**, in **Y**, and in **Z** common to all of the appended matrices in the original data set. In the case of different HPLC-DAD data runs analyzed simultaneously, the decomposition of the three-way array gives an **X** matrix with chromatographic profiles, a **Y** matrix with pure spectra, and a **Z** matrix with the quantitative information about the amount of each compound in the different chromatographic runs. In this case, a trilinear structure would imply that the shape of the pure spectrum and the pure chromatogram of a compound remain invariant in the different chromatographic runs. If the experimental conditions in the different runs are similar enough, the UV spectrum of a pure compound should not change; however, small run-to-run differences in peak shape and position are commonly found in practice. Assuming that the elution process of the same compound in different runs always yields an identically shaped chromatographic profile does not make sense from a chemical point of view and, therefore, the data set should be considered nontrilinear. In the example related to the fluorimetric monitoring of a kinetic process, the decomposition of the original data set gives a matrix **X** with pure excitation spectra, a matrix **Y** with pure emission spectra, and a matrix **Z** with the kinetic profiles of the process. A trilinear structure would indicate that the shapes of the excitation and emission spectra of a compound do not change at the different reaction times of the kinetic process. This invariability of the spectra is an acceptable statement if the experimental conditions during the process are not modified. Therefore, this data set can be considered trilinear.

In practice, however, most of the systems are nontrilinear, due to either the underlying chemical process (e.g., UV-reaction-monitoring coupling experiments with different reagent ratios) or to the instrumental lack of reproducibility in the response profiles (e.g., chromatographic profiles in different HPLC-DAD runs). Therefore, multivariate curve-resolution methods are mainly focused on the study of real examples lacking the trilinear structure. Despite the higher abundance of nontrilinear data sets, many of the algorithms proposed to study three-way arrays rely on the assumption of trilinear structure. This is the case with the generalized rank annihilation method (GRAM) [79], designed to work with two matrices, or its natural extension, direct trilinear decomposition (DTD) [80], which can handle larger data sets with more appended matrices. Both GRAM and DTD are noniterative methods and use latent variables to resolve the profiles in **X**, **Y**, and **Z**. When these methods are applied to nontrilinear data sets, the profiles obtained often contain imaginary numbers (see Chapter 12, Section 12.6). Iterative methods are also used in three-way data, and the scheme followed in their application is the same as for a single data matrix, i.e., determination of number of components, use of initial estimates, application of constraints, and iterative optimization until convergence. Most of the iterative algorithms are based on least-squares calculations. As for two-way data sets, three-way iterative methods are more flexible in how constraints can be applied and can deal with data sets that are more diverse.

We have noted that three-way resolution methods generally work with the unfolded matrices. Depending on the algorithm used, all three types of unfolded matrices may be used, or only some of them. In the PARAFAC decomposition of a trilinear data set, all three types of unfolded data matrices are used, whereas in the resolution of a nontrilinear data set by the MCR-ALS method, only one type of unfolded matrix is used.

Three-way data sets have been presented as cubes of data formed by appending several matrices together. This means implicitly that all of the data matrices in a tensor should be equally sized; otherwise, the cube cannot be constructed. Additionally, the information in the rows, in the columns, and in the third direction of the array must be synchronized for each of the layers of the cube. For example, in the HPLC-DAD example, if the columns are wavelengths, all of the runs should span the same wavelength range, and if the rows are retention times, the elution time range should also coincide. In experimental measurements, it may not be easy or convenient to fulfill these two requirements. Synchronization can be difficult when the parameter that changes in one of the directions of the array cannot be controlled in a simple manner. Obtaining equally sized matrices may also be inconvenient if synchronization forces the inclusion of irrelevant information in some of the two-way appended arrays. An example of difficult synchronization is the combination of experiments in which a pH-dependent process is monitored by UV spectroscopy. In this case, the pH variations may not be easily reproducible among the experiments. The inconvenience of appending equally sized two-way arrays is also evident when several HPLC-DAD runs of a mixture and matrices of its single standards are treated together. If the standard runs cover the same elution time range as the mixture runs, then most of the information in the standard matrices will be formed by baseline spectra that are not relevant for the resolution of the mixture.

When building a typical three-way data set is not possible, there is no need to give up the simultaneous analysis of a group of matrices that have something in common. Some methods, such as MCR-ALS, are designed to work with only one of the three possible unfolded matrices. This operating procedure greatly relaxes the demands in how the two-way arrays are combined. Indeed, MCR-ALS requires only one common direction in all of the matrices to be analyzed. In both of the previous examples, the common direction is the wavelength range of the spectra collected (see Figure 11.11).

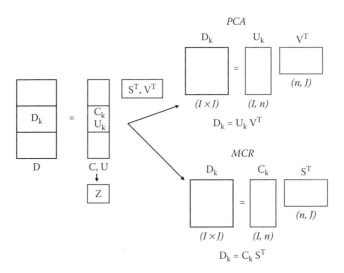

FIGURE 11.11 Bilinear models for three-way data, unfolded PCA and unfolded MCR.

The MCR-ALS decomposition method applied to three-way data can also deal with nontrilinear systems [81]. Whereas the spectrum of each compound of the columnwise augmented matrix is considered to be invariant for all of the matrices, the unfolded **C** matrix allows the profile of each compound in the concentration direction to be different for each appended data matrix. This freedom in the shape of the **C** profiles is appropriate for many problems with a nontrilinear structure. The least-squares problems solved by MCR-ALS, when applied to a three-way data set, are the same as those in Equation 11.11 and Equation 11.12; the only difference is that **D** and **C** are now augmented matrices. The operating procedure of the MCR-ALS method has already been described in Section 11.5.4, but some particulars regarding the treatment of three-way data sets deserve further comment.

In the resolution of a columnwise augmented data matrix, the initial estimates can be either a single S^T matrix or a columnwise augmented **C** matrix. The columnwise concentration matrix is built by placing the initial **C**-type estimates obtained for each data matrix in the three-way data set one on top of each other. The appended initial estimates must be sorted into the same order as the initial data matrices in **D**, and they must keep a correct correspondence of species, i.e., each column in the augmented **C** matrix must be formed by appended concentration profiles related to the same chemical compounds. When no prior information about the identity of the compounds in the different data matrices is available, the correct correspondence of species can be estimated from the two-way resolution results of each single matrix.

The same constraints used in the resolution of a two-way data matrix can be applied to three-way data sets [21, 42]. Selectivity and nonnegativity affect the spectrum and the augmented concentration profile of each species, whereas unimodality is applied separately to each of the profiles appended to form the augmented concentration profile. The closure constraint operates by applying the corresponding closure constant to each of the single matrices in the columnwise concentration matrix. Another constraint specific of three-way data sets is the so-called correspondence among species. In each single matrix of a three-way data set, the concentration profiles of absent compounds are set equal to zero after each iterative cycle.

Although MCR-ALS is especially able to cope with nontrilinear data sets formed by matrices of varying sizes, it can also work with trilinear data sets. Because of the inherent freedom in the modeling of the profiles of the augmented **C** matrix, trilinear structure can be included in the MCR-ALS method as an optional constraint [81, 82]. The application of this constraint is performed separately on the concentration profile of each species. To implement this type of constraint, profiles of a certain species are placed one beside each other to form a new augmented concentration profile matrix, and PCA is performed on it. If the system is trilinear, the score vector related to the first PC will show the real shape of the concentration profile, and the rest of PCs must be related to noise contributions. The loadings related to the first PC are scaling factors accounting for the species concentration level in the different appended matrices. Therefore, the new single profiles can be calculated as the product of the score vector by their corresponding scaling factors. The constrained single profiles are finally appended

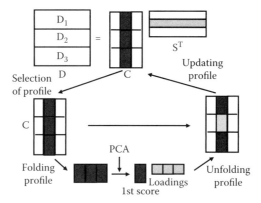

FIGURE 11.12 Implementation of a trilinearity constraint in MCR bilinear models for three-way data.

to form the new augmented concentration profiles. This process is shown graphically in Figure 11.12. In contrast to other three-way resolution methods specially designed to work with trilinear systems, the implementation of this constraint in MCR-ALS need not necessarily be all inclusive, i.e., some or all of the compounds can be forced to have common profiles in the **C** matrix. This flexibility allows a more representative modeling of some real situations, such as in (a) systems with trilinear profiles related to the evolution of chemical compounds common to each experiment and (b) freely modeled profiles related to important background contributions that differ in each experiment.

The information in the third direction of the array, i.e., the **Z** matrix, is directly extracted from the augmented matrix **C** in MCR-ALS. This dimension of the data set is usually the smallest in size and represents scaling differences among the appended matrices. Because the S^T profile of each compound is common to all of the appended data matrices, the area of the concentration profile of each compound is scaled according to the concentration level of the species in each single data matrix. Thus, the profile of a compound in the **Z** matrix accounts for the relative concentration of a particular compound in each of the appended matrices and can be obtained from (a) the ratio between the area of its concentration profile in a given matrix and (b) the area related to the concentration profile of the same compound in a matrix taken as a reference.

11.7 UNCERTAINTY IN RESOLUTION RESULTS, RANGE OF FEASIBLE SOLUTIONS, AND ERROR IN RESOLUTION

The main sources of uncertainty associated with the resolution results are the ambiguity of the recovered profiles and the experimental noise of the data. Providing methodologies to quantify this uncertainty is not only a topic of interest in the current literature, but a necessary requirement to enable the use of resolution methods in standard analytical procedures.

The possible existence of ambiguity in resolution results has been known since the earliest research in this area [1, 2, 23]. After years of experience, it has been possible to set resolution theorems that indicate clearly the conditions needed to recover uniquely the pure concentration and signal profiles of a compound in a data set. These conditions depend mainly on the degree of selectivity in the column mode or row mode of the measurements. The degree of selectivity, in turn, depends on (a) the amount of overlap in the region of occurrence for the compound of interest with the rest of constituents and (b) the general distribution of the different compound windows in the data set [21, 22]. Therefore, in the same system, some profiles can be recovered uniquely, and some others will necessarily be affected by a certain ambiguity. When ambiguity exists, a compound is represented by a band of feasible solutions instead of a unique profile. Calculating the boundaries of these bands is not straightforward, and the first attempts proposed were valid only for systems with two or three components [1–4]. More recent approaches extended their applicability to systems with no limit in the number of contributions [83]. The most recent methods use optimization strategies to find the minimum and maximum solution boundaries by minimizing and maximizing objective functions subject to selected constraints. The objective functions represent the ratio between the signal contributed by the compound of interest and the total signal from all compounds in the data set [84, 85]. These strategies are more powerful than previous ones and allow for an accurate study of the effect of the different constraints in the magnitude of the bands of feasible solutions (see Figure 11.13).

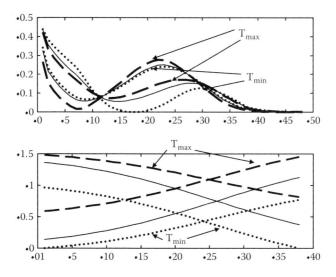

FIGURE 11.13 Effect of constraints in feasible bands. T_{max} and T_{min} define maximum and minimum feasible ranges/bands (dashed and dotted lines, respectively) around the resolved solutions (solid lines).

Even in the absence of ambiguity, the experimental error contained in real data propagates into the resolution results. This source of uncertainty affects the results of all kinds of data analysis methods and, in simpler approaches, like multivariate or univariate calibration, is easily quantified with the use of well-established and generally accepted figures of merit. Although some figures of merit have been proposed for higher-order calibration methods [86], finding analytical expressions to assess the error associated with resolution results is an extremely complex problem because of the huge number of nonlinear parameters that are calculated, as many as the number of elements in all of the pure profiles resolved. To overcome this problem and still give a reliable approximate estimation of the error propagation in resolution, other strategies known under the general name of resampling methods are used [87, 88]. In these strategies, an estimate of the dispersion in the resolution results is obtained by the resolution of a huge number of replicates. To simulate these replicates, the complete data set can be resolved multiple times after adding a certain amount of noise on top of the experimental measurements (noise-added method), or the replicates of the data set can be constructed by addition of a certain amount of noise to a noise-free simulated or reproduced data set (Monte Carlo simulations). Finally, a data matrix can be resolved repeatedly after removing different rows or columns, or in the case of a three-way data set, by removing complete data matrices (jackknife). These strategies provide an enormous number of results from the different resolution runs to allow for an estimation of the uncertainty due to the noise in the resolved profiles. The estimate of uncertainty in resolved profiles in turn allows for the computation of the accuracy of parameters estimated from the resolved profiles, such as rate constants or equilibrium constants [64]. Resampling and Monte Carlo simulation methods have been recently proposed for estimating uncertainty of multivariate curve-resolution profiles and of the parameters derived from them [89].

Although ambiguity and noise are two distinct sources of uncertainty in resolution, their effect on the resolution results cannot be considered independently. For example, the boundaries of the compound windows can be clearly blurred due to the effects of noise, and this can give rise to ambiguities that would be absent in noise-free data sets. A definite advance would be the development of approaches that can consider this combined effect in the estimation of resolution uncertainty.

11.8 APPLICATIONS

Within the variety of multicomponent systems, "processes" and "mixtures" can be placed at the two extremes. The term "process" holds for reaction data, where the compositional changes respond to a known physicochemical model, or for any evolving chemical system (e.g., a chromatographic elution) whose sequential compositional variation is caused by physical or chemical changes and whose underlying physicochemical model, if any, is too complex or simply unknown. "Mixtures" would have a completely random variation along the compositional direction of the data set. An example could be a series of spectra collected from independent multicomponent samples. Other data sets lie between these two extremes because they lack the global continuous compositional evolution of a process, although they can present it locally. For example, spectroscopic images can have a smooth compositional variation in neighboring pixels. In this respect, environmental data sets are similar, where close geographical sampling points can be compositionally related.

The examples that are given in the following subsections show the power of multivariate curve resolution to resolve very diverse chemical problems. Different strategies adapted to the chemical and mathematical features of the data sets are chosen, and resolution of two-way or three-way data sets is carried out according to the information that has to be recovered. Because MCR-ALS has proved to be a very versatile resolution method, able to deal with two-way and three-way data sets, this is the method used in all of the following examples.

11.8.1 BIOCHEMICAL PROCESSES

Biochemical processes are among the most challenging and interesting reaction systems. Due to the nature of the constituents involved, macromolecules such as nucleic acids or proteins, the processes to be analyzed do not follow a simple physicochemical model, and their mechanism cannot be easily predicted. For example, well-known reactions for simple molecules, e.g., protonation equilibria, increase in complexity for macromolecules due to the presence of polyelectrolytic effects or conformational transitions. Because the data analysis cannot be supported in a model-fitting procedure (hard-modeling methods), the analysis of these processes requires soft-modeling methods that can unravel the contributions of the process without the assumption of an *a priori* model.

Examples of biochemical processes successfully studied by spectroscopic monitoring and multivariate resolution techniques include protonation and complexation of nucleic acids and other events linked to these biomolecules, such as drug intercalation processes and salt, solvent, or temperature-induced conformational transitions [90–97]. In general, any change (thermodynamic or structural) that these biomolecules undergo is manifested through a distinct variation in an instrumental signal (usually spectroscopic) and can be potentially analyzed by multivariate resolution techniques.

A relevant example in this field is the study of protein folding processes [94]. They have an intrinsic biochemical interest linked to the relationship between protein structure and biological activity. Protein structure is organized into four hierarchical levels, including primary structure (the sequence of amino acids in the polypeptide chains), secondary structure (the regular spatial arrangements of the backbone of the polypeptide chain stabilized by hydrogen bonds that give rise to helical or flat sheet arrangements), tertiary structure (spatial arrangement of the secondary structure motifs within a polypeptide chain, responsible for the globular or fibrillar nature of proteins), and quaternary structure (union of several polypeptide chains by weak forces or disulfide bridges). All proteins must adopt specific three-dimensional folded structures to acquire the so-called native conformation and be active, i.e., their secondary and tertiary structure should be fully organized.

Protein folding can take place as a one-step process, where only the folded or native (N) and the unfolded (U) states are detected, or as a multistep process, where intermediate conformations occur. The so-called molten globular state has often been reported as a well-characterized intermediate conformation that shows organized secondary structure motifs and unordered tertiary structure [94–96]. Indeed, protein folding events most often follow one of the two following mechanisms: (a) one-step process: Native (N) ↔ Unfolded (U); or (b) two-step process: Native (N) ↔ Intermediate ("molten globule") (I) ↔ Unfolded (U). The detection and characterization of intermediate conformations is not easy because either the lifetime of these transient intermediates is frequently too short to be detected, or it is not possible to separate and isolate them from other protein conformations simultaneously present.

The mechanism and the identity of the protein conformations involved in a folding process can be studied by monitoring spectrometrically the changes in the protein tertiary and secondary structures. Far-UV circular dichroism has long been used to elucidate the secondary structure of proteins and to follow specifically changes in this structural level. Near-UV circular dichroism is known to be sensitive to changes in the protein tertiary structure. These two techniques were used to monitor the thermal unfolding of bovine α-apolactalbumin, a globular protein present in milk with major alpha helical content in the secondary structure (see Figure 11.14).

Protein folding can be studied by recording a complete spectrum at each stage in the monitored process. The spectra recorded in a thermal-dependent protein folding process are organized in a data matrix $\mathbf{D}$, where rows represent spectra recorded at each temperature and columns represent the melting curves (absorbance vs. temperature profiles) at each wavelength. Recalling the general expression of resolution methods (Equation 11.1), $\mathbf{D} = \mathbf{C}\,\mathbf{S}^T + \mathbf{E}$, the columns in matrix $\mathbf{C}$ represent

FIGURE 11.14 Structure of α-apolactalbumin.

Multivariate Curve Resolution 451

the thermal-dependent concentration profiles of the detected protein conformations, and the rows in matrix S^T represent their corresponding pure spectra. Matrix E describes the experimental error.

The results presented below were obtained by multivariate curve resolution–alternating least squares (MCR-ALS). MCR-ALS was selected because of its flexibility in the application of constraints and its ability to handle either one data matrix (two-way data sets) or several data matrices together (three-way data sets). MCR-ALS has been applied to the folding process monitored using only one spectroscopic technique and to a row-wise augmented matrix, obtained by appending spectroscopic measurements from several different techniques.

In general, MCR-ALS resolution analyses of protein folding have been performed using initial estimates obtained by evolving-factor analysis (EFA), the method most suitable to describe the evolution of the contributions present in processes. The concentration profiles in C were constrained to be nonnegative and unimodal with closure. Unimodality was used because the evolution of each protein conformation can be appropriately represented by an emergence-decay profile having a single peak maximum as long as the temperature changes during the experiment always increase or always decrease. The condition of closure is appropriate in these systems because the total concentration of protein remains constant during the unfolding/folding process. Selectivity constraints were also applied at the lowest temperatures, where only the native conformation of the protein is supposed to be present. Circular dichroism (CD) spectra in S^T are not forced to be nonnegative because negative ellipticities can naturally occur in CD spectra.

Different arrangements of the data from these experiments have allowed the study of several aspects linked to protein folding, namely: (a) changes in the protein secondary structure, (b) changes in the protein tertiary structure, and (c) global mechanistic and structural description of the protein-folding process. The results obtained are briefly presented in the following subsections.

11.8.1.1 Study of Changes in the Protein Secondary Structure

Changes in protein secondary structure were studied by the analysis of the far-UV (190 to 250 nm) circular dichroism spectra from a single data matrix. Resolved concentration and spectra profiles are shown in Figure 11.15a. The resolved CD spectrum associated with the native conformation shows the typical spectral features associated with a major contribution of the α-helix motif in its secondary structure, i.e., an intense negative band with two shoulders located around 220 and 210 nm [95, 96]. The resolved spectrum for the unfolded conformation shows the typical spectral features associated with a random coil motif (a sharper negative band at short wavelengths and weaker features around 220 nm). The resolved concentration profiles show the thermal evolution of the concentration of the different protein conformations in the process. The crossing point in the plot of these concentration profiles gives the temperature at which 50% of the native protein has acquired its native secondary structure (T_{sec}), which is about 60°C.

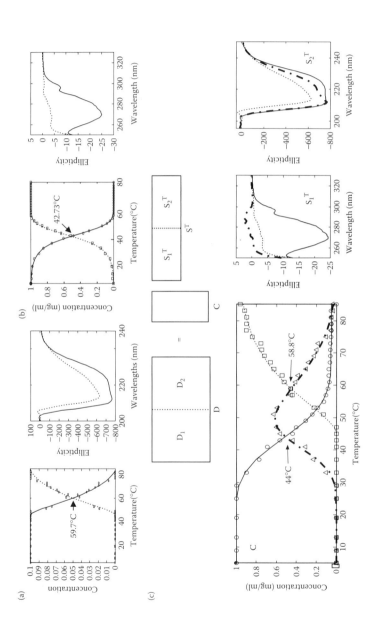

FIGURE 11.15 Resolution of the protein folding of α-apolactalbumin. (a) Detection of changes in protein secondary structure (far-UV circular dichroism measurements). (b) Detection of changes in protein tertiary structure (near-UV circular dichroism measurements). (c) Complete description of protein folding. Resolution of the row-wise data set formed by near-UV (D_1) and far-UV (D_2) circular dichroism measurements. Solid line: native conformation, dash-dotted line: intermediate conformation, dotted line: unfolded conformation.

11.8.1.2 Study of Changes in the Tertiary Structure

Changes in tertiary structure during protein folding processes have been studied using near-UV (250 to 330 nm) circular dichroism. Figure 11.15b shows the MCR-ALS resolved concentration profiles and spectra related to the different tertiary structures present during the folding of α-apolactalbumin. The CD spectrum of the folded protein conformation shows a very intense negative band [95], whereas the denatured/unfolded species shows a nearly flat signal in the CD spectrum. In this experiment, the crossing point in the plot of concentration profiles represents the temperature at which 50% of the native protein has acquired its initial tertiary structure (T_{tert}), which happens at about 40°C for α-apolactalbumin.

11.8.1.3 Global Description of the Protein Folding Process

To check for the presence of an intermediate in a protein folding process, the temperatures at which the secondary structure (T_{sec}) and the tertiary structure (T_{tert}) of the folded conformation are half-formed can be compared. If both coincide, the protein loses the tertiary and the secondary structures simultaneously, and only a native conformation with secondary and tertiary structures ordered or an unfolded conformation with both structural levels unordered describe the process. If significant differences are observed in the crossing temperatures of concentration profiles, a new, intermediate third species with the secondary structure ordered and the tertiary unordered may be needed to explain the shift in the appearance of the tertiary and secondary structures. The difference of almost 20°C found between T_{sec} and T_{tert} in the above two experiments seems to guarantee the presence of an intermediate conformation in the folding of α-apolactalbumin, but only the multivariate resolution analysis of the suitable measurements (far-UV and near-UV CD spectra) together can confirm this hypothesis and model the appearance of the intermediate conformation.

Figure 11.15c shows the resolved concentration profiles and spectra coming from the row-wise appended matrix containing data from the three techniques mentioned previously. The need for one additional intermediate conformation has been proven to be necessary to explain the protein folding process of α-apolactalbumin. Additionally, the thermal range of occurrence and the evolution of this intermediate can now be known. The resolved spectrum obtained for the α-apolactalbumin intermediate shows that it has an ordered secondary structure similar to the native folded protein and an unordered tertiary structure similar to the unfolded protein at high temperatures. These spectral features match the spectral description attributed to the molten globular state and provide additional evidence to confirm the presence of this species as a real intermediate conformation.

As has been shown, complex protein folding processes involving the presence of intermediate conformations can be successfully described combining multispectroscopic monitoring and multivariate curve resolution. The detection and modeling of intermediate species that cannot be isolated either by physical or chemical means is fully achieved. The fate of the intermediate during the process, i.e., when it is present and in what amount, is unraveled from the original raw measurements.

The spectral information obtained by using appropriate deconvolution approaches, particularly the resolved pure CD spectrum of the intermediate, is the essential starting point, unobtainable by other methods, for deducing the secondary structure of the intermediate.

11.8.2 ENVIRONMENTAL DATA

Principal component analysis and multivariate curve resolution are powerful tools for the investigation and modeling of large multivariate environmental data arrays measured over long periods of time in environmental monitoring programs [98]. The goals of these studies are the computation, resolution, modeling, screening, and graphical display of patterns in large environmental data sets, looking for possible data groupings and sources of environmental pollution, as well as for their temporal and geographical distribution. The fundamental assumption in these studies is that variance in the measured concentrations of contaminants (or properties) can be attributed to a small number of contamination sources of different origin (industrial, agricultural, etc.) and that they can be modeled by profiles describing their chemical composition, their geographical distribution, and their temporal distribution. Large environmental analytical data arrays containing concentration information of multiple chemical compounds collected at different sampling sites and at different sampling periods are arranged in large data tables or matrices, or in more complex data arrays according to different dimensions, ways, modes, orders, or directions of experimental measurement. In the chemometrics literature, these complex ordered data arrays are commonly called multiway data arrays or higher-order tensor data.

As discussed in Chapter 4, principal component analysis, PCA [99] is one of the multivariate data analysis methods more frequently used in exploratory analysis and modeling of two-way data arrays (data tables or data matrices). PCA allows the transformation and visualization of complex data sets into a new perspective in which the most relevant information is made more obvious. In environmental studies, by use of PCA, main contamination sources can be identified, and their geographical and temporal distributions can be interpreted and further investigated. Multivariate curve resolution using alternating least squares (MCR-ALS) has also been proposed to achieve similar goals [15, 16, 98, 100, 101]. Although this method is traditionally used for curve-resolution purposes, i.e., to resolve spectra and concentration profiles having a smooth curved shape, there is no fundamental reason why this method cannot also be used for resolution and modeling of noncurved and nonsmooth types of profiles, including those describing geographical and temporal pollution patterns observed in environmental data studies.

Both PCA and MCR-ALS can be easily extended to complex data arrays ordered in more than two ways or modes, giving three-way data arrays (data cubes or parallelepipeds) or multiway data arrays. In PCA and MCR-ALS, the multiway data set is unfolded prior to data analysis to give an augmented two-way data matrix. After analysis is complete, the resolved two-way profiles can be regrouped to recover the profiles in the three modes. The current state of the art in multiway data analysis includes, however, other methods where the structure of the multiway data array is explicitly built into the model and fixed during the resolution process. Among these methods, those based in parallel factor analysis (PARAFAC) and Tucker multiway models have been developed in recent years [77, 78] and have been used for the analysis of environmental data sets [102, 103]. A complete description of these methods is given in Chapter 12.

In the example described here, multiway principal component analysis and multivariate curve resolution are compared in the analysis of a data set obtained in an exhaustive and systematic monitoring program in Portugal [16]. The study included measurements of 19 priority semivolatile organic compounds (SVOC) in a total of 644 surface-water samples distributed among 46 different geographical sites during a period of 14 months. These data arrays were organized into 14 data matrices, each one corresponding to one month of the sampling campaign (Figure 11.16). The resulting data set was arranged as one single columnwise augmented large concentration data matrix of dimensions 644 × 19. In the columnwise augmented data matrix, the individual concentration data matrices from different months were stacked consecutively one on top of the other. Only one of the three modes is unambiguous in the columns, the composition mode (19 SVOCs). The other two modes (geographical and temporal) are mixed in the rows of the columnwise data matrix.

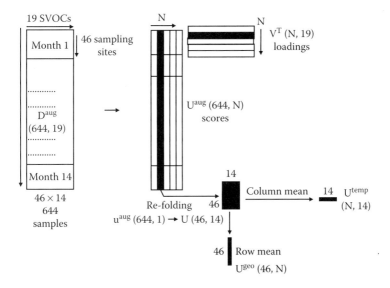

FIGURE 11.16 Data matrix augmentation arrangement and bilinear model for PCA or MCR-ALS decompositions. Resolved loadings, $\mathbf{V}^T$, provide the identification of the main sources of data variance (contamination sources). Resolved scores, $\mathbf{U}^{aug}$, provide the identification of the temporal and geographical distribution of these sources after appropriate refolding: $\mathbf{U}^{geo}$ geographical distribution scores and $\mathbf{U}^{temp}$ temporal distribution scores. For each component, a scores matrix is obtained by refolding its resolved long augmented score vector. Taking averages of the rows or columns of this score matrix, temporal and geographical distributions are obtained for this component.

SVOC concentration values below the limit of detection and nondetected values were set to half the detection limit value [104]. Missing values were estimated by PCA using the PLS Toolbox "missing" function (Eigenvector Research, Inc., Manson, WA) for MATLAB. Different data pretreatment methods were tested and compared. They included column mean centering, column autoscaling, and log transformation and they were performed on the columns of the columnwise augmented data matrix. Column mean centering removed constant background contributions, which usually were of no interest for data variance interpretation. However, in this particular case, mean centering caused little changes in the results, since most of the values of the different variables (SVOC) were so low that their averages were also low and close to zero. Column scaling to unit variance increased the weight of variables that initially had lower variances. In some cases, this effect may distort significantly the results, making interpretation more difficult, especially for those noisy variables having only very few values higher than the detection limit. Log transformation of experimental data is another procedure that has been frequently recommended in the literature for skewed data sets, like those in environmental studies where the majority of the values are low values with a minor contribution of high values. With log data pretreatment, a more symmetrical distribution of experimental data is expected. To remove large negative values, a constant value equal to 1 was added to all of the variable entries. In this way, log values were always nonnegative [105].

Principal component analysis (PCA) and multivariate curve resolution–alternating least squares (MCR-ALS) were applied to the augmented columnwise data matrix $\mathbf{D}^{aug}$, as shown in Figure 11.16. In both cases, a linear mixture model was assumed to explain the observed data variance using a reduced number of contamination sources. The bilinear data matrix decomposition used in both cases can be written by Equation 11.19:

$$d_{ij}^{aug} = \sum_{n=1}^{N} u_{in}^{aug} v_{jn}^{aug} + e_{ij} \qquad (11.19)$$

In this equation, d^{aug}_{ij} is the concentration of SVOC j in sample i in the augmented experimental data matrix $\mathbf{D}^{aug}$. The variable u^{aug}_{in} (the score of component n on row i) is the contribution of contamination source n in sample i. The variable v_{jn} (the loading of variable j on component n) is the contribution of SVOC j in contamination source n. The residual, e_{ij}, is the variance in sample i and variable j of d_{ij} not modeled by the N environmental contamination sources. The same equation can be written in matrix form as:

$$\mathbf{D}^{aug} = \mathbf{U}^{aug} \mathbf{V}^{T} + \mathbf{E}^{aug} \qquad (11.20)$$

where $\mathbf{D}^{aug}$ is the whole data array arranged in an augmented data matrix of dimensions 46 × 14 rows (46 sampling sites, 14 months) and 19 columns (SVOC), as shown in Figure 11.16.

Equation 11.20 describes the factorization of the experimental data matrix into two factor matrices, the loadings matrix $\mathbf{V}^{T}$ and the augmented scores matrix $\mathbf{U}^{aug}$. The loadings matrix $\mathbf{V}^{T}$ identifies the nature and composition of the N main contamination sources defined by means of their chemical composition (SVOC concentrations) profiles. The augmented scores matrix U^{aug} gives the geographical and temporal distribution of these contamination sources. This geographical and temporal information is intermixed in the columns of the resolved augmented scores matrix U^{aug}, and it is not directly available from it. A relatively easy way to recover geographical (U^{geo}) and temporal (U^{temp}) information is by properly rearranging columns of scores (U^{aug}) into matrices having dimensions 46 × 14 (46 geographical sites × 14 months), as shown in Figure 11.16. Analysis of the resulting matrices by taking the row and column averages or by computing the SVD gives the average contribution of the resolved source profiles as a function of geographical location (46 × N) or by month (N × 14), as shown in Figure 11.16. In this way, U^{geo}, U^{temp}, and V^T (the three mode components) were estimated and directly compared with those resolved by three-way model-based methods like PARAFAC and Tucker3 [77, 78]. Finally, E^{aug} gives the residual part of D^{aug} not modeled by the N contamination sources, i.e., the unexplained variance associated with noise and minor nonmodeled environmental contamination sources. The proper complexity of the PCA model or MCR-ALS model, i.e., the number of components or contamination sources included in the model, is a compromise between different goals: model simplicity (few components), maximum variance explained by the model (more components), and model interpretability.

Whereas PCA models provide a least-squares solution of Equations 11.19 and 11.20 under orthogonal constraints and maximum variance explained by each successively extracted component, MCR-ALS models give a nonnegative least-squares solution of the same equation without use of orthogonal constraints or maximum explained variance. Also, whereas PCA orthogonal solutions of Equations 11.19 and 11.20 for a two-way data matrix are unique, MCR-ALS solutions of the same equation are not unique, and they may be rotationally ambiguous [1, 2, 21–23]. However, MCR-ALS models provide solutions that hopefully are more similar to the real sources of pollution than PCA solutions. MCR-ALS solutions can be seen as oblique rotated PCA solutions fulfilling nonnegativity constraints. MCR-ALS models provide a complementary insight to the problem under study by helping to resolve and interpret real environmental sources of data variance. In other works, MCR-ALS has been shown to be a powerful tool for resolving species profiles in different chemical systems [7, 10–12, 17, 20, 21, 42, 45, 46, 51, 55, 62, 67, 68, 74, 81, 82, 90–92] and, more recently, it has also been applied for the resolution of environmental contamination sources [15, 16, 98, 100, 101].

MCR-ALS solutions can be additionally constrained to fulfill a trilinear model [82]. When this trilinearity constraint is applied, the profiles in the three different modes (U^{geo}, U^{temp}, and V^T) are directly recovered and can be compared with the profiles obtained using PARAFAC- or Tucker-based model decompositions. MCR-ALS results have already been compared with Tucker3-ALS and PARAFAC-ALS results in the resolution of different chemical systems [81].

Table 11.1 gives the results from the application of PCA on column mean-centered data, on column autoscaled data, and on log-transformed column mean-centered data. Using five PCs, the amount of variance explained was 84.0, 46.1, and 69.6%, respectively. The results for the column mean-centered data were nearly identical to those obtained for the nonmean-centered raw data. The reason for this is that the means of

TABLE 11.1
Explained Variances by PCA for Three Data Pretreatment Methods(%)

Data Pretreatment Method	PC Number				
	1	2	3	4	5
Mean-centered data	29.1 (29.1)	25.1 (54.2)	13.9 (68.1)	8.7 (76.8)	7.2 (84.0)
Autoscaled data	11.8 (11.8)	11.4 (23.2)	9.2 (32.4)	7.1 (39.6)	6.5 (46.1)
Log-transformed mean-centered data	28.3 (28.3)	15.7 (44.0)	9.9 (53.9)	8.5 (62.4)	7.2 (69.6)

Note: Explained variances are given individually for each PC and for the total number of considered components (in parentheses).

the concentrations of the different analyzed compounds are always low and very close to zero. To avoid bias in the interpretation toward one of these treatments, the complete study was performed using these three data pretreatments. However, to keep this summary concise, the PCA results for the column mean-centered data are given first and described in detail. Only the main trends are summarized for the results obtained with other pretreatment methods. PCA results for the column mean-centered data were selected first because they provided a simpler interpretation of the main contamination sources. The PCA results (loadings and scores) for the first five components are given in Figure 11.17 and Figure 11.18. Environmental interpretation of these results is given in detail in a previous work [16].

When the data were autoscaled, the variance was more evenly distributed among principal components. The variance explained by the PC1 and PC2 was relatively lower than for mean-centered data (Table 11.1). Industrial emission point contamination sources were easily perceived using autoscaled data because individual events and point contamination sources counted much less in the final results. On the other hand, the weighting of variables corresponding to organic compounds present at low uncertain concentration levels increased considerably, and they gave large contributions to the loadings of the first principal component. The autoscaled PCA results were more difficult to interpret, and they did not provide further relevant information apart from that revealed previously using the mean-centered data. The log-transformed mean-centered data gave PCA results very similar to those obtained using mean-centered data. Only some differences in the order of the components explaining small amounts of variance were observed. As indicated in Table 11.1, the percentage of variance explained by each component in the log-transformed results was lower than for the raw mean-centered data, but still considerably higher than for the autoscaled data.

Figure 11.19 shows the results obtained after applying MCR-ALS with nonnegativity constraints to the same SVOC data set shown in Figure 11.16. Using five components, the explained data variance was 84.1%, very close to the value obtained by PCA (84.4%) for mean-centered data. MCR-ALS was directly applied to raw data, without any further data pretreatment apart from imputation of missing data (PLS Toolbox *missing.m* function) and setting values below the detection limit to

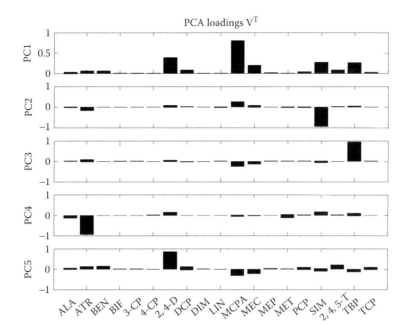

FIGURE 11.17 PCA loadings for raw mean-centered augmented data matrix; from top to bottom, first to fifth (PC1 to PC5) principal components. Compound names and abbreviations are as follows: alachlor (ALA), atrazine (ATR), bentazone (BEN), biphenyl (BIF), 3-chlorophenol (3-CP), 4-chlorophenol (4-CP), (2,4-dichlorophenosy)acetic acid (2,4-D), dichloroprop (DCP), dimethoate (DIM), linuron (LIN), h-chloro-z-methyphenoxyacetic acid (MCPA), mecoprop (MEC), 4-chloro-3-methylphenol (MEP), metholachlor (MET), pentachlorophenol (PCP), simazine (SIM), (2,4,5-trichlorophenoxy)acetic acid (2,4,5-T), tributylphosphate (TBP), 2,4,6-trichlorophenol (TCP).

half the detection limit value, as previously described. The components resolved using MCR-ALS were not orthogonal and, therefore, the variances explained by each component overlap and their sums will not be equal to the total variance explained by all of them simultaneously. This fact might be considered to make identification and interpretation of contamination sources more difficult; however, on the other hand, real sources of contamination are not orthogonal and they do overlap. MCR-ALS attempts their resolution. Environmental interpretation of these results is given in detail in a previous publication [16].

In general, interpretation of the results obtained by MCR-ALS is similar to that obtained by PCA. Geographical and temporal distributions were deduced from resolved scores, once they were properly rearranged and averaged (See Figure 11.16). Geographical and temporal distributions were quite similar to those observed by PCA scores, with a similar interpretation. If trilinearity was also applied as a constraint during MCR-ALS resolution [82], the three mode profiles, the loadings $\mathbf{V}^T$, and scores $\mathbf{U}^{geo}$ and $\mathbf{U}^{temp}$ would be obtained directly from the data analysis. However, due to this trilinearity constraint, the amount of variance explained for the same

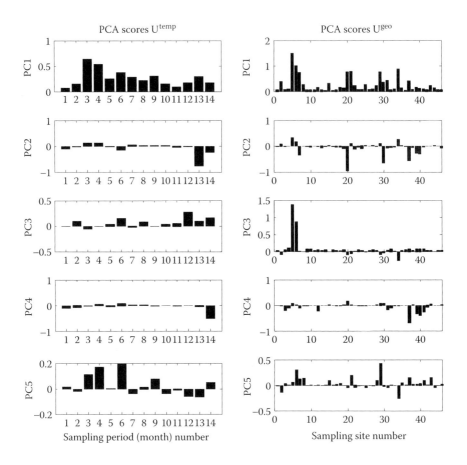

FIGURE 11.18 PCA scores for raw mean-centered augmented data matrix. Temporal (left) and geographical (right) distribution of resolved composition profiles obtained after appropriate augmented scores matrix refolding. From top to bottom, first to fifth (PC1 to PC5) principal components. Lower code numbers are for northern samples and higher code numbers are for southern areas. Sampling time periods are numbered from 1 to 14 for each month from April 1999 to May 2000.

number of components decreased to 62%. Loadings of the five resolved components were rather similar to those obtained in previous analysis when the trilinearity constraint was not applied. Larger differences were obtained for resolved geographical and temporal score profiles, U^{geo} and U^{temp}, which were now simpler to interpret, having a lower number of large score values. The main trends were similar to those previously found in the results obtained without applying the trilinearity constraint. Results obtained using the trilinearity constraint were, in fact, practically identical to those obtained using the PARAFAC-ALS algorithm.

In summary, using either principal component analysis or multivariate curve-resolution methods, the main contamination sources of semivolatile organic compounds present in the surface waters of Portugal were identified, and their geographical

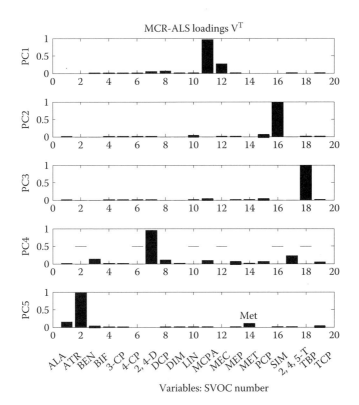

FIGURE 11.19 Composition (loading) profiles of resolved components in the MCR-ALS analysis of raw augmented data matrix. Top components explain more variance; bottom components explain less variance. Names for the compounds are defined in the caption to Figure 11.17.

and temporal distributions were estimated. See Tauler et al. [16] for a more detailed description of the chemometric and environmental interpretation of the results obtained in this study. A similar interpretation of the contamination sources (loadings) and of their geographical and temporal distributions (scores) was possible using these two different methods. The proposed method for averaging unfolded PCA and MCR-ALS scores was found to be very useful for recovering geographical and temporal information from score profiles as an alternative to fitting three-way models like PARAFAC and Tucker3. Similar multiway data analysis approaches are being proposed [106, 107] for resolution and apportionment of environmental sources in air contamination studies.

11.8.3 Spectroscopic Images

A spectroscopic image consists of a group of spectra collected along localized points (pixels) spread on a sample surface [108–110]. These measurement can be displayed as a data cube with two dimensions related to the x, y surface coordinates of the pixels and a third dimension related to the spectral wavelengths recorded (see Figure 11.20).

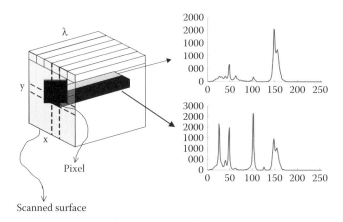

FIGURE 11.20 Structure of a spectroscopic image.

An image data set can easily contain hundreds of thousands of measurements, and a common concern is how to handle such a huge data set. Classical approaches tend to work with reduced representations of the image, where only one measurement per pixel, most commonly the cumulative intensity or, alternatively, the signal at a particular selective wavelength, is used [111–113]. These methods, while simple, do not take advantage of the quality and amount of information contained in the raw measurements.

Resolution methods adapt easily to the huge size and complex spatial structure of spectroscopic images, and they work with the complete information in the data set [46, 114]. Most techniques in spectroscopic imaging provide signals for which the intensity is proportional to the concentration of compounds in each pixel, following a relationship equivalent to the Beer-Lambert law in spectroscopy. Therefore, the measurement variation in an image data set follows a bilinear model, where the mixture spectra recorded at each pixel are described by the concentration-weighted sum of the pure signals of the chemical compounds present. The pure spectra and distribution maps for the different compounds are then resolved as in any other two-way data set. The only additional operations required are purely formal and consist of unfolding the original image cube into a matrix of pixel spectra and then refolding the elements in the resolved concentration profiles according to the original spatial structure of the image to produce the distribution map (see Figure 11.21).

The use and potential of multivariate resolution for the analysis of spectroscopic images is illustrated using a real example from pharmaceutical quality control, where the detection of minor impurities in a tablet or of an uneven distribution of compounds are relevant issues. The measurements used are a series of Raman images of tablets, with different active-compound to excipient-composition ratios ranging from 0 to 80% of the active ingredient. All images were 45×45 pixels in size and contained 576 readings per pixel spectrum recorded in the spectral range of 609 to 1173 cm^{-1}.

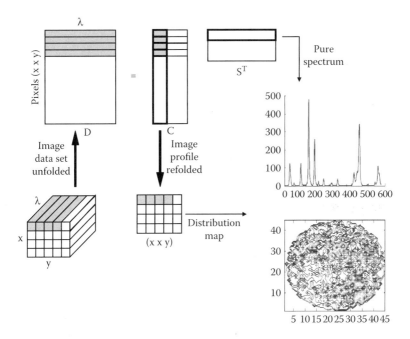

FIGURE 11.21 Resolution of a spectroscopic image.

Exploration of a data set before resolution is a golden rule fully applicable to image analysis. In this context, there are two important domains of information in the data set: the spectral domain and the spatial domain. Using a method for the selection of pure variables like SIMPLISMA [53], we can select the pixels with the most dissimilar spectra. As in the resolution of other types of data sets, these spectra are good *initial estimates* to start the constrained optimization of matrices **C** and **S**^{T}. The spatial dimension of an image is what makes these types of measurement different from other chemical data sets, since it provides local information about the sample through pixel-to-pixel spectral variations. This local character can be exploited with chemometric tools based on local-rank analysis, like FSMW-EFA [30, 31], explained in Section 11.3.

Fixed-size image window–evolving-factor analysis (FSIW-EFA) [46] is a modification of FSMW-EFA and works by analyzing small windows constructed taking the spectra of a pixel and its most immediate neighbors. PCA analysis is performed on the areas around each of the pixels in the image. Joining all of the results obtained, we can plot a local-rank map, which represents the number of components (image constituents) needed to describe each of the pixel areas of the image. The local-rank maps are displayed as two-dimensional plots with the *x* and *y* directions corresponding to those of the scanned surface. In Figure 11.22, the maps of several tables are presented.

The local-rank maps provide information on the local complexity of the image. Areas with higher ranks indicate a more complex chemical composition with more

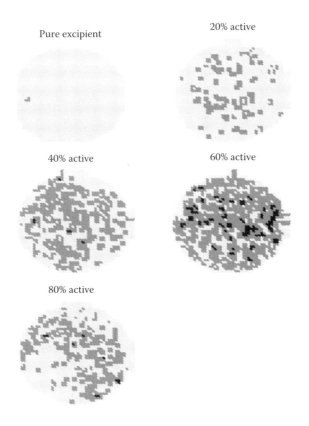

FIGURE 11.22 Local-rank maps of the different pill images (the percent refers to content of active compound). Light, medium, and dark grey refer to rank 1, 2, and 3, respectively.

constituents overlapping, whereas areas with low local ranks, e.g., 1, contain only one chemical compound, i.e., a selective pixel. In this example, the presence and location of an unexpected third compound (impurity) in the tablets containing 40, 60, and 80% of active component is detected through the occurrence and position of some rank-3 pixels. It is important to note that detecting the presence of this minor component would be very difficult by visual inspection of the thousands of spectra collected or by using classical data analysis approaches.

As in other chemical problems, the data sets of all five tablet images are combined to form a three-way data set to be resolved. The modeling of images with low impurity content is improved due to the information included in images with higher concentrations of this compound. As expected, exploiting the three-way character of the data set significantly decreases the ambiguity of the resolved pure-compound profiles compared to an individual analysis of each tablet image.

The information obtained in the FSIW-EFA exploratory analysis is used in the resolution step. The total number of compounds in the three-way data set has been found to be equal to three (active, excipient, and unknown). The iterative optimization process starts with a matrix S^T containing the initial estimates found by

SIMPLISMA. Nonnegativity constraints are applied to constrain concentration profiles and Raman spectra. The concentration submatrices related to each individual image are constrained, taking into account the information available about the presence or absence of the different compounds in the data set. This information is expressed in a binary-coded data matrix, where 1 represents the present compounds and 0 represents the absent compounds. The column labels A, E, and U refer to active, excipient, and unknown, respectively, and the percentages show the abundance of the active compound in the corresponding tablet.

$$\begin{array}{c} \\ 0\% \\ 20\% \\ 40\% \\ 60\% \\ 80\% \end{array} \begin{pmatrix} A & E & U \\ 0 & 1 & 0 \\ 1 & 1 & 0 \\ 1 & 1 & 1 \\ 1 & 1 & 1 \\ 1 & 1 & 1 \end{pmatrix} \qquad (11.21)$$

The concentrations of absent components are set to zero in the corresponding submatrices of **C**.

Figure 11.23 shows the pure-component distribution maps, using mesh plots to display more clearly the distribution of the impurity. The corresponding pure-component spectra for each of the resolved components are also shown.

The variance explained in the raw data reconstruction is 96.6%. The distribution maps match qualitatively the expected trend of relative abundance of major compounds according to the nominal content in the different pills, and they show the heterogeneity in the compound distribution of both excipient and active compound along the surface of the different pills. In each pill, the contribution of each compound in each pixel is straightforwardly obtained from the distribution maps of the pure compounds. The degree of overlap among compounds and, therefore, the chemical complexity of the sample, is locally known. The resolution results confirm the need to include a third compound to describe the composition of the pills, which appears with a very distinct spectral shape. The modeling and location of this impurity is successfully achieved, even for the pills where this compound is present in a very low proportion.

11.9 SOFTWARE

Implementations of curve-resolution algorithms and methods in public domain and commercial software remain scarce, reflecting the intrinsic difficulties of developing robust and user-friendly methods for curve resolution. Solving curve-resolution problems still typically requires strong user interaction and knowledge of the problem under study. There are some exceptions, including the MATLAB software that we have been developing for years and offer as a free download at our Web page, http://www.ub.es/gesq/eq1_eng.htm. This software includes a set of MATLAB m-files that facilitate the exploration and resolution of two- and three-way data sets.

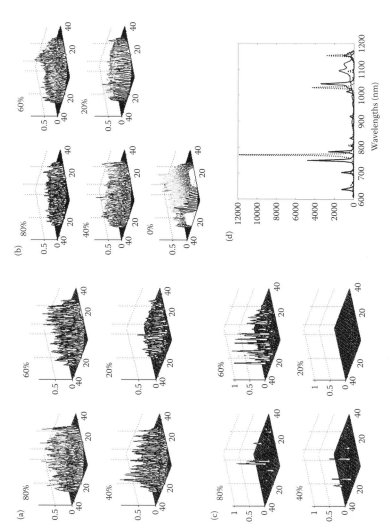

FIGURE 11.23 Mesh representation of the pure distribution maps for: (a) active compound, (b) excipient, (c) impurity (the percent in a, b, and c always refers to content of active component; hot and cold colors refer to high and low concentrations, respectively), and (d) pure resolved spectra (thick solid line: active compound, thin solid line: excipient, dotted line: impurity).

These routines come with detailed instructions of use along with data examples of diverse application (chromatographic examples, reaction systems, etc.). Our research group offers free support for the users of this software and appreciates suggestions about possible improvements in the routines and in the contents of the Web page.

REFERENCES

1. Lawton, W.H. and Sylvestre, E.A., Self modeling curve resolution, *Technometrics*, 13, 617–633, 1971.
2. Sylvestre, E.A., Lawton, W.H., and Maggio, M.S., Curve resolution using a postulated chemical reaction, *Technometrics*, 16, 353–368, 1974.
3. Borgen, O.S. and Kowalski, B.R., An extension of the multivariate component-resolution method to three components, *Anal. Chim. Acta*, 174, 1–26, 1985.
4. Borgen, O.S., Davidsen, N., Zhu, M.Y., and Oyen, O., The multivariate n-component resolution problem with minimum assumptions, *Mikrochim. Acta*, 2, 1, 1986.
5. Hamilton, J.C. and Gemperline, P.J., An extension of the multivariate component-resolution method to three components, *J. Chemom.*, 4, 1–13, 1990.
6. Windig, W., Self-modeling mixture analysis of spectral data with continuous concentration profiles, *Chemom. Intell. Lab. Sys.*, 16, 1–16, 1992.
7. de Juan, A., Casassas, E., and Tauler, R., Soft modeling of analytical data, in *Encyclopedia of Analytical Chemistry: Instrumentation and Applications*, John Wiley & Sons, New York, 2000, pp. 9800–9837.
8. Liang, Y. and Kvalheim, O.M., Resolution of two-way data: theoretical background and practical problem-solving, Part 1: Theoretical background and methodology, *Fresenius J. Anal. Chem.*, 370, 694–704, 2001.
9. Jiang, J.H. and Ozaki, Y., Self-modeling curve resolution (SMCR): principles, techniques, and applications, *Appl. Spectr. Rev.*, 37, 321–345, 2002.
10. de Juan, A. and Tauler, R., Chemometrics applied to unravel multicomponent processes and mixtures: revisiting latest trends in multivariate resolution, *Anal. Chim. Acta*, 500, 195–210, 2003.
11. Navea, S., de Juan, A., and Tauler, R., Detection and resolution of intermediate species in protein folding processes using fluorescence and circular dichroism spectroscopies and multivariate curve resolution, *Anal. Chem.*, 64, 6031–6039, 2002.
12. Jaumot, J., Escaja, N., Gargallo, R., González, C., Pedroso, E., and Tauler, R., Multivariate curve resolution: a powerful tool for the analysis of conformational transitions in nucleic acids, *Nucl. Acid Res.*, 30, e92, 2002. http://nar.oxfordjournals.org/content/vol 30/issue 17/index.
13. Hopke, P.K., *Receptor Modeling in Environmental Chemistry*, John Wiley & Sons, New York, 1985.
14. Xie, Y.L., Hopke, P.K., and Paatero, P., Positive matrix factorization applied to a curve resolution problem, *J. Chemom.*, 12, 357–364, 1998.
15. Tauler, R., Interpretation of environmental data using chemometrics, *Tech. Instrum. Anal. Chem.*, 21, 689, 2000.
16. Tauler, R., Lacorte, S., Guillamón, M., Céspedes, R., Viana, P., and Barceló, D., Resolution of main environmental contamination sources of semivolatile organic compounds in surface waters of Portugal using chemometric compounds, *Environ. Toxicol. Chem.*, 23, 565–575, 2004.

17. Esteban, M., Ariño, C., Díaz-Cruz, J.M., Díaz-Cruz, M.S., and Tauler, R., Multivariate curve resolution with alternating least squares optimization: a soft-modeling approach to metal complexation studies by voltammetric techniques, *Trends Anal. Chem.,* 19, 49–61, 2000.
18. Zampronio, C.G., Moraes, L.A.B., Eberlin, M.N., and Poppi, R.J., Multivariate curve resolution applied to MS/MS data obtained from isomeric mixtures, *Anal. Chim. Acta,* 446, 495–502, 2001.
19. Cruz, B.H., Díaz-Cruz, J.M., Sestakova, I., Velek, J., Ariño, C., and Esteban, M., Differential pulse voltammetric study of the complexation of Cd(II) by the phytochelatin (gamma-Glu-Cys)(2)Gly assisted by multivariate curve resolution, *J. Electroanal. Chem.,* 520, 111–118, 2002.
20. López, M.J., Ariño, C., Díaz-Cruz, S., Díaz-Cruz, J.M., Tauler, R., and Esteban, M., Voltammetry assisted by multivariate analysis as a tool for complexation studies of protein domains: competitive complexation of α and β-metallothionein domains by cadmium and zinc, *Environ. Sci. Technol.,* 37, 5609–5616, 2003.
21. Tauler, R., Smilde, A.K., and Kowalski, B.R., Selectivity, local rank, three-way data analysis and ambiguity in multivariate curve resolution, *J. Chemom.,* 9, 31–58, 1995.
22. Manne, R., On the resolution problem in hyphenated chromatography, *Chemom. Intell. Lab. Sys.,* 27, 89–94, 1995.
23. Martens, H., Factor analysis of chemical mixtures, *Anal. Chim. Acta,* 112, 423–448, 1979.
24. Malinowski, E.R. and Howery, D.G., *Factor Analysis in Chemistry,* 1st ed., John Wiley & Sons, New York, 1980.
25. Malinowski, E.R., *Factor Analysis in Chemistry,* 3rd ed., John Wiley & Sons, New York, 2002.
26. Federation of Analytical Chemistry and Spectroscopy Society. Multivariate Curve Resolution: The Sleeping Giant of Chemometrics, Proceedings of 30th FACSS 2003, Ft. Lauderdale, FL, October 19–23, 2003.
27. Gampp, H., Maeder, M., Meyer, C.J., and Zuberbühler, A.D., Calculation of equilibrium constants from multiwavelength spectroscopic data, III: model-free analysis of spectrophotometric and ESR titrations, *Talanta,* 32, 1133–1139, 1985.
28. Maeder, M. and Zuberbühler, A.D., The resolution of overlapping chromatographic peaks by evolving factor-analysis, *Anal. Chim. Acta,* 181, 287–291, 1986.
29. Maeder, M., Evolving factor analysis for the resolution of overlapping chromatographic peaks, *Anal. Chem.,* 59, 527–530, 1987.
30. Keller, H.R. and Massart, D.L., Peak purity control in liquid chromatography with photodiode array detection by fixed-size-moving-window–evolving factor analysis, *Anal. Chim. Acta,* 246, 379–390, 1991.
31. Keller, H.R., Massart, D.L., Liang, Y.Z., and Kvalheim, O.M., Evolving factor analysis in the presence of heteroscedastic noise, *Anal. Chim. Acta,* 263, 29–36, 1992.
32. McCue, M. and Malinowski, E.R., Rank annihilation factor-analysis of unresolved LC peaks, *J. Chromatog. Sci.,* 21, 229–234, 1983.
33. Gampp, H., Maeder, M., Meyer, C.J., and Zuberbühler, A.D., Quantification of a known component in an unknown mixture, *Anal. Chim. Acta,* 193, 287–292, 1987.
34. Malinowski, E.R., Window factor analysis: theoretical derivation and application to flow injection analysis data, *J. Chemom.,* 6, 29–40, 1992.
35. Malinowski, E.R., Automatic window factor analysis: a more efficient method for determining concentration profiles from evolutionary spectra, *J. Chemom.,* 10, 273–279, 1996.

36. Kvalheim, O.M. and Liang, Y.Z., Heuristic evolving latent projections — resolving two-way multicomponent data, 1: selectivity, latent projective graph, datascope, local rank and unique resolution, *Anal. Chem.,* 64, 936–946, 1992.
37. Liang, Y.Z., Kvalheim, O.M., Keller, H.R., Massart, D.L., Kiechle, P., and Erni, F., Heuristic evolving latent projections —- resolving two-way multicomponent data, 2: detection and resolution of minor constituents, *Anal. Chem.,* 64, 946–953, 1992.
38. Manne, R., Shen, H., and Liang, Y., Subwindow factor analysis, *Chemom. Intell. Lab. Sys.,* 45, 171–176, 1999.
39. Jiang, J.H., Šašic′, S., Yu, R., and Ozaki, Y., Resolution of two-way data from spectroscopic monitoring of reaction or process systems by parallel vector analysis (PVA) and window factor analysis (WFA): inspection of the effect of mass balance, methods and simulations, *J. Chemom.,* 17, 186–197, 2003.
40. Gampp, H., Maeder, M., Meyer, C.J., and Zuberbühler, A., Calculation of equilibrium constants from multiwavelength spectroscopic data, IV: model free least squares refinement by use of evolving factor analysis, *Talanta,* 33, 943–951, 1986.
41. Tauler, R. and Casassas, E., Principal component analysis applied to the study of successive complex formation data in the Cu(II) ethanolamine systems, *J. Chemom.,* 3, 151–161, 1988.
42. Tauler, R., Multivariate curve resolution applied to second order data, *Chemom. Intell. Lab. Sys.,* 30, 133–146, 1995.
43. Toft, J. and Kvalheim, O.M., Eigenstructure tracking analysis for revealing noise patterns and local rank in instrumental profiles: application to transmittance and absorbance IR spectroscopy, *Chemom. Intell. Lab. Sys.,* 19, 65–73, 1993.
44. Whitson, A.C. and Maeder, M., Exhaustive evolving factor analysis (E-EFA), *J. Chemom.,* 15, 475–484, 2001.
45. de Juan, A., Navea, S., Diewok, J., and Tauler, R., Local rank exploratory analysis of evolving rank-deficient systems, *Chemom. Intell. Lab. Sys.,* 70, 11–21, 2004.
46. de Juan, A., Dyson, R., Marcolli, C., Rault, M., Tauler, R., and Maeder, M., Spectroscopic imaging and chemometrics: a powerful combination for global and local sample analysis, *Trends Anal. Chem.,* 23, 71–79, 2004.
47. Vandeginste, B.G.M., Derks, W., and Kateman, G., Multicomponent self-modeling curve resolution in high performance liquid chromatography by iterative target transformation factor analysis, *Anal. Chim. Acta,* 173, 253–264, 1985.
48. Gemperline, P.J., A priori estimates of the elution profiles of the pure components in overlapped liquid chromatography peaks using target factor analysis, *J. Chem. Inf. Comp. Sci.,* 24, 206–212,1984.
49. Manne, R. and Grande, B.V., Resolution of two-way data from hyphenated chromatography by means of elementary matrix transformations, *Chemom. Intell. Lab. Sys.,* 50, 35–46, 2000.
50. Mason, C., Maeder, M., and Whitson, A.C., Resolving factor analysis, *Anal. Chem.,* 73, 1587–1594, 2001.
51. Tauler, R., Izquierdo-Ridorsa, A., and Casassas, E., Simultaneous analysis of several spectroscopic titrations with self-modeling curve resolution, *Chemom. Intell. Lab. Sys.,* 18, 293–300, 1993.
52. Malinowski, E.R., Obtaining the key set of typical vectors by factor analysis and subsequent isolation of component spectra, *Anal. Chim. Acta,* 134, 129–137, 1982.
53. Windig, W. and Guilment, J., Interactive self-modeling mixture analysis, *Anal. Chem.,* 63, 1425–1432, 1991.
54. Sánchez, F.C., Toft, J., van den Bogaert, B., and Massart, D.L., Orthogonal projection approach applied to peak purity assessment, *Anal. Chem.,* 68, 79–85, 1996.

55. de Juan, A., Vander Heyden, Y., Tauler, R., and Massart, D.L., Assessment of new constraints applied to the alternating least squares method, *Anal. Chim. Acta,* 346, 307–318, 1997.
56. Lawson, C.L. and Hanson, R.J., *Solving Least-Squares Problems,* Prentice-Hall, Upper Saddle River, NJ, 1974.
57. Hanson, R.J. and Haskell, K.H., Two algorithms for the linearly constrained least-squares problem, *ACM Trans. Math. Softw.,* 8, 323–333, 1982.
58. Haskell, K.H. and Hanson, R.J., An algorithm for linear least-squares problems with equality and non-negativity constraints, *Math. Prog.,* 21, 98–118, 1981.
59. Bro, R. and de Jong, S., A fast non-negativity-constrained least squares algorithm, *J. Chemom.,* 11, 393–401, 1997.
60. Bro, R. and Sidiropoulos, N.D., Least squares algorithms under unimodality and non-negativity constraints, *J. Chemom.,* 12, 223–247, 1998.
61. Van Benthem, M.H., Keenan, M.R., and Haaland, D.M., Application of equality constraints on variables during alternating least squares procedures, *J. Chemom.,* 16, 613–622, 2002.
62. Díaz-Cruz, J.M., Tauler, R., Grabaric, B., Esteban, M., and Casassas, E., Application of multivariate curve resolution to voltammetric data, 1: study of Zn(II) complexation with some polyelectrolytes, *J. Electroanal. Chem.,* 393, 7–16, 1995.
63. Gampp, H., Maeder, M., Meyer, C.J., and Zuberbühler, A.D., Quantification of a known component in an unknown mixture, *Anal. Chim. Acta,* 193, 287–293, 1987.
64. Bijlsma, S. and Smilde, A.K., Estimating reaction rate constants from a two-step reaction: a comparison between two-way and three-way methods, *J. Chemom.,* 14, 541–560, 2000.
65. Antunes, M.C., Simao, J.E.J., Duarte, A.C., and Tauler, R., Multivariate curve resolution of overlapping voltammetric peaks: quantitative analysis of binary and quaternary metal mixtures, *Analyst,* 127, 809–817, 2002.
66. Bijlsma, S. and Smilde, A.K., Application of curve resolution based methods to kinetic data, *Anal. Chim. Acta,* 396, 231–240, 1999.
67. de Juan, A., Maeder, M., Martínez, M., and Tauler, R., Combining hard- and soft-modeling to solve kinetic problems, *Chemom. Intell. Lab. Sys.,* 54, 123–141, 2000.
68. Díaz-Cruz, J.M., Agulló, J., Díaz-Cruz, M.S., Ariño, C., Esteban, M., and Tauler, R., Implementation of a chemical equilibrium constraint in the multivariate curve resolution of voltammograms from systems with successive metal complexes, *Analyst,* 126, 371–377, 2001.
69. Bezemer, E. and Rutan, S.C., Study of the hydrolysis of a sulfonylurea herbicide using liquid chromatography with diode array detection and mass spectrometry by three-way multivariate curve resolution-alternating least squares, *Anal. Chem.,* 73, 4403–4409, 2001.
70. Bezemer, E. and Rutan, S.C., Multivariate curve resolution with non-linear fitting of kinetic profiles, *Chemom. Intell. Lab. Sys.,* 59, 19–31, 2001.
71. Jandanklang, P., Maeder, M., and Whitson, A.C., Target transform fitting: a new method for the non-linear fitting of multivariate data with separable parameters, *J. Chemom.,* 15, 511–522, 2001.
72. Windig, W., Hornak, J.P., and Antalek, B., Multivariate image analysis of magnetic resonance images with the direct exponential curve resolution algorithm (DECRA) — Part 1: algorithm and model study, *J. Magnet. Res.,* 133, 298–306, 1998.
73. Diewok, J., de Juan, A., Maeder, M., Tauler, R., and Lendl, B., Application of a combination of hard and soft modeling for equilibrium systems to the quantitative analysis of pH-modulated mixture samples, *Anal. Chem.,* 75, 641–647, 2003.

74. de Juan, A., Maeder, M., Martínez, M., and Tauler, R., Application of a novel resolution approach combining soft- and hard-modeling features to investigate temperature-dependent kinetic processes, *Anal. Chim. Acta,* 442, 337–350, 2001.
75. de Juan, A., van den Bogaert, B., Cuesta Sánchez, F., and Massart, D.L., Application of the needle algorithm for exploratory analysis and resolution of HPLC-DAD data, *Chemom. Intell. Lab. Sys.,* 33, 133–145 ,1996.
76. Amhrein, M., Srinivasan, B., Bonvin, D., and Schumache, M.M., On the rank deficiency and rank augmentation of the spectral measurement matrix, *Chemom. Intell. Lab. Sys.,* 33, 17–33, 1996.
77. Bro, R., PARAFAC: tutorial and applications, *Chemom. Intell. Lab. Syst.,* 38, 149–171, 1997.
78. Kiers, H. and Smilde, A.K., Constrained three-mode factor analysis as a tool for parameter estimation with second-order instrumental data, *J. Chemom.,* 12, 125–147, 1998.
79. Sanchez, E. and Kowalski, B.R., Generalized rank annihilation factor analysis, *Anal. Chem.,* 58, 499–501, 1986.
80. Sanchez, E. and Kowalski, B.R., Tensorial resolution: a direct trilinear decomposition, *J. Chemom.,* 4, 29–45, 1990.
81. de Juan, A. and Tauler, R., Comparison of three-way resolution methods for non-trilinear data sets, *J. Chemom.,* 15, 749–771, 2001.
82. Tauler, R., Marqués, I., and Casassas, E., Multivariate curve resolution applied to three-way trilinear data: study of a spectrofluorimetric acid-base titration of salicylic acid at three excitation wavelengths, *J. Chemom.,* 12, 55–75, 1998.
83. Wentzell, P.D., Wang, J., Loucks, L.F., and Miller, K.M., Direct optimization of self modeling curve resolution: application to the kinetics of the permanganate-oxalic acid reaction, *Can. J. Chem.,* 76, 1144–1155, 1998.
84. Gemperline, P.J., Computation of the range of feasible solutions in self-modeling curve resolution algorithms, *Anal. Chem.,* 71, 5398–5404, 1999.
85. Tauler, R., Calculation of maximum and minimum band boundaries of feasible solutions for species profiles obtained by multivariate curve resolution, *J. Chemom.,* 15, 627–646, 2001.
86. Booksh, K.S. and Kowalski, B.R., Theory of analytical chemistry, *Anal. Chem.,* 66, 782A–791A, 1994.
87. Faaber, K. and Kowalski, B.R., Propagation of measurement errors for the validation of predictions obtained by principal component regression and partial least squares, *J. Chemom.,* 11, 181–238, 1997.
88. Faber, K., Comment on a recently proposed resampling method, *J. Chemom.,* 15, 169–188, 2001.
89. Jaumot, J., Gargallo, R., and Tauler, R., Noise propagation and error estimations in multivariate curve-resolution alternating least squares using resampling methods, *J. Chemom.,* 18, 324–340, 2004.
90. de Juan, A., Izquierdo-Ridorsa, A., Tauler, R., Fonrodona, G., and Casassas, E., A soft modeling approach to interpret thermodynamic and conformational transitions of polynucleotides, *Biophys. J.,* 73, 2937–2948, 1997.
91. Vives, M., Gargallo, R., and Tauler, R., Study of the intercalation equilibrium between the polynucleotide poly(adenylic)-poly(uridylic) acid and the ethidium bromide dye by means of multivariate curve resolution and the multivariate extension of the continuous variation and mole ratio methods, *Anal. Chem.,* 71, 4328–4337, 1999.
92. Navea, S., de Juan, A., and Tauler, R., Three-way data analysis applied to multispectroscopic monitoring of protein folding, *Anal. Chim. Acta,* 446, 187–197, 2001.

93. Pain, R.H., *Mechanisms of Protein Folding,* Oxford University Press, Oxford, 1994.
94. Creighton, T.E., *Proteins: Structures and Molecular Properties,* 2nd ed., W.H. Freeman and Co., New York, 1997.
95. Berova, N., Nakanishi, K., and Woody, R.W., *Circular Dichroism: Principles and Applications,* John Wiley & Sons, New York, 2000.
96. Greenfield, N.J., Methods to estimate the conformation of proteins and polypeptides from circular dichroism data, *Anal. Biochem.,* 235, 1–10, 1996.
97. Manderson, G.A., Creamer, L.K., and Hardman, M.J., Effect of heat treatment on the circular dichroism spectra of bovine beta-lactogobulin A, B, and C, *J. Agric. Food Chem.,* 47, 4557–4567, 1999.
98. Salou, J., Tauler, R., Bayona, J., and Tolosa, I., Input characterization of sedimentary organic chemical markers in the northwestern Mediterranean Sea by exploratory data analysis, *Environ. Sci. Technol.,* 31, 3482–3490, 1997.
99. Wold, S., Esbensen, K., and Geladi, P., Principal component analysis, *Chemom. Intell. Lab. Syst.,* 2, 37–52, 1987.
100. Tauler, R., Barceló, D., and Thurman, E.M., Multivariate correlations between concentration of selected herbicides and derivatives in outflows from selected U.S. midwestern reservoirs, *Environ. Sci. Technol.,* 34, 3307–3314, 2000.
101. Peré-Trepat, E., Petrovic, M., Barceló, D., and Tauler, R., Identification of main microcontaminant sources of non-ionic surfactants, their degradation products and linear alkylbenzene sulfonates in coastal waters and sediments in Spain by means of principal component analysis and multivariate resolution, *Anal. Bioanal. Chem.,* 378, 642–654, 2004.
102. Leardi, R., Armanino, C., Lanteri, S., and Albertotanza, L., Three-mode principal component analysis of monitoring data from Venice lagoon, *J. Chemom.,* 14, 187–195, 2000.
103. Barbieri, P., Andersson, C.A., Massart, D.L., Predozani, S., Adami, G., and Reisenhofer, E., Modeling bio-geochemical interactions in the surface waters of the Gulf of Trieste by three-way principal component analysis (PCA), *Anal. Chim. Acta,* 398, 227–235, 1999.
104. Farnham, I.M., Singh, A.K., Stetzenbach, K.J., and Johannesson, K.H., Treatment of non-detects in multivariate analysis of groundwater geochemistry data, *Chemom. Intell. Lab. Syst.,* 60, 265–281, 2002.
105. Zitko, V., Principal component analysis in the evaluation of environmental data, *Mar. Pollut. Bull.,* 28, 718–722, 1994.
106. Hopke, P.K., An introduction to receptor modeling, *Chemom. Intell. Lab. Sys.,* 10, 21–43, 1991.
107. Paatero, P., Least squares formulation of robust non-negative factor analysis, *Chemom. Intell. Lab. Syst.,* 37, 23–35, 1997.
108. Hornak, J.P., *Encyclopedia of Imaging Science and Technology,* John Wiley & Sons, New York, 2002.
109. Colarusso, P., Kidder, L.H., Levin, I.W., Fraser, J.C., Arens, J.F., and Lewis, E.N., Infrared spectroscopic imaging: from planetary to cellular systems, *Appl. Spectrosc.,* 52, 106A–120A, 1998.
110. Geladi, E.N., *Multivariate Image Analysis in Chemistry and Related Areas: Chemometric Image Analysis,* John Wiley & Sons, Chichester, U.K., 1996.
111. Orsini, F., Ami, D., Villa, A.M., Sala, G., Bellottu, M., and Doglia, S.M., FT-IR microspectroscopy for microbiological studies, *J. Microbiol. Methods,* 42, 17–27, 2000.

112. Fischer, M. and Tran, C.D., Investigation of solid phase peptide synthesis by the near-infrared multispectral imaging technique: a detection method for combinatorial chemistry, *Anal. Chem.,* 71, 2255–2261, 1999.
113. Baldwin, P.M., Bertrand, D., Novales, B., Bouchet, B., Collobert, G., and Gallant, D.J., Chemometric labeling of cereal tissues in multichannel fluorescence microscopy images using discriminant analysis, *Anal. Chem.,* 69, 4339–4348, 1997.
114. Wang, J.H., Hopke, P.K., Hancewicz, T.M., and Zhang, S.L., Application of modified alternating least squares regression to spectroscopic image analysis, *Anal. Chim. Acta,* 476, 93–109, 2003.

12 Three-Way Calibration with Hyphenated Data

Karl S. Booksh

CONTENTS

12.1	Introduction	475
12.2	Background	476
12.3	Nomenclature of Three-Way Data	478
12.4	Three-Way Models	478
12.5	Examples	481
12.6	Rank Annihilation Methods	482
	12.6.1 Rank Annihilation Factor Analysis	482
	12.6.1.1 RAFA Application	483
	12.6.2 Generalized Rank Annihilation Method	485
	12.6.2.1 GRAM Application	486
	12.6.3 Direct Trilinear Decomposition	489
	12.6.3.1 DTLD Application	490
12.7	Alternating Least-Squares Methods	491
	12.7.1 PARAFAC / CANDECOMP	491
	12.7.1.1 Tuckals	493
	12.7.1.2 Solution Constraints	493
	12.7.1.3 PARAFAC Application	494
12.8	Extensions of Three-Way Methods	495
12.9	Figures of Merit	496
12.10	Caveats	497
References		499
Appendix 12.1	GRAM Algorithm	502
Appendix 12.2	DTLD Algorithm	503
Appendix 12.3	PARAFAC Algorithm	504

12.1 INTRODUCTION

Three-way calibrations methods, such as the generalized rank annihilation method (GRAM) and parallel factor analysis (PARAFAC), are becoming increasingly prevalent tools to solve analytical challenges. The main advantage of three-way calibration is estimation of analyte concentrations in the presence of unknown, uncalibrated spectral interferents. These methods also permit the extraction of analyte, and often interferent, spectral profiles from complex and uncharacterized mixtures. In this chapter, a theoretical and practical progression and overview of three-way calibration methods from the simplest rank annihilation factor analysis (RAFA) to the more flexible PARAFAC is presented. Extensions of many three-way methods are covered to highlight the paradigm's flexibility in solving particular analytical calibration problems.

12.2 BACKGROUND

One challenge faced by analytical chemists interested in instrumental calibration for estimation of analyte concentrations in unknown solutions is the resolution of highly convolved signals returned from modern analytic instrumentation. Chromatograms do not always have baseline separation of adjacent peaks, and molecular spectroscopic techniques are seldom 100% selective. For quantitative applications involving convolved signals, it is imperative to explicitly or implicitly extract the target analyte signature from the overlapping instrumental responses of any interfering species. For many qualitative and exploratory applications, important trends and effects can be better noticed if the tapestry of data can be selectively unwoven and viewed one piece at a time by deconvolving the overlapping signals.

The advent of inexpensive microprocessors has simultaneously led to more complicated, and powerful, analytical instruments and more powerful, and complicated, methods for resolving and interpreting data gathered from such instruments. In an abstract sense, the power and complexity of instrumentation increases with the number of "ways" the data is collected and treated [1]. This, of course, assumes all other factors, such as technique appropriateness and signal-to-noise ratio, are equal.

"One-way" data consists of either one sample and multiple variables, e.g., a single chromatogram generated over time, or multiple samples and one variable, e.g., calibration of an ion-selective electrode. A single gas chromatogram (GC) collected with a flame-ionization detector is considered one-way data, as the data is only collected in one way, over time. Analyzing multiple solutions with an ion-selective electrode is also one-way, over samples. In these applications, the mathematical and statistical tools applicable to one-way data analysis limit the analyst. For qualitative analysis, e.g., GC peak resolution, Gaussian curves or orthogonal polynomials are often fit to the chromatogram to elicit the number of overlapping species [2]. Quantitative applications, e.g., in electrochemistry, require the assumption of complete selectivity of the detector [3]. The results derived from these analyses are only as accurate and reliable as the assumptions made.

"Two-way" data consists of multiple samples, each represented by multiple variables, or one sample represented by two sets of interacting variables. The type of two-way data determines the amount of quantitative and qualitative information that can be extracted [4]. Digitized ultraviolet (UV) spectra collected for each of I samples is an example of multiple samples, the first way, and multiple variables, the second way, forming "two-way" data. Such data is usually employed for quantitative applications where the independent variables, the spectra, are related to a property (e.g., concentration) of the samples, the dependent variable. A number of linear calibration methods described in Chapter 6, such as principal component regression (PCR) and partial least squares regression (PLS) [5], and nonlinear regression methods, such as predictive alternating conditional expectations (PACE), multivariate adaptive regression splines (MARS), and artificial neural networks (ANN) [6, 7], are available to facilitate determination of the dependent variable's value in an unknown sample. However, these quantitative methods extract little to no qualitative information.

On the other hand, data such as a single sample represented by a collection of digitized spectra at discrete time intervals during a chromatographic run, or a collection of digitized spectra at discrete pH values during a titration, forms two-way data that is best suited for qualitative applications. It is important to note that the instrument in the first way, e.g., the chromatograph, must modulate the signal from the instrument in the second way, e.g., the spectrometer. For example, a liquid chromatography-UV/Vis diode-array spectroscopy (DAS) system forms two-way data, but an FTIR-UV/Vis DAS does not form two-way data, since the FTIR does not modulate the UV/Vis DAS signal. Spectral deconvolution methods described in Chapter 11 — such as self-modeling curve resolution (SMCR), iterative target factor analysis (ITFA), evolving factor analysis (EFA), and their descendants — are successfully employed to deduce the number and qualitative identity of species in a complex mixture [8–11].

"Three-way" data, in chemical applications, is usually visualized as multiple samples, each consisting of two sets of interacting variables. A collection of excitation-emission matrix fluorescence spectra, or multiple diode-array absorption spectra, collected at regular intervals during each of several chromatographic runs fit the traditional image of three-way chemical data that has one "object" way and two "variable" ways. Hirschfeld listed 66 "hyphenated" methods to produce three-way data that are feasible with state-of-the-art technology [12]. Applications of one-object–two-variable, three-way data are often quantitative in nature. The ability of three-way data analysis to uniquely deconvolve overlapped spectral signatures is harnessed to extract the analyte signal from an unknown and uncalibrated background prior to quantitation [3, 13]. The data structure requirements and algorithms that permit accurate calibration with three-way data is the primary focus of this chapter.

However, three-way data can also be formed with two object ways and one variable way and by one sample with three variable ways. Environmental data where several distinct locations are monitored at discrete time intervals for multiple analytes exemplifies three-way data with two object ways and one variable way. Excitation-emission-time decay fluorescence or gas chromatography with a tandem mass spectroscopic detector are instrumental methods that form three-way data with three variable ways. These data types are employed mostly for qualitative application. Herein, the desire of the analyst to elicit underlying factors that influence the ecosystem or to deconvolve highly overlapped spectral profiles to deduce the number, identity, or relaxation coefficients of constituents in a complex sample can be realized. The same procedures employed for quantitation lend themselves to the extraction of qualitative information.

In truth, there is no limit to the number of "ways" that can form a data set. The open-ended description for data with more than three ways is "N-way" data. For example, a collection of excitation-emission-time decay fluorescence spectra forms "four-way" data. Add reaction kinetics or varying experimental conditions, and "five-way" data, or greater, could easily be formed. Many of the techniques discussed below, in particular the PARAFAC-based algorithms, are readily extended to N-way applications [14, 15].

12.3 NOMENCLATURE OF THREE-WAY DATA

There are two competing and equivalent nomenclature systems encountered in the chemical literature. The description of data in terms of "ways" is derived from the statistical literature. Here a "way" is constituted by each independent, nontrivial factor that is manipulated with the data collection system. To continue with the example of excitation-emission matrix fluorescence spectra, the three-way data is constructed by manipulating the excitation-way, emission-way, and the sample-way for multiple samples. Implicit in this definition is a fully blocked experimental design where the collected data forms a cube with no missing values. Equivalently, hyphenated data is often referred to in terms of "orders" as derived from the mathematical literature. In tensor notation, a scalar is a zeroth-order tensor, a vector is first order, a matrix is second order, a cube is third order, etc. Hence, the collection of excitation-emission data discussed previously would form a third-order tensor. However, it should be mentioned that the "way-based" and "order-based" nomenclature are not directly interchangeable. By convention, "order" notation is based on the structure of the data collected from each sample. Analysis of collected excitation-emission fluorescence, forming a second-order tensor of data per sample, is referred to as second-order analysis, as compared with the "three-way" analysis just described. In this chapter, the "way-based" notation will be arbitrarily adopted to be consistent with previous work.

Furthermore, traditional notation for scalars, vectors and variables will be adopted. A scalar of fixed value, e.g., the number of factors in a model, is represented by an italicized capitol, N. An italicized lowercase letter, e.g., the nth factor, represents a scalar of arbitrary value. All vectors are column vectors designated by lowercase bold, e.g., $\mathbf{x}$. Matrices are given by uppercase bold, e.g., $\mathbf{X}$, and cubes (third-order tensors by uppercase open-face letters, e.g., $\mathbb{R}$. Transposes of matrices and vectors, defined by switching the row and column indices, is designated with a superscript T, e.g., $\mathbf{x}^T$. The transpose of a cube need not be defined for this chapter. Subscripts designate a specific element of a higher-order tensor, where the initial order is inferred by the number of subscripts associated with the scalar.

12.4 THREE-WAY MODELS

There are six classes of three-way data, and four of these classes can be appropriately modeled with the basic trilinear, or PARAFAC (PARAllel FACtor), model, where the data cube is decomposed into N sets of triads, $\hat{\mathbf{x}}$, $\hat{\mathbf{y}}$, and $\hat{\mathbf{z}}$ [16]. The trilinear model can be presented equivalently in statistical

$$\hat{R}_{ijk} = \sum_{n=1}^{N} \hat{X}_{ni} \hat{Y}_{nj} \hat{Z}_{nk} + E_{ijk} \qquad (12.1a)$$

or mathematical forms

$$\hat{\mathbf{R}}_k = \sum_{n=1}^{N} \hat{\mathbf{x}}_n \hat{Z}_{nk} \hat{\mathbf{y}}_n^t + \mathbf{E}_k \qquad (12.1b)$$

Here N refers to the number of factors employed by the model to describe the $I \times J \times K$ data cube, or the rank of the model.

In general, the number and form of factors in a model are not constrained to be representative of any physical reality. With self-modeling curve resolution (SMCR) as the two-way data model, this is referred to as the rotational ambiguity of the factors; there is a continuum of factors that satisfy the SMCR model and equivalently describe the data [17]. However, in problems with an underlying trilinear structure, when the proper number of factors is chosen for a trilinear model, the factors are accurate estimates of the true underlying factors. In other words, if a LC-UV/Vis DAS formed $\mathbb{R}$, then each $\hat{\mathbf{x}}_n$ would correspond to one of the true N chromatographic profiles, each $\hat{\mathbf{y}}_n$ to one of the true spectroscopic profiles, and each $\hat{\mathbf{z}}_n$ to the relative concentrations in the K samples. Therefore, in three-way analysis, when, for an isolated chemical component (1) the true underlying factor in each of the three modes is independent, except for scale, from the state of the other two modes; (2) the true underlying factor in any of the three modes cannot be expressed by linear combinations of the true underlying factors of other components in the same mode; (3) there is linear additivity of instrumental responses among the species present and; (4) the proper number of factors is chosen for the model, then the factors $\hat{\mathbf{x}}$, $\hat{\mathbf{y}}$, and $\hat{\mathbf{z}}$ are unique to a scaling constant and are accurate estimates of the true underlying factors, $\mathbf{x}$, $\mathbf{y}$, and $\mathbf{z}$. This is shown graphically in Figure 12.1, where the data set $\mathbb{R}$ is, in reality, composed of N triads. If the proper number of factors N is chosen, the estimates of the factors in each of the triads, $\hat{\mathbf{x}}$, $\hat{\mathbf{y}}$, and $\hat{\mathbf{z}}$, will be accurate estimates of the true underlying factors.

The trilinear model is actually a specific case of the Tucker3 model. The Tucker3 model is best understood by viewing a graphical representation such as in Figure 12.2. A data cube, $\mathbb{R}$, is decomposed into three sets of factors, $\hat{\mathbf{x}}$, $\hat{\mathbf{y}}$, and $\hat{\mathbf{z}}$, as with PARAFAC. However, the Tucker3 model differs from the PARAFAC model in three key ways. First, the number of factors in each order of the Tucker3 model is not constrained to be equal. Second, the Tucker3 model employs a small core cube, $\mathbb{C}$, that governs the interactions among the factors. A nonzero element in the pth, qth, and rth position of the core $\mathbb{C}$ dictates an interaction between the pth factor in the X-way, the qth factor in the Y-way, and the rth factor in the Z-way. This permits modeling of two or more factors that might have, for example, the same chromatographic profile

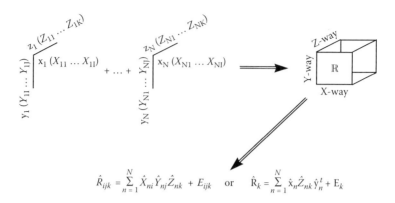

FIGURE 12.1 Construction and decomposition of a three-way array via the trilinear PARAFAC model.

but different spectral and concentration profiles [18, 19]. Third, the interactions in the Tucker3 model make it a nonlinear model, and it is not really appropriate for problems that possess trilinear structure unless special structure is imposed upon the core matrix, $\mathbb{C}$. If there is the same number of factors in each way, and $\mathbb{C}$ is constrained to only have nonzero elements on the superdiagonal, then the Tucker3 model is equivalent to the PARAFAC model.

With the exception of the Tucker3 model, the models discussed here are intrinsically linear models, and a straightforward application thus assumes linear interactions and behavior of the samples. While many of the systems of interest to chemists contain nonlinearities that violate the assumptions of the models, the PARAFAC model forms an excellent starting point from which many subsidiary methods are

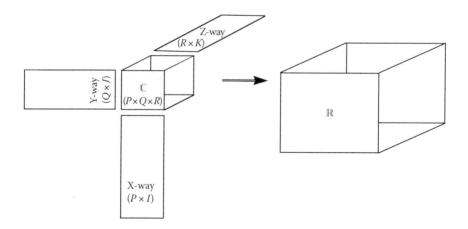

FIGURE 12.2 Construction of a three-way array according to the unconstrained Tucker model.

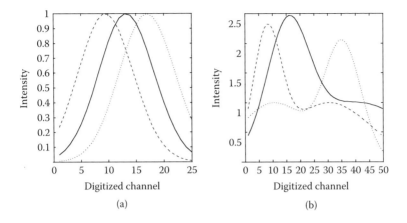

FIGURE 12.3 X-way (A) and Y-way (B) profiles used to construct the data sets for each example. Solid line represents the analyte spectra, while the dashed and dotted lines correspond to the two interferents.

constructed to incorporate nonlinear behavior into calibration models formed from three-way data collected with hyphenated methods.

12.5 EXAMPLES

Seven simulated LC-UV/Vis DAS data matrices were constructed in MATLAB® 5.0 (MathWorks Inc., Natick, MA). Each sample forms a 25×50 matrix. The simulated LC and spectral profiles are shown in Figure 12.3a and Figure 12.3b, respectively. Spectral and chromatographic profiles are constructed to have a complete overlap of the analyte profile by the interferents. Three of the samples represent pure standards of unit, twice-unit, and thrice-unit concentration. These standards are designated S1, S2, and S3, respectively. Three three-component mixtures of relative concentrations of interferent 1: analyte: interferent 2 are 1:1.5:0.5, 2:0.5:2, and 2:2.5:1, and these are employed for all examples. These mixtures are designated M1, M2, and M3, respectively. An additional two-component mixture, 2:2.5:0, is employed as an example for rank annihilation factor analysis. This sample is designated M4. In most applications, normally distributed random errors are added to each digitized channel of every matrix. These errors are chosen to have a mean of zero and a standard deviation of 0.14, which corresponds to 10% of the mean response of the middle standard. In the rank annihilation factor analysis (RAFA) examples, errors were chosen to simulate noise levels of 2.5 and 5% of the mean response of S2. In some PARAFAC examples, the noise level was chosen to be 30% of the mean response of the second standard.

Data analysis algorithms were also constructed in the MATLAB programming language. These programs are included in the appendices of this chapter (Appendices 12.1, GRAM Algorithm; 12.2, DTLD Algorithm, 12.3, PARAFAC Algorithm) and are available on the Web at www.public.asu.edu/~booksh. The Web server also contains the MATLAB script for regenerating the data employed in the examples.

12.6 RANK ANNIHILATION METHODS

Rank annihilation methods employ eigenvalue-eigenvector analyses for direct determination of analyte concentration with or without intrinsic profile determination. With the exception of rank annihilation factor analysis, these methods obtain a direct, noniterative solution by solving various reconstructions of the generalized eigenvalue-eigenvector problem.

12.6.1 RANK ANNIHILATION FACTOR ANALYSIS

The original rank annihilation factor analysis (RAFA) postulates determination of analyte concentration from the successive eigenanalysis of the equation $\mathbf{R}_1 - \alpha \mathbf{R}_2$, where $\mathbf{R}_2$ is an $I \times J$ data matrix from a mixture of the analyte and other interferents, $\mathbf{R}_1$ is an $I \times J$ matrix from a pure analyte of known concentration, and α is an estimate of relative analyte concentration in the standard compared with the mixture [20]. Determination of the eigenvalues of the matrix $(\mathbf{R}_1 - \alpha \mathbf{R}_2)(\mathbf{R}_1 - \alpha \mathbf{R}_2)^T$ or $(\mathbf{R}_1 - \alpha \mathbf{R}_2)^T(\mathbf{R}_1 - \alpha \mathbf{R}_2)$ yields N eigenvalues that correspond to "significant" factors and $I - N$ (or $J - N$ if $J < I$) "insignificant" factors that correspond to random noise in the samples. A plot of the calculated eigenvalues vs. α will show a minimum in the Nth eigenvalue when 1 is equal to the relative analyte concentration between the standard and the unknown. An iterative line search method is employed to find the optimum 1 that yields the minimum value for the Nth eigenvalue.

Lorber [21, 22] recognized that the roots of

$$|(\mathbf{R}_2 - \alpha \mathbf{R}_1)| = 0 \qquad (12.2)$$

as iteratively determined by RAFA, can be directly determined by reconfiguring the RAFA equation into the form of the generalized eigenvalue-eigenvector problem (GEP),

$$\mathbf{R}_1 \Psi = \Lambda \mathbf{R}_2 \Psi \qquad (12.3)$$

Here Λ is the diagonal matrix of eigenvalues, and the columns of Ψ are the eigenvectors of $\mathbf{R}_1$ and $\mathbf{R}_2$. However, Equation 12.3 cannot be solved directly as a GEP, since $\mathbf{R}_1$ and $\mathbf{R}_2$ are not square, full rank matrices.

One method of solving Equation 12.3 is to transform it into an eigenvalue-eigenvector problem with one square, nonsingular matrix. The solution can be derived by first expressing $\mathbf{R}_1$ as the outer product of two vectors $\mathbf{x}$ and $\mathbf{y}$ where $\mathbf{R}_1 = \mathbf{xy}^T$. This is equivalent to Equation 12.1b with $\mathbf{x}$ and/or $\mathbf{y}$ scaled such that z is 1 for the standard sample. Therefore, the quantity of signal due to analyte in $\mathbf{R}_2$ is given by $\mathbf{xy}^T/\alpha$ where α gives the concentration ratio of analyte in $\mathbf{R}_1$ and $\mathbf{R}_2$, e.g., $\alpha = C_1/C_2$. While $\mathbf{x}$ and $\mathbf{y}$ cannot be known without *a priori* information, the matrix $\mathbf{R}_2$ can be decomposed via the singular-value decomposition (SVD) [23] into orthogonal rotation components, **U** and **V**, that span the row and column space of $\mathbf{R}_2$, respectively (see Chapter 5). By considering only the N most significant orthogonal components, $\mathbf{R}_2$ is approximated as

$$\mathbf{R}_2 \approx \overline{\mathbf{U}}\overline{\mathbf{S}}\overline{\mathbf{V}}^T \qquad (12.4)$$

where **S** is the diagonal matrix of singular values of $\mathbf{R}_2$, and the subscript indicates that only the N most significant orthogonal rotation components are employed in reconstructing $\mathbf{R}_2$. Thus the minimum dimensional space required to represent the information content of $\mathbf{R}_2$, with the assumption that $\mathbf{R}_1$ is contained in this N dimensional space, is determined. Projecting $\mathbf{R}_1$ and $\mathbf{R}_2$ into the subspace defined by Equation 12.4 gives

$$\overline{\mathbf{S}}^{-1}\overline{\mathbf{U}}^T\mathbf{R}_1\overline{\mathbf{V}}\Psi = \Lambda\overline{\mathbf{S}}^{-1}\overline{\mathbf{U}}^T\overline{\mathbf{U}}\overline{\mathbf{S}}\overline{\mathbf{V}}^T\overline{\mathbf{V}}\Psi \qquad (12.5a)$$

Since $\mathbf{U}^T\mathbf{U}$ and $\mathbf{S}^{-1}\mathbf{S}$ are identity matrices and $\mathbf{V}^T\mathbf{V}\mathbf{X} = \mathbf{X}$, Equation 12.5a reduces to

$$\overline{\mathbf{S}}^{-1}\overline{\mathbf{U}}^T\mathbf{R}_1\overline{\mathbf{V}}\Psi = \Lambda\Psi \qquad (12.5b)$$

with $\overline{\mathbf{S}}^{-1}\overline{\mathbf{U}}^T\mathbf{R}_1\overline{\mathbf{V}}$ being an $N \times N$ full rank matrix and any nonzero eigenvalue equal to α. Concurrently, α may also be determined from the trace of $\overline{\mathbf{S}}^{-1}\overline{\mathbf{U}}^T\mathbf{R}_1\overline{\mathbf{V}}$ where

$$a = \mathrm{trace}(\overline{\mathbf{S}}^{-1}\overline{\mathbf{U}}^T\mathbf{R}_1\overline{\mathbf{V}})/q \qquad (12.6)$$

The trace of a matrix is defined as the sum of the diagonal elements, and q is given as the rank of $\mathbf{R}_1$. Please note in the above derivations, as in RAFA, that $\mathbf{R}_1$ must be a rank 1 matrix representing the pure standard.

Considering the algorithmic simplicity and efficiency in this formulation, RAFA is an attractive method of estimating analyte concentration in a complex mixture. Compared with the other three-way methods discussed below, very few floating-point operations (FLOPs) are required. For small values of N, abandoning the SVD for the more efficient kernel NIPALS (nonlinear iterative partial algorithm least squares) can further reduce the required FLOP budget. However, RAFA is limited in applicability by two restrictions: (1) a pure standard of known concentration is required, and (2) only one standard and one unknown sample can be simultaneously analyzed. These two deficiencies are corrected in the generalized rank annihilation method (GRAM) and direct trilinear decomposition (DTLD), respectively.

12.6.1.1 RAFA Application

The predictive limitations of the iterative RAFA method are evident from Figure 12.4. Figure 12.4a demonstrates the sharp minimum in the plot of the singular values versus α for an errorless S3 and M3. The third singular value contains the minimum, since M3 is, ideally, a rank 3 matrix. The minimum accurately occurs at a predicted concentration of 2.500 when the concentration of the standard is accounted. However, adding random errors of 1.667% and 3.333% of the S3 response

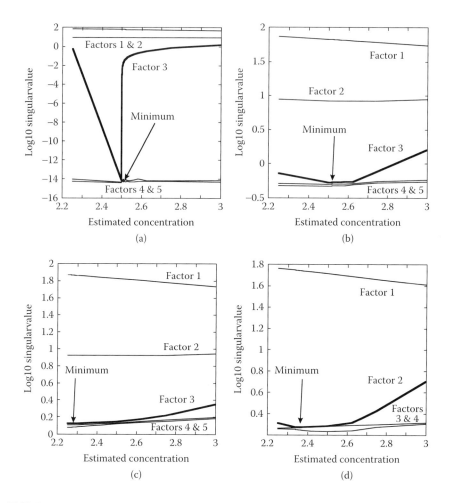

FIGURE 12.4 Plots of log10 of singular values vs. estimated concentration for RAFA. Bold line represents correct factor for quantitation: (a) errorless, ideal data, (b) 2.5% relative error added to each sample, (c) 5.0% relative error added to each sample, (d) only one interferent with 5.0% relative error added to each sample.

(2.5% and 5.0% of the mean S2) to the mixture and standard matrices broadens the minimum in this plot (Figure 12.4b and Figure 12.4c). The minimum in Figure 12.4b, albeit broader, still accurately occurs at 2.500. However, with errors as small as 3%, there is evidence in Figure 12.4c that RAFA becomes unreliable for this application. Here the analyte concentration in the mixture is estimated to be 2.26, significantly lower than the true value of 2.500. More accurate quantitation and a sharper minimum in the plot of singular values vs. α occur when fewer factors are required to model the mixture spectrum. Figure 12.4d is derived from RAFA on S3 and M4, with M4 being, ideally, a rank 2 matrix. Added errors have a standard deviation of 3.333% of the S3 response. Although the minimum is sharper and the estimated mixture concentration is more accurate (2.34) than in Figure 12.4c, the level of precision and accuracy demonstrated by RAFA at this noise level would surely not be acceptable in many applications.

12.6.2 GENERALIZED RANK ANNIHILATION METHOD

An alternative method to solving Equation 12.3 is to reduce both $\mathbf{R}_1$ and $\mathbf{R}_2$ to square, nonsingular, nonidentity matrices by projecting each matrix independently onto the space formed jointly by the two matrices. This permits calculation of Λ and Ψ, via the QZ algorithm [23] and, by extension, relative concentration estimates, $\hat{\mathbf{Z}}$, and estimates of the true underlying factors in the X- and Y-ways. This is known as the generalized rank annihilation method (GRAM) [24, 25].

The first step in GRAM is projecting $\mathbf{R}_1$ and $\mathbf{R}_2$ into the joint row and column spaces of the two matrices. Determination of the row and column spaces can be performed with the singular value decomposition of $\mathbf{R}_1 + \mathbf{R}_2$ where

$$\mathbf{R}_1 + \mathbf{R}_2 \approx \bar{\mathbf{U}}\bar{\mathbf{S}}\bar{\mathbf{V}} \quad (12.7)$$

Here the N columns of $\bar{\mathbf{U}}$ define the significant factors in the joint row space of $\mathbf{R}_1$ and $\mathbf{R}_2$, and the N columns of $\bar{\mathbf{V}}$ define the joint column space. The generalized eigenvalue problem

$$\bar{\mathbf{U}}^T\mathbf{R}_1\bar{\mathbf{V}}\Psi = \bar{\mathbf{U}}^T\mathbf{R}_2\bar{\mathbf{V}}\Psi\Lambda \quad (12.8)$$

is solved by the QZ algorithm, where the estimates of the true underlying factors in the X and Y orders are found by

$$\hat{\mathbf{Y}} = \bar{\mathbf{V}}(\Psi^{-1})^T \quad (12.9a)$$

and

$$\hat{\mathbf{X}} = (\mathbf{R}_1 + \mathbf{R}_2)\bar{\mathbf{V}}\,\Psi \quad (12.9b)$$

The relative concentration estimates are found in the diagonal elements of Λ, e.g., $\Lambda_{kk} = 0$ if the kth species is absent in $\mathbf{R}_2$ and $\Lambda_{kk} = \infty$ if the kth species is absent in $\mathbf{R}_1$.

In the intervening decade since the inception of GRAM, numerous refinements to the basic algorithm have been published. Wilson, Sanchez, and Kowalski [25] proposed three initial improvements. Inserting $\mathbf{R}_1 + \mathbf{R}_2$ for $\mathbf{R}_1$ in Equation 12.8 solves stability problems encountered when $\mathbf{R}_2$ contains components absent in $\mathbf{R}_1$. Here, the diagonal matrix Λ now contains the fractional contribution, e.g., $\Lambda_{kk} = 0$ if the kth species is absent in $\mathbf{R}_2$ and $\Lambda_{kk} = 1$ if the kth species is absent in $\mathbf{R}_1$. Second, the significant joint row and column spaces of $\mathbf{R}_1$ and $\mathbf{R}_2$ can be more rapidly calculated with a NIPALS-based algorithm than with the SVD. Finally, the joint row

TABLE 12.1A
Estimated Concentrations in Mixture Samples from GRAM with Concatenated Matrices

	Mixture M1 (Truth: 1.5)	Mixture M2 (Truth: 0.5)	Mixture M3 (Truth: 2.5)
Standard S1	1.412	0.533	2.256
Standard S2	1.504	0.565	2.394
Standard S3	1.514	0.565	2.384

TABLE 12.1B
Estimated Concentrations in Mixture Samples from GRAM with Added Matrices

	Mixture M1 (Truth: 1.5)	Mixture M2 (Truth: 0.5)	Mixture M3 (Truth: 2.5)
Standard S1	1.260	0.627	2.151
Standard S2	1.538	0.653	2.356
Standard S3	1.610	0.639	2.401

and column spaces of $\mathbf{R}_1$ and $\mathbf{R}_2$ can alternatively be determined from the concatenated matrices $[\mathbf{R}_1 | \mathbf{R}_2]$ and $\frac{\mathbf{R}_1}{\mathbf{R}_2}$, respectively. However, Poe and Rutan [26] have demonstrated that determination of the joint row and column spaces from $\mathbf{R}_1 + \mathbf{R}_2$ yields the most robust estimates of X, Y, and Z when model errors are introduced in the form of spectral shifts (e.g., slight changes in retention indices) in the X- and Y-ways.

Li and Gemperline [27] solved the problem of GRAM returning X- and Y-way estimates with imaginary components when two or more factors were collinear in the Z-way, or concentration. Here, similarity transformations are employed to rediagonalize Λ and eliminate any imaginary components in Ψ that do not correspond to a degenerate solution for Equation 12.8. The need for similarity transformations can be circumvented by judicious choice of the QZ algorithm [28].

12.6.2.1 GRAM Application

The estimated analyte concentrations in each of the three mixtures, with added random errors of a standard deviation equal to 10% of the mean value of S2, are summarized in Table 12.1. The results are presented for prediction with each of the three standards for each of the three mixtures utilizing the GRAM algorithm with concatenated (Table 12.1a) and added (Table 12.1b) samples. Similarly, the squared correlation coefficients between the true and estimated X-way and Y-way profiles are presented in Table 12.2. Figure 12.5 presents the best-case and worst-case examples for estimating the X-way and Y-way profiles with GRAM. In the worst-case scenario ($\rho^2_x = 0.9938$, $\rho^2_y = 0.9987$), there is very little discernible variation between the true and estimated profiles. In most cases, the true and estimated profiles

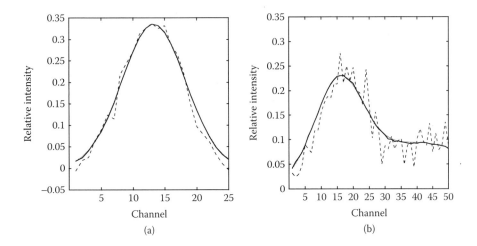

FIGURE 12.5 True (solid), best (gray), and worst (dashed) estimated (a) X-way and (b) Y-way analyte profiles from GRAM. There is insufficient difference between the best and true analyte profiles to discriminate them in these plots.

are indistinguishable. GRAM is incapable of estimating the true profiles of the interferents, since the two interfering spectral concentrations are collinear between samples.

Two trends in the predictive performance of GRAM are apparent from Table 12.1a and Table 12.1b and from Table 12.2a through Table 12.2d. First, it is readily evident that GRAM is a more robust algorithm for concentration estimation than the iterative RAFA. At its worst, GRAM yields a 30% error for prediction of mixture M2. While this error seems superficially very large, it should be tempered with the realization that the random noise added to each sample is equivalent to 40% of the analyte concentration in the mixture. For a prediction with more concentrated standards and mixtures, the prediction error is closer to 4%. Second, performing GRAM

TABLE 12.2A
Squared Correlation Coefficient (angle cosign) of Predicted and True Y-Way Spectra from GRAM with Concatenated Matrices

	Mixture M1	Mixture M2	Mixture M3
Standard S1	0.9993	0.9999	1.0000
Standard S2	0.9995	1.0000	1.0000
Standard S3	0.9996	1.0000	1.0000

TABLE 12.2B
Squared Correlation Coefficient (angle cosign) of Predicted and True Y-Way Spectra from GRAM with Added Matrices

	Mixture M1	Mixture M2	Mixture M3
Standard S1	0.9987	0.9994	0.9997
Standard S2	0.9988	0.9996	0.9998
Standard S3	0.9987	0.9996	0.9998

TABLE 12.2C
Squared Correlation Coefficient (angle cosign) of Predicted and True X-Way Spectra from GRAM with Concatenated Matrices

	Mixture M1	Mixture M2	Mixture M3
Standard S1	0.9985	0.9996	0.9992
Standard S2	0.9998	0.9999	0.9998
Standard S3	0.9999	0.9999	0.9999

TABLE 12.2D
Squared Correlation Coefficient (angle cosign) of Predicted and True X-Way Spectra from GRAM with Added Matrices

	Mixture M1	Mixture M2	Mixture M3
Standard S1	0.9939	0.9959	0.9982
Standard S2	0.9971	0.9963	0.9983
Standard S3	0.9986	0.9971	0.9986

with concatenated matrices generally yields lower prediction errors than performing GRAM by adding the two matrices to determine the reduced sample space (Equation 12.7). Although these results may seem counter to the conclusions of Poe and Rutan [26], recall that Poe and Rutan employed correlated errors as spectral and chromatographic shifts, while in these examples, uncorrelated random errors were employed. The logical conclusion of these two observations is that, on average, GRAM with added matrices is best employed when errors in profile reproducibility predominate, while GRAM with concatenated matrices is best employed when random, uncorrelated errors predominate.

12.6.3 DIRECT TRILINEAR DECOMPOSITION

RAFA and GRAM are limited to simultaneous analysis of one standard and one unknown. One way around this limitation is the direct trilinear decomposition (DTLD) [29]. Aside from being a rapid method for solving Equation 12.1, DTLD also provides an excellent starting solution for least-squares and iterative methods.

At the heart of DTLD lies the GRAM algorithm. DTLD functions by judiciously combining the K samples in the data matrix into two pseudosamples, applying GRAM, then reinflating the $I \times N$, $J \times N$, and $K \times N$ matrices to present the X-way profiles, Y-way profiles, and Z-way profiles, respectively.

The first steps in DTLD are unfolding the K $I \times J$ matrices into a $K \times (I*J)$ matrix and employing the SVD algorithm to calculate the first two vectors, $\mathbf{u}_1$ and $\mathbf{u}_2$, of the row space in this unfolded matrix. These two vectors serve as mixing vectors to construct the two pseudosamples, $\mathbf{G}_1$ and $\mathbf{G}_2$, from linear combinations of the K matrices where

$$\mathbf{G}_1 = \sum_{k=1}^{K} \mathbf{u1}_k \mathbf{R}_k \quad (12.10a)$$

and

$$\mathbf{G}_2 = \sum_{k=1}^{K} \mathbf{u2}_k \mathbf{R}_k \quad (12.10b)$$

with $\mathbf{R}_k$ being the Kth sample and $\mathbf{u}i_k$ being the kth element in the ith vector of the row space of the set of unfolded samples. Another alternative to constructing the two pseudosamples is to add half of the sample matrices to generate $\mathbf{G}_1$ and add the other half of the sample matrices to generate $\mathbf{G}_2$. However, potential problems can result from arbitrary summation of samples through inducing collinearity. For example, consider four unknown samples with concentrations [1,2], [2,1], [2,2], and [1,1]. If the first two samples are added together and the second two samples are added together, the two resulting pseudosamples are collinear, with apparent concentrations [3,3] and [3,3]. Hence, the calculated model would be unstable. By choosing orthogonal vectors that lie in the sample space of the data as mixing guidelines for constructing the two pseudosamples, all possibility of introducing unnatural collinearity into the system is eliminated.

As with GRAM, the matrices $\mathbf{G}_1$ and $\mathbf{G}_2$ need to be projected into the joint row and column spaces of the K samples. This can be accomplished by augmenting the K samples row- and columnwise prior to singular-value decomposition (SVD), or by summing the K samples prior to SVD. Equation 12.8 is solved for Ψ with $\mathbf{G}_1$ and $\mathbf{G}_2$ replacing $\mathbf{R}_1$ and $\mathbf{R}_2$. The estimates of the true X-way and Y-way factors are found by Equation 12.9a and Equation 12.9b. The difference from GRAM is that the concentration estimates, or the columns of $\mathbf{Z}$, are found by a least-squares fitting of the columns of $\hat{\mathbf{X}}$ and $\hat{\mathbf{Y}}$ to $\mathbf{R}_k$. The least-squares fit is performed to obtain the N values of the kth row, $\hat{\mathbf{Z}}$.

TABLE 12.3
Predicted Standard and Mixture Concentrations from DTLD and PARAFAC

Sample	True	DTLD	PARAFAC	PARAFAC ($\times 3$ noise)
S1	1.0	1.0006	1.0007	0.9582
S2	2.0	1.9987	1.9987	2.0279
S3	3.0	3.0006	3.0007	2.9955
M1	1.5	1.5090	1.5025	1.5440
M2	0.5	0.4838	0.4997	0.6587
M3	2.5	2.5385	2.5257	2.5782

It is important to note that problems may arise if $\hat{\mathbf{X}}$ and $\hat{\mathbf{Y}}$ are calculated from Y values that are obtained from solutions to

$$\bar{\mathbf{U}}^T \mathbf{G}_1 \bar{\mathbf{V}} \Psi = \bar{\mathbf{U}}^T \mathbf{G}_2 \bar{\mathbf{V}} \Psi \Lambda \tag{12.11a}$$

and

$$\bar{\mathbf{V}}^T \mathbf{G}_1 \bar{\mathbf{U}} \Psi = \bar{\mathbf{V}}^T \mathbf{G}_2 \bar{\mathbf{U}} \Psi \Lambda \tag{12.11b}$$

as was proposed in the original reference [30]. Instead, it is advisable to use the most numerically stable, best-conditioned solution from Equation 12.11a and Equation 12.11b.

12.6.3.1 DTLD Application

DTLD yields even more accurate and precise results than GRAM. Table 12.3 presents the predicted analyte concentrations for the three standards and three unknowns. The estimated standard concentrations are accurate to less than 0.1% of the true analyte concentration. Prediction errors for the mixture samples are, in general, a factor of 2 to 4 less than the prediction errors realized by GRAM. The largest prediction error, for the least concentrated sample, is only 3.25%. This is compared with an average prediction error of 11% with the best application of GRAM.

DTLD allows estimation of the interferent X-way and Y-way profiles, as well estimation of the analyte profiles. The squared correlation coefficients of these estimated profiles with the true, noiseless profiles are presented in Table 12.4. For the analyte, which is present in all six samples, the true and estimated profiles are indistinguishable. The interferent profiles, which are present in only half of the samples, are comparable in accuracy with the estimated analyte profiles from GRAM.

TABLE 12.4
Squared Correlation Coefficient (angle cosign) between True and Estimated Profiles from DTLD and PARAFAC

	DTLD	PARAFAC	PARAFAC (× 3 noise)
X-way (analyte)	1.000	1.000	0.999
Y-way (analyte)	1.000	1.000	0.9997
X-way (interferent 1)	0.9976	0.9998	0.9987
X-way (interferent 2)	0.9990	0.9998	0.9988
Y-way (interferent 1)	0.9996	0.9999	0.9985
Y-way (interferent 2)	0.9985	0.9998	0.9976

12.7 ALTERNATING LEAST-SQUARES METHODS

Alternating least squares (ALS) methods are both slower, due to their numeric intensity, and more flexible than eigenvalue–eigenvector problem-based methods for solving Equation 12.1a and Equation 12.1b. The basic PARAFAC model of Equation 12.1 is expanded to include cross-terms. This designates interactions between factors in different modes. Simultaneously, the ALS solutions to Equation 12.1 can be constrained to be positive while, or while not, having unimodal factors in one or all ways.

12.7.1 PARAFAC / CANDECOMP

PARAFAC refers both to the parallel factorization of the data set **R** by Equation 12.1a and Equation 12.1b and to an alternating least-squares algorithm for determining $\hat{\mathbf{X}}$, $\hat{\mathbf{Y}}$, and $\hat{\mathbf{Z}}$ in the two equations. The ALS algorithm is known as PARAFAC, emanating from the work by Kroonenberg [31], and as CANDECOMP, for canonical decomposition, based on the work of Harshman [32]. In either case, the two basic algorithms are practically identical.

The PARAFAC/CANDECOMP algorithm begins with an initial guess of the X-way and Y-way starting profiles. The initial Z-way profiles are determined by solving

$$\mathbf{R}_C = \mathbf{A}\mathbf{Z}^T \qquad (12.12)$$

where $\hat{\mathbf{Z}} = \mathbf{C}^+\mathbf{R}_C$, and $\mathbf{C}^+$ is the generalized inverse of $\mathbf{C}$ that can be calculated from the normal equations or singular-value decomposition of $\mathbf{C}$. In Equation 12.12, $\mathbf{R}_C$ is an $(I*J) \times K$ matrix. The matrix is constructed by unfolding the K slices of $\mathbb{R}$ in the I–J plane, where $R_{C(j-1)I+i,k} = R_{i,j,k}$. Similarly, $\mathbf{C}$ is an $(I*J) \times N$ matrix formed from the N columns of $\hat{\mathbf{X}}$ and $\hat{\mathbf{Y}}$ where $C_{(j-1)I+i,n} = X_{i,n}Y_{j,n}$.

Updated estimates of the X-way and Y-way profiles are found by solving

$$\mathbf{R}_A = \mathbf{A}\mathbf{X}^T \qquad (12.13)$$

and

$$\mathbf{R}_B = \mathbf{B}\mathbf{Y}^T \quad (12.14)$$

such that $\hat{\mathbf{X}} = \mathbf{A}^+\mathbf{R}_A$, and $\hat{\mathbf{Y}} = \mathbf{B}^+\mathbf{R}_B$, respectively. $\mathbf{R}_A$ and $\mathbf{R}_B$ are constructed analogously to $\mathbf{R}_C$. Their construction requires the respective unfolding of $\mathbb{R}$ in the Y–Z and X–Z planes. This forms a $(J * Z) \times I$ matrix for $\mathbf{R}_A$, and an $(I * K) \times J$ matrix for $\mathbf{R}_B$. Similarly to $\mathbf{C}$, $A_{(k-1)J+k,n} = Y_{j,n}Z_{k,n}$ and $B_{(k-1)I+k,n} = X_{i,n}Z_{k,n}$. The algorithm proceeds iteratively, cycling through Equation 12.12 through Equation 12.14 until the convergence criterion is satisfied. For each cycle, the most recent estimates of $\mathbf{X}$ and $\mathbf{Y}$ are used to determine $\hat{\mathbf{Z}}$ in Equation 12.12; the most recent estimates of $\mathbf{Y}$ and $\mathbf{Z}$ are used to determine $\hat{\mathbf{X}}$ in Equation 12.13; and the most recent estimates of $\mathbf{X}$ and $\mathbf{Z}$ are used to determine $\hat{\mathbf{Y}}$ in Equation 12.14. There are, consequently, two important factors that influence the final estimates of the X-way, Y-way, and Z-way profiles. They are the starting guess for $\hat{\mathbf{X}}$ and $\hat{\mathbf{Y}}$, and the convergence criterion.

The PARAFAC algorithm is sensitive to the initial guess of the solutions for $\hat{\mathbf{X}}$ and $\hat{\mathbf{Y}}$. Inaccuracy of the initial guess may cause PARAFAC to become trapped in local minima and, hence, not to converge to the global optimum least-squares solution. To further the problem, the PARAFAC algorithm can become delayed in "swamps" far from the optimum solution [33]. Consequently, the speed of the algorithm is sensitive to the initial guess for $\hat{\mathbf{X}}$ and $\hat{\mathbf{Y}}$. The starting iteration of $\hat{\mathbf{X}}$ and $\hat{\mathbf{Y}}$ can be provided by a random number generator [34], DTLD [35], or *a priori* knowledge of analyte profiles. Multiple initial guesses should be considered when employing a random starting value, even though these efforts markedly increase the analysis time. Although the solution for each starting value will be different, if all, or most, of the solutions are similar, it is safe to assume that PARAFAC has converged near the global optimal solution. The convergence time for PARAFAC can be improved by initializing the algorithm with guesses near the optimal solution. These guesses can come from DTLD or reference spectra of species known, or highly suspected, to be in the data set. Care should be employed when utilizing the DTLD solutions, since DTLD often yields significant imaginary components in predicting X-way and Y-way factors. The problems caused by initializing PARAFAC with imaginary components can be circumvented by applying the real components of $\hat{\mathbf{X}}$ and $\hat{\mathbf{Y}}$ generated from DTLD. Alternatively, the absolute value, or complex modulus, of $\hat{\mathbf{X}}$ and $\hat{\mathbf{Y}}$ from DTLD may be employed.

Two popular convergence criteria for the PARAFAC algorithm are based on (1) the changes in the residuals, or unmodeled data, between successive iterations, and (2) changes in the predicted profiles between successive iterations. In the first case, the algorithm is terminated when the root averages of the squared residuals between successive iterations agree to within an absolute or relative tolerance; for example, they may be within 10^{-6} of one another. While such fit-based stopping criteria are conceptually easy to visualize, a faster method for determining convergence relies on the correlation between the predicted X-, Y-, and Z-way profiles between successive iterations. When the product of the cosines between successive iterations in the X-, Y-, and Z-modes approaches arbitrarily close to 1, say within 10^{-6}, the algorithm is terminated. The cosine in the X-way is determined by unfolding the $I \times N$ matrices, $\hat{\mathbf{X}}_{old}$ and $\hat{\mathbf{X}}_{new}$, into the column vectors $\hat{\mathbf{x}}_{old}$ and $\hat{\mathbf{x}}_{new}$. The cosine θ_X is then defined as

$$\cos \theta_X = \frac{\mathbf{x}_{old}\mathbf{x}_{new}}{\sqrt{(\mathbf{x}_{old}\mathbf{x}_{old})(\mathbf{x}_{new}\mathbf{x}_{new})}} \tag{12.15}$$

The other two terms, $\cos \theta_Y$ and $\cos \theta_Z$, are defined equivalently. Convergence, when $\cos \theta_X * \cos \theta_Y * \cos \theta_Z > 1 - 10^{-6}$, implies that successive iterations in all three modes are correlated to at least $1 - 10^{-6}$. Mitchell and Burdick [32] cite, besides speed, an additional benefit to correlation-based convergence. In cases when two factors are highly correlated in one or more of the three ways, ALS methods may become mired in "swamps," where the fit of the model changes slightly but the correlation between the predicted X-, Y-, and Z-ways changes significantly between successive iterations. Following numerous iterations, the ALS algorithm will emerge from the "swamp," and the residuals and estimated profiles will then both rapidly approach the optimum. Hence, correlation-based convergence is more resistant to inflection points in the error response surface when optimizing the model.

12.7.1.1 Tuckals

The generalization of the PARAFAC model is the Tucker3 model. As with PARAFAC, the Tucker3 model decomposes a data cube $\mathbb{R}$ into three matrices: **X**, **Y**, and **Z**. In addition, it also generates a core of reduced dimensions, $\mathbb{C}$, from $\mathbb{R}$ (Figure 12.2). One alternating least-squares algorithm for estimating the parameters of the Tucker3 model is Tuckals, or TUCK Alternating Least Squares. This iterative Tuckals algorithm proceeds similarly to the PARAFAC/CANDECOMP algorithm. However, instead of cycling through three sets of parameters, four sets of parameters must be successively updated, $\hat{\mathbf{X}}, \hat{\mathbf{Y}}, \hat{\mathbf{Z}}$, and $\hat{\mathbb{C}}$. Furthermore, while PARAFAC preassumes N, or the number of factors in the model, Tucker3 requires that the three dimensions of the core array, P, Q, and R, be assumed.

12.7.1.2 Solution Constraints

ALS algorithms are more flexible than rank-annihilation-based algorithms because constraints can be placed onto the solutions derived from ALS methods. Ideally, constraints are not needed to achieve accurate, meaningful estimates of concentration and spectral profile. However, the presence of slight nonlinear interactions among the true underlying factors, of highly correlated factors, or of low SNR will often result in profile estimates that are visually unsatisfying and that contain significant quantitative errors derived from the model. These effects can often be minimized by employing constraints to the solutions that are based on *a priori* knowledge or assumptions of the data structure. Common *a priori* constraints include prior knowledge of sample concentrations, spectral profiles, or analyte characteristics. ALS algorithms implicitly constrain the estimated profiles to lie in real space as opposed to the rank annihilation methods, which may fit factors with imaginary components to the data.

Perhaps the most common constraint consciously placed on the PARAFAC or Tucker3 models is nonnegativity. When one of the modes represents concentrations, chromatographic profiles, or spectra, constraining the solutions to yield only nonnegative profile estimates often improves the quantitative and qualitative accuracy of the models. Care should be taken when applying nonnegativity constraints to spectra, such as absorbance and quenching in fluorescence, which can be manifested, detected, and modeled as negative profiles. Nonnegative estimates of the three-way profiles can be obtained by replacing the least-squares update of any given profile with the nonnegative least squares (NNLS) solution that is well defined in the mathematics literature [36]. The method described in [36] is readily available as a MATLAB function. The downside of this method is that it is numerically intensive compared with computing the regular least-squares solution for each update. Alternatively, nonnegativity can be more rapidly enforced by setting all negative parts of each profile to zero, or its absolute value, prior to updating. Empirically, convergence is achieved with fewer floating-point operations compared with calculating the true NNLS solution. However, the relative efficacy of setting all negative values to zero compared with NNLS is unknown.

A second constraint often applied in three-way calibration of chromatographic data is unimodality. This constraint exploits the knowledge that chromatographic profiles have exactly one maximum. Unlike NNLS, there is a method to calculate the true unimodal least-squares update during each iteration. With unimodal constraints, a search algorithm is implemented to find the maximum of each profile and to ensure that, from that maximum, all values are monotonically nonincreasing. Values found to be not monotonically decreasing can be suppressed with equality constraints.

The third common constraint is based on *a priori* knowledge of the three-way profiles. In this case, the known relative concentrations of the standards, or the known spectral profiles of one or more components, can be fixed as part of the solution. In the Tucker3 model, it is common to restrict some of the potential interactions between factors when they are known not to exist. Constraint values, again, lend themselves to careful selection, as the scaling of the factors must still be taken into account.

12.7.1.3 PARAFAC Application

Table 12.3 compares the estimated analyte concentrations for DTLD, PARAFAC, and PARAFAC × 3 noise (PARAFAC with the addition of a factor of three greater random errors) applied to the same calibration problem. Table 12.4 is analogous to Table 12.3, except that it also presents the squared correlation coefficients between the true and estimated X-way and Y-way profiles for all three species present in the six samples. It is first evident that PARAFAC slightly outperforms DTLD when applied to the same calibration problem. However, the improvement often lies in the third or fourth decimal place and is hardly significant when compared with the overall precision of the data. This near equivalence of DTLD and PARAFAC is rooted in the fact that DTLD performs admirably, and there is little room for improvement to "refine" the DTLD solution. A direct visual comparison of the true and estimate profiles is not shown; however, the sets of two curves are identical to the resolution of the plots.

Increasing the magnitude of the added random errors by a factor of three has surprisingly little effect on the accuracy of PARAFAC. The largest prediction errors are associated with the samples that include the least-significant analyte spectra, S1 and M1. In these two samples, the standard deviations for the added errors are 60% and 120% of the mean analyte signal, yet the associated prediction errors are only 4% and 30%, respectively. This is even more impressive when considering that the absolute prediction errors are only 0.04 and 0.15 units over a spread of 2.5 units for all samples.

12.8 EXTENSIONS OF THREE-WAY METHODS

GRAM is applicable in more areas than just calibration of two samples: one standard and one unknown, where multiple measurements are collected in two interlinked "ways." Kubista [37] developed a three-way DATAN (DATa ANalysis) method applied to calibration of multiple samples with fluorescence measurements, each at two excitation wavelengths and multiple digitized emission wavelengths. In this application, the X-way contains the concentration information and Λ contains the relative excitation cross sections of the fluorescent species present. Many of the limitations of this procrustean rotation-based method can be circumvented by employing GRAM instead [38]. It is easy to see that in this application it is unnecessary to be limited to two excitation wavelengths and analysis by GRAM. If more excitation wavelengths are desired, either DTLD or least-squares fitting of the PARAFAC model is appropriate.

In special applications, GRAM or any least-squares solution to the PARAFAC model can also be applied to qualitative analysis of one sample. Windig and Antalek applied GRAM in the direct exponential curve-resolution algorithm (DECRA) to facilitate signal resolution with pulsed gradient spin echo (PGSE) NMR data [39]. Here, the exponential signal decay rate in the X-way was exploited to reconstruct the data set into two matrices, where the signal intensity in the second matrix differed by a factor of the decay constant from the signal intensity in the first matrix. From the original $I \times J$ matrix, $\mathbf{R}_1$ is constructed from the first $I - 1$ NMR spectra and $\mathbf{R}_2$ is constructed from the last $I - 1$ NMR spectra. The estimated X-way factors, scaled by Λ, yield information regarding the diffusion coefficients of the N species present, and the Y-way factors are estimates of the NMR spectra for each species. MATLAB functions for DECRA can be found on the Chemolab archive: sun.mcs.clarkson.edu.

Although the PARAFAC model is a "trilinear" model that assumes linear additivity of effects between species, the model can be successfully employed when there is a nonlinear dependence between analyte concentration and signal intensity. Provided that the spectral profiles in the X- and Y-ways are not concentration dependent, the resolved Z-way profiles will be a nonlinear function of analyte concentration. By utilizing multiple standards with DTLD or PARAFAC, the nonlinear relationship between $\hat{\mathbf{Z}}$ and analyte concentration can be determined, and the analyte concentration can be estimated using univariate nonlinear regression of the appropriate column of $\hat{\mathbf{Z}}$ onto concentration [30].

The PARAFAC model is often applicable for calibration when a finite number of factors cannot fully model the data set. In these traditionally termed "nonbilinear" applications, the additional terms in the PARAFAC model successively approximate the variance in the data set. This approximation is analogous to employing additional factors in a PLS or PCR model [5]. Nonbilinear rank annihilation (NBRA) exploits the property that, in many cases when the PARAFAC model is applied to a set consisting of a pure analyte spectrum and mixture spectrum, some factors will be unique to the analyte, some will be unique to the interferent, and some factors will describe both analyte and interferent information [40]. Accurate calibration and prediction can be accomplished with the factors that are unique to the analyte. If these factors can be found by mathematically multiplying the pure spectrum by α, then the estimated relative concentrations that decrease by $1/\alpha$ are unique to the analyte [41]. In Reference [41] the necessary conditions required to enable accurate prediction with nonbilinear data are discussed.

As with univariate and multivariate calibration, three-way calibration assumes linear additivity of signals. When the sample matrix influences the spectral profiles or sensitivities, either care must be taken to match the standard matrix to those of the unknown samples, or the method of standard additions must be employed for calibration. Employing the standard addition method with three-way analysis is straightforward; only standard additions of known analyte quantity are needed [42]. When the standard addition method is applied to nonbilinear data, the lowest predicted analyte concentration that is stable with respect to the leave-one-out cross-validation method is unique to the analyte.

12.9 FIGURES OF MERIT

Analytical figures of merit, for example sensitivity, selectivity, and signal to SNR, are useful tools for comparing different analytical techniques. The connection of figures of merit from univariate to three-way analysis has been extensively reviewed and critiqued [3, 43]. With two-way and three-way calibration, the figures of merit are based on the "net analyte signal," the NAS. The NAS is loosely defined as the portion of the analyte signal that is employed for calibration. This is contrasted to the full analyte signal that is used in univariate applications. With multivariate data analysis, the NAS is the portion of the pure analyte signal that is orthogonal to all interferents present in the data set, where

$$\text{NAS} = \mathbf{r}_a^T(\mathbf{I} - \mathbf{R}_i\mathbf{R}_i^+) \qquad (12.16)$$

Here $\mathbf{r}_a$ is the instrumental response of the analyte, and $\mathbf{R}_i$ is the collection of instrumental responses of the interferents. The remaining figures of merit are then derived from the NAS.

In three-way calibration, as with two-way calibration, the figures of merit are similarly derived from the three-way NAS. Assuming that all calculations are performed at unit analyte concentration, the selectivity, sensitivity, and SNR are the magnitude of the NAS divided by the magnitude of the analyte signal, concentration, and noise, respectively. Mathematically, these equations can be found from

$$\text{SEL} = \|\text{NAS}\|_F / \|\mathbf{R}_A\|_F \tag{12.17a}$$

$$\text{SEN} = \|\text{NAS}\|_F / c \tag{12.17b}$$

$$\text{S/N} = \|\text{NAS}\|_F / \|\mathbf{E}\|_F \tag{12.17c}$$

Here $\mathbf{R}_A$ is the response of the analyte at unit concentration, c; $\mathbf{E}$ is a matrix of expected, or estimated, errors; and $\|\ \|_F$ is the Froebus norm, or root sum of the squared elements, of a matrix. It should be noted that while the NAS is a matrix quantity, selectivity (SEL), sensitivity (SEN), and signal-to-noise (S/N) are all vector quantities. The limit of detection and the limit of quantitation can also be determined via any accepted univariate definition by substituting $\|\text{NAS}\|_F$ for the analyte signal and $\|\mathbf{E}\|_F$ for the error value.

There is still debate over the proper manner to calculate the NAS. In the earliest work by Ho et al. [20], the three-way NAS is calculated as the outer product of the multivariate NAS from the resolved X-way and Y-way profiles, such that

$$\mathbf{x}_{\text{NAS}} = \mathbf{x}_a^T (\mathbf{I} - \mathbf{X}_i \mathbf{X}_i^+) \tag{12.18a}$$

and

$$\mathbf{y}_{\text{NAS}} = \mathbf{y}_a^T (\mathbf{I} - \mathbf{Y}_i \mathbf{Y}_i^+) \tag{12.18b}$$

Therefore,

$$\text{NAS} = \mathbf{x}\mathbf{y}^T \tag{12.18c}$$

Similarly, Messick et al. [44] suggested that the NAS can be found by orthogonal projection of Equation 12.16 following unfolding each $I \times J$ sample and interferent matrix into an $IJ \times 1$ vector. The three-way NAS is the consequent NAS of Equation 12.16 refolded into an $I \times J$ matrix. The third alternative, propounded by Wang et al. [41], is to construct the NAS from the outer products of the X-way and Y-way profiles that are unique to the analyte. In this method, no projections are explicitly calculated.

12.10 CAVEATS

There are numerous other considerations not covered in this chapter that a thorough treatment of three-way analysis would demand. Perhaps the most important of these is the choosing of the optimal number of factors, N, to include in the three-way model. In truth, there is no single, best way to decide on N and to consequently validate and justify the choice of N. Initial estimates of N can be derived from PCA analysis of the unfolded data matrix $\mathbb{R}$ where any, or all, of the statistical, empirical, or *a priori* methods for deducing the optimal number of factors in a PCA model can be employed [11]. Similarly, visual inspection of the estimated factors is often beneficial. Inclusion of too few factors yields overly broad and featureless factors. On the other hand, inclusion of too many factors often yields nonsensical or redundant factors.

Visual inspection of the estimated factors is not to be trusted in the presence of degenerate factors, which occur when two or more factors are collinear in one or more of the three "ways." When this is the case, in the concentration way or Z-way, the PARAFAC model is still valid, but the rotational uniqueness of the X-way and Y-way profiles of the degenerate factors is lost. This often results in estimated profiles that are hard to interpret. If the collinearity occurs in the X-way or Y-way, the PARAFAC model may not be appropriate, and the constrained Tucker3 model should be used instead. Collinearity in the X-way or Y-way can be checked by successively performing PCA on data unfolded to an $I \times (J*K)$ matrix, and then to a $J \times (I*K)$ matrix. If there are no collinearities in the X- or Y-ways, the optimal number of factors determined by both unfoldings will be the same.

Once the choice of N, or potential range of N, is determined, the next concern is the choice of model and algorithm. As discussed previously, DTLD is considerably faster than ALS algorithms for determining model parameters; however, ALS algorithms are more flexible and robust to small model errors. Similarly, two alternatives for nonnegative least-squares fitting of model parameters were discussed. Table 12.5 lists the speeds, in FLOPs, for the algorithms and data employed as examples in this chapter. GRAM is easily the fastest algorithm, but it is incapable of handling four, then two, samples concurrently. The FLOPs required for a complete GRAM analysis increase geometrically when all combinations of multiple samples are to be included in the analysis. DTLD is slower than GRAM when fewer than four or five samples are analyzed, but for larger data sets GRAM will be considerably faster.

TABLE 12.5
Relative Speed (in GigaFLOPs) for the Discussed Three-Way Methods

Method	G FLOPs
RAFA (mean of 6)	4.2
GRAM (mean of 18)	1.8
DTLD	8.3
PARAFAC (DTLD start)	41.9
PARAFAC ($\times$ 3 noise, DTLD start)	43.5
PARAFAC ($\times$ 3 noise, DTLD start, NNLS)	1111
PARAFAC (random start; 5 replicates)	$\mu = 36.4$; $\sigma = 8.1$

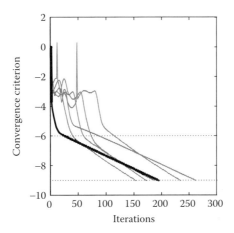

FIGURE 12.6 Convergence progress for PARAFAC from DTLD initialization (bold) and random initialization (gray). Dotted lines represent common convergence thresholds.

PARAFAC is much slower than all other alternatives. Employing DTLD as an initial guess of the X-way and Y-way profiles often reduces the computation time for PARAFAC. This is evident in Figure 12.6. When a low convergence criterion is employed, demonstrated as the dotted line at –6, PARAFAC with DTLD is much faster than PARAFAC with random initial guesses. However, when a more conservative stopping criterion is employed, such as $\cos \theta_X * \cos \theta_Y * \cos \theta_Z > 1 - 10^{-9}$ from Equation 12.15, refining the DTLD model shows no improvement in speed over PARAFAC with random starting values. This is also shown in Table 12.5, where PARAFAC with the DTLD start converges in 43.5 gigaFLOPs, and the means of five replicate random starting points converge on an average of 36.4 gigaFLOPs, with a standard deviation of 8.1 gigaFLOPs. However, it must be noted that this is only one example, and it should be viewed as a potential trend, not a hard rule of thumb. Finally, when constraints are placed on the PARAFAC solution, such as nonnegative least squares, the number of FLOPs required to achieve the final solution increases rapidly. It is a judgment call, best left up to the individual users, to decide on what is an acceptable speed/performance trade.

REFERENCES

1. Esbensen, K.H., Wold, S., and Geladi, P., Relationships between higher-order data array configurations and problem formulations in multivariate data analysis, *J. Chemom.*, 3, 33–48, 1988.
2. Smit, H.C., Signal processing and correlation techniques, in *Chemometrics in Environmental Chemistry*, Einax, J., Ed., Springer, Berlin, 1995.
3. Booksh, K. and Kowalski, B.R., Theory of analytical chemistry, *Anal. Chem.*, 66, 782A–791A, 1994.

4. Gerritsen, M., van Leeuwen, J.A., Vandeginste, B.G.M., Buydens, L., and Kateman, G., Expert systems for multivariate calibration, trendsetters for the wide-spread use of chemometrics, *Chemom. Intell. Lab. Syst.,* 15, 171–184, 1992.
5. Martens, H. and Naes, T., *Multivariate Calibration,* Wiley, Chichester, U.K., 1989.
6. Frank, I.E., Modern nonlinear regression methods, *Chemom. Intel. Lab. Syst.,* 27, 1–9, 1995.
7. Widrow, B. and Sterns, S.D., *Adaptive Signal Processing,* Prentice-Hall, New York, 1985.
8. Lawton, W.H. and Sylvestre, E.A., Self modeling curve resolution, *Technometrics,* 13, 617, 1971.
9. Windig, W., Self-modeling mixture analysis of spectral data with continuous concentration profiles, *Chemom. Intel. Lab. Syst.,* 16, 1–16, 1992.
10. Gampp, H., Maeder, M., Meyer, C.J., and Zuberbuhler, A.D., Calculation of equilibrium constants from multiwavelength spectroscopic data, III: model-free analysis of spectrophotometric and ESR titrations, *Talanta,* 32, 1133, 1985.
11. Malinowski, E., *Factor Analysis in Chemistry,* 2nd ed., John Wiley & Sons, New York, 1991.
12. Hirschfeld, T., The hyphenated methods, *Anal. Chem.,* 52, 297A, 1980.
13. Smilde, A.K., Three-way analyses: problems and perspectives, *Chemo. Intel. Lab. Sys.,* 15, 143–157, 1992.
14. Burdick, D.S., An introduction to tensor products with applications to multiway analysis, *Chemom. Intell. Lab. Syst.,* 28, 229, 1995.
15. Bro, R., PARAFAC, tutorial and applications, *Chemom. Intel. Lab. Syst.,* 38, 149–171, 1997.
16. Booksh, K.S. and Kowalski, B.R., Calibration method choice by comparison of model basis functions to the theoretical instrument response function, A*nal. Chim. Acta,* 348, 1–9, 1997.
17. Jackson, J.E., *A User's Guide to Principal Components,* John Wiley & Sons, New York, 1991.
18. Smilde, A.K., Tauler, R., Henshaw, J.M., Burgess, L.W., and Kowalski, B.R., Multicomponent determination of chlorinated hydrocarbons using a reaction based sensor, 3: medium-rank second-order calibration with restricted Tucker models, *Anal. Chem.,* 66, 3345–3351, 1994.
19. Smilde, A.K., Wang, Y., and Kowalski, B.R., Theory of medium-rank second-order calibration with restricted-Tucker models, *J. Chemom.,* 8, 21–36, 1994.
20. Ho, C.-H., Christian, G.D., and Davidson, E.R., Application of the method of rank annihilation to quantitative analysis of multicomponent fluorescence data from the video fluorometer, *Anal. Chem.,* 50, 1108–1113, 1978.
21. Lorber, A., Quantifying chemical composition from two-dimensional data arrays, *Anal. Chim. Acta,* 164, 293–297, 1984.
22. Lorber, A., Features of quantifying chemical composition from two-dimensional data arrays by the rank annihilation factor analysis method, *Anal. Chem.,* 57, 2395–2397, 1985.
23. Golub, G.H. and Van Loan, C.F., *Matrix Computations,* 3rd ed., Johns Hopkins University Press, Baltimore, 1996.
24. Sanchez, E. and Kowalski, B.R., Generalized rank annihilation factor analysis *Anal. Chem.,* 58, 496–499, 1986.

25. Wilson, B.E., Sanchez, E., and Kowalski, B.R., An improved algorithm for the generalized rank annihilation method, *J. Chemom.*, 3, 493–498, 1989.
26. Poe, R. and Rutan, S., Effects of resolution, peak ratio, and sampling frequency in diode-array fluorescence detection in liquid chromatography, *Anal. Chim. Acta,* 283, 245–253, 1993.
27. Li, S. and Gemperline, P.J., Generalized rank annihilation method using similarity transformations, *Anal. Chem.,* 64, 599–607, 1992.
28. Faber, K., On solving generalized eigenvalue problems using MATLAB, *J. Chemom.,* 11, 87–91, 1997.
29. Sanchez, E. and Kowalski, B.R., Tensorial resolution: a direct trilinear decomposition, *J. Chemom.,* 4, 29–45, 1990.
30. Booksh, K.S., Lin, Z., Wang, Z., and Kowalski, B.R., Extension of trilinear decomposition method with an application to the flow probe sensor, *Anal. Chem.,* 66, 2561–2569, 1994.
31. Kroonenberg, P.M., *Three-Mode Principal Component Analyses: Theory and Applications,* DSWO Press, Leiden, The Netherlands, 1983.
32. Harshman, R.A., Foundations of the PARAFAC procedure, UCLA Working Paper on Phonetics, 16, 1–84, 1970.
33. Mitchell, B.C. and Burdick, D.S., Slowly converging PARAFAC sequences: swamps and two-factor degeneracies, *J. Chemom.,* 6, 155, 1992.
34. Harchman, R.A. and Lundy, M.E., The PARAFAC model for three-way factor analysis and multidimensional scaling, in *Research Methods for Multimode Data Analysis,* Law, H.G. et al., Eds., Praeger, New York, 1984.
35. Burdick, D.S., Tu, X.M., McGown, L.B., and Millican, D.W., Resolution of multicomponent fluorescent mixtures by analysis of the excitation-emission-frequency array, *J. Chemom.,* 4, 15–28, 1990.
36. Lawson, C.L. and Hanson, R.J., *Solving Least Squares Problems,* Prentice-Hall, Upper Saddle River, NJ, 1974.
37. Scarmino, I. and Kubista, M., Analysis of correlated spectral data, *Anal. Chem.,* 65, 409–416, 1993.
38. Booksh, K.S. and Kowalski, B.R., Comments on the DATa ANalysis (DATAN) algorithm and rank annihilation factor analysis for the analysis of correlated spectral data, *J. Chemom.,* 8, 287–292, 1994.
39. Windig, W. and Antelek, B., Direct exponential curve resolution algorithm (DECRA): a novel application of the generalized rank annihilation method for single spectral mixture data set with exponentially decaying contribution profiles, *Chemom. Intel. Lab. Sys.,* 37, 241–254, 1997.
40. Wilson, B.E. and Kowalski, B.R., Quantitative analysis in the presence of spectral interferents using second-order nonbilinear data, *Anal. Chem.,* 61, 2277–2284, 1989.
41. Wang, Y., Borgen, O.S., Kowalski, B.R., Gu, M., and Turecek, F., Advances in second order calibration, *J. Chemom.,* 7, 117–130, 1993.
42. Booksh, K.S., Henshaw, J.M., Burgess, L.W., and Kowalski, B.R., A second-order standard addition method with application to calibration of a kinetics-spectroscopic sensor for quantitation of trichloroethylene, *J. Chemom.,* 9, 263–282, 1995.

43. Faber, K., Lorber, A., and Kowalski, B.R., Analytical figures of merit for tensorial calibration, *J. Chemom.*, 11 419–462, 1997.
44. Messik, N.J., Kalivas, J.H., and Lang, P.M., Selectivity and related measures for nth-order data, *Anal. Chem.*, 68, 1572–1579, 1996.

APPENDIX 12.1 GRAM ALGORITHM

```
function [X,Y,c_est]=gram_demo(STAN,UNKN,rank,opts)
%Generalized Rank Annihilation Method as per Wilson, Sanchez, and Kowalski.
%
%INPUT
%    STAN: Standardized matrix of known analyte concentration.
%    UNKN: Mixture matrix of indeterminate constitution.
%    rank: Estimated rank of the concatenated STAN and UNKN matrices
%    opts: By default GRAM_WSK employs the concatenated matrices and
%          [STAN;UNKN]
%    Setting options to any non-zero value employs the additive matrix
%          [STAN+UNKN] for GRAM.
%
%Output:
%    X: Estimated, unit length, intrinsic profiles in the X order.
%    Y: Estimated, unit length, intrinsic profiles in the Y order.
%    c_est:Estimated relative constituent concentrations in UNKN.
%

% Initialization
if nargin == 3
     opts = 0;
end

%Compute row space and column space
if opts == 0
     [v,s,u]=svd([STAN,UNKN]',0); col_sp=u(:,1:rank);
     [u,s,v]=svd([STAN;UNKN],0);  row_sp=v(:,1:rank);
else
     [u,s,v]=svd([STAN+UNKN],0); col_sp=u(:,1:rank);
   row_sp=v:,1:rank);
end

%Reduce STAN and UNKN into square, full rank matrices and solve GEP
STAN=col_sp'*STAN*row_sp; UNKN=col_sp'*UNKN*row_sp;
[STAN_t,UNKN_t,q,z,Eig_vec]=qz(STAN,UNKN);

%Calculate X, Y, and c_est
Y=row_sp*pinv(Eig_vec)'; Y=Y./(ones(length(Y),1)*sum(Y.^2).^.5);
X=col_sp*(STAN+UNKN)*Eig_vec; X=X./(ones(length(X),1)*sum(X.^2).^.5);
c_est=diag(UNKN_t)./diag(STAN_t);
```

APPENDIX 12.2 DTLD ALGORITHM

```
function [X,Y,Z] =dtld(DATA,nsam,npc)
%Direct Trilinear Decomposition as per Booksh, Lin, Wang, and Kowalski
%
%INPUT
%      DATA: Samples concatenated [S1;S2; ... ;Sn].
%      nsam: Number of samples in DATA.
%      rank: Number of factors to be employed in the model.
%
%OUTPUT
%      X: Estimate row intrinsic profiles.
%      Y: Estimate column intrinsic profiles.
%      Z: Estimate sample intrinsic profiles (relative concentrations).

%Initialization
[i,j]=size(DATA); i=i/nsam; row_X=[]; tube_Z=[]; Q=[];
% UNFOLD KEEPING COLUMN SPACE INTACT
col_Y=DATA;
% UNFOLD KEEPING ROW SPACE INTACT
for r=0:nsam-1
     row_X=[row_X,DATA(i*r+1:i*(r+1),:)];
end
% UNFOLD KEEPING TUBE SPACE INTACT
for z=0:nsam-1
     DATA_temp=DATA(i*z+1:i*(z+1),:);
     tube_Z = [tube_Z,DATA_temp(:)];
end

%COMPUTE REDUCED SPACES IN THREE ORDERS
%COMPUTES ECONOMY SIZE SVD TO SAVE SPACE
[u,s,v]=svd(col_Y,0);              V=v(:,1:npc);
[u,s,v]=svd(row_X',0);             U=v(:,1:npc);
[u,s,v]=svd(tube_Z,0);             W=v(:,1:2);

%PROJECT DATA TO UVW BASIS SET
G1=zeros(npc);           G2=zeros(npc);
for g = 1:nsam
     G2=G2+W(g,1).*U'*DATA(i*(g-1)+1:i*g,:)*V;
     G1=G1+W(g,2).*U'*DATA(i*(g-1)+1:i*g,:)*V;
end

%SOLVE QZ
[G1_t,G2_t,q,z,Eig_vec]=qz(G1,G2);

%CALCULATE X, Y, and c_est
Y=V*pinv(Eig_vec)'; Y=Y./(ones(length(Y),1)*sum(Y.^2).^.5);
X=U*(G1+G2)*Eig_vec; X=X./(ones(length(X),1)*sum(X.^2).^.5);
```

```
%Estimate Sample Concentrations
for i=1:npc
     xy=X(:,i)*Y(:,i)';   Q=[Q;xy(:)'];
end
Z=tube_Z'*Q'*inv(Q*Q');
```

APPENDIX 12.3   PARAFAC ALGORITHM

```
function
[X,Y,Z,stats,X_dtld,Y_dtld,Z_dtld]=als_3d(DATA,nsam,rank,in_opt,x_opt,y_opt,z_opt);
%INPUT
%     DATA:Column augmented samples e.g. [SAMP1;SAMP2;SAMP3]
%     nsam:Number of samples in DATA.
%     rank:Number of factors to use in the model.
%     in_opt: Initialization options (1 for random X,Y vectors; default
%     for DTLD).
%     x_opt: X profile constraint options (1 for non-negativity; 2 for
%     unimodality; 3 for both; default for none).
%     y_opt: Y profile constraint options (1 for non-negativity; 2 for
%     unimodality; 3 for both; default for none).
%     z_opt: Z profile constraint options (1 for non-negativity; 2 for
%     unimodality; 3 for both; default for none).
%
%OUTPUT
%     X: Estimate of the normalized X order intrinsic profiles.
%     Y: Estimate of the normalized Y order intrinsic profiles.
%     Z: Estimate of the normalized Z order intrinsic profiles.
%     stats: correlation between [X,Y,Z,product of the 3 correlations]
%      Terminates algorithm when 1-product is less than 10e-6.
%      Initial Divide-by-0 warning is a byproduct of this step. Ignore it!
%     X_init:Initial X vector guess.
%     y_init:Initial Y vector guess.

% Initialization
if nargin < 4, in_opt = 0; end
if nargin < 5, x_opt = 0; end
if nargin < 6, y_opt = 0; end
if nargin < 7, z_opt = 0; end
UCCold = 0; UCCnew = 1e-4; Zold=ones(nsam,rank); stats=[];
[x_size,y_size]=size(DATA); x_size=x_size/nsam;
reps=0; row_X=[]; tube_Z=[]; Q=[];

%Find initial X and Y vectors
if in_opt == 1
     x_init=rand(x_size,rank);
     y_init=rand(y_size,rank);
else
     [x_init,y_init] = dtld(DATA,nsam,rank);
end
Xold=real(x_init); Yold=real(y_init);
if x_opt==1 | x_opt==3, Xold=abs(Xold); end
```

Three-Way Calibration with Hyphenated Data

```
if y_opt==1 | y_opt==3, Yold=abs(Yold); end
if z_opt==1 | z_opt==3, Zold=abs(Zold); end

%UNFOLD KEEPING COLUMN SPACE INTACT
col_Y=DATA;
%UNFOLD KEEPING ROW SPACE INTACT
for r=0:nsam-1
     row_X=[row_X,DATA(x_size*r+1:x_size*(r+1),:)];
end
%UNFOLD KEEPING SAMPLE SPACE INTACT
for z=0:nsam-1
     DATA_temp=DATA(x_size*z+1:x_size*(z+1),:);
     tube_Z = [tube_Z,DATA_temp(:)];
end
%Major iterative loop
while UCCnew > 1e-9 & reps < 2000

     %CALCULATE NEW Z
     for i=1:rank
          xy=Xold(:,i)*Yold(:,i)';              Q=[Q;xy(:)'];
     end
     if z_opt==1 | z_opt==3 %Apply non-negativity constraints
          for i=1:nsam
               Znew(i,:)=nnls(Q',tube_Z(:,i))';
          end
     else %UNCONSTRAINED SOLUTION
          Znew=tube_Z'*Q'*inv(Q*Q');
     end
     Q=[ ];
     if z_opt==2 | z_opt==3 %APPLY UNIMODALITY CONSTRAINTS
          [val,index]=max(abs(Znew));
          for i = 1:rank
               for j = index(i):-1:2
                    if ((Znew(j,i)-Znew(j-1,i))*...
                         Znew(index(i),i))<0,
                         Znew(j-1,i)=Znew(j,i);
                    end
               end
               for j = index(i):1:nsam-1
                    if ((Znew(j,i)-Znew(j+1,i))*...
                         Znew(index(i),i))<0,
                         Znew(j+1,i)=Znew(j,i);
                    end
               end
          end


     % CALCULATE NEW X
     for i=1:rank
          yz=Yold(:,i)*Znew(:,i)';              Q=[Q;yz(:)'];
     end
     if x_opt==1 | x_opt==3 %APPLY NON-NEGATIVITY CONSTRAINTS
          for i=1:x_size
```

```
                        Xnew(i,:)=nnls(Q',row_X(i,:)')';
        end
else %UNCONSTRAINED SOLUTION
        Xnew=row_X*Q'*inv(Q*Q');
end
Q=[ ];
if x_opt==2 | x_opt==3 %APPLY UNIMODALITY CONSTRAINTS
        [val,index]=max(abs(Xnew));
        for i = 1:rank
            for j = index(i):-1:2
                if ((Xnew(j,i)-Xnew(j-1,i))*...
                    Xnew(index(i),i))<0,
                    Xnew(j-1,i)=Xnew(j,i);
                end
            for j = index(i):1:x_size-1
                if ((Xnew(j,i)-Xnew(j+1,i))*...
                    Xnew(index(i),i))<0,
                    Xnew(j+1,i)=Xnew(j,i);
                end
            end
        end
Xnew=Xnew./(ones(x_size,1)*sum(Xnew));
% CALCULATE NEW Y
for i=1:rank
        xz=Xnew(:,i)*Znew(:,i)';              Q=[Q;xz(:)'];
end
if y_opt==1 | y_opt==3 %Apply non-negativity constraints
     for i=1:y_size
              Ynew(i,:)=nnls(Q',col_Y(:,i))';
     end
else %UNCONSTRAINED SOLUTION
     Ynew=col_Y'*Q'*inv(Q*Q');
end
Q=[ ];
if y_opt==2 | y_opt==3 %APPLY UNIMODALITY CONSTRAINTS
     [val,index]=max(abs(Ynew));
     for i = 1:rank
        for j = index(i):-1:2
          if ((Ynew(j,i)-Ynew(j-1,i))*...
             Ynew(index(i),i))<0,
             Ynew(j-1,i)=Ynew(j,i);
          end
        end
        for j = index(i):1:y_size-1
             if ((Ynew(j,i)-Ynew(j+1,i))*...
                Ynew(index(i),i))<0,
                Ynew(j+1,i)=Ynew(j,i);
             end
        end
     end
end
Ynew=Ynew./(ones(y_size,1)*sum(Ynew));;
%TEST FOR CONVERGENCE
```

```
    UCCX = (Xnew(:)'*Xold(:))/((Xnew(:)'*Xnew(:))*...
           (Xold(:)'*Xold(:)))^.5;
    UCCY = (Ynew(:)'*Yold(:))/((Ynew(:)'*Ynew(:))*...
           (Yold(:)'*Yold(:)))^.5;
    UCCZ = (Znew(:)'*Zold(:))/((Znew(:)'*Znew(:))*...
           (Zold(:)'*Zold(:)))^.5;
    UCCnew = 1 - (UCCX*UCCY*UCCZ);
    Xold=Xnew;        Yold=Ynew;            Zold=Znew;
    reps=reps+1;
    if reps == 1
        UCCnew = 1;
    end

    stats = [stats;1-UCCX, 1-UCCY, 1-UCCZ, UCCnew];
end
X=Xnew; Y=Ynew; Z=Znew;
```

13 Future Trends in Chemometrics

Paul J. Gemperline

CONTENTS

13.1 Historical Development of Chemometrics 510
      13.1.1 Chemometrics — a Maturing Discipline 511
13.2 Reviews of Chemometrics and Future Trends 511
      13.2.1 Process Analytical Chemistry 512
      13.2.2 Spectroscopy 512
      13.2.3 Food and Feed Chemistry 512
      13.2.4 Other Interesting Application Areas 513
13.3 Drivers of Growth in Chemometrics 513
      13.3.1 The Challenge of Large Data Sets 514
      13.3.2 Chemometrics at the Interface of Chemical and Biological Sciences 514
13.4 Concluding Remarks 516
References 516

The term "chemometrics" was coined more than 30 years ago in 1971 by Svante Wold. Through his collaborations with Bruce Kowalski, they recognized the importance of this new field and formed the International Chemometrics Society in 1974. Together, they are considered to be the founders of a new subdiscipline called chemometrics. The term "chemometrics" first appeared in the chemical literature in 1975 [1, 2]. These early pioneers recognized the power of multivariate methods for uncovering hidden relationships between variables and objects. In the beginning, their research in the area of chemometrics focused on pattern recognition methods and applications [3–5], principal component analysis [6], and partial least-squares [7], among other topics. In the previous 12 chapters of this book, various authors have covered the most important topics of chemometrics, including those listed above, that were used by the earliest pioneers of the field. The methods described by these early authors and the applications they addressed are just as relevant, useful, and important today as they were 30 years ago.

    Over the last 30 years, chemometrics has evolved into an interdisciplinary subdiscipline of chemistry that combines mathematical modeling, multivariate statistics, and chemical measurements. There have been numerous definitions of the term "chemometrics," and a recent paper by Workman cites no fewer than ten [8]. All ten definitions share elements in common with the above definition. The ultimate goal of research in chemometrics is to find new and powerful ways to extract information from chemical data. It has been exciting to participate in this field of research during the past 25 years because advances in computer technology have made it possible to make physical/chemical measurements more rapidly than it is possible to process them. At the same time, advances in computer technology and graphics have opened up new possibilities for data analysis.

13.1 HISTORICAL DEVELOPMENT OF CHEMOMETRICS

Many of the developments in chemometrics can be attributed to the successful introduction and application of methods known for a long time and developed in other fields. The methods of principal component analysis (PCA) and partial least squares (PLS) are good examples. Principal component analysis was originally developed by Pearson in 1901 [9] and independently developed later by Hotelling in 1933 [10]. Applications of principal component analysis can be found in a work by Rao in 1964 [11] and in a book by Gnanadesikan in 1977 [12]. The application of principal component analysis in analytical chemistry can be found as early as 1972 in the deconvolution of two overlapped GC (gas chromatography) peaks [13], in 1973 for estimating the distributions of different dyes in photographic mixtures [14], and in 1974 for the detection of overlapping GC-MS (gas chromatography–mass spectrometry) peaks [15].

Partial least squares was developed in the 1960s by Herman Wold, working in the field of econometrics. His son, Svante Wold, introduced PLS to the field of chemistry and further developed the method [16]. Application of these powerful methods to interesting chemical problems produced innovative solutions to chemical problems previously thought to be intractable. In many ways, the introduction of multivariate methods of analysis into the discipline of chemistry was revolutionary.

The historical development of near-infrared spectroscopy (NIR) nearly parallels the historical development of chemometrics. In fact, the development of chemometrics, especially multivariate calibration methods such as PLS, has been an important enabling technology for the development of quantitative NIR applications. Prior to the widespread availability of desktop computers and powerful multivariate calibration software, the NIR spectral region (700 to 3000 nm) was considered useless for most routine analytical analysis tasks because so many chemical compounds give broad overlapping absorption bands in this region. During the past 30 years, however, NIR spectroscopy has rapidly replaced many time-consuming conventional methods of analysis such as the Karl Fisher moisture titration, the Kjeldahl nitrogen titration method for determining total protein [17], and the ASTM gasoline engine method for determining motor octane ratings of gasoline [18]. These applications would be impossible without chemometric methods like multivariate calibration. These advanced calibration methods can be used to "unmix" complicated patterns of broad overlapping absorption bands observed in the information-rich NIR spectral region.

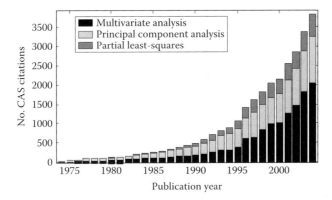

FIGURE 13.1 Citations for selected topics in chemometrics in *Chemical Abstracts* covering the period 1974 to 2004.

13.1.1 CHEMOMETRICS — A MATURING DISCIPLINE

From its humble beginnings in the early days more than 30 years ago, chemometrics has evolved into a mature discipline. Today, a search of *Chemical Abstracts* using the term "multivariate analysis" turns up more than 16,000 citations, with more than 2000 citations in 2004 alone. A significant number of the citations for the term "multivariate analysis" from 2004 can be attributed to work at the interface between chemistry, biology, medicine, and the treatment of human disease, with such topics as pharmacology, immunology, and toxicology, including the analysis of tumor markers and other biomarkers. A search of *Chemical Abstracts* using the term "principal component analysis" turns up nearly 12,000 citations, with about 1200 citations in 2004 (Figure 13.1). The application areas in which these PCA publications appear are diverse and include examples from the study of protein chemistry, air pollution, IR spectroscopy, food analysis, mass spectrometry, environmental and water pollution, process control, gas chromatography, soil chemistry, DNA microarray technology, sensors, NMR (nuclear magnetic resonance) spectroscopy, HPLC (high-performance liquid chromatography), and QSAR (quantitative structure–activity relationships). The diversity of application areas and these signs of significant growth over the past 30 years ago suggest that chemometrics is a mature field, well accepted by the scientific community and firmly entrenched in a wide range of interdisciplinary studies.

13.2 REVIEWS OF CHEMOMETRICS AND FUTURE TRENDS

The chemometrics community has done an excellent job of providing frequent general literature reviews and application-specific reviews that provide excellent sources of information on past trends and allow one to make extrapolation to areas of future growth. The most important regular series of reviews are the "Fundamental Reviews in Chemometrics," published every 2 years in the journal, *Analytical Chemistry* [19–30]. Topical areas in chemometrics are also frequently reviewed. Examples of specialized methods in chemometrics have been the topic of reviews, including the wavelet transform [31], partial least squares [32], self-modeling curve resolution [33], genetic algorithms [34], and calibration transfer [35, 36], among others.

13.2.1 PROCESS ANALYTICAL CHEMISTRY

One of the major application areas of chemometrics is process analytical chemistry. Reviews of this topic are also published every 2 years in *Analytical Chemistry* [37–40]. The goal of research in this field is to use measurements from chemical analyzer systems and chemometrics to monitor, optimize, and control chemical processes. These applications are typically oriented toward increasing the yield of a process, improving product quality, reducing waste, reducing processing time, and improving safety. Chemical analyzers include systems based on temperature, pressure and flow sensors, gas or liquid chromatography, flow-injection analysis, electrochemistry, x-ray spectrometry, NMR and microwave spectroscopy, mass spectrometry, ultrasonic methods, and spectroscopic methods including UV/Visible, near-infrared, mid-infrared, and Raman spectroscopy. Typical chemometrics methods for process analytical applications include multivariate calibration methods such as time-series analysis, partial least-squares (PLS), and multivariate statistical process control (MSPC).

13.2.2 SPECTROSCOPY

Chemometrics finds widespread use in spectroscopy, and there are a number of reviews that describe advances in this area. In a review by Geladi [41], some of the main methods of chemometrics are illustrated with examples. A series of three reviews addresses the topic of chemometrics in spectroscopy [42–44]. Part 1 has 199 references and focuses specifically on chemometric techniques applied to spectroscopic data [42]. Part 2 has 68 references and focuses on data-preprocessing methods and data transformations aimed at reducing noise, removing effects of baseline offsets, and filtering to remove noise [43]. Part 3 focuses on multiway methods of analysis applied to spectroscopic data [44].

The application of chemometrics in near-infrared spectroscopy is finding widespread use in many different industries for monitoring the identity and quality of raw materials and finished products in the food and agricultural industry [46], polymer, pharmaceutical, and organic chemical manufacturing industries [18, 47].

13.2.3 FOOD AND FEED CHEMISTRY

Chemometrics and multivariate methods are increasingly being used in the food and feed chemistry industries for exploratory analysis of large data sets, for multivariate quality assurance and quality control, for detecting adulteration, and for estimating chemical, physical, and sensory properties of food. Multivariate approaches to these kinds of applications are essential for constructing mathematical models of sensory properties that are inherently multivariate. Historically, chemometrics has been an important enabling technology in this field, and a recent review on this topic provides a good starting point for individuals interested in learning more about this important application area of chemometrics [48]. Rapid spectroscopic/chemometric methods for detecting adulteration and confirming the authenticity of food products have been reviewed [49]. Such methods are being increasingly demanded by food processors, consumers, and regulatory bodies. Another review [50] describes the use of chemometrics and low-resolution time-domain NMR measurements (relaxation profiles) for the characterization of food products. Reviews of other application-specific chemometrics/spectroscopic methods in the food industry cover postharvest evaluation of fruit and vegetable quality [51], characterization of essential oils [52], and the history and methods of chemometrics in the wine industry [53].

13.2.4 OTHER INTERESTING APPLICATION AREAS

Many other important application areas of chemometrics have been the subject of reviews and are too numerous to list here. A sampling of reviews in this category illustrates the breadth and diversity of chemometrics application areas. A review of applications in smart sensors [54] describes how chemometrics is an important enabling technology for the development of smart and reliable sensor systems. A review of environmental forensics [55] describes how numerical methods are critical in the process of identifying the chemical fingerprints of complex contaminant sources in environmental systems. Often, multiple sources are present at different geographic sites. By use of appropriate chemometric methods, these mixtures of different sources can be mathematically resolved to identify them and map their temporal and spatial distributions.

13.3 DRIVERS OF GROWTH IN CHEMOMETRICS

Chemometrics is a child of the information age. Inexpensive computers, beginning with minicomputers in the 1970s and microcomputers in the 1980s, fueled enormous growth in the development of computerized scientific instrumentation. New instruments are now available that were impossible to envision 20 or 30 years ago. For example, confocal hyperspectral imaging microscopes are capable of recording images of live microorganisms [56]. A recently described three-dimensional hyperspectral imaging microscope has 250-nm resolution in the x–y-plane and 750-nm resolution in the z-direction [57, 58]. The instrument is capable of measuring a 512-point fluorescence spectrum in the visible range at each voxel in the image at a rate of up to 8300 spectra per second, giving a data-acquisition rate of more than 4.2 Mb/s. A recently described hyperspectral image of a 20 × 60-mm yeast genome microarray slide at 10-μm resolution represented more than 3 Gbyte of data [58, 59]. The challenge with such a flood of data is to develop fast, efficient, and effective chemometric methods aimed at extracting meaningful information from the mountain of data produced [59]. The availability of inexpensive, powerful computers capable of performing sophisticated numerical data-analysis tasks has also helped fuel the development of chemometrics. The same drivers that pushed the development of chemometrics forward during the past 30 years are still at work today in many fields, including hyperspectral image analysis.

13.3.1 THE CHALLENGE OF LARGE DATA SETS

There are many challenging data analysis problems yet to be solved that will drive the development of chemometrics in the future. One of these is the very large data sets that will require development of specialized methods and algorithms for treating them. Some research groups have already begun developing methods for handling huge data sets in specialized applications [57, 60, 61]. One method for dealing with these kinds of problems includes the use of PCA or wavelet compression. This approach works well for data sets when huge amounts of the same kind of variables are measured. For example, compression in the x–y-plane of hyperspectral images can dramatically reduce the size of such data arrays, with a corresponding loss of x–y-resolution in the image. Estimation of spectroscopically unique features can easily be accomplished in these low-resolution images to produce pure-component chemical images or "maps." Once the pure-component spectra of these features are extracted, they can easily be used to estimate a chemical composition map at full resolution. An alternative approach to coping with huge data sets is to segment the problem or divide the problem into smaller-sized problems. This approach can also work well for hyperspectral images, where pure-component features are extracted from small segments of the image and then used to estimate the chemical image at full resolution [57].

For many types of data sets, it is not possible to use compression or segmentation. In such cases, multiblock models [62] and hierarchical models can prove helpful [63]. Multiblock models work by dividing variables into meaningful blocks that are measured on similar parts of a process or system [64]. Multiblock projection methods based on PCA and PLS can be used in situations where processes can be naturally blocked into subsections [62, 65, 66]. The multiblock projection methods allow establishment of monitoring charts for individual process subsections as well as for entire processes. Computational efficiency can be dramatically improved by blocking large-scale problems into smaller blocks. By this approach, meaningful information can be extracted from very large historical databases [67]. Special events or faults are also generally detected earlier using these approaches. When a special event or fault occurs in a subsection of the process, multiblock methods can generally detect the event earlier and reveal the subsection within which the event has occurred. Hierarchical models typically employ blocking on two levels, an upper level, in which the relationships between blocks are modeled, and a lower level, which shows the details of each block. On each level, PLS or PCA scores and loadings are available for model interpretation. For more complex problems, hierarchical models can be extended to several hierarchical levels, thereby providing a scalable approach to modeling very large data sets [64].

13.3.2 CHEMOMETRICS AT THE INTERFACE OF CHEMICAL AND BIOLOGICAL SCIENCES

Some of the most exciting opportunities for innovation and new developments in the field of chemometrics lie at the interface between chemical and biological sciences. These opportunities are made possible by the exciting new scientific advances and discoveries of the past decade that have been made possible by the rise of "quantitative biology." The advances made possible by the molecular biology revolution can be attributed to collaborations between physical scientists and biological scientists, and between experimentalists and modelers.

Quantitative biology refers to recent developments in high-throughput quantitative experimental methods and the application of modeling approaches to understand relationships between (a) structure and function in molecules of biological interest, (b) cellular biochemical processes, (c) evolutionary processes, and (d) networks of biological molecules and populations. For example, technological advances in the area of high-throughput technologies have made it possible to elucidate entire genome sequences. Other technological advances, such as live-cell hyperspectral imaging, open up the prospects for modeling biochemical networks using spectroscopic measurements at the cellular and organelle level. Chemometrics is already having an impact on these fields.

A recent review of metabonomics describes how the use of chemometrics in genomics, proteomics, and metabonomics is enabling the pharmaceutical industry to expand drug pipelines [68]. The field of metabonomics uses multivariate methods of analysis to extract information about toxicity or the diagnosis of disease from complex profiles measured by spectroscopic instruments on biological systems. For example, high-field NMR measurements of biofluids, tissues, and cell cultures produce huge amounts of data and complex patterns that are difficulty to interpret, but that contain useful information. Chemometrics can help in extracting this information.

Analysis of complex genomics-based data in systems biology requires multivariate data-analysis methods to obtain useful information. The variation observed in these data sets occurs simultaneously on different levels, such as variation between organisms and variation in time. In conventional two-way methods like PCA, the different types of variation present in these data sets is mathematically confounded. A new method called multilevel component analysis (MCA) was recently introduced to separate these different types of variation [69]. The method was recently demonstrated using a data set containing ^{1}H NMR spectra of urine collected from 10 monkeys at 29 time points during a 2-month study. In this application, MCA was used to generate different submodels for different types of variation that are easier to interpret.

Multiway methods such as Tucker3 models and N-way PLS are especially useful for identifying and separating different sources of variation in different modes that would otherwise be masked by high levels of variability naturally found in biological populations. By use of multiway methods, it becomes easier to identify chemically meaningful information and find interesting correlations from huge tables of uninteresting measurements. Metabolomic profiling experiments often contain underlying structure, such as time × dose, that cannot be exploited by current biostatistical methods. Multiway chemometrics methods can appropriately incorporate the structure of these data sets into multiway models that allow for easier interpretation of the variation induced by the different factors of the design [70]. For example, a recent study applied a Tucker3 model to a one-dimensional high-field ^{1}H NMR spectrum measured from rat urine samples after the animals were exposed to a toxic substance [71]. The three-way data array consisted of NMR spectra of urine × time × rats. Three groups of rats were studied: a control group, a low-dose group, and a high-dose group [71]. The use of Tucker3 analysis on this data set permitted the investigators to recover variance from low-dose rats as well as earlier detection of biochemical markers that were otherwise masked in conventional two-way multivariate methods of analysis. Another study used multiway PLS and PARAFAC on two-dimensional diffusion-edited ^{1}H NMR spectra to quantify lipoprotein fractions in human plasma samples determined by ultracentrifugation [72]. PLS calibration models were developed for the four main fractions of lipoprotein as well as 11 subfractions with high correlations (R = 0.75–0.98). Further developments at the interface between chemometrics and bioinformatics can be expected to improve our ability to extract meaningful information from these kinds of measurements.

Another recent review describes the use of chemometrics in NIR and Raman spectroscopy of biological materials, another active area of growth for research in chemometrics [73]. Clinical applications of NIR spectroscopy as well as biological and biomedical applications for Raman spectroscopy are rapidly growing areas of research, enabled by chemometrics and multivariate methods of analysis. Also of significant biological importance are studies of vibrational optical activity and Raman optical activity for multicomponent qualitative and quantitative analysis. Time-resolved step-scan methods as well as surface-enhanced methods and imaging approaches are yielding unprecedented amounts of data about systems of biological interest. New chemometrics techniques will be required to obtain useful information from the flood of data provided by these new applications.

One of the emerging biological and biomedical application areas for vibrational spectroscopy and chemometrics is the characterization and discrimination of different types of microorganisms [74]. A recent review of various FTIR (Fourier transform infrared spectrometry) techniques describes such chemometrics methods as hierarchical cluster analysis (HCA), principal component analysis (PCA), and artificial neural networks (ANN) for use in taxonomical classification, discrimination according to susceptibility to antibiotic agents, etc. [74].

13.4 CONCLUDING REMARKS

The future of chemometrics lies in the development of innovative solutions to interesting problems. This problem-oriented approach is required because relatively few advances can be expected in the area of new mathematical and numerical methods.

REFERENCES

1. Kowalski, B.R., Chemometrics: views and propositions, *J. Chem. Inf. Comput. Sci.*, 15, 201–203, 1975.
2. Kowalski, B.R., Measurement analysis by pattern recognition, *Anal. Chem.*, 47, 1152A–1162A, 1975.
3. Kowalski, B.R. and Bender, C.F., Application of pattern recognition to screening prospective anticancer drugs: Adenocarcinoma 755 biological activity test, *J. Am. Chem. Soc.*, 96, 916–918, 1974.
4. Duewer, D.L., Kowalski, B.R., and Schatzki, T.F., Source identification of oil spills by pattern recognition analysis of natural elemental composition, *Anal. Chem.*, 47, 1573–1583, 1975.

5. Wold, S. and Sjostrom, M., SIMCA: a method for analyzing chemical data in terms of similarity and analogy, *ACS Symp. Ser.*, 52, 243–282, 1977.
6. Carey, R.N., Wold, S., and Westgard, J.O., Principal component analysis: an alternative to "referee" methods in method comparison studies, *Anal. Chem.*, 47, 1824–1829, 1975.
7. Gerlach, R.W., Kowalski, B.R., and Wold, H.O.A., Partial least-squares path modelling with latent variables, *Anal. Chim. Acta*, 112, 417–421, 1979.
8. Workman, J., The state of multivariate thinking for scientists in industry: 1980–2000, *Chemom. Intell. Lab. Syst.*, 60, 13–23, 2002.
9. Pearson, K., On lines and planes of closest fit to systems of points in space, *Philos. Mag.*, 6, 559–572, 1901.
10. Hotelling, H., Analysis of a complex of statistical variables into principal components, *J. Educ. Psychol.*, 24, 417–441, 498–520, 1933.
11. Rao, C.R., The use and interpretation of principal component analysis in applied research, *Sankhya A*, 26, 329–358, 1964.
12. Gnanadesikan, R., *Methods for Statistical Data Analysis of Multivariate Observations*, John Wiley & Sons, New York, 1977.
13. MacNaughtan, D., Jr., Rogers, L.B., and Wernimont, G., Principal component analysis applied to chromatographic data, *Anal. Chem.*, 44, 1421–1427, 1972.
14. Ohta, N., Estimating absorption bands of component dyes by means of principal component analysis, *Anal. Chem.*, 45, 553–557, 1973.
15. Davis, J.E., Shepard, A., Stanford, N., and Rogers, L.B., Principal component analysis applied to combined gas chromatographic-mass spectrometric data, *Anal. Chem.*, 46, 821–825, 1974.
16. Geladi, P. and Kowalski, B.R., Partial least-squares regression: a tutorial, *Anal. Chim. Acta*, 185, 1–17, 1986.
17. Watson, C.A., Near infrared reflectance spectrophotometric analysis of agricultural products, *Anal. Chem.*, 49, 835A–840A, 1977.
18. Workman, J., Jr., A brief review of near infrared in petroleum product analysis, *J. Near Infrared Spectrosc.*, 4, 69–74, 1996.
19. Kowalski, B.R., Chemometrics, *Anal. Chem.*, 52, 112R–122R, 1980.
20. Frank, I.E. and Kowalski, B.R., Chemometrics, *Anal. Chem.*, 54, 232R–243R, 1982.
21. Ramos, L.S., Beebe, K.R., Carey, W.P., Sanchez, M.E., Erickson, B.C., Wilson, B.E., and Wangen, L.E., Chemometrics, *Anal. Chem.*, 58, 294R–315R, 1986.
22. Brown, S.D., Barker, T.Q., Larivee, R.J., Monfre, S.L., and Wilk, H.R., Chemometrics, *Anal. Chem.*, 60, 252R–273R, 1988.
23. Brown, S.D., Chemometrics, *Anal. Chem.*, 62, 84R–101R, 1990.
24. Brown, S.D., Bear, R.S., Jr., and Blank, T.B., Chemometrics, *Anal. Chem.*, 64, 22R–49R, 1992.
25. Brown, S.D., Blank, T.B., Sum, S.T., and Weyer, L.G., Chemometrics, *Anal. Chem.*, 66, 315R–359R, 1994.
26. Brown, S.D., Sum, S.T., Despagne, F., and Lavine, B.K., Chemometrics, *Anal. Chem.*, 68, 21–61, 1996.
27. Lavine, B.K., Chemometrics, *Anal. Chem.*, 70, 209R–228R, 1998.
28. Lavine, B.K., Chemometrics, *Anal. Chem.*, 72, 91R–97R, 2000.
29. Lavine, B.K. and Workman, J.J, Jr., Chemometrics, *Anal. Chem.*, 74, 2763–2769, 2002.
30. Lavine, B.K. and Workman, J.J., Jr., Chemometrics, *Anal. Chem.*, 76, 3365–3372, 2004.
31. Chau, F.-T. and Leung, A.K.-M., Application of wavelet transform in processing chromatographic data, *Data Handling Sci. Technol.*, 22, 205–223, 2000.

32. Wold, S., Sjostrom, M., and Eriksson, L., PLS-regression: a basic tool of chemometrics, *Chemom. Intell. Lab. Syst.*, 58, 109–130, 2001.
33. Jiang, J.-H., Liang, Y., and Ozaki, Y., Principles and methodologies in self-modeling curve resolution, *Chemom. Intell. Lab. Syst.*, 71, 1–12, 2004.
34. Leardi, R., Genetic algorithms in chemometrics and chemistry: a review, *J. Chemom.*, 15, 559–569, 2001.
35. Feudale, R.N., Woody, N.A., Tan, H., Myles, A.J., Brown, S.D., and Ferre, J., Transfer of multivariate calibration models: a review, *Chemom. Intell. Lab. Syst.*, 64, 181–192, 2002.
36. Fearn, T., Standardisation and calibration transfer for near infrared instruments: a review, *J. Near Infrared Spectrosc.*, 9, 229–244, 2001.
37. Blaser, W.W., Bredeweg, R.A., Harner, R.S., LaPack, M.A., Leugers, A., Martin, D.P., Pell, R.J., Workman, J., Jr., and Wright, L.G., Process analytical chemistry, *Anal. Chem.*, 67, 47R–70R, 1995.
38. Workman, J., Jr., Veltkamp, D.J., Doherty, S., Anderson, B.B., Creasy, K.E., Koch, M., Tatera, J.F., Robinson, A.L., Bond, L., Burgess, L.W., Bokerman, G.N., Ullman, A.H., Darsey, G.P., Mozayeni, F., Bamberger, J.A., and Greenwood, M.S., Process analytical chemistry, *Anal. Chem.*, 71, 121R–180R, 1999.
39. Workman, J., Jr., Creasy, K.E., Doherty, S., Bond, L., Koch, M., Ullman, A., and Veltkamp, D.J., Process analytical chemistry, *Anal. Chem.*, 73, 2705–2718, 2001.
40. Workman, J., Jr., Koch, M., and Veltkamp, D.J., Process analytical chemistry, *Anal. Chem.*, 75, 2859–2876, 2003.
41. Geladi, P., Sethson, B., Nystroem, J., Lillhonga, T., Lestander, T., and Burger, J., Chemometrics in spectroscopy, *Spectrochimica Acta, B: At. Spectrosc.*, 59B, 1347–1357, 2004.
42. Workman, J.J., Jr., Mobley, P.R., Kowalski, B.R., and Bro, R., Review of chemometrics applied to spectroscopy: 1985–95, part 1, *Appl. Spectrosc. Rev.*, 31, 73–124, 1996.
43. Mobley, P.R., Kowalski, B.R., Workman, J.J., Jr., and Bro, R., Review of chemometrics applied to spectroscopy: 1985–95, part 2, *Appl. Spectrosc. Rev.*, 31, 347–368, 1996.
44. Bro, R., Workman, J.J., Jr., Mobley, P.R., and Kowalski, B.R., Review of chemometrics applied to spectroscopy: 1985–95, part 3: multi-way analysis, *Appl. Spectrosc. Rev.*, 32, 237–261, 1997.
45. Wetzel, D.L.B., Analytical near infrared spectroscopy, *Dev. Food Sci.*, 39, 141–194, 1998.
46. Geladi, P. and Dabakk, E., An overview of chemometrics applications in near infrared spectrometry, *J. Near Infrared Spectrosc.*, 3, 119–132, 1995.
47. Forina, M. and Drava, G., Chemometrics [in food chemistry], *Food Sci. Technol. (NY)*, 77, 21–58, 1996.
48. Downey, G., Food and food ingredient authentication by mid-infrared spectroscopy and chemometrics, *Trends Anal. Chem.*, 17, 418–424, 1998.
49. Rutledge, D.N., Chemometrics and time-domain nuclear magnetic resonance, *Analusis*, 25, M9–M14, 1997.
50. Massantini, R., Carlini, P., and Anelli, G., NIR spectroscopy in evaluation of fruit and vegetable quality during the postharvest period, *Industrie Alimentari (Pinerolo, Italy)*, 36, 321–326, 1997.
51. Hibbert, D.B., Chemometric analysis of data from essential oils, *Mod. Meth. Plant Anal.*, 19, 119–140, 1997.
52. Forina, M. and Drava, G., Chemometrics for wine: applications, *Analusis*, 25, M38–M42, 1997.

53. Vogt, F., Dable, B., Cramer, J., and Booksh, K., Recent advancements in chemometrics for smart sensors, *Analyst (Cambridge, U.K.)*, 129, 492–502, 2004.
54. Johnson, G.W. and Ehrlich, R., State of the art report on multivariate chemometric methods in environmental forensics, *Environ. Foren.*, 3, 59–79, 2002.
55. Zimmermann, T., Rietdorf, J., and Pepperkok, R., Spectral imaging and its applications in live cell microscopy, *FEBS Lett.*, 546, 87–92, 2003.
56. Timlin, J.A., Haaland, D.M., Sinclair, M.B., Aragon, A.D., Martinez, M.J., and Werner-Washburne, M., Hyperspectral microarray scanning: impact on the accuracy and reliability of gene expression data, *BMC Genomics*, 6, 2005.
57. Sinclair, M.B., Timlin, J.A., Haaland, D.M., and Werner-Washburne, M., Design, construction, characterization, and application of a hyperspectral microarray scanner, *Appl. Opt.*, 43, 2079–2089, 2004.
58. Martinez, M.J., Aragon, A.D., Rodriguez, A.L., Weber, J.M., Timlin, J.A., Sinclair, M.B., Haaland, D.M., and Werner-Washburne, M., Identification and removal of contaminating fluorescence from commercial and in-house printed DNA microarrays, *Nucleic Acids Res.*, 31, e18/11–e18/18, 2003.
59. Eriksson, L., Antti, H., Gottfries, J., Holmes, E., Johansson, E., Lindgren, F., Long, I., Lundstedt, T., Trygg, J., and Wold, S., Using chemometrics for navigating in the large data sets of genomics, proteomics, and metabonomics (gpm), *Anal. Bioanal. Chem.*, 380, 419–429, 2004.
60. Eriksson, L., Johansson, E., Lindgren, F., Sjoestroem, M., and Wold, S., Megavariate analysis of hierarchical QSAR data, *J. Comp.-Aided Mol. Des.*, 16, 711–726, 2002.
61. MacGregor, J.F., Jaeckle, C., Kiparissides, C., and Koutoudi, M., Process monitoring and diagnosis by multiblock PLS methods, *AIChE J.*, 40, 826–838, 1994.
62. Wold, S., Kettaneh, N., and Tjessem, K., Hierarchical multiblocks PLS and PC models for easier model interpretation and as an alternative to variable selection, *J. Chemom.*, 10, 463–482, 1996.
63. Wold, S., Berglund, A., and Kettaneh, N., New and old trends in chemometrics: how to deal with the increasing data volumes in R&D&P (research, development, and production), with examples from pharmaceutical research and process modeling, *J. Chemom.*, 16, 377–386, 2002.
64. Lopes, J.A., Menezes, J.C., Westerhuis, J.A., and Smilde, A.K., Multiblock PLS analysis of an industrial pharmaceutical process, *Biotechnol. Bioeng.*, 80, 419–427, 2002.
65. Westerhuis, J.A., Kourti, T., and Macgregor, J.F., Analysis of multiblock and hierarchical PCA and PLS models, *J. Chemom.*, 12, 301–321, 1998.
66. Kourti, T., Nomikos, P., and MacGregor, J.F., Analysis, monitoring and fault diagnosis of batch processes using multiblock and multiway PLS, *J. Proc. Cont.*, 5, 277–284, 1995.
67. Holmes, E. and Antti, H., Chemometric contributions to the evolution of metabonomics: mathematical solutions to characterizing and interpreting complex biological NMR spectra, *Analyst (Cambridge, U.K.)*, 127, 1549–1557, 2002.
68. Jansen, J.J., Hoefsloot, H.C.J., van der Greef, J., Timmerman, M.E., and Smilde, A.K., Multilevel component analysis of time-resolved metabolic fingerprinting data, *Analytica Chimica Acta*, 530, 173–183, 2005.
69. Smilde, A.K., Jansen, J.J., Hoefsloot, H.C.J., Lamers, R.-J.A.N., van der Greef, J., and Timmerman, M.E., ANOVA-simultaneous component analysis (ASCA): a new tool for analyzing designed metabolomics data, *Bioinformatics*, 21, 3043–3048, 2005.
70. Dyrby, M., Baunsgaard, D., Bro, R., and Engelsen, S.B., Multiway chemometric analysis of the metabolic response to toxins monitored by NMR, *Chemom. Intell. Lab. Syst.*, 76, 79–89, 2005.

71. Dyrby, M., Petersen, M., Whittaker, A.K., Lambert, L., Norgaard, L., Bro, R., and Engelsen, S.B., Analysis of lipoproteins using 2D diffusion-edited NMR spectroscopy and multi-way chemometrics, *Analytica Chimica Acta*, 531, 209–216, 2005.
72. Ozaki, Y. and Murayama, K., Infrared and Raman spectroscopy and chemometrics of biological materials, *Pract. Spectros.*, 24, 515–565, 2001.
73. Mariey, L., Signolle, J.P., Amiel, C., and Travert, J., Discrimination, classification, identification of microorganisms using FTIR spectroscopy and chemometrics, *Vibra. Spectrosc.*, 26, 151–159, 2001.

Index

A

absorption spectroscopy, 218
accuracy
　defined, 10, 12
active-experiment approach, 266
active experiments, presumptions for, 265
　control of values and, 265
　measurement of output variables and, 265
　noise variables and, 265
algorithms, 4
algorithms for realizable optimal experimental designs
　exact (or N-Point) D-Optimal designs, 307–309
　sequential composite D-Optimal designs, 313–315
　sequential D-Optimal designs, 310–311
aliasing, 388
all possible subsets, 138
ALS algorithm
　PARAFAC and, 491
alterative hypothesis, 48–49, 122
alternating least-squares (ALS) methods, 491
alternating least-squares (ALS) methods, solution constraints, 493
　known profiles, 494
　nonnegativity, 494
　unimodality, 494
alternating least-squares methods
　PARAFAC / CANDECOMP, 491–494
American Society for Testing Materials (ASTM), 107, 114
analysis of emission spectroscopic data, *241*
analysis of multivariate data in kinetics, *228*
analysis of variance (ANOVA), 28
　analysis of residuals, 30–32
　ANOVA to test for differences between means, 28
　between-sample variation (between-treatment variation), 29–30
　within-sample variation (within-treatment variation), 29
analyte concentrations in unknown solutions, 476
analytical chemistry, 2, 512
ANOVA (analysis of variance), 125
　calculations, summary approach, 31
　table, *32*

apodization
　transfer function and, 397
application of DOE in multivariate calibration, 321–329
　improving quality from historical data, 330–335
application of DOE in multivariate calibration, construction of a calibration set, 321–329
　defining the bounds of the factor space, 327–328
　estimating extinction coefficients, 329–330
　identifying the number of significant factors, 322–324
application of DOE in multivariate calibration, improving quality from historical data
　improving the numerical stability of the data set, 333
　prediction ability, 334
applications, chemometric
　biochemical processes, 449, 449–459
　environmental data, 454
archaeological artifacts
　obsidian, composition of, 356
artificial neural networks (ANN), 477
Association of Official Analytical Chemists (AOAC), 114
AT&T Bell Labs, 294
autocatalysis, 246–248
autocatalytic reaction scheme, *246*
autoscaled data with E-optimal subset (ref Fig 8.31), ***335***
autoscaling, 79, 458

B

backward elimination, 137
　Efroymson's stepwise regression algorithm and, 137
　forward selection and, 137
bandwidth-limited discrete Fourier transform, 388
barycentric coordinate systems, 319
baseline correction, 80
　of Raman emission spectra, *80*
basis vectors, 71–72, 96, 141, 189
　column basis vectors, 71
　row basis vectors, 71
Beer-Lambert's Law, 219, 237, 241, 324, 328

521

matrix notation, result of data-fitting procedure and, *238*
spectroscopy and, 462
Beer's law, 107
Belousov-Zhabotinsky (BZ) Reaction, 251–253, 256
  FKN mechanism, 252
  four stages of, 252
  Oregonator model, 252
  oscillating reactions and, 251–252
  oxidation of organic species in, 251
  represented b the Oregonator model, *263*
between-run, defined, 10
bias
  laboratory, 18
  method, 18
  systematic error and, 18
biased methods of calibration, 139
bilinear data matrix, 71
bivariate normal distribution
  scatter plots of, *56–58*
"black box" principle, *265*, 265, 266, 269
BP Amoco naphtha processing plant, 330
Butterworth filter equation, 396

C

calculation of the concentration profiles: case I, simple mechanisms, 220
calculation of the concentration profiles: case II, complex mechanisms, 241–252
  Fourth-Order Runge-Kutta Method in Excel, 242–245
  Interesting kinetic examples, 246–252
calculation of the concentration profiles: case III, very complex mechanisms, 253
  Arrhenius theory and, 254
  changes in ionic strength and, 254–255
  changes in pH and, 254–255
  experimental inconsistencies and, 253
  Eyring theory and, 254
  temperature fluctuations and, 253
calibration, 105–166
  curvilinear, 112
  data sets and, 107–108
  internal standards, 156
  introduction to, 109
  measurement error and measures of prediction error, 114
  multivariate, 111
  non-zero intercepts and, 110
  preproccssing techniques, 156, 156–157
  sample sets, 264
  selection of calibration and validation samples, 113

sets, 322
software, 159
statistical evaluation of calibration models obtain by least squares, 121–134
univariate, 109, 211
calibration, a practical example, 116
  graphical survey of NIR water-methanol data, 116
  multivariate, 119–120
  univariate, 118, 118–119
calibration, biased methods of, 139–153
  example regularization results, 153
  other calibration methods, 150–151
  partial least squares, 147–149
  principal component regression, 140–153
  regularization, 151
calibration line, fulcrum and, 129
calibration models, 330
  spectral data and, *324*
  spectroscopic, 321
calibration models, statistical evaluation obtained by least squares, 121–134
  coefficient of determination and multiple correlation coefficient, 130
  confidence interval and hypothesis tests for regression coefficients, 126
  hypothesis testing, 122
  interference effects and selectivity, 234
  interpreting regression in ANOVA tables, 125
  leverage and influence, 128
  model departures and outliers, 129
  partitioning of variance in least-squares solutions, 123–124
  prediction confidence intervals, 127
  sensitivity and limit of detection, 131–132
calibration, multivariate, 119–120
  chemometrics and, 264
calibration, software and, 159
  DeLight, 159
  GRAMS, 159
  MATLAB and, 159–160
  Pirouete, 159
  PLS_Toolbox, 159
  SIMCA, 159
  Unscrambler, 159
calibration, standard addition method, 153–154, 153–155
  multivariate standard addition method, 155
  univariate standard addition method, 154, 154–155
calibration, standardization, 157, 157–159
  alternative standardization (transfer) methods, 157
  calibration-transfer methods, 157
  of instrument response, 158

Index

of predicted values, 157–158
  with preprocessing techniques, 159
calibration steps
  calibration prediction step, 143
  compute a regression vector using calibration samples, 143
  compute the projected calibration spectra, 142
  estimate the RMSEC, 143
calibration, three-way, 4
  hyphenated data and, 475–508
calibration, variable selection, 135–139
  all possible subsets, 138
  backward elimination, 137
  Efroymson's stepwise regression algorithm, 136–137
  forward selection, 135
  recommendations and precautions, 138
  sequential-replacement algorithms, 138
  simulated annealing andgeneic algorihm, 138
calibration work, UV range and, 330
cascade of wavelet coefficient vectors from wavelet tree filter banks defining the discrete wavelet transform, *411*
catalog of four eact D-Optimal experimental designs
  for the spectroscopic calibration problem, *327*
central composite designs (CCD), 293
  three-factor with axial values and four center points, *294*
  for two factors, *293*
central limit theorem, *13*, 45
  implications of, 45
Chebyshev polynomials, 396, *404*
chemical analysis
  interferences and, 134–135
chemical measurements
  chemical properties and, 2
  decision making and, 1, 210
  physical properties and, 2
  statistical properties and, 2
  as three-legged platform, 2
chemical properties, 2
  equilibria, 2
  kinetics, 2
  mass balance, 2
  stoichemetry, 2
chemometrics, 1–6, 509, 510, 516
  analytic chemistry and, 510
  biomedical applications of, 514
  calibration, validation, and significance testing, 2
  in Chemical Abstracts, 511
  chemical and biological sciences, 511, 514
  chemical data and, 510
  chemical measurements and, 2, 509
  computing and, 2, 510, 512, 513
  environmental forensics and, 512
  extraction of chemical information from analytical data, 2
  future trends, 509–520
  gas chromatography (GC) and, 510
  gas chromatography-mass spectrometry (GC-MS), 510
  historical development of, 50, 509
  hyper spectral images, 514
  large data sets and, 514
  live cell hyperspectral imaging, 514
  mathematical modeling and, 509
  as maturing discipline, 511
  multivariate calibration methods, 510
  multivariate methods of analysis, 510
  multivariate statistics and, 509
  near infrared spectroscopy and, 510
  new instruments and, 513
  NIR spectroscopy and, 510, 514
  partal least-squares, 510
  partial least-squares, 509
  pattern recognition and, 509
  principal component analysis (PCA), 509
  principle component analysis (PCA) and, 510
  quantitative biology and, 514
  Raman spectroscopy and, 514
  smart sensors and, 512
chemometrics citations in *Chemical Abstracts* from 1974–2004, 511
chemometrics, drivers of growth in, 513
  chemical and biological sciences, 514
  large data sets, 514
chemometrics, reviews of and future trends, 511
  food and feed chemistry, 512–513
  interesting application areas, 511, 513
  process analytical chemistry, 512
  spectroscopy, 512
chi-square distribution, 47, 58–59
  degrees of freedom and, *48*
chromatography, 140, 477
chromographic profile optimized by ITTFA, *438*
classical and robust tolerance ellipse of the phosphorus data set, *173*
classical PCA (CPCA), 186, 188, 194
classical PCR (CPCR), 196
classical quadratic discriminant analysis (CQDA), 207
classification, 207, 351–354
  discriminant analysis and, 207
  K-Nearest neighbor, 352
  partial least squares (PLS), 352
  SIMCA, 353
  supervised learning and, 207
classification and pattern recognition, 339–378

clustering and, 347
   data preprocessing, 341
   introduction to, 339–340
   mapping and display, 342–346
   practical considerations, 354
classification in high dimensions, 211
   dimension-reduction procedure (PCA), 211
   MCD as uncomputable, 211
   SIMCA method (soft independent modeling of class analogy), 211
classification in low dimensions, 207–210
   classical and robust discriminant rules, 207
   evaluating the discriminant rules, 208
   an example, 209
classification methods, 351
   K-nearest neighbor (KNN), 351
   partial least squares (PLS), 351
   soft independent modeling by class analogy (SIMCA), 351
classification of jet fuels, *364*, 365
closure, 434
cluster analysis, 347
clustering, 345, 347, 363
   classification with PCA score plots and, 98
   hierarchical, 351
   single linkage and, 351
clustering algorithms, 348
   FCV, 348
   hierarchical, 348
   K-means, 348
   Patrick Jarvis, 348
   similarity matrix, 348
clusters
   determining numbers of, *348*
   elliptical, 351
   identifying, 347
   linear, 351
   spherical, 351
   validity, 347
CM-estimators, 183
coefficient of determination, 131
   multiple correlation coefficient and, 130
combined canonical models
   independent (process) variables and, 284
   mutually dependent (mixture variables) and, 284
combined mixture and process factor space
   constructed from three mixture variables and one process variable, *286*
components included in product P1, with UV-Inactive components shaded, *328*
composition (loading) profiles of resolved components in the MCR-ALS analysis of raw augmented data matrix, *460*
compound maps, 462

compounds, kinetic or thermodynamic process and, 435
computation for distance between data cluster and a point, *350*
concentration of Components C1 to C9 used to measure pure-component spectra, *324*
concentration profiles
   for a first-order reaction at constant and increasing temperature, *254*
   of an HPLC-DAD data set, *425*
   for a reaction as calculated by MATLAB, 221
   for a two-component system, 421
concentration windows, 426
confidence intervals, 14, 299
   compared with nonoptimal design and D-optimal design, *301*
   critical value of the Student statistic, 300
   estimate of the measurement error and, 300
   hypothesis tests for regression coefficients and, 126–127
   influence of on the error of the prediction, *300*
   of the mean, equation for, 46
   for the regression line and for predictions, illustration of, *128*
   variance of the prediction, 300
confidence level, 133
consrained region
   in three-component mixtures, *281*
   of three mixture variables with lower bounds only, 280
constrained mixture region with upper and lower bounds, 282
constrained mixture spaces, 279–280
constraints, 433–437
   application of, 433
   chemical properties and, 433
   closure, 434
   effect on the shape of resolved profiles, *433*
   equality and inequality and, 433
   hard-modeling constraints: physicochemical models, 435
   implementation of, 433
   iterative optimization and, 433
   known profiles, 435
   local rank, 435
   mathematical properties and, 433
   nonnegativity, 433, 434, 465
   selectivity, 435
   unimodality, 434
   zero-concentration, 435
constraints, hard-modeling, 435
   concentration profiles and, *436*
   physicochemical models, 435
construction and decomposition of a three-way array via the trilinear PARAFAC model, *480*

Index   525

construction of a three-way array according to the unconstrained Tucker model, *480*
continuous sine-wave signal with aliased curve reconstructed, *387*
controlled and noise variables, 297
  examples of, **295**
convergence progress for PARAFAC from DTLD initialization and random initialization, 499
convolution, 397
Cook's distance
  distance measures and, 129
coordinate systems
  barycentric, 271
  barycentric and Descartes, three-dimensional example, *271*
  barycentric and Descartes, two-dimensional example, *270*
  Descartes, 271
  simplex, 271
correlation coefficients between absorbence at selected pairs of wavelengths for sulfamethoxazole training set, **63**
CPCR
  RPCR and, 199
CQDR, 209
critical values for F
  for a one-tailed test, *20*
  for a two-tailed test, 21
curvature effect, 113
curve resolution (CR) method, 421, 423
curvilinear calibration, 112–113
cutoff frequency, 395

D

D-optimal design, 301
data analysis algorithms, 481
  DTLD Algorithm, 481
  GRAM algorithm, 481
  MATLAB programming language and, 481
  PARAFAC algorithm, 481
data matrix
  augmentation arrangement and bilinear model for PCA or MCR-ALS decompositions, *455*
  plotting, 431
data points in subspace of original measurement space, *344*
data preprocessing, 341
  normalization, 342
  raw data into computer ready form, 341
  scaling, 341
data sets, 55
  fundamental modes of vibration, overtones, and combinations, 108
  near infrared spectroscopy and, 107
  solvent interactions, 108
  water-methanol mixtures, 108
data sets from multicomponent systems, *418*
data space
  dividing, 351
data vector, *78*
DATAN data analysis method, 495
datascope, 431
Daubechies wavelets, 407
decision limits, graphical representation of, 132
degrees of freedom, 46, 123–124, 125, 235
dendogram, single-linkage
  of the obsidian data, 357
dendograms, 351
  as intuitive and, *349*
  using single-linkage method developed from gs chromatograms of scent marks of Marmoset monkeys, *349*
denoised signal, *414*
denoising, 413
  of wavelet components using entropy threshhold, *413*
Descartes coordinate systems, 319
design efficiency, 303
Design Expert, 319
design measures, 303
design of experiment (DOE), 264
  software, 320
designer transfer function, 395
designs, full-factorial and simplex-lattice, **283**
determination limit, 134
DETMAX algorithm, 308
DFT, 400
direct exponential curve-resolution algorithm (DECRA), 495
direct methods,simplex, 225
  simplex algorithm and, 225
direct trilinear decomposition (DTLD), 489
  GRAM algorithm and, 489
discrete, bandwidthlimited Fourier transform of noisy spectrum, *390*
discrete convolution of two functions, *393*
discrete Fourier transforms (DFT), 389, 409
discrete wavelet, 409
discrete wavelet transform (DWT), 409, 410
discriminant analysis
  classical quadratic discriminant analysis (CQDA), 207
  cross-validation and, 208
  evaluating the discriminant rules, 208
  probability of misclassification, 208
distribution
  functions, 41–68
  normal, 14
  volume of, 53

distribution curves
  normal and standard normal, 44
distribution of error in Eigenvalues, 93
Dixon Q test, 33, *33*
DTLD, 498
  algorithm (Appendix 12.2), 503
  application, 490, 494–495
  application, GRAM and, 490

E

E-Optimal calibrations, *336*
effect of using a 30-point E-optimal subset during the latent-variable extraction, 334
effects of constrains in feasible bands, *447*
Efroymson's stepwise regression algorithm, 136–137
  convergence of algorithm, 137
  variable-addition step, 136, 136–137
  variable-deletion step, 137
eigenvalues, 70, 74, 75, 76, 141
  information and, 345
  primary, 74
  secondary, 75
eigenvectors, 70, 74, 75, 76, 141, 189
  complete set of, 74
  eigenspectra and, 73
environmental data
  environmental monitoring and, 454
  MCR-ALS and, 454
  multivariate curve resolution, 454
  PCA and, 454
  principle component analysis and, 454
environmental monitoring program in Portugal, 455
environmental study of SVOc in surface waters of Portugal, 460
error, type I, 50, 123, 133
error, type II, 50, 123
error, types of
  gross error, 10
  random error (noise), 10
  systematic error, 10
errors, estimating, 235
estimated concentration in mixture samples from GRAM with concatenated matrices, *486*
estimated concentrations in mixture samples from GRAM with added matrices, *486*
estimated factors
  visual inspection and, 498
estimating extinction coefficients, 329
Euclidean distance between two data points, *350*
Euler's method for numerical integration, *242*
  simple, 242
evolving factor analysis (EFA), 423, 426
  backward EFA, 424
  emergence and decay of compounds and, 424
  forward EFA, 424
  plot, 424
exact (or N-Point) D-Optimal designs
  DETMAX algorithm, 308
  Fedorov's algorithm, 307
  MD Galil and Kiefer's algorithm, 309
  Wynn-MItchell and van Schalkwyk algorithms, 308
Excel, 218
Excel's Solver, 227
  algorithm, 234
  using for nonlinear fitting of a first-order reaction, *228*
experimental design, 4, 265
  catalogs of, 320
  controlled factors and, 294
  desired properties for, 298–299
  full factorial design (FFD), 285
  inomplete lattice, 285
  for lithium lubricant study, *278*
  mixture and process design, 285
  noise factors and, 294
  simplex-lattice designs, 272
experimental design, off-the-shelf software packages, 316–320
  Design Expert, 319
  MATLAB, 316–318
  other packages, 319
experimental error
  Monte Carlo simulation methods and, 448
  resampling methods and, 448
experimental region, 268
experiments, passive, 330
explained varianced by PCA for three data pretreatment methods, *458*
extensions of three-way methods, 495
extraction of analyte, 476
extreme values
  outliers, 34
  stragglers, 34

F

F-test for determining the number of factors, 93
Factor Analysis in Chemistry, Malinowsky
  history of chemometrics and, 422
factor spaces
  constrained mixture spaces, 279–282
  defining bonds of, 327
  mixture and process spaces, 283–286
  one-dimensional, *269*
  process, 268
  simplex-centroid designs, 275–278

Index 527

simplex-lattice designs, 272–274
two and three-dimensional in coded
  variables, *269*
two-dimensional, *268*
factor spaces, mixture, 269–271
  Euclidean space and, *269*
  mixture variables and, 269
factorial designs
  advantages of, 290
  disadvantages of, 290
factorial designs, full, 290
  three-level, three factor, *292*
  three or more levels in, 291
  two-level three factor, ***291***
fast Fourier transforms (FFT), 389
  algorithms, 389
  Cooley-Tukey transform and, 389
  Sande-Tukey transform and, 389
FAST-MCD method, 203
Fearn's algorithm for OSC smoothing, 385
Fedorov's algorithm, 307
FFT, 400
figures of merit, analytical, 496
  "net analyte signal" (NAS) and, 496
filter
  lag, 400
  window, *402*
filtering, 380, 400
finite impulse response (FIR) filter, 401
  signal processing with a simple moving average
    filter, *401*
first and second derivatives, 82
  spectra of water-methanol mixtures, *117*
first order kinetic single-wavelength experiment
  and result of a least-squares fit, *223*
fitting parameters to a given measurement, 225
  direct methods, 225
  Newton-Gauss methods, 225
fitting, with Excel's Solver, *239*
fixed-size image window-evolving-factor
  analysis(FSIW-EFA), 463
fixed-size moving-window--evolving-factor
  analysis (FSMW-EFA), 423, 426
flash photolysis, 218
flow reactors, 256
food and feed chemistry, 512
Ford Motor CO., 294
forecasting, 127
forward selection, 136
Fourier domain, 388, 399
Fourier series, 385
Fourier transform, 386, 397, 400
  convolution and, 391
  correlation and, 391
  Parseval's theorem and, 391

  properties of, 389–394, ***391***
  window function and real part, *393*
Fourier transform infrared spectrometer
  (FTIR), 70
frequency-domain signal processing, 385–393
  bandwidth-limited discrete Fourier
    transform, 388
  Fourier transform, 386
  properties of the Fourier transform, 389
  sampling theorem and aliasing, 386–387
frequency-domain smoothing, 394, 394–397
  with designer transfer functions, 395–397
frequency response of noisy signal and its
  matched smoother, *398*
FSIW-EFA analysis, 464
fuel spill identification, 358
  gas chromatograms, 358
full factorial design (FFD), 283–284
Fundamental Reviews in Chemometrics, 512
fundmental modes of vibration, overtones, and
  combinations, 108

G

gas chromatography (GC), 70, 330
general equivalence theorem, 304
generalized eigenvalue-eigenvector problem
  (GEP), 482
generalized M-estimators (GM-estimators), 183
generalized multivariate squared distance, 52
generalized or mahalanobis distances, 52
generalized rank annihilation method (GRAM),
  4, 475, 485
generic plot of a variance indicator against a bias
  measure, *145*
genetic algorithm (GA), 138
genome sequences, chemometrics and, 514
GRAM, 489, 495, 498
  algorithm (Appendix 12.1), 502
  application, 486–488
  trends in predictive performance, *487*
graphical survey of NIR water-methanol data, 116
Grubbs' tests
  application to chemical data, 35
  critical values of G for, 35
  outliers, 34

H

Hadamard transform function, 317
hard-modeling constraints: physicochemical
  models, 435
harmonious PCR plot for methanol, *149*
Hawkins-Bradu-Kass data set, 187
  artificial data set, 186

Heisenberg uncertainty principle, 409
Hessian matrix, 234
heuristic evolving latent projecions (HELP), 429–430
   HELP method, 431
hierarchical models, 514
high-performance liquid chromatography (HPLC), 70
high-speed gas chromatographic profiles of jet fuels, *360*
histogram showing absorbence of sulfamethoxazole training set, 62
Hotelling, 510
HPLC-DAD
   data set, 431
   system, 430
"hyphenated" mass spectrometry-chromatographic methods, 423
hyphenated methods, 477
hypothesis test
   comparison of multivariate means, 59
hypothesis testing, 122–123
   alterative hypothesis, 48–49
   null hypothesis, 48–49
   principles of, 48–49
   statistical inference and, 48–49

I

impulse response function, 399
impulse response function of the seven-point rectangular smoother window function, *396*
incept, nonzero, 119, 462
initial estimates, chemometric methods and, 432
input variables, 268
   mixture variables, 268
   process variables, 268
interferences, 134–135
   chemical, 134–135
   physical, 134–135
   spectral, 134–135
International Chemometric Society, 509
International Union of Pure and Applied Chemistry (IUPAC), 12
Interpreting regression ANOVA tables, 125
iterative resolution methods, 431
iterative target transformation factor analysis (ITFA), 432, 437–438, 477
   process and, 438–439
   target factor analysis (TFA) and, 437
   target vector and, 439

J

Jacobian J
   representation of, *232*
Jacobian J and Marquardt parameter, *234*
Jacobian J matrix, 232–233

K

K-nearest neighbor (KNN), 352, 354
   Bayes classifier and, 352
   Euclidean distance and, 352
kinetic data, analysis of, 246–253
   additional methods, 258
   autocatalysis, 246
   Belousov-Zhabotinsky (BZ) Reaction, 251–253
   flow reactors, 258
   globalization of the analysis, 258–259
   Lotka-Volterra (Sheep and Wolves), 250–251
   measurement techniques and, 256
   model parser, 256
   related issues, 255–258
   soft-modeling methods, 257–258
   Zeroth-Order reaction, 248–249
kinetic modeling of multivariate measurements, nonlinear regression, 217–262
   Beer-Lambert's Law, 219
   calculation of the concentration profiles: case I, simple mechanisms, 220
   calculation of the concentration profiles: case II, complex mechanisms, 241–252
   calculation of the concentration profiles: case III, very complex mechanisms, 253–254
   matrix notation, 219
   model-based non-linear fitting and, 222–240
   multivariate data, 219–220
   related issues, 255–258
kinetic modeling of multivariate measurements, related issues
   flow reactors, 256
   globalization of the analysis, 256
   measurement techniques, 256
   model parser, 256
   other methods, 258
   soft-modeling methods, 257
kinetics of chemical processes
   light-absorption spectroscopy and, 218
kinetiic reactions, 246–252
   autocatalysis, 242–246
   Belousov-Zhabotinsky (BZ) Reaction, 251–252
   Lotka-Volterra (Sheep and Wolves), 250
   zeroth-order reaction, 248–249
Kjeldahl nitrogen titration method, 107

known profiles, 435
Kohonen neural network, 345–346
Kowalski, Bruce
  chemometrics and, 509

L

least median of squares (LMS), 183
least squares, 121
  classic, 107
  estimates, 111
least-squares fit, 222, 237
  of emission spectroscopic data, *240*
leave-multiple-out CV (LM) (CV)
  Monte Carlo CV (MCCV), 115
leave-one-out cross-validation (LOOCV), 115
leverage and influence, 128
  calibration samples and, 128
leverage, in statistics, 128
leverage points, 184, 191
light-absorption spectroscopy
  CD (circular dichroism), 218
  NIR (near-infrared_ reflectance spectra, 218
  UV/Visible spectroscopy, 218
limit of detection, 132
  determination limit, 134
  multivariate detection limit, 134
  univariate decision limit, 132
  univariate detection limit, 133
limitation of quantity (LOQ), 114, 134
linear and non-linear parameters, *228*
linear discriminant rule, 209
linear effect, 113
linear models with n standards, diagram of three different types, *109*
linear parameter fitting, 241
linear regression in low dimensions, 176
  linear regression with one response variable, 176
linear regression with one response variable, 176
  the classical least-squares estimator, 177
  the multiple linear regression model, 176
  other robust regression estimators, 182, 182–183
  an outlier map, 180, 180–181
  the robust LTS estimator (least trimmed squares), 178–179
linear regression with several response variables, 183
  an example, 185, 194
  the multivariate linear regression model, 183–184
  principle component regression (PCR), 195
  the robust MCD-regression estimator, 184
  selecting the number of principle components, 193

local rank analysis methods, 423
  evolving factor analysis (EFA), 423
  fixed-size moving-window--evolving-factor analysis (FSMW-EFA), 423
  images and, 463
local rank constraints, 435–436
local rank maps, 426
  of pill images, *464*
location and covariance estimation in low dimensions, 173
  the empirical mean and covariance matrix, 173
  other robust estimators of location and covariance, 176
  the robust MCD estimator, 174–176
location and scale estimation
  mean and standard deviation, 169–171
  median and median absolute deviation, 171
  other robust estimators of location and scale, 171–173
logarithm of the square sum (ssq) of the residual vector r as a function of the rate constant k, *224*
Lotka-Volterra (Sheep and Wolves), 250–251
  dynamics of population of predators and preys in closed systems, 250
  kinetic population profiles and, 250
  MATLAB's Runge-Kuta solver and, 250
  predator and prey kinetics, *250*
LTS estimator
  regression, scale, and affine equivariant, 179
LTS regression, 183, 197, 198
  errors and, 179

M

M-estimators, 183
Mahalanobis, 52, 192
  distances, 52, 59, 64
Mahalobis Distances and Probability Densities from Hotelling's T^2 Statistic for Test Samples of Sulfamethoxazole compared with the Training Set, *64*, **66**
Malinowski's empirical indicator function (IND), 91
Malinowski's RE function, *91*
Manne's resolution theorems, 421, 427
mapping and display, 341
  data compression, 343
  dimensionality reduction with PCA and, *343*
  principle component analysis (PCA) and, 342, 343
  variance and, 342
Marquardt parameter mp, 234
mass spectrometer (MS), 70
matched smoothing of noisy signal, *398*

mathematical handling of multiwave absorption data sets, 218
MATLAB, 74, 202, 232, 241, 456, 465, 481
  backslash notation, 234
  chemometrics and, 218
  code, 59, 218
  corrocoef functions, 62
  cov functions, 62
  D-optimal designs and, 317–318
  diff command, 60
  example 3.1, 61
  example 3.2, 61
  example 3.3, 62
  example 3.4, *63*
  example 3.5, 65
  example 3.6, 66
  example 4.1: eigenvectors and eigenvalues, 75
  example 4.2: principal component analysis of a spectroscopic data set, 75–76
  example 4.3: principal component analysis using the SVD), 76–77
  example 4.4 autoscale of a matrix, 79–80
  example 4.5: PCA procedure, 86
  example 4.6: function for calculating Malinowski's RE, IND, and REV functions, 92–93
  example 4.7: determining the number of significant principal components in a data matrix, 95–96
  example 4.8: computation of residual spectra, 100
  example 4.9: residual variance analysis, 102
  example 7.1, 222, 223, 254
  example 7.2, 222
  example 7.3a, 226, 227
  example 7.3b, 226, 227
  example 7.4a, 229
  example 7.4b, 230
  example 7.5a, 235, 236
  example 7.5b, 235, 236–237
  example 7.5c, 235, 237
  example 7.6a (autocatalysis), 247
  example 7.6b, 248
  example 7.7a (Zeroth-order kinetics), 249
  example 7.7b (Zeroth-order kinetics), 249
  example 7.8a, 250
  example 7.8a lotka volterra, 251
  example 7.8b ode lotka volterra, 251
  find command, 61, 65
  functions for DECA, 495
  hist command, 61
  m_dist.m function, 64
  mean functions, 62
  plot command, 60
  Runge-Kutta method and, 242, 246
  Statistics Toolbox, 317
  zoom command, 60, 65
matrix notation, 218, 219
maximum-likelihood rule, 207
MCD estimates, 184, 187, 193
MCD estimator, 198
MCD regression, 197, 198, 203
  an example, 185
MCR-ALS decomposition method, 448
  biochemical processes and, 449–452
MD Galil and Kiefer's algorithm, 309
mean
  arithmetic, 11
  defined, 10
  inferences about, 49
  population, *13*
mean absolute deviation (MAD), 37
mean centering, 78, 159
  effect on a bivariate distribution of data points, graphical illustration, *78*
measurement error and measures of prediction error, 114
measurement techniques, 256
median absolute deviation (MAD), 171
membership probability, 207, 208, 209
mesh representation of pure distribution maps, 465
methanol model at without an intercept term, graphical displays, *120–121*
methanol model with a non-zero intercept, *124–125*
methanol model with a nonzero incept, graphical displays, *122–123*
methanol result from univariate and multivariate calibration of water-methanol mixture, **118**
middle point of the window, *402*
misclassification probabilities for RQDR and CQDR applied to the fruit data set, **209**
misclassification probability(MP), 208, 209
misclassification rates, 210
mixture and process factor space for incomplete cubic simplex-lattice design combined with a full factorial design, **284**
mixture and process factor spaces, 283
  symmetric experimental designs and, 283
mixture factor spaces
  coordinates in, shown in barycentric and Descartes coordinate systems, **271**
model
  calibration and validation, 198
  parser, 256
  validation, 114
model-based fitting of kinetic data, 218
  methods for nonlinear least-squares fitting, 218

Index

modeling the concentration profiles of the reacting components, 218
model-based non-linear fitting, 222
   direct methods, simplex, 225–226
   linear and non-linear parameters, 228–229
   Newton-Gauss-Levenberg/Marquardt (NGL/M), 230
   non-linear fitting using Excel's Solver, 227
   nonwhite noise, 237–239
model departures and outliers, 129
   residual plot and, 129
model determination showing the bias/variance tradeoff with selection of the metaparameter, *144*
Monte Carlo CV (MCCV)
   leave-multiple-out CV (LM)(CV), 115
mother Daubechies wavelet and two shifted and scaled children fitting a noisy signal, *407*
mother wavelet functions, 407, *407*
moving-average filter, noise reduction and, 403
multiblock models, large data sets and, 514
multicollinearity, 194
multicomponent systems
   mixtures, 448
   processes, 448
multiple correlation coefficient, 131
multiple linear regression (MLR), 169
multiple overlapping absorption bands in the NIR spectrum f water, *89*
multiple rank analyses, 423
multiplicative scatter correction (MSC), 83, *85*
multivariate
   distribution, 52
   measurements, 225
multivariate absorbence data
   Beer-Lambert's Law and, 241
multivariate adaptive regression splines (MARS), 477
multivariate advantages, 3, 52
multivariate calibration, 3, 111–112, 195
   methods, 512
multivariate classification and pattern recognition, 4
multivariate curve resolution, 4, 417–467, 417–474
   ambiguities and resolution theorems, 418–422
   environmental data, 454–460
   error in resolution, 446–447
   historical background of, 422
   local rank and resolution: evolving factor analysis and related techniques, 423–425
   MCR-ALS Simultaneous analysis of multiple correlated data matrices, 440–445
   range of feasible solutions, 446–447
   resolving diverse chemical problems and, 448
   self-modeling curve resolution extended to multiway data, 440–445
   software, 465–467
   spectroscopic images, 461–465
   uncertainty in resolution results, 446–447
   useful methods and, 441
multivariate curve resolution-alternating least squares (MCR-ALS), 432
multivariate curve resolution-alternating least squares MCR-ALS, 439
multivariate curve resolution-alternating least squares (MCR-ALS), 439
multivariate curve resolution-alternating least squares MCR-ALS, 456–459
   procedure, 439
multivariate curve resolution, applications, 448–453
   biochemical processes, 449–450
   global description of the protein folding process, 453
   study of changes in the protein secondary structure, 451–452
multivariate curve resolution, iterative methods, 431–439
   constraints based on chemical or mathematical properties, 433–437
   constraints, classification, 433–437
   constraints, definition, 433–437
   constraints equality and inequality, 433–437
   generation of initial estimates, 432
   iterative target transformation factor analysis (ITFA), 437–438
   multivariate curve resolution-alternating least squares (MCR-ALS), 439
multivariate curve resolution, noniterative resolution methods, 426–429
   Heuristic Evolving Latenet Projections (HELP), 429–430
   other techniques, 429–430
   Subwindow Factor Analysis (SFA), 429–430
   window factor analysis (WFA), 427–428
multivariate data, 219
   parameters and, 229
multivariate dataset, 218
   matrix form and, *54*
multivariate detection limit, 134
multivariate distances, 59
   "acceptable" and "unacceptable" objects, 64
   calculate Mahalanobis distances and probability densities, 64
   compute training set mean and variance-covariance matris`, 62   Au: matris not clear
   example of, 59–69
   graphical review of smx.mat data file, 69

histograms of variables (wavelengths), 69
  selection of, 60
multivariate distances, example
  step 1: graphical review of smx.mat data file, 60
  step 2: selection of variables (wavelengths), 61
  step 3: view histograms of selected variables, 61–62
  step 4: compute the training set mean and variance-covariance matrix, 62
  step 5: calculate Mahalanobis distances and probability densities, 64
  step 6: find "acceptable" and "unacceptable" objects, 65
multivariate limit of detection (LOD), 134
multivariate normal distribution, 3, 41–68, 51, 51–52
  assumptions, 55
  bell-shaped curve and, 51
  chi-square distribution, 56
  estimation of population parameters from small samples, 54
  generalized or Mahalanobis distances, 52
  generalized sample variance, 55
  graphic illustration of selected bivariate normal distributions, 56
  variance-covariance matrix, 53
multiway data analysis approaches, 461

N

N-way data, 478
"net analyte signal" (NAS)
  calibration and, 496–497
net analyte signal (NAS) vector, 134
Newton-Gauss algorithm, 233
  flow diagram of, *233*
Newton-Gauss-Levenberg/Marquardt (NGL/M), 230–237
  algorithm, 230, 234, *235*, 235
  fitting procedure, *235*
  routine, 235
Newton-Gauss method, 233
NIR absorbence spectra of water-methanol mixtures, *109*
NIR (near-infrared-reflectance spectra), 60, 510
  of sulfamethoxazole powder, 59, *60*
NIR spectra projected ino a three-dimensional space formed by the first three principle components, *335*
NIR water-methanol data set, 141
noise
  amplitude, 380
  variables, 266
noise removal and the problem of prior information, 380–381

signal estimation and signal detection, 381
noise removal, matched smoothers and, 397
noisy spectrum to be enhanced, *381*
nomadization applied to Raman spectra, *83*
non-zero intercepts, *109*, 110
non-zero mean noise effects, 399
noniterative resolution methods, HELP method, 431
nonlinear fitting, using Excel's Solver, 227
nonlinear multivariate model estimation, chemical kinetics and, 4
nonlinear parameter fitting, 241
nonlinear regression, 477
nonnegativity, 433
nonsymmetric optimal designs, 298–306
  optimal vx. equally distanced designs, 299–301
  optimality criteria, 298
nonsymmetric optimal designs, design optimality and design efficiency criteria, 302–302
  A-Optimality, 306
  D-Optimality and D-Efficiency, 304
  design measures, 303
  E-Optimality, 306
  G-Optimality and G-Efficiency, 305
nonwhite noise, 237
normal distribution
  plot of, *14*
normalization, 83
null hypothesis, 19, 48–49, 122, 133
  acceptance and rejection errors, 122
numerical approximation of the first derivative on near-infrared spectra of water-methanol mixtures, 82
Nyquist criterion, 388

O

octane data set with NIR absorbence spectra
  outlier maps and, 205
one variable at a time approach, 288–290
  advantages of, 290
  applied to a two-factor problem, *289*
  disadvantages of, 290
  illustration of, *289*
one variable at a time vs. optimal design
  bivariate (multivariate) example, 288–289
  one-variable at a time approach, advantages of, 290
  one-variable at a time approach, disadvantages, 290
one-way data, 476
operating region, 268
optimal composite sequential designs (OCSD), 314
  schematic representation of, *314*

Index 533

for three mixture variables and one process variable, optimal for models having structures in equations, *321*
for two process variables and two polynomial models, *315*
optimal sequential designs (OSD), 314
ordinary differential equation solvers (ODE-solvers), 245
orthogonal projection approach (OPA), 432
orthogonal signal correction (OSC), 384
other calibration methods, 150
    common basis vectors and a generic model, 150, 150–151
    continuum regression (CSR), 150
    cyckuc subspace regression (CSR), 150
    generalized RR (GRR), 150
    ridge regression (RR), 150
    ridge variations of PCR, PLS, etc., 150
outlier maps, 185, 194, 199
    of artificial regression data, *181*
    of the biscuit dough data set obtained with RPCR, *201*
    of glass data set based on three principal components computed with CPCA and ROBPCA, *195*
    PCA analysis results and, 191–192
    of robust residuals vs. robust distances of the carriers for the foam data set, ***186***
    for the stars data set, *182*
outlier maps, three-dimensional
    PCA and regression outlier maps and, 202
outliers, *32*, 168–169
    experimental, 33
    Grubbs' tests, 34
    orthogonal, 191, 194
    and outlier map of a three-dimensional data set projected on a robust two-dimensional PCA subspace, *192*
    vertical, 184

P

PARAFAC, 481, 494–495, 495, 496
    Algorithm (Appendix 12.3), 504
    application, 494
    relative speed of, 499
PARAFAC / CANDECOMP, 491–494
    algorithm, 491
    PARAFAC application, 494
    solution constraints, 493
    tuckals, 493
PARAFAC model, 491, 498
    as trilinear, 495
parallel factor analysis (PARAFAC), 4, 455, 475, 478–480

parallel vector analysis (PVA0, 428
parameters
    linear, *228*
    molar absorptivities, 228
    non-linear, *228*
    rate-constants, 228
Parseval's theorem, 391
partial least squares (PLS), 112, 140, 147, 147–149, 196
    applications, 352
    comparison with PCR, 149, 149–150
    mathematical procedures, 148–149
    method summarized, 352
    modeling method, 352
    number of basis vectors selection, 149
    regression method, 352
    software, 352
partial least squares (PLS) regression, 202, 477
    classical PLSR and, 202
    an example, 204
    model calibration and validation, 204
    robust PLSR and, 202
partitioning of variance in least-squares solutions, 123
pattern-recognition, 355
pattern-recognition methodology, 341
    classification, 341
    clustering, 341
    mapping and display, 341
pattern-recognition techniques, applications of, 355, 355–371
    archaeological artifacts, 356–357
    class-membership problem, 355
    classification, 355
    data preprocessing, 355
    feature selection, 355
    fuel spill identification, 358–364
    mapping and display, 355
    PLS models, 355
    principle component analysis, 355
    self-organizing map (SOM), 355
    sorting plastics for recycling, 365–370
    taxonomy based on chemical constitution, 371–373
PCR, PLS, and RR harmonious plots for methanol, *152*
PCR, PLS, and simplex harmonious plots for NI analysis of moisture in soy samples, *153*
peak-matching program, 361
pharmaceutical manufacturing process, 43
piecewise direct standardization (PDS), 158
pixels, 461, 462
    PCA analysis and, 463
plots

log 19 of singular values vs. estimated
    concentration for RAFA, 484
NIR spectra of water-methanol mixtures, 86
residual spectra from the training set and the
    residual spectrum for an unknown w
    spectrum, *99*
of second and third largest principal
    components for jet fuel samples, *362*
simulated two-component mixture spectra
    (solid line) and a spectrum contaminated
    with a third unknow component, *99*
two largest PLS components in the training set
    (bees), *373*
two largest principle components for jet fuel
    samples, *361*
two largest principle components of 10 metals
    in obsidian samples, *356*
two largest principle components of the training
    set, *369*
PLS, 514
    map from training-set data (honeybees), *374*
    models, 336
point, measurement and, 302
polynomial least-squares filtering, *404*
polynomial moving-average (Savitsky-Golay)
    filters, 403
polynomial smoothing, 81–82
    polynomial smoothing on near-infrared spectra
    of water-methanol mixtures, *81*
polynomials
    canonical form of, 272
population, 42
    in chemistry, 43
    finite, 42
    infinite, 43
    mean, 42
    mean vector, 54
    standard deviation, 42
population parameters, estimation of
    from small samples, 54
population variance, 47
population variance-covariance matrix, 54
possible residual plots with the estimated residual
    on the y-axis and he estimated concentration
    on the x-axis, 130
practical considerations in experiment design, 354
    choice of classifier, 354
    choice of training set, 354
    controlling experimental variables, 354
    feature selection, 354–355
    validity set, 354
precision and accuracy
    four common scenarios, *13*
precision, defined, 10, 12
predicted calibration spectra for product P1, 330

predicted standard and mixture concentrating
    from DTLD and PARAFAC, *490*
predicted sum of squares (PRESS) statistic,
    193–194
prediction confidence intervals, 127–128
prediction set
    for honeybee example, *373*
    for jet fuel example, *359*
    for plastics sample, *368*
prediction-set results (F-values) for Jet Fuel
    sample, 366
prediction-set samples projected onto the
    principle component map, *369*
predictive alternating conditional expectations
    (PACE), 477
preprocessing options, 77
    baseline correction, 80
    first and second derivatives, 82
    mean centering, 77, 78
    multiplicative scatter correction (MSC) and, 83
    normalization, 83
    smoothing and filtering, 81
    standard normal variate (SNV) transforms,
        83–85
    variance scaling, 77, 78
PRESS statistic, 198
principal component analysis (PCA), 69–104, 70
    chemometrics and, 3
principal component regression (PCR), 112
principle component analysis (PCA), 70–102,
    140–153, 189, 347, 351, 422, 456–459, 514
    alternataive formulations of, 77
    applied to chromatographic-spectroscopic data,
        70
    axes defining basis vectors, *344*
    basis vectors, 141, 144
    classical PCA (CPCA), 185
    conclusions, 102
    data exploration procedures, 86
    example PCR results, 145–146
    introduction to, 70
    loadings for raw mean-centered augmented
        data matrix, *459*
    mathematical procedures, 142
    outlier map, 199
    outliers, 193, 198, 199
    PCA data exploration procedure, 86
    preprocessing options and, 77–85
    score plots of the water-methanol mixtures, *88*
    scores for raw mean-centered augmented data
        matrix, *460*
principle component analysis (PCA), basic
    vectors, 96–98
    clustering and classification with PCA score
        plots, 98

Index   535

principle component analysis (PCA), influencing
    factors, 87–88
  distribution of error in Eigenvalues, 93
  F-test for determining the number of factors, 93
  variance and residual variance, 89–92
principle component analysis (PCA) outlier map
    applying RPCR to the biscuit dough data set
  regression outlier map and, *201*
principle component analysis (PCA) outlier map
    of octane data set obtained with RSIMPLS
    and SIMPLES
  regression outlier map obtained with
    RSIMPLES and SIMPLES, 205–206
principle component analysis (PCA), residual
    spectra, 98–101
  residual variance analysis, 100–101
principle component map, jet fuel data, *362*, 363
principle component model, 73–76
  alternative formulations of, 77
  Eigenvectors and Eigenvalues, 74–75
  singular-value decomposition, 76
principle component model, diagram for the
    chromatographic-spectroscopic data set
    shown in Figure 4.1, *73*
principle component regression (PCR), 140–141,
    195, 477
  basic vectors, 141
  classical PCR, 194
  linear regression models and, 194
  methanol, *145–146*
  method, 196
  model calibration and validation, 198
  number of principal components and, 198
principle component regression (PCR)
    applications, multivariate calibration and,
    195
principle component regression (PCR),
    mathematical procedures, 141–142
  calibration steps, 142
  example PCR results, 145
• number of basic vectors, 144
  unknown prediction steps, 143
principle component regression (PCR) models,
    336
  prediction error, estimating, 198
  validation of, 198
principle components, selecting the number of,
    193
probability distribution function, 43
process analytical chemistry, applications, 512
process variables, transformation formula, 269
products with the respective quantities of the
    components in each, *323*
projection of a three-dimensional vector a onto a
    two-dimensional subspace, *96*

projection pursuit (PP) algorithm
  RAPCA (relection algorithm for principal
    components analysis), 188–189
projection pursuit (PP) techniques, 188
protein folding, *451*, 451
  changes in tertiary structure and, 453
  global description of the protein folding
    process, 453
protocols for planning experiments, 3
pulse radiolysis, 218
pulsed gradient spin echo (PGSE), 495
pure-component NIR spectra for water and
    methanol, *116*

Q

quadratic discriminant rule (RQDR), 209
quality, illustrated with distributions of important
    quality characteristics, *296*
QZ algorithm, 486

R

RAFA, 489
  application, 483
Raman
  images, 462
  spectra of plastic, *367*
rank annihilation factor analysis (RAFA), 476
rank annihilation methods, 482–490
  direct trilinear decomposition (DTLD), 489
  direct trilinear decomposition (DTLD)
    application, 490
  generalized rank annihilation method (GRAM),
    485
  generalized rank annihilation method (GRAM)
    application, 486–488
  rank annihilation factor analysis (RAFA), 482
  rank annihilation factor analysis (RAFA)
    application, 482–484
RAPCA (relection algorithm for principal
    components analysis), 189
rate constants, 228
reactions, physical properties and, 2
  energy transfer, 2
  phase transitions, 2
  temperature, 2
recovery of concentration compound by window
    factor analysis, *428*
"reduced" product formulations after removal of
    UV-Inactive components, *326*
reexpressing data in alternate bases to analyze
    structure, 381–384
  projection-based signal analysis as signal
    processing, 383–384

regression
  deepest, 183
  depth, 183
  outliers, 198
regression analysis notation, 286
regression coefficients, 326
regression model
  identifying the type of, 325–326
regularization, 151–152
  Tikhonov regularization, 152, 153
regularization results: example, 153
relative speed (in GIgaFLOPs) for the discussed three-way methods), 498
repeatability, defined, 10
reproducibility, defined, 10
residual spectra, 98
residual sum of squares (RSS), 136
residual variance analysis, 100
  example, *102*
residuals, 128
  studentized, 129
resolution of a multicomponent chromatographic HPLC-DAD, *420*
resolution results, ambiguity in, 447
response-surface modeling (RSM), 264, 265, 265–286
  design of experiments (DOE) and, 265
  general scheme of, 265–267
  methodology, 267
  model fitting and, 265
  process or product optimization, 265
  regression-analysis-related notation and, 286–287
response-surface modeling (RSM) and experimental design, 256, 263–339
  algorithms for realizable optimal experimental designs, 306–315
  example: application of DOE in multivariate calibration, 321–329
  nonsymmetric optimal designs, 298–306
  off-the-shelf software and catalogs of designs of experiments, 316–320
  one variable at a time vs.optimal design, 288–290
  symmetric optimal designs, 290–294
  Taguchi Experimental Design Approach, 294
response-surface modeling (RSM), factor spaces, 268–285
  constrained mixture spaces, 279–282
  mixture and process spaces, 283–286
  mixture factor spaces, 269–271
  process factor spaces, 268
  simplex-centroid designs, 275–278
  simplex-lattice designs, 272–274

results from factor analysis of simulated chromatographic data, **94**
results from univariate calibration of the water-methanol mixture, ***118***
ridge regression (RR), 112, 140
RMSEC values, 118, 119
RMSEV values, 118, 119
ROBPCA, 193, 197, 199, 202, 203, 211
  algorithm, 190, 198
  outlier map and, 194
  projection pursuit and robust covariance estimation, 189
robust calibration, 159–160, 166–216
  introduction to, 168
  principle component analysis, 185
  software availability, 211
robust calibration, classification, 207–208, 207–211
  classification in high dimensions, 211
  classification in low dimensions, 207–208
  an example (spectra of three cultivars of cantaloupe), 209
  in high dimensions, 211
  in low dimensions, 207–210
robust calibration, linear regression in low dimensions
  linear regression with one response variable, 176
  linear regression with several response variables, 183
robust calibration, location and covariance estimation in low dimensions, 173–176
  the empirical mean and covariance matrix, 173
  other robust estimators of location and covariance, 176
  the robust MCD estimator, 174
robust calibration, location and scale estimation, 162–172
  the mean and standard deviation, 169
  the median and the median absolute deviation, 169
  other robust estimators of location and scale, 169
robust calibration, partial least-squares regression, 202–206
  classical PLSR, 202
  an example, 204
  robust PLSR, 203
robust calibration, principle components analysis, 185–194
  classical principle components analysis, 185
  an example, 194
  an outlier map, 191
  robust PCA based on a robust covariance estimator, 187

Index

robust PCA based on projection pursuit, 188
robust PCA based on projection pursuit and he MCD, 189
selecting the number of principle components, 193
robust calibration, principle components regression, 194–199
   classical PCR, 194
   an example, 199
   model calibration and validation, 198
   robust PCR, 197
robust calibration, software availability, 211
   FAST-LTS, 211
   FAST-MCD, 211
   LIBRA (Library for Robust Analysis), 211
robust discriminant analysis, 210
robust linear discriminant analysis, 208
   MCD estimates and, 208
robust LTS-subspace estimator, 191
robust multivariate methods, 4
robust PCA, 4
   based on a robust covariance estimator, 187
   based on projection pursuit, 188
   based on projection pursuit and he MCD, 189–190
robust PCA methods
   outliers and, 187
robust PCR, an example (biscuit data set), 198
robust PCR (RPCR) method, 197
   regression techniques and, 197
   robust PCA and, 197
robust PLC, 4
robust principal component analysis (RAPCA), 209
robust quadratic discriminant analysis (RQDA), 207
robust R-RMSECV curve for biscuit dough data set, 199
robust tolerance ellipses for the fruit data, 210
root mean square eror of cross-validation (RMSECV), 114, 198
root mean square error of calibration (RMSEC), 114
root mean square error of prediction (RMSEP)
   computing, 114
   RMSEV for validation and, 114
root mean squared error of estimation (RMSEE), 382
RPCR algorithm, 198
RSIMPLS
   ROBPCA and, 202
RSIMPLS approach, 203
   octane data set with NIR absorbence spectra, 204
Runge-Kutta algorithms

chemical kinetics and, 242
Runge-Kutta method, fourth order
   in Excel, 242–245, *244*
   in Excel, concentration profiles modeled, *246*

S

S-estimators, 183
S-PLUS software, 202
sample variance, 47
   generalized, 55
sample variance-covariance matrix, 54
samples, 43
   influential, 128
   projected onto two-dimensional space, *343*
sampling and sampling distributions, 42, 42–44
   normal distribution, 43–45
   simple random sampling, 42
   standard normal distribution, 45
sampling theorem, 387
sampling theorem and aliasing, 386–388
sampling theory, 41–68
Savitsky-Golay filtering, *404*
Savitzky-Golay smoothing, 81
scale plots, 410
scatter plot
   of absorbence at 1912 nm vs. 2264 for he sulfamethoxazole training set, *63*
   of the principal component (PC) scores from the SVD analysis of the water-methanol NIR mixture data, *142*
   of the principal component scores from the analysis of the HPLC-UV/visible data set, *96*
score plot and 97.5% tolerance ellipse of the Hawkins-Bradu-Kass data obtained with CPCA and MCD, 187
second loading vector and calibration vector of sucrose for the biscuit dough data set computed with CPCR and RPCR, *200*
selected bivariate normal distributions
   graphical illustration of, 56
selection of calibration and validation samples, 113
selectivity, 135
self-modeling curve resolution extended to multiway data
   MCR-ALS Simultaneous analysisof multiple correlated data matrices, 440–441
self-modeling curve resolution (SMCR), 477, 479
self-modeling methods, 257–258
self-organizing map (SOM), 345, 351
   disadvantages of, 347
   distance not preserved in, 347
   generated from the training set, *370*

semivolatile organic compounds (SVOC), 455, 456
sensitivity, 131
  analytic, 131
  calibration, 131
  effective, 131
  and limit of detection, 131
sequential D-Optimal experimental design for two process variables and a second-order polynomial, 312
sequential-replacement algorithms, 138
  forward-selection algorithm and, 138
  obtaining, 138
Shell foam data, 185
short-term Fourier transform (STFT), 406
signal estimation and signal detection, 381
  known signal and, 381
  root mean squared error of estimation (RMSEE), 382
signal processing
  convolution property and, 391
  digital filtering and, 4, 379–416
  serial, 400
signal processing and digital filtering, 379–416
  frequency-domain signal processing, 385–393
  frequency-domain smoothing, 394–397
  introduction to, 379
  noise removal and the problem of prior information, 380–381
  reexpressing data in alternate bases to analyze structure, 382–384
  time-domain filtering and smoothing, 398–405
  wavelet-based signal processing, 406–413
signal-to-noise ratio (SNR), 297, 322, 381, 411
  formulas for, 298
significance level, 50
significance testing, 18
  comparison of a sample mean with a certified value, 24–25
  comparison of the means from two samples, 24, 25
  comparison of two methods with different test objects or specimens, 26
  F-test for comparison of variance (precision), 19–21
  one-tailed or two-tailed tess, 24
  student t-test, 21, 22–23
significant factors, identifying the number of, 322
SIMCA method (soft independent modeling of class analogy), 211
SIMCA training set results for Jet Fuel Sample, 365
similarity values, 348
simple moving average filter ("boxcar averager"), 401

simple-to-use interactive self-modeling analysis (SIMPLISMA), 432
simplex, 280
  algorithm, 226, 228, 234
  defined, 226
  unconstrained, 281
simplex-centroid designs, 275–276
  advantages of, 277
  disadvantages of, 277
  example (lithium lubricant study), 278
simplex-lattice designs, 272
  advantages of, 274
  component contour plots and, 274
  as composite designs, 274
  as D-optimal, 274
  disadvantages of, 274
  example, 275
  examples of for linear, quadratic, full cubic, and special cubic models, *273*
  homogenous mixtures required for, 275
simplex-lattice mixture design, four-component for a cubic polynomial model, **276**
simplex minimization with three parameters, principle of, *226*
SIMPLISMA, 463, 465
SIMPLS algorithm, 202
simulated annealing (SA) and genetic algorithm, 138
simulated HPLC-UV/visible chromatographic data set showing two overlapping peaks with different UV/visible spectra, 72
singular-value decomposition (SVD), 75–76, 188, 430
"slow-fast" ambiguity, 237
small sample distributions, 46
  chi-square distribution, 47
  t-distribution, 46
  t-distribution, the, 46
smoothed data
  after inverse Fourier transformation, *394*
  from application of trapezoidal smoother, 397
smoothing, 380, 394
  with designer transfer functions, 395–398
  in the frequency domain with a trapezoidal smoothing function, *396*
  of noisy signal with wavelet transform, *412*
  operation, 383
  time-domain and, 398
  transfer function and, 394
smoothing and filtering, *81*
  signal-to-noise ratio and, 81
soft independent modeling by class analogy (SIMCA), 354, 364
  classification in high dimensions, 353
  quadratic discriminant analysis and, 354

Index   539

soft modeling, 345
software, 465
  curve-resolution algorithms and methods, 465
software applications, 4
  ANOVA using Excel, 39–40
  MATLAB and, 4
  Microsoft Excel, 37
solution for the integration of ODEs, 222
solvent interactions, 108–109
sorting plastics for recycling, 365
  pattern recognition techniques and, 365
  principle component analysis, *368*
  Raman spectroscopy, 365
  self-organizing map (SOM), *368*
spectra, 462
  first-derivative, 116–117
  pixels and, 463
  second derivative, 116–117
  of the UV-active components, *324*
  of the UV-inactive components, *325*
spectral profiles, 476
spectral windows, 423
spectroscopic-chromatographic data, 70–72
  basic vectors, 71
spectroscopic images, 461–462, 462
  multivariate resolution and, 462
  pharmaceutical qualiity control and, 462
  resolution of, 463
  structure of, *462*
spectroscopic measurement, 256
spectroscopy, 477
  applications of, 512
  chemometrics and, 512
  near infrared spectroscopy and, 512
spectrum, *78*, 195
spectrum, techniques for obtaining, 195
  energy-dispersive x-ray fluorescence spectrometry (ED-XRF), 195
  florescence spectrometry, 195
  NIR (near-infrared_ reflectance spectrometry), 195
  nuclear magnetic resonance (NMR), 195
  ultraviolet sperometry (UV), 195
square sum ssq of the residuals r as a function of two parameters, *224*
squared correlation coefficient (angle cosign) between True and Estimated Profiles from DTLD and PARAFAC, 491
squared correlation coefficient (angle cosign) of predicted and true X-Way Spectra from GRAM with added matrices, ***488***
squared correlation coefficient (angle cosign) of predicted and true X-Way Spectra from GRAM with concatenated matrices, ***488***
squared correlation coefficient (angle cosign) of predicted and true Y-Way Spectra from GRAM with added matrices, ***488***
squared correlation coefficient (angle cosign) of predicted and true Y-Way spectra from GRAM with concatenated matrices, ***487***
standard deviation, 10
  population, 11, *13*
  relative standard (or coefficient of variation), 11
  sample, 11
standard normal distribution, 45
standard normal variate (SNV) preprocessing, *85*, 87
standard normal variate (SNV) transforms, 83
standard residuals of the stars data set based on a classical MLR and a robust LTS estimator, *179*
stars data, 180–181
stars regression data set with classical and robust fit, 177
statistical evaluation of data, 7–40
  analysis of variance and, 27–30
  common terms, 10
  introduction to, 8
  normal distribution, properties of, 14
  outliers, 33–35
  precision and accuracy, 12
  robust estimates of central tendency and spread, 36–38
  significance testing, 18–26
  software and, 38–40
  sources of error, 9
statistical evolution of calibration models obtained by least squares, 121
statistical modeling, 265
statistical properties, 3
  calibration of response signals, 2
  control of interfering factors, 2
  measurement process errors, 2
  modeling of complex multivariate signals, 2
strong filtering of noisy data using quadratic polynomials, *405*
structure of he extended design matrix F for a second-order model with two process variables, ***287***
structure of protein, *450*
student t-test
  paired t-test, 21
  standard t-test, 21
  t-test with nonequal variance, 21
subwindow factor analysis (SFA), 428–430, 429–430
  algorithm, 431
  application of, *430*
sum of squares (ssq), 123–124, 234

summary data for the analysis of Pb in orchard leaves, 49
summary statistics for NIR calibration of water in water-methanol mixtures using one wavelength and a nonzero intercept, *125*
sums of squares, 125
symmetric optimal designs, 290
symmetric optimal designs, central composite designs, 293–294
symmetric optimal designs, three or more levels in full factorial designs, 291292
    advantages, 290symmetric optimal designs, two-level factorial designs, 290-294
    disadvantages, 291
symmetry matrix, *349*
systematic error or bias, 17

T

t-distribution, *16*
    degrees of freedom and, *47*
t-test
    comparison of a sample mean with a certified value, 24
    comparison of the means from two samples, 24
    comparison of the means of two methods with different samples, 24
Taguchi, 296
    Experimental Design Approach, 294
    methods, 295
    orthogonal designs, 319
Taguchi design, 297
    for two controlled factors and three noise factors, *297*
Taguchi, Genichi, *294*
    quality engineering and, 294
target factor analysis (TFA) and, 437
taxonomy based on chemical constitution
    Africanized and European honeybees, 371
    gas chromatography and, 371
    principle component analysis, 372
three-factor principal component model for NIR spectra of water-methanol mixtures, diagram of, *142*
three-way calibration, advantages of, 475
three-way calibration with hyphenated data, 475–499
    alternating least-squares methods, 491–494
    background, 476
    caveats, 497
    extensions of three-way methods, 495
    figures of merit, 496
    introduction to, 475
    nomenclature of three-way data, 478
    rank annihilation methods, 482–490

three-way data, 477
    models of, 478–479
three-way data analysis, 443–444, 464, 477
    calibration with, 477
    constraints and, 445
    examples of, 481
    MCR-ALS decomposition method, 445
    nontrilinear, 442
    trilinear, 442
    unfolding and, 441
    unfolding or matricization, *441*
three-way data, bilinear models for unfolded PCA and unfolded MCR, *444*
three-way data, nomenclature of
    data as orders, 478
    data as ways, 478
time-domain filtering and smoothing, 398–405
    filtering, 400–402
    polynomial moving-average (Savitsky-Golay) filters, 403–405
    smoothing, 398–399
time-domain processing of noisy data with moving-average filter, *402*
time-domain smoothing of noisy data, *399*
training set
    for the honeybee sample, *372*
    jet fuel sample, *358*
    for plastics sample, *368*
transfer functions for FIR filters as a function of window size, *403*
translated experimental design for product P1, *329*
translation of coded factor levels to real experimental levels for product P1, *328*
trilinearity constraint in MCR bilinear models for three-way data, *446*
true, best, and worst estimated X-way and Y-way analyte profiles from GRAM, *487*
"true values", 114
TUCK alternating least squares, 493
Tuckals, 493
Tucker3 model, 479–480, 493
two-way data, 476
two-way data matrix, augmentation of, *441*
types of observations based on heir robust distance (RD) and their Residual Distance (RD, *185*)
types of observations based on score distance (SD) and orthogonal distance (OD), *193*

U

uncertainty in resolution results, experimental error and, 448
uncertainty in resolution results, sources of, 446

Index   541

ambiguity of recovered profiles, 446
experimental noise of the data, 446
unimodality, 434
Union of Pure and Applied Chemistry (IUPAC), 131
univariate calibration, 109, 118, 118
   with an intercept term, 119
   without an intercept term, 118, 118–119
univariate decision limit, 132
univariate detection limit, 133
univariate hypotheses testing, 48
   inferences about means, 49
   inferences about variance and the
      F-distribution, 51
univariate MCd, 193
univariate normal distribution, 3
unknown prediction steps
   prediction step, 143
   validation step, 143
UV absorption, 323
UV-inactive components, 323
UV spectra, 323
   of all nine pure components, *324*
UV spectroscopy, 322
UV/visible spectrometer, 70

V

v-fold CV, 115
validation set, 208
variables, 10, 11, 44
   between-sample means, 28
   collinear, 344
   continuous random, 43
   discrete random, 43
   heteroscedastic, 130
   homoscedastic, 130
   residual variance and, *89*
   scaling, 78–79
   selection, 135
   within-sample means, 28
variance and the f-distribution
   inferences about, 51
variance-covariance matrix, 53
viscose data set and standardized residuals
     obtained with different estimators of location
     and scale, ***170***
Visual Basic, 218

W

water-methanol mixtures, 108
wavelengths, 462

wavelet, 406
   components resulting for discrete wavelet
     transform, *410*
   sine basis functions for orthogonal transforms
     and, 408
   truncation and, 412
wavelet analysis
   Fourier analysis and, 408
wavelet-based signal processing, 406–413
   discrete wavelet transform, 409–410
   smoothing and denoising with wavelets,
     411–413
   wavelet function, 406–407
   wavelet function, time and frequency
     localizations of, 408
wavelet family, 407
   vanishing moments and, 407
wavelet functions, 411
   coiflet wavelet families, 411
   Daubechies wavelet families, 411
   symlet wavelet families, 411
   time and frequency localizations of,
     408
weak filtering of noisy data using quadratic
     polynomials, *405*
white noise, 222, 397
   defined, 237
Wilson-Hilferty transformation, 192
window factor analysis (WFA), 427–428
   steps of, 427–428
within-run, defined, 10
Wold, Herman
   partial least squares (PLS) and, 352, 510
Wold, Svante
   chemometrics and, 509
   partial least squares (PLS) and, 510
Wynn-MItchell and van Schalkwyk algorithms,
     308

X

X-way and Y-way profiles used to construct data
   sets for each example, *481*
Xerox, 294

Z

Zeroth-Order reaction, 248–249
   catalyst and, 248
   macroscopically observed, 248